European Communities Oil and Gas Technological Development Projects

Fourth Status Report

European Communities Oil and Gas Technological Development Projects

Fourth Status Report

Compiled by

J. P. JOULIA

D. VAN ASSELT

P. ARGYRIS

J. PASQUIER

Commission of the European Communities

Directorate-General for Energy,

Brussels

Published by

Graham & Trotman

A member of the Kluwer Academic Publishers Group

LONDON/DORDRECHT/BOSTON

for the Commission of the European Communities

Graham & Trotman Limited Kluwer Academic Publishers Group
Sterling House 101 Philip Drive
66 Wilton Road Assinippi Park
London SW1V 1DE Norwell MA 02061
UK USA

First published in 1990 for the Commission of the European
Communities, Directorate-General Telecommunications, Information
Industries and Innovation.

EUR 13070

ISBN 1 85333 478 2

British Library and Library of Congress Cataloging-in Publication
data is available from the publisher.

Printed in Great Britain by Bell and Bain Ltd., Glasgow

CONTENTS

PREFACE

Energy is of vital importance to the Community, which is currently engaged in the largest single challenge in its history, namely the completion of the 1992 Internal Market.

This fundamental change can only be fully achieved if a secure energy supply is guaranteed to the Community. Of all the energy sources, it is the hydrocarbons which have the dominant role as a consequence of their major share in the Community energy supply, and it is unlikely that there will be any significant change in their position before the end of the century.

Since the crisis of 1973, the Community industries have largely succeeded in ensuring a secure supply of oil and gas to the Member States. A programme of Community support has been directed towards enabling these industries to develop innovative techniques for exploiting reserves economically and at the highest possible safety levels, both for personnel and the environment. The purpose of this report is to show the most recent achievements of this programme.

It is important for the Community to continue to cooperate with industry to promote effective energy technologies in the market place. Security of energy supplies, including of course hydrocarbons, and efficient use of energy are the major axes of a new Community programme called THERMIE for the promotion of energy technology in Europe. THERMIE will run from 1990 to 1994 and will extend and build upon the activities started in 1973.

I am convinced that the Community's enterprises, operating in the hydrocarbon sector, would benefit from the possibilities offered by this new programme to reinforce the importance of the energy technology in Europe.

C.S. MANIATOPOULOS
Director-General for Energy

INTRODUCTION

At the beginning of the 1990s the European Community is a major force on the world scene, having a significant involvement in all sectors of economic life.

The Community is seeking to further improve its economic and social development through the single market, which is to be established by 1992. Energy is of strategic importance for growth, for improved competitiveness, for job creation and for this drive towards the 1992 single market.

The last two decades have shown that, because of its dependence on hydrocarbons, and particularly on imported oil, the Community is highly vulnerable in terms of energy supplies. This situation has led the Community to develop an energy with three basic elements :
- energy objectives to reduce dependence,
- financial support, particularly for technological development, to improve security of supply,
- action in specific areas such as stocks and crisis mechanisms.

In view of the importance of having adequate technologies to enhance the security of the oil and gas supplies, in 1973 the Community launched a programme to support the development of innovative technologies in this sector. This programme operated under two Regulations[1], the second and final one expiring at the end of 1989.

This is the fourth report to be published on the programme and its results. It deals with the projects supported from the ninth (1983) to the fifteenth (1989) round. Projects which have been completed since the publication of the previous report[2], and those which are ongoing, are the subject of a detailed summary. Projects completed before 1987, and also 1989 projects for which contracts have not yet been signed, are not covered by this report.

COMMUNITY RESEARCH AND DEVELOPMENT STRATEGY

Technology is of the highest importance for securing energy supplies, particularly in the oil and gas sector, where the exploitation of indigeneous reserves has been possible only as a consequence of technological progress. Indeed, the North Sea only emerged as a major oil resource in the 1970s through new techniques for the discovery and economic exploitation of its reserves under the best possible safety conditions.

Some of these technologies have been validated with the support of the Community, and European oil-related companies have proved their capability to deliver to the oil and gas industry successful techniques and processes, not only for application inside the Community, but also on a worldwide scale, thus giving their promoters strong positions in foreign markets.

---

(1) regulations 3056/73 and 3639/85
(2) European Communities Oil and Gas Technological Development Projects
  — Third Status Report — 1987

The basic instrument for encouraging research and development throughout the Community is the framework programme for science, research and development. The first framework programme which followed the Single Act and which was operative from 1987 to 1990, has encouraged research in the hydrocarbons sector. A specific sub-programme called "JOULE" focused among other things on the exploration and production of hydrocarbons. It enabled research in upstream activities to be supported as a complement to the programme for supporting technological development, which is closer to the market.

Research and development is needed to promote innovative ideas, but penetration of the market is a step which is essential in order to achieve effectiveness. Promotion of energy technologies is required in order to establish the strong and stable energy base which is necessary for the Internal Market. The Council has therefore now adopted "THERMIE", a new programme which will provide financial support to projects aiming to promote innovative technologies, particularly where market penetration still entails a large element of technical and financial risk. This programe will run for five years, from 1990 to 1994, and will include technology promotion in the hydrocarbons sector.

PROGRAMME CHARACTERISTICS

The financial support provided under Regulations 3056/73 and 3639/85 was intended to support technological development directly related to those activities in exploration, production, transportation and storage of hydrocarbons which would be likely to improve the security of Community oil and gas supplies. The support granted was in the form of a subsidy which was repayable by the beneficiary in the case of the commercial success of the project. This support did not exceed 40% of the eligible costs.

The projects in the hydrocarbons sector which have been supported within the programme are those which :

- develop innovatory techniques, processes or products, or exploit a new application of techniques, processes or products for which the research stage is completed;

- offer prospects of industrial, economic and commercial viability;

- present financing difficulties because of the technical and economic risks involved, so that most probably they would not be carried out without Community financial support.

The projects were submitted by promoters according to technical priorities given in the annual call for proposals. The selection of projects was carried out by the Commission after consulting an Advisory Committee consisting of representatives from the Member States.

In the selection, a preference was given to projects involving at least two independent companies established in different Member States, provided that the contribution of each of the undertakings was effective and significant, and to projects submitted by small and medium-size firms.

Once support is granted, it is followed by the signing of a detailed contract. During the execution of the project, the Commission scrutinises the progress of the work both technically and financially. The technical coordination of the programme is the responsibility of Directorate D, "Energy Technology" of the Directorate General for Energy, whilst the Division "Contracts and Management of Resources" is responsible for all administrative matters. Reports on the state of advancement of the programme are made every 2 years to the European Parliament and the Council.

STATUS OF IMPLEMENTATION

From 1974 to 1989, 244 companies submitted 1087 projects in response to the Commission's annual call for proposals. The Council (Regulation 3056/73) and the Commission (Regulation 3639/85) granted financial support to 689 projects totalling 537 million Ecus. This resulted in 587 contracts, a figure which will be increased by the 1989 contracts which have still to be signed.

In the period covered by this report (1987-1989), it is obvious that the status of R&D in the hydrocarbons sector was still being influenced by the consequences of the 1986 reduction in oil prices. Budget reductions undoubtedly affected the investment levels of the industry and more particularly of oil-related companies. Nevertheless, European companies have continued their efforts to develop new equipment and techniques, and these efforts have resulted in some significant improvements.

In the EXPLORATION sector, the emphasis still remained on seismic techniques, and there were significant advances. A new set of tools will be made available for exploration :

- AGIP has obtained good results in a project which focused on the resolution power of seismic data by developing new methods for the calculation of primary static corrections to improve deep reflexions (TH. 01.037/84).

- GERTH has developed a simulation model for the spatial distribution of heterogeneities in the reservoirs. This 3D simulation tool should be very useful for the geological fluvio-deltaic environments which are found in the North Sea (TH. 01.070/86).

- MIDLAND VALLEY has brought to a commercial stage its two-dimensional-balanced section software. This tool should allow the seismic interpreter to validate underground geometrics through balanced cross sections (TH. 01.073/86).

Regarding PRODUCTION techniques, there are several aspects :

- Offshore structures and foundation engineering have been disappointing. Nevertheless, some results can be highlighted:

    . The Tripod tower steel platform has been developed by HEEREMA with a view to its application in the Troll field. A similar structure was used like a wellhead platform in 27 m of water in the Netherlands Helder field. Dynamic and fatigue tests, material and welding tests and foundation design proved the feasibility of the concept (TH. 03.139/83).

    . A project led by MENCK, the leading contractor in the field of pile hammers. In this project an underwater hammer and its power pack were intensively tested in order to determine thermal behaviour and to study reliability. This development has been followed by the installation of the Menck hammer on board the Micoperi 7000 heavy lift vessel, for application in the Campos basin (TH. 06.044/88).

    . The Hydralok system for connection between piles and structures is a significant achievement, which might be used for deeper applications. It has been applied on one of the Ravenspurn structures (TH. 06.019/85).

- Subsea technology has been the subject of intensive development in the last few years. The Community has supported several major projects :

    . The design by SUBSEA INTERVENTION SYSTEMS with the support of oil companies, of an integrated system involving a christmas tree and its associated remotely operated vehicle for maintenance and repair. This system is significant in the evolution of the industry towards reliable systems, fully capable of operating in deep water. Recently the system has successfuly completed underwater tests (TH. 07.059/86).

    . Since 1983 AGIP has been promoting its SAF — Sistemi per alti Fondali —. This system does not require divers or guidelines for installation, operation and maintenance. The new christmas tree, purposely designed, is particularly adapted to deep and very deep waters. Its integrated approach, including an intervention tool, should improve the reliability of the system, which is to be installed on a well for offshore testing in the near future (TH. 03.201/86).

    . B.P.'s concept of DISPS — "Diverless subsea production system" — is also in line with the important technological evolution to design diverless systems. This system, to be used in water depths of 350 to 750 m, is basically composed of a template as supporting structure, production equipment and the Remote Guidance Vehicle, which performs installation and recovery of modules without diver intervention. Underwater testing has been successfully conducted at 150 m(TH. 03.233/87).

- Underwater production systems using subsea separation have been developed in recent years, and this should have economic prospects for small field production:

    . The BOET concept of a subsea separator, tested in ARGYLL field, is one of them. The results of this test should demonstrate the reliability of a technique which is suitable for satellite field exploration (TH. 03.182/85).

    . The GASP approach, promoted by Goodfellow & Associates. This system design, using the modularity principle, is suitable for exploitation of small fields in the vicinity of existing facilities, and economic studies have shown that it could be applied to 85 of these satellite fields, within a 50 km of radius (TH. 03.217/86).

- Multiphase pumping is a field of technology which could produce a technical revolution in the next decade. The "Poseidon" project by IFP/TOTAL is being supported with respect to its French part, its Norwegian part being carried out by STATOIL. A phased approach has led to the design of a concept which is particularly cost-effective, since it will avoid the use of a platform for treatment of the effluent, and it could also enable exploitation of satellite fields located far away from existing facilities. A key item in this concept is the helico-axial pump, of which an industrial prototype began long-term tests in 1989.

- In the field of human intervention in deep waters, the past two years have seen important progress :

    . The SAGA, the first civil submarine able to put divers at work without a surface vessel, developed by COMEX, made its first operational mission (TH. 07.050/83).

    . The COMEX divers extended the world record for human intervention to a depth of 520 m and COMEX continued to explore the promising use of hydrogen as a breathing mixture (TH. 15.110/87).

    . MAN and BRUKER associated proved the technical validity of a closed cycle diesel motor using Argon with an innovative regeneration system of exhaust gases (TH. 13.06/85 and TH. 13.07/85).

TRANSPORT technology is receiving more and more attention due to its safety implications, both for personnel and for the environment. Repair systems and inspection methods have been the subject of numerous projects, among which it is worth noting :

- SNAM development for a complete repair system (TH. 10.47/85) ;

- RTD success in developing a pipeline and riser inspection tool which enables the measurement of both interior and exterior corrosion (TH. 15.052/84).

EVALUATION OF THE PROGRAMME

The whole programme was evaluated by an independent consultant, Smith & Rea Energy Analysts Limited, with the assistance of nine European independent experts.

The overall conclusions of this evaluation were that through certain projects, such as the horizontal drilling, the application of enhanced oil recovery methods and the development of the SWOPS ("Single Well Oil Production System"), the programme contributed to the Community's security of supply by means of improved techniques and processes. SREA found that service companies had generally been the most successful, as opposed to design companies. Projects aimed at producing a manufactured product have had a high success rate, particularly where the potential end-users, namely the oil companies, participated. In addition, the evaluation brought to light the fact that Community industry in the oil and gas sector, and particularly the oil-related industries which provide equipment and services to the oil companies is important in terms of employment, technical capabilities, potential for innovation and economic power on a global scale. In the past few years this industry, which has been traditionally largely orientated towards international exchanges, has begun to develop a Community identity. This evolution has been undoubtedly favoured by the programme for supporting technological development in the hydrocarbons sector.

During the application of the second Regulation from 1985 to 1989 the links between companies of different Member States have been reinforced and the number of applications involving several companies has increased, to reach a point where more than 50 % of the support went to cooperative projects in 1989.

DISSEMINATION OF THE RESULTS

The dissemination of project results is mainly carried out by the contractor, who remains the owner of the developed technology. Circulation of information within the scientific and technical community has been achieved by project promoters through papers at technical conferences, articles in specialized newspapers and participation at oil industry exhibitions.

Nevertheless, the role of the Commission in this context has been essential:

a) As in 1979 and 1984, the Commission organised in 1988 in Luxembourg a Symposium on "New Technologies for the Exploration and Exploitation of Oil and Gas Resources". Attended by several hundred participants it demonstrated the dynamism of this industrial sector and the need for technology transfer in order to exchange experience and promote collaboration.

b) The "SESAME" data base was launched in 1983. Available since 1987 to external users by way of commercial hosts, it is a useful means of following the progress of all the energy technology projects.

c) Participation of the Commission in the important exhibitions of this sector, such as the Deep Offshore Technology Conference held in October 1989.

FUTURE ACTION

Considering the achievements of the past and the requirements of the future — among which the completion of the Internal Market by 1992, with its important energy component, remains the major objective of the Community — the Council has adopted the Commission's proposal for a new programme for the promotion of European energy technologies, called "THERMIE".

This new programme will have two lines of action :

- financial support for innovation and dissemination projects;

- associated measures for the dissemination of project results and successful technologies.

The implementation of the "THERMIE" programme will give concrete substance to Community policy for ensuring the availability of reliable and efficient energy technologies and their promotion within the market.

Further information on Community action in the field of energy technology may be obtained from
    Directorate for Energy Technology
    Directorate General for Energy
    Commission of the European Communities
    rue de la Loi, 200
    B - 1049 Brussels

BREAKDOWN OF SUPPORT BY SECTOR OF ACTIVITY FROM 1974 TO 1989

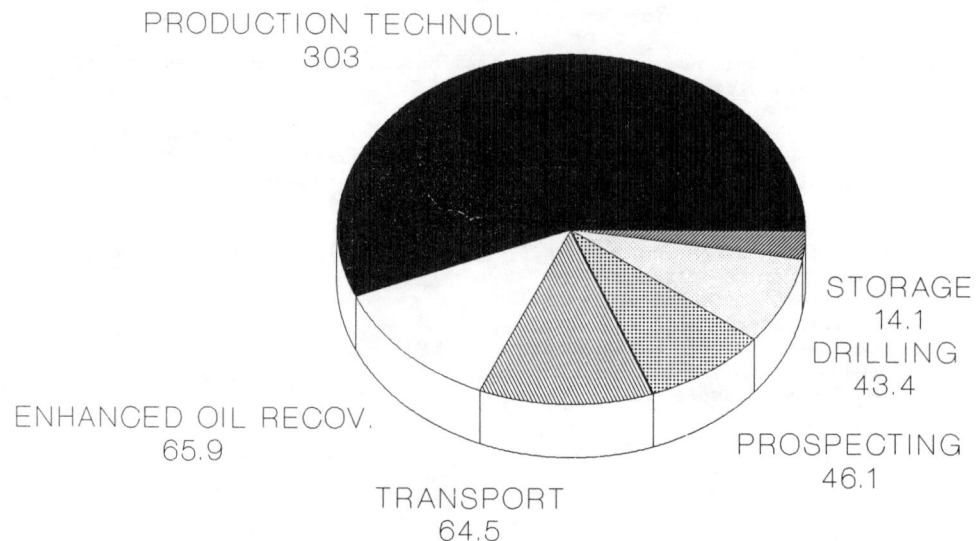

PRODUCTION TECHNOL.
303

STORAGE
14.1
DRILLING
43.4

ENHANCED OIL RECOV.
65.9

PROSPECTING
46.1

TRANSPORT
64.5

TOTAL : 536.7 MILLIONS OF ECUS

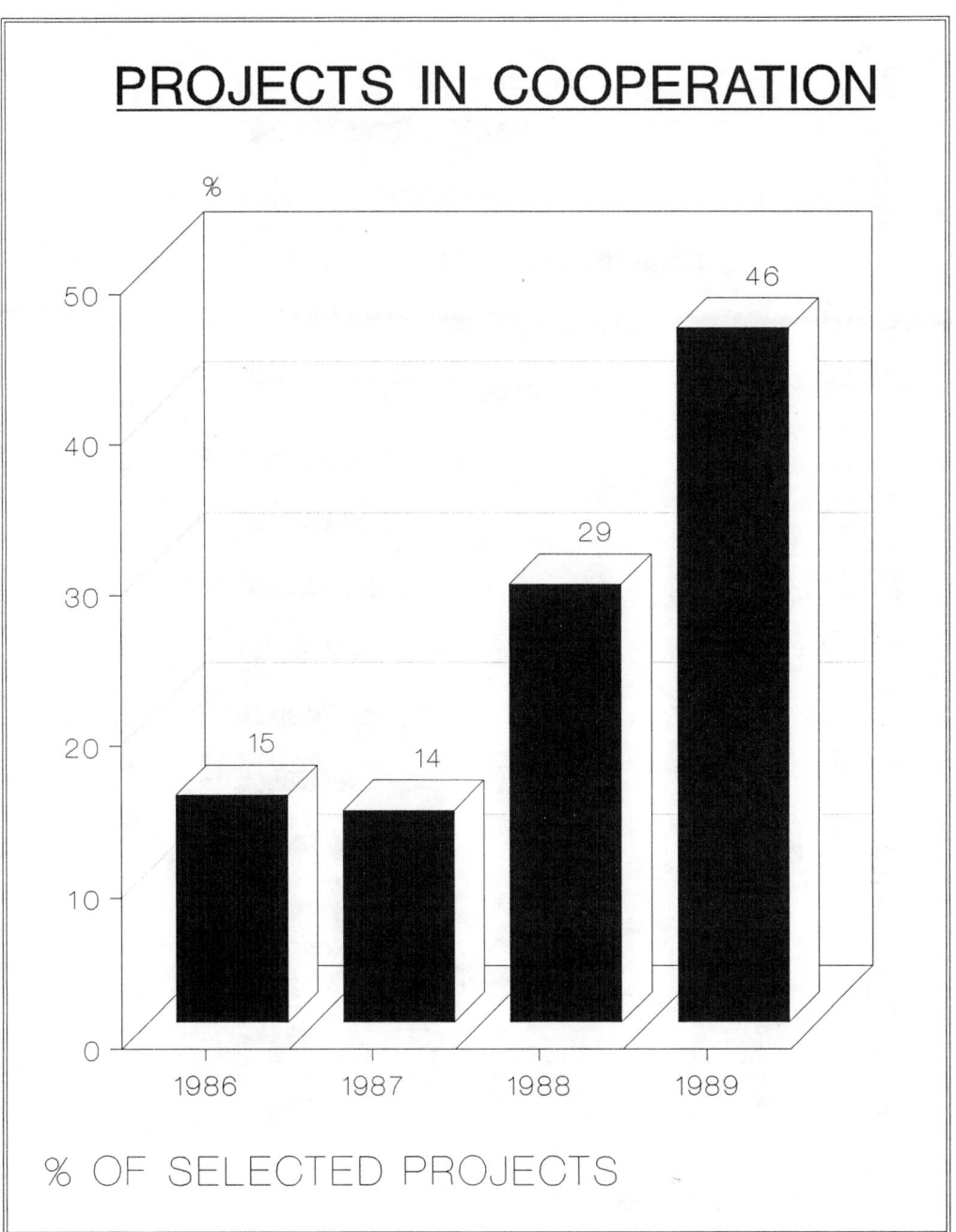

PROJECTS IN COOPERATION

% OF SELECTED PROJECTS

APPLICATIONS SUBMITTED TO THE COMMISSION AND PROJECTS SUPPORTED

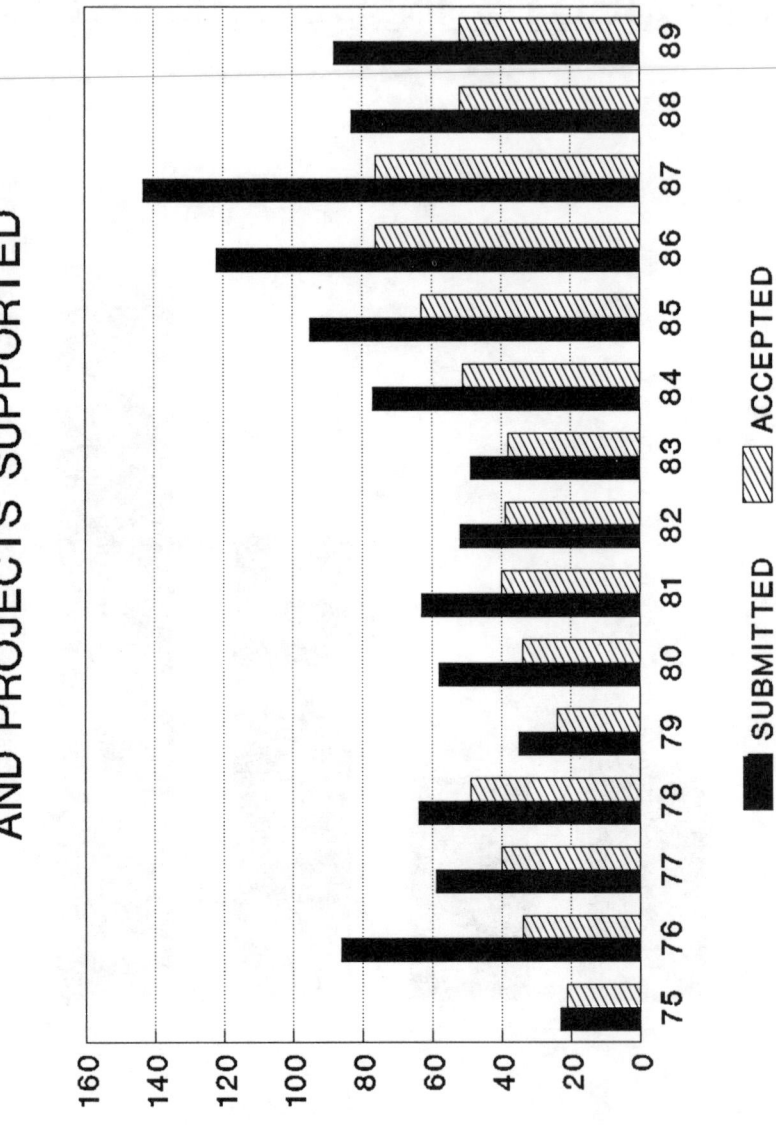

EEC

■ SUBMITTED   ▨ ACCEPTED

GEOPHYSICS AND PROSPECTING

```
****************************************************************************
* TITLE : ACOUSTIC MEASUREMENTS ON RESERVOIR ROCK.        *    PROJECT NO    *
*                                                         *                  *
*                                                         *  TH./01035/83/NL/..  *
*                                                         *                  *
****************************************************************************
* CONTRACTOR :                                            *  TELEPHONE NO    *
*    TECHNISCHE HOGESCHOOL DELFT, AFD MIJNBOUWKUNDE        *                  *
*                                                         *  015 78 16 17    *
*    POSTBUS 5028                                         *                  *
*    NL - 2600 GA DELFT                                    *                  *
*                                                         *  TELEX NO        *
*                                                         *                  *
* TECHNICAL DIRECTOR :                                    *  38151           *
*    IR. J.P. VAN BAAREN                                   *                  *
*                                                         *                  *
****************************************************************************
                                                    VERSION : 31/12/88
```

AIM OF THE PROJECT :
****
The objective of the research is to determine more rock properties from the
microseismogram obtained via borehole measurements with the Sonic tool.

PROJECT DESCRIPTION :
****
Microseismograms are measured under atmospheric pressure and room temperature on
artificially created rock. The synthetic rock closely resembles "natural"
sandstone but it has the advantage that all rock parameters are known exactly
and can be changed independently. Thus by changing one rock parameter it is
attempted to establish a correlation between that particular parameter and the
received microseismogram. To investigate the influence of pressure and
temperature on the recorded wave train the acoustic measurements will also be
performed in a triaxial cell under conditions of up to 400 bar pressure and 150
deg. C.

STATE OF ADVANCEMENT :
****
After a one year extension of the contract, the project was completed by
31/12/87.

RESULTS :
****
- An atmospheric acoustic measurement set-up has been constructed consisting of :
  XYZ system, programmable waveform generator, oscilloscope, transducers and
amplifiers. The equipment is controled by an IBM PC and allows the storage,
plotting and calculation of the data. Both refraction and transmission
measurements can be performed.
- A borehole simulator (triaxial cell) has been designed and built allowing
acousting refraction measurements at 435 bar and 150 degrees C at conditions
corresponding to 4000 m depth. The sample size is 400 mm in diameter with a
height of 600 mm.
- Establishing a relation of rock properties with acoustic properties from field
microseismograms is extremely difficult because too many unknown factors are
present.
- Reservoir samples can be constructed in the laboratory with the silica-lock
process or with araldite in which one rock property can be changed keeping the
other properties constant. The rock behaves as real subsurface reservoir rock.
- The software developed for this project gives theoretical estimates that
highly improves the interpretation of the experiments. The computer models are
based on the Biot theory and Biot theory with a frequency limit.
- Using the transmission measurements with Plona's method the attenuation
constants of the rock are determined with three methods. The methods compare
very well. It is therefore advised to use the highest amplitude method.
-The computer program using the 'Biot theory with frequency limit' is very fast
and gives accurate wave mode velocitites. The attenuation of shear and fast
compressional are predicted accurately. The attenuation of the slow
compressional mode is not correct.
During the field applications new evaluation methods have been developed:
1. The shale fraction present in a reservoir can consist of a varying mixture of
different clay minerals. Each mixture is characterized by its own shale point,
resulting in a range of five shale points valid for the reservoir interval in
our shaly sandstone example.
2. A method is proposed to obtain the five shale points from the logs.
3. As a result the hydrocarbon conten of the reservoir is determined more
accurately.

REFERENCES :
****
- VISSER R., MODELLING OF ACOUSTIC MEASUREMENTS ON POROUS ROCKS, DELFT PROGRESS
REPORT, VOLUME 13, NUMBER 1/2, 1988.
-ROUSCHOP J., H.K.J. HELLER AND J.P. VAN BAAREN, HIGH PRESSURE/TEMPERATURE
BOREHOLE SIMULATOR, DELFT PROGRESS REPORT, VOLUME 13, NUMBER 1/2, 1988.
-VISSER R., H.K.J. HELLER AND J.P. VAN BAAREN, ARTIFICIAL RESERVOIR SAMPLES,

DELFT PROGRESS REPORT, VOLUME 13, NUMBER 1/2, 1988.
-VISSER R., H.K.J. HELLER AND J.P. VAN BAAREN, A MEASUREMENT SET-UP FOR
DETERMINING ULTRASONIC VELOCITY IN POROUS ROCKS, DELFT PROGRESS REPORT, VOLUME
13, NUMBER 1/2, 1988.

```
*****************************************************************************
* TITLE : NEW METHODOLOGY AIMED TO ENHANCE DEEP      *      PROJECT NO       *
*         SEIMIC REFLECTIONS DEFINING THE STATIC     *                       *
*         CORRECTIONS DUE TO A COMPLEX SHALLOW       *   TH./01037/84/IT/..   *
*         GEOLOGY                                    *                       *
*****************************************************************************
* CONTRACTOR :                                       *   TELEPHONE NO         *
*     AGIP SPA                                        *                       *
*                                                     *   02 - 52023227        *
*   ELGE DEPT                                         *                       *
*   C.P. 12069                                        *                       *
*   IT - 20120 MILANO                                 *   TELEX NO             *
*                                                     *                       *
* TECHNICAL DIRECTOR :                                *   310246               *
*     ING. D. BILGERI                                 *                       *
*                                                     *                       *
*****************************************************************************
                                              VERSION : 29/08/89
```

AIM OF THE PROJECT :
****
The goal of this research project consists in increasing the resolutive power of
seismic data by searching and developing new methodologies for the determination
of both field and residual static corrections and for the broadening of the
frequency content of the data.

PROJECT DESCRIPTION :
****
The project will develop through different steps of studies and experimental
works in the filed and in the processing :
1. Data acquisition
2. Data processing
3. Special processing (surface static correction)
4. Synthesis of the tests to get an estimate of velocities and thickness and
   to draw conclusions from the seismic results and from the developed
   methodology.

STATE OF ADVANCEMENT :
****
Completed

RESULTS :
****
The enhancement of seismic signal resolution through the exploitation of two
complementary seismic surveys is obtained with techniques of static calculation
from geological model in areas of complex surface geology and of a multi wavelet
investigation in areas of stratigraphic problems. In field data acquisition new
techniques have been experimented with satisfactory results on eight Hydrapulse,
four Vibroseis, two Dynamite seismic lines and three geological surveys have
been carried on. Static calculation in areas characterised by a particularly
complex surface geology is performed within the project through a detailed
definition of the geological model of the near surface layers along the
acquisition profile.High resolution enhancement for stratigraphic delineation is
achieved by merging records acquired with different sources which illuminate the
exploration target with wavelets of complementary characteristics better
detailing the geological feature.

REFERENCES :
****
- HAGEDOORN, J.G. 1959, THE PLUS-MINUS METHOD...: G.PROSP.,7,158-182.- PALMER, D.
, 1980, THE GENERALIZED RECIPROCAL METHOD....:SOC.EXPLOR.GEOPHYS.
-SCHNEIDER, W.A., AND SHIN-YEN KUO, 1985, REFRACTION MODELING FOR STATIC
CORRECTIONS :
   55TH SEG,WASHINGTON,D.C.
- FOSTER, M.R.AND GUINZY,J.,1967,THE COEFFICIENT OF COHERENCE...GEOPHYSICS 32,
602-616.-WALDEN,A.T. AND WHITE, R.E.,1984,ON ERRORS OF FIT AND ACCURACY IN
MATCHING.G.PROSP.32,871-891
-WHITE,R.E.,1980,PARTIAL COHERENCE MATCHING...,G.PROSP.28,333-358.

```
**************************************************************************
* TITLE : HIGHER RESOLUTION MARINE SEISMIC DATA      *      PROJECT NO        *
*         ACQUISITION AND PROCESSING.                *                         *
*                                                    *   TH./01039/84/NL/..    *
*                                                    *                         *
**************************************************************************
* CONTRACTOR :                                       *    TELEPHONE NO         *
*    TECHNISCHE HOGESCHOOL DELFT, AFD MIJNBOUWKUNDE   *                         *
*                                                    *    015 785190           *
*    POSTBUS 5028                                    *                         *
*    NL - 2600 GA DELFT                              *                         *
*                                                    *    TELEX NO             *
*                                                    *                         *
* TECHNICAL DIRECTOR :                               *    38151                *
*    PROF. A.M. ZIOLKOWSKI                           *                         *
*                                                    *                         *
**************************************************************************
```
VERSION : 01/01/87

AIM OF THE PROJECT :
The aim of the project is to improve the resolution of marine seismic data by improved data acquisition methods and corresponding improvements in the data processing. In particular, an accurate description of the angular-dependent source wavefield must be provided and new data processing software developed to utilize this information.

PROJECT DESCRIPTION :
A data set already exists consisting of a seismic line shot over a logged well in the North Sea. The line has been shot with a number of different source configurations and in each case the source wavefield can be determined. The well logs provide a check on the processing of the data in two ways: first, synthetic seismic data can be calculated from the well logs and compared with the real data; second, the real data can be deconvolved and inverted to obtain density and velocity logs that can be compared with the real well logs.

STATE OF ADVANCEMENT :
Completed.

RESULTS :
The validity of three new methods for specifying the source wavefield for marine seismic data has been tested. The interaction method and the source decomposition method had already been validated before the start the project with the result that the interaction method was viable, succesful and likely to be easy to use, and the source decomposition method was unsuccesful. The third method, the scaling method, has been tested within the project, and was very succesful, surviving an experiment in which it stood at risk of being refuted. However, it is not easy to use. The conclusion is that the interaction method can be used in production to determine the full wavefield of the source. Also, software has been developed and tested for taking account of source directivity during processing. The source directivity can be specified if the interaction method is used in acquisition. The angular dependent deconvolution can be performed on CMP gathers in the frequency wavenumber domain if the dip of the layers is less than 10 degrees, and if the target depth is very large compared with the dimensions of the source array. With these assumptions, the software that has been developed performed well in tests and was subsequently applied to real data. The software was applied to the data from the Delft Air Gun Experiment, and the expected improvement in resolution over current acquisition and processing methods was tested. It is shown that, even when the far field source signature in the vertical direction is known from measurements, the optimun signature deconvolution of the data is inferior to the angular dependent deconvolution of the data. The increase in resolution at the target depth is noticeable, as expected. It is shown that the conventional statistical approaches to deconvolution fail to extract the known wavelet, probably because the whitness assumption for the earth reflection response is not valid for seismic data.
The whole improved data acquisition and processing route was tested by comparing the results with offset-dependent synthetic seismograms calculated from well data. The forward modelling is done by calculating individual plane wave responses using the reflectivity methodand a matrix extension for the elastic case, superposing the plane waves to create the point source response, and then superposing the point source responses using the notional source concept. In this scheme the biggest problem has been the superposition of the plane waves via the Fourier-Bessel transform. Six different schemes have now been developed with various trade-offs between accuracy and speed. Despite this very large effort into the modelling of synthetic seismograms from the well log data, we did not succeed in getting a very good match with the measured data. However, we feel that the real reason for our lack of success here is that the well logs are not accurate.

REFERENCES :

6

****
- ZIOLKOWSKI A.M. REPORT NO. 1987/6: FINAL REPORT
- FOKKEMA J.T. AND ZIOLKOWSKI A.M. THE CRITITCAL REFLECTION THEOREM. GEOPHYSICS 52, NO 7.
ZIOLKOWSKI ET AL, THE SIGNATURE OF AN AIR GUN ARRAY: COMPUTATION FROM NEAR FIELD MEASUREMENTS, INCLUDING INTERACTIONS. GEOPHYSICS 47, NO 10.
DRIJKONINGEN G.G. AND FOKKEMA J.T., THE EXACT SEISMIC RESPONSE OF AN OCEAN AND A N-LAYER CONFIGURATION. GEOPHYSICAL PROSPECTING 35, 33-61.

```
***************************************************************************
* TITLE : INVERSION OF SEISMOGRAMS              *      PROJECT NO        *
*                                               *                        *
*                                               *   TH./01042/84/FR/..   *
*                                               *                        *
***************************************************************************
* CONTRACTOR :                                  *   TELEPHONE NO         *
*    GERTH                                       *                        *
*                                               *   1 47 52 61 39        *
*    AVENUE DE BOIS PREAU 4                      *                        *
*    FR - 92502 RUEIL-MALMAISON                  *                        *
*                                               *   TELEX NO             *
*                                               *                        *
* TECHNICAL DIRECTOR :                           *   203 050              *
*    MR KOLB                                      *                        *
*                                               *                        *
***************************************************************************
```

VERSION : 18/04/88

AIM OF THE PROJECT :
****
This project deals with the inversion of reflection seismic data, in other words
attempting to find the subsurface geological parameters that best explain the
surface recorded seismic traces.
The subsurface is approximated as an acoustic medium where wave propagation is
described by only two parameters: velocity of propagation and acoustic impedance
(for 1D-models only the latter is involved).
These two parameters are the unknowns for the inverse problem.
The first objective of this project is to study the feasibility of several
inversion type problems (depending on the geometrical hypotheses : 1D, 2D
horizontally stratified, heterogeneous 2D).
The second goal is to write software for processing large field data volume.

PROJECT DESCRIPTION :
****
The reflection seismic method can be briefly described by a source near the
surface which produces a propagation wave in the substratum. The resulting
perturbation which is reflected, refracted or diffracted is measured at the
surface by hydrophones (Gi) as pressure (or displacement) Pobs (Gi,t) as a
function of time.
The forward problem involves the computation of synthetic seismograms Pm (Gi,t)
that correspond to a given subsurface model m. Acoustic wave propagation is
assumed.
The inverse problem, briefly explained, will involve the search of an earth
model m that produces seismograms Pm (Gi,t) as close as possible to the observed
seismograms Pobs (Gi,t).
The first step is then to construct a tool that efficiently resolves the forward
problem, inversion requiring the computation of J(m) for a large number of
models m.
The next step will require the construction of a second basic tool which can
calculate the gradient of J with respect to the model m.
An optimization strategy must then be defined, entirely automatic or at the
contrary interactive, conveniently integrating a priori information of the model.
A sensitivity study has had to be done, with two objectives: one was to
determine the space of admissible models in which the search of the solution was
stable; the second objective was to find the best optimization variables along
with a correct preconditioning of the problem to increase the stability and the
convergence rate.
Many software tests were required, first with noisy synthetic data, and later
with field data.
Three inversion problems have been studied:
A 1D inversion : plane waves modelling in an horizontally stratified media.
Acoustic impedance as a function of travel-time is sought.
B 2D inversion of horizontally stratified media: waves propagation is 2D, the
model to be found is 1D. The information available for different offsets allows
the search for both velocity and acoustic impedance.
C Heterogeneous 2D inversion : the geometry of the entire problem is 2D, which
means that there will be a considerable number of unknowns, and that a large
number of seismic records is required to supply enough information.
This problem is therefore very complex as well as very costly to study.

STATE OF ADVANCEMENT :
****
Completed

RESULTS :
****
Part A of the project, involving 1D inversion, has been developed. Numerous
theoretical as well as pratical results can be found in the literature,
describing the possibilities and the difficulties of this problem.
From these results, after the construction of a solid modelling and gradient
calculation software package we were able to quickly set a path towards its
practical application.
A software package was created for stratigraphic extrapolation of well data

using 1D inversion. A seismic profile of zero offset traces is constructed, after which each trace is inverted by using the low frequency impedance trend of the previous trace. The first inversed trace uses the trend from the well. This tool appears to be a very useful on for nearly horizontal stratigraphies. Nevertheless, field acquisition problems have been highlighted. These problems have to be solved before the tool can acquire a certain credibility.

Parts B and C of this project related to 2D inversion have needed the invention of new techniques to solve the problems associated with finding a velocity field. These techniques (use of travel time instead of depth, progressive downward continuation in time, progressive increase in frequencies, adapted unknowns) were successfully implemented in the problem B. This has led to a satisfying solution for synthetic data of large dimensions.

For heterogenous 2D media, these new techniques are much too heavy to be implemented at the moment. There is however, one aspect that has been solved in a very successful fashion : that is finding the impedance reflectors when the velocity field is approximately known. The results consist in a refinement of a migration before stack using the wave equation.

```
*******************************************************************************
* TITLE : UTILISATION OF DRILLING NOISES IN      *      PROJECT NO        *
*          SEISMICS.                              *                       *
*                                                 *    TH./01043/84/FR/..  *
*                                                 *                       *
*******************************************************************************
* CONTRACTOR :                                    *    TELEPHONE NO        *
*    GERTH                                         *                       *
*                                                 *    1 47.52.61.39       *
*    AVENUE DE BOIS PREAU 4                        *                       *
*    FR - 92502 RUEIL-MALMAISON                   *                       *
*                                                 *    TELEX NO            *
*                                                 *                       *
* TECHNICAL DIRECTOR :                            *    203 050             *
*    MR. P. GROS                                  *                       *
*                                                 *                       *
*******************************************************************************
                                                       VERSION : 01/01/87
```

AIM OF THE PROJECT :

This project is based on the utilisation of the drill bit as an acoustic source. Seismic signals emitted during the abrasion of the rock by the drill bit are recorded via seismic pick-ups placed in a special manner on the surface or in nearby wells.
After these signals have been processed and interpreted in real time, it is possible to obtain data comparable to that provided presently off-line by seismic measurements taken in wells, for example the equivalent of a "speed log" in relation to the depth, the seismic profile of a well, or a 3D image of formations crossed in the vicinity of the well. Thus, it is a matter of studying and finalizing appropriate methods and equipment, thus providing access to a better instant acknowledgement of the crossed formations, and allowing to improve drilling procedures.

PROJECT DESCRIPTION :

To achieve this goal, it is necessary to use both, the seismic signals received at the surface on a seismic type apparatus, or on special sensors placed in nearby wells, and the signals transmitted by the drill string and received at its head, as the latter signals can be representive of a pseudo-signature of the seismic signals transmitted by the drill bit.
This project has been divided in five major phases :
Phase 1 - Measures at the drill string
Phase 2 - Measures at the surface geophone
Phase 3 - Combination of both types of measures
Phase 4 - Well seismic source
Phase 5 - Special data acquisition and processing device on the site.
Phase 1, 2 and 3 are closely related, as phase 3 can only be executed once Phases 1 and 2 have been carried out on the same site and at the same time.
In this regard, measures were made at three different sites in France :
LE MAYET DE MONTAGNE, in July 1984, while drilling in granite for the purpose of geothermals studies. Measures were taken at a depth between 200 and 800 m, and recorded with a 47-trace surface device over a length of about 700 m and placed radially to the well, and on a set of sensors placed in bores near the well. An accelerometer was fitted at the end of the drill string. Drilling was executed out in the open with a bottomhole drill hammer.
LACQ SUPERIOR, in May 1985, during a vertical drilling operation, very near a deviated well. This particularity has allowed us to record with the help of well geophones, the signals transmitted by the bit drilling at depths of about 400 m, and measured at horizontal distances inferior to 50 m. The purpose was to compare the signals recorded at the upper end of the drill string with those recorded at the drill bit.
SOUDRON 116, in October 1985, when drilling a production well on the Soudron field, south of Chalons sur Marne in the Paris Basin. During this field data acquisition, measures were taken between 1000 and 2100 m deep (one level of measures for every string adjunction) on seismic devices comprising sensors placed, both, at the surface and in bores at a depth of about 40 m.
Phase 3 consisted mainly of the acquisition via numerical processing of seismic profiles of the transposed and multi-offset wells, or of data on the vertical seismic speeds by using the records provided by surface devices and drill strings, and the repeated cross-correlations.
Phase 4 consisted in a feasibility study and in the definition of specifications in view of designing a prototype seismic source while drilling in case the feasibility results of phases 1 to 3 proved insufficient at the level of the signal to noise ratio transmitted only by the drill bit.
Phase 5 of the project depended mainly on the positive results obtained during the previous phases. The purpose of phase 5 was to study and make an in-situ acquisition and processing prototype unit, capable of providing in real time the data required to run the drilling operation and to pursue exploration.

STATE OF ADVANCEMENT :

Phases 1, 2 and 3 have been completed while using only the drill bit as seismic source. For phase 4, a number of manufacturers of drilling equipment, both

French and American, have been investigated. American equipment is not available and it is not possible, within the scope of this project to study and develop a seismic source while drilling. Studies interrupted on 31.12.86.

RESULTS :
****

Studied carried out during phase 3, that is the combination of drill string measurements and surface measurements, have led to the following conclusions :
1. Seismic profiles for transposed wells can be obtained from the sole recording of drill bit noises, but the quality of the results remain poor, and depends partly on the hardness of the formation and on the drilling mode (ordinary rotation or percussion rotation).
2. The signal to noise ratio remaining low, it would be useful to improve it by using a seismic source on the drill string near the drill bit.
3. The spectrum of seismic signals, emitted at the bit during drilling and transmitted in the soil is relatively high frequency and may reach 300 to 400 Hz.
4. In most cases, seismic signals recorded at the top of the drill string are not faithfully representative of the signals transmitted at the bit.
For a better correlation with recordings on devices spread at the surface, it is advised, during cross-correlations of traces, to use the "pseudo-signatures" recorded near the bit. However, this raises the problem of their transmission up to the surface, along the drill string.
5. Obtaining an evaluation of the mean speeds of the formation sections between the bit and the surface, and in a correlative manner, of the speeds of seismic waves in vertical sections (section speed) is possible through a statistical study of the curvatures on seismic profiles without the use of plots at the top of the drill string. However, the results obtained like the previous ones are closely linked to the quality of the signal to noise ratio (soft terrains providing hardly any positive result). Their improvement implies the elaboration of a seismic source.
6. Different drilling equipment have been studied both in Europe and in the U.S. A inherent to the study of a seismic source while drilling. There are very few studies on this subject and because of the present context, there is no real incentive for manufacturers to pursue research in this domain. Consequently studies have been interrupted at this stage of the project.

```
********************************************************************************
* TITLE : DEVELOPMENT OF A TECHNIQUE TO EXPLORE      *     PROJECT NO          *
*         SUBSURFACE GAS-WATER CONTACTS.             *                         *
*                                                    *   TH./01046/84/IR/..    *
*                                                    *                         *
********************************************************************************
* CONTRACTOR :                                       *   TELEPHONE NO          *
*    KISH DEVELOPMENT LTD                            *                         *
*                                                    *   01/332211             *
*    162 CLONTARF ROAD                               *                         *
*    CLONTARF                                        *                         *
*    IR - DUBLIN 3                                   *   TELEX NO              *
*                                                    *                         *
* TECHNICAL DIRECTOR :                               *   33438                 *
*    DR. D. NAYLOR                                   *                         *
*                                                    *                         *
********************************************************************************
                                                    VERSION : 29/08/89
```

AIM OF THE PROJECT :
****
To examine the geological and geophysical problems of detecting and validating
subsurface gas-water interfaces in hydrocarbon reservoirs as an aid to the
direct detection and mapping of gas reserves. A number of known gasfield
examples in northwest Europe were examined and compared to other possible, but
unproven, examples. In particular a horizontal subsurface seismic feature in the
Kish Basin, offshore Ireland, was detailed by a seismic survey, and later
penetrated by an exploration well, at which time geological-geophysical studies
were made of the critical rock section.

PROJECT DESCRIPTION :
****
Phase 1: carefully monitored seismic acquisition (with large volume water guns)
and processing of reflection seismic data over the Kish Basin example and the
production of time and depth maps of the principal reflecting horizons.
Delineation of the gas water contact. Maturation studies of existing well
samples.
Phase 2: detailed lithological/petrological studies of the two existing Kish
Basin wells and analysis of the possible gas-water contact in the Basin.
Documentation of known gas-water examples from other areas world-wide. Drilling
of a deep exploration well in the Kish Basin and study of the results.
Phase 3: complete analysis of Kish Basin results.
Studies of three gas-water contacts with reprocessing of seismic data and
detailed analysis of electric log and petrological data.
Phase 4: final report preparation.

STATE OF ADVANCEMENT :
****
A study of European gas-water contacts. The ability of the seismic method to
detect gas-water contacts has been demonstrated in recent years, although the
precise parameters for optimum recording are not clearly understood. Other
subsurface phenomena may also mimic a gas-water contact on the seismic record.
The project considered the optimum technical package available to the industry
in gas-prone areas.

RESULTS :
****
100 km of additional reflection seismic data were obtained to examine a possible
gas-water interface in the Kish Basin. At the same time a maturation study of
samples from the existing Amoco 33/22-1 well and related areas was aimed at a
study of source potential. Processing of the Kish Basin lines demonstrated that
the reflection was probably not a fluid-contact reflection. Petrographic and
electron microscope studies of samples from the nearby shell 33/21-1 well showed
that the potential Triassic reservoir demonstrated high average porosities with
only limited secondary mineral development to inhibit permeability.
As part of a wider documentation of gas-water contacts about 20 published and
unpublished examples have been collected worldwide.
These examples are from a wide variety of geological settings and show that for
recognition of a fluid contact refection it must be a positive reflection,
observed in horizontal or near-horizontal attitude, in a trappint position,
against a background of moderate dip.
Drilling of the Kish Basin 33/17-1 well took place in May-June 1986. The well
was plugged and abandoned as a dry hole at 6600 feet (2012 m). Results from the
well substantially confirmed the geological prognosis for the structure and
confirmed that the anomalou flat refector on the prospect was not a fluid-
contact reflection.
The study showed that where seismic quality is good and reservoir thickness/gas
column are adequate, then careful acquisition and processing can allow a
determination of whether the gas-liquid contqct is horizontal and meaningful
estimates of reservoir porosity and gas column. Confirmation of porosity by
amplitude study is worth a test and may on occasion indicate, by comparison with
velocity data, a component from a porosity step caused by differential
cementation. Where such a step is separated from the present gas-water contact
it represents a fossil gas-liquid contact, a study ofd which can reveal the

B

history of folding and tilting since gas emplacement and may allow definition of the commercial base of the gas accumulation. Improving seismic techniques allow increasing recognition of gas-liquid reflections in poor-quality data areas (for example from reservoir porosities of the order of 15 %). Nevertheless, hoaxes can occur, and the report describes a number of these. In terms of exploration philosophy the report points out that a gas-liquid reflection is the most significant indication of hydrocarbons that is available in exploration and that it should be exhaustively investigated prior to drilling. The absence of a reflector may also be of utmost importance and factors influencing this are analysed. The study of gas-liquid reflections is useful not only in the positive sense of confirming the presence and volume of gas, but also in the negative sense of avoiding dry holes.

```
*************************************************************************
* TITLE : ACCESS (PHASE 1)                        *    PROJECT NO       *
*                                                 *                     *
*                                                 *  TH./01047/85/FR/.. *
*                                                 *                     *
*************************************************************************
* CONTRACTOR :                                    *  TELEPHONE NO       *
*    GERTH                                         *                     *
*                                                 *  1 47.52.61.39      *
*    AVENUE DE BOIS PREAU 4                        *                     *
*    FR - 92502 RUEIL-MALMAISON                    *                     *
*                                                 *  TELEX NO           *
*                                                 *                     *
* TECHNICAL DIRECTOR :                            *  203 050            *
*    MR. P. LALOUEL                                *                     *
*                                                 *                     *
*************************************************************************
```
                                                     VERSION : 31/12/88

AIM OF THE PROJECT :
****
The aim of the project was to develop computer aided interpretation stations to
meet all the requirements of explorers.
The development of an integrated as well as a portable package was initiated
during the Phase I of the ACCESS project.
Basis tools and routines were built up to be used by subsequent applications.

PROJECT DESCRIPTION :
****
The phase I of the project was subdivided into two sub-phases. Sub-phase 1 deals
with the Kernel of the software. Specialized routines to be used by any
applications were developed and thoroughly tested to ensure an user friendly
dialog between man and machine. Other subroutines are dedicated to the
management of the data (extensive use of a relational data base structure). A
graphic Editor was developed to manipulate the graphic objects (ACCESS uses
graphics extensively). A numeric Interpretor was set up to allow the user to
customize the software whenerver requested.
Sub-phase 2 deals with some of the applications in relation to the
interpretation, The topographical module will manage constant data i.e. limits
of concessions, well locations, seismic profiles... The single well module will
carry out the log interpretation while multiwell module will work on several
wells to allow cross sections, correlations, average value calculations for
instance.
Geophysical data will be used through several applications modules such as
Horizon (Management of any kind of horizon), Velocity (time vs depth conversion),
 Trace Attributes (processing of signals), Synthetic Seismograph, Modelling
through Ray Tracing.

STATE OF ADVANCEMENT :
****
All the above mentioned modules are under betatesting or user testing at the end
of ACCESS Phase I.

RESULTS :
****
It is definetely too early to conclude on a project which is still going on (see
ACCESS Phase II). Most of the application modules give satisfactory results.
However, some improvements as far as performance is concerned are to be
implemented.ref

```
******************************************************************************
* TITLE : MULTIPLEXED STREAMER OF SECOND        *      PROJECT NO          *
*         GENERATION.                           *                          *
*                                               *                          *
*                                               *   TH./01049/85/FR/..     *
*                                               *                          *
******************************************************************************
* CONTRACTOR :                                  *    TELEPHONE NO          *
*    GERTH                                       *                          *
*                                               *                          *
*    4 AVENUE DE BOIS PREAU                      *    1 47.52.61.39         *
*    FR - 92502 RUEIL-MALMAISON                 *                          *
*                                               *    TELEX NO              *
*                                               *                          *
* TECHNICAL DIRECTOR :                          *    203 050               *
*    MR. BEAUDUCEL                               *                          *
*                                               *                          *
******************************************************************************
                                                    VERSION : 30/03/89
```

AIM OF THE PROJECT :
****
Feasibility study of a multiplexed streamer with very high number of traces
(1500), through realisation of subassemblies.

PROJECT DESCRIPTION :
****
Important aspects of this streamer are the following :
- Very low traction noise has been obtained by improving the buoyancy balance of
the immersed part. This has led to the study :
    - the weight reduction of boxes containing the electronics and/or the
distribution of the electronics along the streamer elements
    - the weight reduction and the homogenization of the lineic density of the
streamer elements.
- An efficient filtering of residual noises :
    - through the utilisation of continuous hydrophones
    - by considering their location at the level of the frequency/number of wave
(fanshaped filtering).

STATE OF ADVANCEMENT :
****
The following are underway: - Measures on the new cylindrical hydrophones with
variation of parameters:
- nature and thickness of support
- nature of absorbing agent
- Definition of the final structure of prototype hydrophones used in test
sections
- Study of hydrophone fixation in the flute section
- Achievement of a preamplifier-filter waterproof module.

RESULTS :
****
Waterproof modules: the new VAC assembly provides the following performances:
- total distorsion < 10-4 on a dynamic range of 50 dB
- consumption < 800 mw that is 100mw/channel
- preamplifier-filter model in process of miniaturization measured with
following results:
 -dynamics > 105 dB
 -distorsion < 10-4
Weight reduced boxes:
End of stage III of specifications for the optical fiber transmission.
Fabrication of titane boxe containing 2 transmitters and 2 receivers, and
realisation of printed circuit : 2 prototypes are being assembled.
Continuous hydrophones :
Measures on new cylindrical hydrophone with acoustic screen following empirical
determination of best support and acoustic screen:
- capacity : 60 nF/meter of hydrophone
- sensitivity: 20 V/bar
- sensitivity versus acceleration : < 1mV/g

REFERENCES :
****
PATENT INHERENT TO VARIABLE GAIN AMPLIFIERS
3 PATENTS RELATIVE TO CONTINUOUS HYDROPHONES
PATENT FOR THE IMPROVED SAMPLE AND HOLD UNIT
PATENT FOR THE ACOUSTIC SCREEN INCREASING THE SENSITIVITY OF THE CONTINUOUS
CYLINDRICAL HYDROPHONE
NO PUBLICATION YET AVAILABLEPATENT FOR THE VAC ASSEMBLY
2 PATENTS FOR THE HYDROPHONE
NO PUBLICATION YET AVAILABLE.

```
****************************************************************************
*  TITLE : MULTIWELL FIELD GEOPHYSICS                *       PROJECT NO     *
*                                                    *                      *
*                                                    *   TH./01050/85/FR/.. *
*                                                    *                      *
****************************************************************************
*  CONTRACTOR :                                      *   TELEPHONE NO       *
*      GERTH                                         *                      *
*                                                    *   1 47.52.61.39      *
*      AVENUE DE BOIS PREAU 4                        *                      *
*      FR - 92502 RUEIL-MALMAISON                    *                      *
*                                                    *   TELEX NO           *
*                                                    *                      *
*  TECHNICAL DIRECTOR :                              *   203 050            *
*      MR. DELVAUX                                   *                      *
*                                                    *                      *
****************************************************************************
                                                        VERSION : 31/12/87
```

AIM OF THE PROJECT :
****
The object of the project is to develop methods allowing to obtain geophysical
data on the nature of the terrains between different wells within a same field.
The aim is to improve our knowledge on the reservoir layer with the help of
methods complementing those already existing and which are generally implemented
from the surface, such as seismics or electric methods or which are implemented
directly in the wells and presenting a limited investigation range, such as logs.
The implementation of new methods in wells would provide, at any point of the
reservoir located between these wells, significant and thorough data allowing:
- a better static comprehension of the field :
  new structural definition, location and identification of accidents
  (faults, reefs, etc...), facies variations, etc...
- to obtain data on the dynamic behaviour of the reservoir : preferential drains,
  water inlets, evolution of contacts between fluids.

PROJECT DESCRIPTION :
****
Two major types of methods are being studied :
- methods based on the propagation of acoustic waves between wells
- methods based on the measurement of phenomena provoked like the
electromagnetic measurements between two wells.
One of the particular features of the methodology envisaged within the scope of
this project, resides in the integrated interpretation of results deriving from
the different methods.
Last of all, the latter includes a feasibility phase and a study, execution and
prototype testing phase.
SEISMIC METHOD
Three orientations have been followed, each corresponding to a different
approach of the waves propagation phenomena in the reservoir environment :
- study by transparency :
while using travel time inversion techniques (first arrivals and selected
reflections). The obtained result is a fine repartition of the geological speeds
between wells. This technique, may, for example, evidence undetectable speed
anomalies thanks to the knowledge of the media crossed at each bore.
- study by reflection :
  very high frequency seismics implemented between wells may lead to a very
performing visual display of the reservoir. However, due to geometrical
particularities at acquisition level, its processing is not yet mastered from
the software standpoint. Thus, it is necessary that we prove our ability in
emitting and recording seismic signals in a path band according to the
objectives of the project, and that we finalize a processing method.
- study on guided propagation :
This phenomenon occurs when a number of conditions concerning speeds, geometry
of layers, absence of heterogeneities, are respected. Here we are with an
original research orientation which merits exploration (modelling).
ELECTROMAGNETIC METHODS
As the electric properties of a rock depend very closely on its fluids contents,
this approach seems, on a conceptual level, paricularly suitable to evidence
fluids and contacts between fluids.
However, the transposition of these methods, already used in surface geophysics,
remains to be studied more thoroughly within a feasibility phase which is triple
:
- study of the reservoir and of the possibilities of these methods (modelling
aspect, sensitivity, resolution force).
- drilling environment study
- technological study to draft a prototype pilot chain.
MEANS
All these different approaches mentioned above can only be studied, in a first
step, through the modelling tools which will have to be designed or adapted, and
can only be achieved once specific equipment will have been implemented in the
wells.
- seismic :
three different methods, but a single acquisition chain which should be very
wide band (up to 500 to 1000 Hz), of rapid utilisation (important number of

traces), non destructive at source level.

STATE OF ADVANCEMENT :
****
Seismic method :completed.
The three methods envisaged have been tested and appraised with the help of
studies performed on models. The effective data obtained with the first
prototypes provided results in harmony with the objectives of the project.
Electric method : the theoretical feasibility has been proved.

RESULTS :
****
SEISMIC METHOD
- by transparency : development, adaptations and tests on synthetic data of
three 2D or 3D inversion programs (collaboration IFP, SNEA(P), IRIGM).
- by reflection : familiarize with this type of information thanks to an
important volume of synthetic data modelling a complete acquisition between two
wells. Utilisation of the latter for special classification, in processing
softwares adapted to the seismic profiles of wells with offset.
Continuation of the modelling for the integration of the tube wave phenomena in
the transmitting wells and their interaction with the volume waves (mechanism
which proved preponderant).
- guided waves : different configurations showing the theoretical interest of
such an approach have been modelled (reservoirs presenting faults, or
compartments with slow or rapid speed anomaly).
- acquisition : an important measurement program with the help of four different
well sources, on two different sites. Recording was executed with a conventional
3-component sonde and a wide band 48-channel vertical streamer specially
developed for this project. The various sources that have been tested are
successfully : the shear gun of CPGF, the weight drop on the packer of IFP, the
explosive and the sparker of SWRI. Recorded signals have a path band of 500 to
1000 Hz.
As far as the first acquisitions are concerned, the signal to noise ratio is
variable and the sensitivity of the streamer has had to be improved (bottom
preamplification).
A new test site with wells distant of 450 m has been chosen, so to transpose the
results obtained in 1986 on to a more realistic frame. But it was impossible to
record any data (problems of source and noise level in the wells).
- processing : ongoing.
   These are difficult owing to the low signal to noise ratio of the recorded
signals particularly with the well streamer. The elimination of tube waves, a
stack coverage and a PSO treatment allows to recover a field of reflected waves.
After migration, the obtained image presents a space resolution of about 1 metre.
 This is a very promising result..
ELECTROMAGNETIC METHOD
- Theoretical and model studies have made it possible to conclude on a
theoretical feasibility.
- Having proved that measures are possible across the tubing, the last studies
showed that stability of the field response has no relation with the tubing.
- A prototype measuring chain is defined on paper.
Nevertheless, each field study will remain a particular application and a wide
experience will be necessary to adapt the different choices with the raised
problems (methods, acquisitions, processing).

REFERENCES :
****
- A PUBLICATION IN THE REVUE DE L`ENERGIE (NR 391)
- A CONFERENCE AT CONGRESSES SPE AND SEG
- DIFFERENT WORKING DOCUMENTS BETWEEN PARTNERS.

```
********************************************************************************
* TITLE : VERTICAL SEISMIC PROFILING AND        *        PROJECT NO          *
*         INTERACTIVE MODELING                   *                            *
*                                                *    TH./01051/85/DE/..      *
*                                                *                            *
********************************************************************************
* CONTRACTOR :                                   *    TELEPHONE NO            *
*     PRAKLA-SEISMOS AG                          *                            *
*                                                *    0511/642 3522           *
*     BUCHHOLZER STR. 100                        *                            *
*     DE - 3000 HANNOVER 1                       *                            *
*                                                *    TELEX NO                *
*                                                *                            *
* TECHNICAL DIRECTOR :                           *    922419 + 922847 PRAK*
*     DR. H.A.K. EDELMANN                        *    L D                     *
*                                                *                            *
********************************************************************************
                                                      VERSION : 30/08/89
```

AIM OF THE PROJECT :
****
The project aims at a better exploitation of VSP data for drillhole planning.
The interactive operation allows to directly include survey data, log data and
VSP data for the decision. The VSP measuring data are directly introduced to
develop a subsurface model.

PROJECT DESCRIPTION :
****
The project consists of two groups of activities : data acquisition and onsite
data processing on one side and sophisticated large computer data processing and
interpretation on the other.
Data acquisition consists of two groups of elements : the sonde the cable and
the seismic sources, and which are lowered into the borehole the recording
equipment and onsite data processing computer at the earth's surface. The sonde
is designed to meet the environmental specifications, and to reduce the round
trip time (lowering the sonde down to the largest depths and measuring when
pulled up back to the surface) to a minimum. For this purpose a 3-component
receiver chain, consisting of up to 5 elements with a predetermined spacing of
25 m was designed. The number of elements allows to reduce the round trip time
by recording simultaneously at different depths. For deviated wells with dip
angles of more than 30 deg. against vetical, a gimbled 3-component geophone
receiver was selected for installations in each of the elements.
The recording time can be appreciably reduced when using multi-source technique.
This technique can be preferably applied together with the VIBROSEIS method.
Multi-source technique can be applied for multi-offset VSP and for simultaneous
P- and S-wave recording. The onsite data processing was designed as part of the
quality control of the recording process. It allows to continuously control the
data quality and thus minimize recording time without loss of data quality.
The software for large computer data processing enables a careful preparation of
the data acquisition by determining the optimum parameter and by combining all
available data and the survey data in a straightforward manner for later
interpretation.
A first way of analyse consists of transferring the seismic results into a 3-
dimensional picture of the subsurface in the environment of the well.
This process, called migration, works sufficiently well in all cases where the
layer velocities are known and layer spacing is not too small compared to the
wavelength of the seismic signal.
For complicated structures, however, these conditions are normally not met. In
this case an interactive approach must be made, starting for the subsurface
model constructed from the available data before the VSP measurement has been
made and then refining this model.
In some cases a dip analysis, made from multi-offset VSP data, can help to
analyse difficult structures. A polarization analysis must be made when shear
waves areinvolved, especially when polarization effects have to be considered.

STATE OF ADVANCEMENT :
****
Ongoing investigations concern the onsite data processing and the azimuthal
orientation tests. Both are in the construction phase.
Test of the receiver chain in a deep well in cristalline rock will be finished
this year.

RESULTS :
****
The receiver chain has been successfully tested in boreholes of up to 3000 m. In
these tests the capabilities for multi-offset VSP could not yet be fully
exploited. The gimbled 3-component element, designed for deviated wells, was
successfully tested for dip angles up to 55 deg.C against vertical. The
application of the Disco-software system was abandoned in favour of an onsite
data processing, based on a personal computer system. The software for modeling
and migration has undergone final acceptance tests. Dip analysis and
polarization analysis were successfully applied to field data.

REFERENCES :

****
KOEHLER, K., KOENITG,M., 1986: IMPROVEMENT OF MULTI-OFFSET VSP DATA BY SPECIAL
DATA PROCESSING. 48 TH EAEG-MEETING, OSTENDE, BELGIUM
WIERCEYKO,E., NOLTE, E., 1987: ERFASSUNG SEISMICHER DATEN IM BOHRLOCH. 47,
JAHRESTAGUNG DER DGG IN CLAUSTHALZELLERFELD

```
************************************************************************
* TITLE : DEVELOPMENT OF A COST-EFFECTIVE          *     PROJECT NO    *
*         AMPHIBIOUS SEISMIC SURVEY SYSTEM          *                  *
*                                                   *  TH./01056/85/UK/.. *
*                                                   *                  *
************************************************************************
* CONTRACTOR :                                      *   TELEPHONE NO   *
*    AMPHIBIOUS SEISMIC LTD                         *                  *
*                                                   *   0202 733703    *
*    36A ORCHARD AVENUE                             *                  *
*    POOLE                                          *                  *
*    UK - DORSET BH14 8AJ                           *   TELEX NO       *
*                                                   *                  *
* TECHNICAL DIRECTOR :                              *                  *
*    MR. R.M. BRASINGTON                            *                  *
*                                                   *                  *
************************************************************************
                                               VERSION : 11/10/89
```

AIM OF THE PROJECT :
****
The aim of the project is to improve the technical capability and cost-
effectiveness of the existing system by improving the speed of operation, the
system capability in bad weather and in hard soils, the water resistance of
equipment, the deep water performance and the system logistics.

PROJECT DESCRIPTION :
****
The project evolved as a result of seismic surveys carried out in the U.K. A
dynamite system based on minimal draft barges, using flushing pumps operated
through a well in the middle of the barge was developed. This method uses
centrifugal pumps and a specially designed flushing tool lowered from an A-frame
over the well. Using geophones positioned on the sea bed and survey lines marked
with wooden posts, the system was used successfully for surveys in the U.K. The
reason for the project was to develop the system into a fully amphibious cost-
effective system able to operate in a wide range of transition-zone prospects.
The project was divided into eight phases as follows :
1. Sub-system review. At this stage the sub-systems were reviewed, and additions
to system capability were agreed.
2. Sub-system development. Equipment was improved and developed.
3. Sub-system field trials. Comparative time trials were conducted for selected
sub-systems under varying conditions and terrain.
4. System integration. Attention was switched from the individual sub-system to
an analysis of the system as a whole.
5. System field trials. System field trials were held where operations were
studied under varying conditions. Problems were identified and means of
streamlining operations examined.
6. System modifications. These were made on the basis of the field trials.
7. System acceptance trials. Contract pressure conditions were simulated in
which the system was transported to a prospect and used under operating
conditions.
8. Final report.

STATE OF ADVANCEMENT :
****
The project is at the final report stage.

RESULTS :
****
STAGE 1 - SUB-SYSTEM REVIEW
Possible additions to system capability were assessed including amphibious
vehicles telemetry systems and bay cables.
A suitable lightweight drill was chosen to form the basis of a new design of
minimal draft drilling barge.
A new design cutting device operated from the flushing jet was considered.
The sub-system; surveying, flushing, geophones, shooting and recording were
reviewed to ensure optimal choice of equipment.
Experiments were conducted with alternative methods of achieving amphibious
capability for the recording sub-system.
Experiments were conducted with alternative flushing pumps.
Alternative methods of marking the line and fastening the cables and geophones
were evaluated.
Alternative boats were reviewed and propulsion units evaluated.
STAGE 2 - SUB-SYSTEM DEVELOPMENT
Equipment has been improved and developed including the flushing tools and the
gear for raising and lowering flushing equipment.
All equipment liable to immersion in sea-water has been evaluated and design
studies are proceeding to minimise the risk of water damage.
Consideration has been given to redesigning the equipment for faster laying and
picking up under water.
The working depths for laying and picking up geophones have been increased.
For low tide conditions special equipment including mud boards are being
developed.
The bad weather capability of the flushing and drilling barges has been improved.

STAGE 3 - SUB-SYSTEM FIELD TRIALS
Trials were conducted for selected sub-systems under varying conditions and terrains, this involved :
- flushing and drilling in different sub-soils
- line marking, laying and picking up in water at different depths and over mud, sand, marsh and reeds
- rough and calm weather conditions
- working in still and fast moving water

STAGE 4 - SYSTEM INTEGRATION
Standardisation of equipment has been considered for compatibility of spares. Transport systems to prospects for equipment and craft are being devised for easy crew moves. Equipment design has taken into account the necessity for dismantling and ease of packing.

Operations from a mother ship or a land base far from hard slip-way facilities have been studied as well as methods for towing and carrying sub-systems close to work site. Methods of lifting sub-systems aboard and launching from the mother ship are also under development.

Systems for the rapid transport of men and equipment from base to work site are being developed, e.g., systems for quickly securing the drilling barge over the shot holes.

The technical capability and cost-effectiveness of the original system have been improved. Further work is needed on some of the additions to the system.

```
*********************************************************************************
* TITLE : DEVELOPMENT OF A TRANSIENT-              *      PROJECT NO         *
*         ELECTROMAGNETIC DEPTH SOUNDING SYSTEM    *                         *
*         FOR HYDROCARBONS AND GEOTHERMAL          *   TH./01059/85/DE/..    *
*         RESOURCES EXPLORATION                    *                         *
*********************************************************************************
* CONTRACTOR :                                     *   TELEPHONE NO          *
*    UNIVERSITY OF COLOGNE                         *                         *
*                                                  *   49/221 4702552        *
*    INST FOR GEOPHYSICS AND METEOROLOGY           *                         *
*    ALBERTUS-MAGNUS-PLATZ                         *                         *
*    DE - 5000 KOELN 41                            *   TELEX NO              *
*                                                  *                         *
* TECHNICAL DIRECTOR :                             *   8882279               *
*    DR K.M. STRACK/PROF. F.M. NEUBAUER            *                         *
*                                                  *                         *
*********************************************************************************
                                                     VERSION : 01/08/89
```

AIM OF THE PROJECT :
****
In 'no seismic record' areas transient electromagnetic soundings can provide
information of the subsurface structure. Our aim is the development of a
transient EM field system including a portable receiver, field application,
development of data processing and interpretation techniques. This work is to
be carry out in conjunction with the oil industry in order to directly evaluate
the potential application of the technique.

PROJECT DESCRIPTION :
****
The project is divided into hardware, data processing, interpretation, and
technology transfer tasks. The hardware development included improvements and/or
completely new developments of parts of the entire field system. A transient
electromagnetic field system consists of a grounded wire, high current
transmitter, a portable receiver and a data processing and interpretation
workstation which is used for the in-field data processing and preliminary data
interpretation. The development of data processing techniques included the
design, building and testing of a complete pre-stack data processing system
which allows us to obtain a sufficient signal-to-noise ratio even in Western
Europe. The research conducted on the interpretation part of transient electromagnetic data included improvements of the inversion to obtain for the
interpreter inversion statistics and error bounds of the individual earth model
parameters, three dimensional numerical modelling, and three dimensional
analogue scale modelling. Strong collaboration with the oil industry and other
scientists helped to obtain the necessary feedback for the system design and
specification. During all phases of the project, field tests were done to check
hardware, data processing and interpretation schemes.

STATE OF ADVANCEMENT :
****
The project has been successfully completed.

RESULTS :
****
The developed hardware consists of a grounded wire transmitter, a portable, DC
operated digital data acquisition system, data processing and interpretation
workstation which is being used to produce the preliminary in-field
interpretation. Major developments and improvements were done for all parts of
the hardware. The entire hardware was tested from the very beginning to the very
during test surveys in the field.
The data processing was completely redesigned and successfully tested using a
pre-stack data processing system. This system significantly increases the signal-
to-noise ratio. During all stages of the processing, quality control routines
can be run, thus testing the reliability of the data.
In order to adjust the developed technology to related exploration problems
within the European Community, several numerical feasibility studies were
conducted for exploration problems (hydrocarbon and geothermal) in several
countries of the European Community.
A new approach of further improving the signal-to-noise ratio has been developed
and tested. We call this approach "local noise compensation" which is being done
by maintaining a base station during an entire survey day as reference station.
Mobile receivers are then moved around in a vicinity of that base station. Under
the assumption that the noise does not very much spatially the field data of the
mobile receivers can be subtracting the noise derived at each particular time of
the day at the base station.
The interpretation of the transient electromagnetic data was improved by adding
a statistical inversion package to the interpretation which gives error bounds
of the individual earth model parameters. Also three dimensional modelling was
done to test particularily anomalous transients. The three dimensional modelling
of specific problems can either be done using numerical modelling techniques
such as integral equations or reduction of the surface anomaly onto a thin
latterally inhomogeneous conductive plate. With these techniques we were able to
simulate several phenomenons which we have observed in the field such a sign
reversals during he measurement time of the signal.

During all phases of the project these techyniques were successfully tested in the field by carriying out field surveys over several geophysical targets. These targets included immediate exploration targets as well as geophysical comparative studies. In almost all cases the results compared very well with other geophysical/geological informations giving us confidence in our development.

REFERENCES :
****
APPROX. 40 PUBLICATIONS AT NATIONAL AND INTERNATIONAL TECHNICAL MEETINGS.
PUBLICATIONS IN INTERNATIONAL JOURNALS :1

```
**********************************************************************
* TITLE : INVESTIGATION OF THE APPLICATION OF THE    *    PROJECT NO     *
*         ABSORPTION OF SEISMIC WAVES IN HC-         *                   *
*         EXPLORATION                                *    TH./01065/86/DE/.. *
*                                                    *                   *
**********************************************************************
* CONTRACTOR :                                       *    TELEPHONE NO    *
*    D.G.M.K.                                         *                   *
*                                                    *    040 326479     *
*    STEINSTRASSE 7                                  *                   *
*    DE - 2000 HAMBURG 1                             *                   *
*                                                    *    TELEX NO       *
*                                                    *                   *
* TECHNICAL DIRECTOR :                               *    211 466        *
*    DR. M. ALBERTSEN                                *                   *
*                                                    *                   *
**********************************************************************
```

VERSION : 11/09/89

AIM OF THE PROJECT :
****
Seismic methods are worldwide the most sophisticated tools in geophysicals
exploration of mineral resources. By these methods, the stratigraphy of the
subsurface is discovered commonly by use of the material specific travel times
of the reflected seismic waves.
It is will known, that seismic waves are additionally affected in their
amplitudes and frequencies by rock materials. The application of the so-called
adsorption or attenuation of seismic waves (ASW) in praxis as key to rock and
pore fluid characterization is still not realized because of non available
interpretation tools.
Fundamentals for the analysis and interpretation of seismograms on absorption
phenomena were investigated in a previous phase, this new project concerns with
the applicability of absorption evaluation of seismic data in practical
exploration.

PROJECT DESCRIPTION :
****
The project is concepted as joint venture of research institutes and 8companies
of the oil and gas industry. The institues are concerned with the experimental
and theoretical work. The companies are contributing the geological and seismic
data of actual reservoirs as well as their experience in practical operation.
Main goal of this project is to evaluate the applicability of seismic
attenuation methods in field exploration on oil and natural gas. This general
purpose includes a number of different objectives and innovations :
1. Comparison of the different theoretical methods for the interpretation of
field seismograms
2. Installation of the best methods into existing interpretation software for
field application
3. Full analysis of data at exemplary test reservoirs in connection with all the
available geologic data of the fields
4. Correlation of ASW with other seismic-, log- and core-data respectively
5. Synthesis of all results to a working schedule for practical use in the
industry
The general intention is to contibute to a quick introduction of ASW-methods
into field exploration.

STATE OF ADVANCEMENT :
****
Completed

RESULTS :
****
Main results are :
- A joint experiment on the applicabiliy of the different methods to evaluate
the absorption of seismic waves is completed.
- The application of the different methods on seismic field data (VSP and
reflection seismic) brought up important experiences in the ASW-evaluation of
real data.
- The method "Analytical Signal" developed by Prof. Engelhard, Tu Braunschweig,
in the foregoing ASW-project has been completed. It has been tested as a routine
in a processing software on a broad range of field data.
- The different ASW-mechanisms have been investigated in a extensive laboratory
program on cores from outcrops and wellbores
- In connection with petrophysical investigations different theories are
evaluated to describe the interelation between petrophysical and seismic
parameters.
The results are actually under final interpretation. As final synthesis of the
project results a sophisticating overview on the applicability of the absorption
of seismic waves in HC-exploration is in preparation.

REFERENCES :
****
BURKHARDT, H., PAFFENHOLZ, J., SCHUETT, R
"ABSORPTION OF SEISMIC WAVES (ASW)"
DGMK RESEARCH REPORT 254

```
*****************************************************************************
* TITLE : BOREHOLE SOURCE                              *    PROJECT NO      *
*                                                      *                    *
*                                                      *  TH./01067/86/FR/..*
*                                                      *                    *
*****************************************************************************
* CONTRACTOR :                                         *   TELEPHONE NO     *
*    GERTH/C.G.G/G D F                                 *                    *
*                                                      *   1 47.52.61.39    *
*    4, AVENUE DE BOIS PREAU                           *                    *
*    FR - 92502 RUEIL-MALMAISON                        *                    *
*                                                      *   TELEX NO         *
*                                                      *                    *
* TECHNICAL DIRECTOR :                                 *   203 050          *
*    MR. LAURENT                                       *                    *
*                                                      *                    *
*****************************************************************************
                                                     VERSION : 30/06/89
```

AIM OF THE PROJECT :
****
The project consists in developing a borehole seismic source able to work down
to oil field depths (around 3000 m). It will allow the use of new borehole
seismic methods (well-to-well measurements, inverse method) and will improve the
knowledge of reservoir structures.

PROJECT DESCRIPTION :
****
The project is to be developed in three phases:
PHASE 1 - Theoretical prototypes
Two theoretical prototypes have been built:
- the first one uses a drill string to lift a weight which is released by the
triggering signal and drops on a striking surface equipped with a suitable
damper. The potential energy can be changed by changing the weight (2000 or 3000
J). The coupling of the source with the wall is provided by a packer.
- the second one uses a conventional logging cable with seven conductors to
transmit energy from the surface to the tool. It uses the mud pressure to
actuate a mobile system and can be adapted to work as an implosion source or a
mechanical impact source. A large-surface hydraulically driven clamping system
has been developed to insure the best coupling conditions. The potential energy
is about 2000 J at a 1000 m depth.
The site (property of Elf Aquitaine) includes four boreholes.
PHASE 2 - Design and construction of prototype. This phase started end 1987.
PHASE 3 - Testing on the well. This phase is scheduled to start during the
second term of 1988.

STATE OF ADVANCEMENT :
****
Ongoing

RESULTS :
****
The first prototype has been carefully tested on the surface before a borehole
test performed in October 1986. The test site (property of Elf Aquitaine)
includes four boreholes. Recording devices (3-components borehole geophones and
48 channel borehole streamers) were lowered in two of them. The source was
placed in a third one. A surface spread was also set. The source was fired
between 550 and 450 m depth. More than 700 shots were recorded without any
failure.
   The second prototype has been surface-tested with a pressure cell simulating
the borehole conditions. More than 300 shots have been achieved during the ast
months 1986.
The second prototype has been implemented during two tests on a test site (ELF
AQUITAINE). The measuring device placed on the surface consisted in 48 traces
with vertical geophones and 48 traces with horizontal geophones along the
profile-traces at intervals of 20 meters that is a total length of 960 meters.
During the tests performed end June good seismic signals were obtained at the
surface, the source being buried between 800 m and 1100 m. The coupling device
was then modified so it could be used as a confinement system : results were not
satisfactory. Considering the results obtained with these prototypes, Phase 2
began with the study of a system including a weigth drop and a hydraulic
coupling device; this source connected to the surface via a logging wire is a
synthesis of both prototypes. The prototype will be available in March 1988. A
first version of the weight drop prototype and hydraulic coupling device has
been built and developed during the first semester. A first seismic test between
2 wells distant of 550 m at depths ranging between 150 m and 450 m has been run
end June with recording on a well stramer. Two other test were completed in July
and December 1988. During the last one 1500 shots were performed in 2 boreholes
at depths ranging from 400 m to 500 m. Records were obtained with a borehole
geophone, a well streamer in two other boreholes and a surface display. This
first version is now available for industrial jobs. The processing performed on
the results of the December 1988 operation produced excellent seismic survey
results between wells. The emission spectrum reached frequencies of 400 Hz.
This version was also used in an April 1985 transposed seismic survey test in a

cased well, at depths between 850 and 950 meters. Initial results indicate a fairly low seismic energy on a surface spread, but this experiment should be repeated. The final version is being perfected and well tests should be carried out befor the end of 1989.
Results obtained with prototypes 1 and 2 are mentioned in the following references.

REFERENCES :
****
- IFP REPORT "PROCESSING AND STUDY OF THE SEISMIC WELL TEST RUN AT CHATEAURENARD FROM 29 SEPTEMBER TO 3 OCTOBER 1986" BY J.L. MARI, REFERENCE IFP 35.206
- COMMUNICATION PRESENTED AT SPE (SOCIETY OF PETROLEUM ENGINEERS) IN SEPTEMBER 1987, ENTITLED "ACQUISITION TECHNIQUES IN CROSS-HOLE SEISMIC SURVEYS
 BY J. DELVAUX, G. NOUAL, L. NICOLETIS, J.F. DUTGER
- IFP REPORT "WELL-SURFACE SEISMICS WITH WELL SOURCE OF IMPLOSION AND SHOCK TYPE" BY J.L. MARI - IFP REFERENCE 35657 OF NOVEMBER 1987.
-COMMUNICATION PRESENTED AT THE EAEG (EUROPEAN ASSOCIATION OF EXPLORATION GEOPHYSICISTS) IN JUNE 1988 ENTITLED "WELL-TO-SURFACE SEISMIC MEASUREMENTS WITH A DOWNHOLE SOURCE" BY J.LAURANT AND J.L. MARI.

```
********************************************************************************
* TITLE : CONTROL OF THE PROPAGATION OF HYDRAULIC    *    PROJECT NO          *
*         FRACTURES                                  *                        *
*                                                    *    TH./01068/86/FR/..  *
*                                                    *                        *
********************************************************************************
* CONTRACTOR :                                       *    TELEPHONE NO        *
*    GERTH                                           *                        *
*                                                    *    1 47.52.61.39       *
*    4, AVENUE DE BOIS PREAU                         *                        *
*    FR - 92502 RUEIL-MALMAISON                      *                        *
*                                                    *    TELEX NO            *
*                                                    *                        *
* TECHNICAL DIRECTOR :                               *    203 050             *
*    MR. SARDA                                       *                        *
*                                                    *                        *
********************************************************************************
```

VERSION : 29/03/89

AIM OF THE PROJECT :
****
Provide methods and techniques allowing a better evaluation of the technical
risk associated with hydraulic fracturing techniques in difficult practiical
cases : reservoirs presenting natural fractures, small productive thickness,
possible extension of the fracture either up into the overburden or down into
the water level. As the values of the horizontal principal stresses in the
reservoir and in the adjacent layers is the principal active parameter, methods
will be developed for measuring these stresses. In order to evaluate the
efficiency of these methods the dimensions of the fractures will be evaluated
using several techniques.

PROJECT DESCRIPTION :
****
The project is to be developed along three trends:
Phase 1 :
Measurement of the in-situ horizontal stresses and fracture propagation. In the
problem of hydraulic fracturing the measurement of the horizontal stresses
requires a precision which is not usual in the field of earth sciences. Indeed
the minimum in-situ horizontal stress has to be evaluated with an absolute error
inferior to one megapascal in deep or very deep geological layers. Presently,
the adequate methods and/or techniques do not exist. In order to set-up these
methods the following technical achievements are foreseen on each experimental
site :
- minifracturations at several levels in the overburden and in the reservoir
rock, measurement of the downhole pressure and of other useful parameters
- minifracturation in the openhole under the casing shoe
- prefracturation (or fracturation with a limited amount of fluid and without
propping agents) in the reservoir and with measurement of the bottomhole
pressure.
Phase 2 :
Determination of the dimensions of hydraulic fractures. In order to evaluate the
dimensions of hydraulic fractures three techniques are envisaged : measurement
with inclinometers of the displacements induced at the earth surface by the
hydraulic fracture, detection of the acoustic emissions accompanying propagation
and/or closure of the hydraulic fracture. An attempt will be made to evaluate
the length and direction of the fracture with the help of the first method, the
length direction and height of the fracture using the second method, the
thickness of the fracture using the third method.
Phase 3 :
Synthesis. All the data obtained on several experimental sites will be
interpreted. So will be compared the various techniques and methods experimented
on sites and the existing models which simulate fracture propagation.
Will be proposed:
- recommended methods for the measurement of stresses and of fracture extension
- recommended criteria for the selection of a numerical model of fracture
propagation depending on the actual mechanical and geological conditions of the
hydraulic fracturing operation
- feasibility criteria for hydraulic fracturing operations
- recommendations for the on-site control of fracture propagation.

STATE OF ADVANCEMENT :
****
Advancement program. Three main activities : drafting of a technical guide book
for the minifracturation operations, minifracturations on sites and associated
measurements, interpretation of the minifracturation tests.

RESULTS :
****
Minifracturation on a large size block:
A new series of tests has been performed during July 1988 based on successive
reopening and propagations of the fracture initiated in the block. Peak pressure
and propagation pressure appear to be linear functions of the minor external
stress. Influence of the major external stress is negligible. Any development of
the existing fracture even at very low flow-rate and without external stresses

needs a non negligible pressure. Conclusions about fracture close-up pressure
are not so clear. The block was extracted and cut for analysis of the fracture
geometry. This analysis is now in progress.
Analysis of microfracturations and prefacturations:
This analysis has been performed on 6 wells for which syntheses have been
written. Different analyses are used depending on the type of test
(microfracturation, prefacturation, flow-back test, fall-off test). A summary
of measured stresses in overburden and reservoir layers is presented.

REFERENCES :
****
PH. CHARLEZ, P. HERAIL, D. DESPAX "DETERMINATION DE PARAMETRES DE FRACTURATION
HYDRAULIQUE PAR INVERSION DES COURBES DE PRESSION.

```
*******************************************************************************
* TITLE : EVALUATION OF LITHOLOGIC PARAMETERS IN      *      PROJECT NO       *
*         HYDROCARBON RESERVOIR BY SIMULTANEOUS        *                       *
*         APPLICATION OF COMPRESSION AND SHEAR         *   TH./01069/86/DE/..  *
*         WAVES                                        *                       *
*******************************************************************************
* CONTRACTOR :                                         *   TELEPHONE NO        *
*    D.G.M.K.                                           *                       *
*                                                      *   040 326479          *
*    STEINSTRASSE 7                                    *                       *
*    DE - 2000 HAMBURG 1                               *                       *
*                                                      *   TELEX NO            *
*                                                      *                       *
* TECHNICAL DIRECTOR :                                 *   211 466             *
*    DR. M. ALBERTSEN                                  *                       *
*                                                      *                       *
*******************************************************************************
```

VERSION : 02/09/89

AIM OF THE PROJECT :
****
Exploration efforts are recently more and more directed to hydrocarbon
reservoirs of the "faziell type". Reservoirs of this type are indicated by a
lateral change in the pay zone lithology independent of geological structures.
For the exploration of those reservoirs - e.g. in South Germany - it is
necessary to obtain the total information of seismic data.
Bedsides the travel times of seismic waves, the amplitudes of the reflexions are
important in the above mentioned sence.
At this background, a project LIPS is scheduled to investigate the potential of
comprehensive analysis of seismic data (p and s) in "seismic lithology".

PROJECT DESCRIPTION :
****
The project is scheduled as joint venture in cooperation with three university
institutes and eight oil companies. It includes theoretical work, experiments
and field investigations.
The main goal is the analysis, further development and combination of methods
for the evaluation of lithologic parameters of hydrocarbon reservoirs.
One of the prerequisites for the application of lithoseismic methods and for
the involution into exploration programs is a comprehensive data base,
especially of those reservoirs with lithologic transitions.
The quality of registrations and the availability of both p- and s- waves are
the main criteria for the selection of test reservoirs.
It is planned to analyse selected seismograms in detail besides the installation
of full sections.
A special objective is the relation between amplitude behaviour of the
reflection and its angle of incidence.
In addition to the seismic registrations, well-based observations by means of
logs and VSP are planned. Wellbore waves like the "Stoneley wave" are to be
investigated on their potential for density and permeability evaluations.
Laboratory investigations on rock materials are directed to obtain correlations
between seismic and petrophysical parameters.
The main technical goals are summarized as follows:
- evaluation of reservoir parameters by the combined use of different types of
seismic waves (p-, s- and Stoneley waves)
- optimization of the analysing methods by wavelet processing with concern to
the reflection amplitudes and phases
- interrelation of logging- and VSP -parameter with seismic data along surface
related profiles
- further improvement of the generation, application and interpretation of shear
waves.

STATE OF ADVANCEMENT :
****
All parts of the project are still ongoing. The project will be finished at the
end of 1990.

RESULTS :
****
Interesting results are obtained in different parts of the project :
- modelling of thin and outcropping layers
- comparison of reflectivity and finite difference models
- approximation techniques for calculating reflection coefficients
- improvements of PS-processing and of signal-noise ratio
- development of a elastic model for borehole interface waves
- theoretical and experimental investigations of amplitudes and phases of p- and
s-waves.
The results have to be verified and completed till the end of the project. A
final syntesis of the results is planned to give more insight into the
possibilities to evaluate lithologic parameters in hydrocarbon reservoirs by
simultaneous application of compression and shear waves.

```
**********************************************************************
* TITLE : SIMULATION MODEL FOR RESERVOIR         *     PROJECT NO     *
*         HETEROGENEITIES                         *                    *
*                                                 *   TH./01070/86/FR/.. *
*                                                 *                    *
**********************************************************************
* CONTRACTOR :                                    *    TELEPHONE NO    *
*     GERTH                                        *                    *
*                                                 *   1 47.52.61.39    *
*     4, AVENUE DE BOIS PREAU                      *                    *
*     FR - 92502 RUEIL-MALMAISON                   *                    *
*                                                 *    TELEX NO        *
*                                                 *                    *
* TECHNICAL DIRECTOR :                            *    203 050         *
*     MR. RAVENNE                                  *                    *
*                                                 *                    *
**********************************************************************
                                                  VERSION : 30/03/89
```

AIM OF THE PROJECT :
****
Elaboration of a geological deposit model shaped by a mathematical model in
order to provide a description of the heterogeneities at metric and hectometric
scale of the physical properties of a sandstone reservoir that can be used by
the reservoir engineer in view of an optimal field exploitation.

PROJECT DESCRIPTION :
****
The project is divided in two successive phases:
PHASE 1:STUDY OF THE FORMATION
The object of the field acquisition in two and three dimensions was to acquire
geological,geophysical and petrophysical data allowing set-up the model.
These studies will be carried out on the middle Jurasic outcrop of the Yorkshire
cliffs in Great Britain.The selected cliff is a remarkable representation over
10 kilometers length and a height of 30 to 150 m of a wide variety fluvial to
fluvio-deltaic deposits,hardly altered by the tectonics and thus providing a
good image of their original representation.The sandstonelike sedimentary
elements (eventual reservoirs) show extensions ranging from dozens of meters
(clay dominant) to several kilometers (sandstone dominant).The light overburden
of these elements allows easy access through small core-drills located at the
back of the cliff,and thus makes it possible to obtain physical properties
measurements less altered than those acquired on the outcrops.Core-drills will
also provide us with well calibrated log shapes.
The sedimentologic study of outcrops will include a photogeological study of the
whole cliff,performed with the help of documents obtained by helicopter.A set of
vertical cross-sections will then made,thus providing a detailed quantitative
description of the different sandstone or clay elements in relation with the
deposit mode at a scale below the kilometer.
The execution of core-drills near the outcrops will allow the passage from a 2-D
heterogeneities description to a 3-D description at a kilometric scale and the
comparison of the obtained results with those of the oil drillings thanks to the
recording of logs.
The physical properties will be measured on samples taken from the outcrops and
on plugs sampled from the cores. A geostatistic interpretation of the values
obtained will then be proposed.
High resolution georadar profiles will connect the core-drill with one another
and with the outcrops in order to improve the threedimensional image of the
studied sedimentary elements.
PHASE 2 : ELABORATION OF THE 3-D GEOSTATISTIC MODEL
This model aims at describing the heterogeneities that can be the object of a
direct exploitation by the reservoir engineer in order to optimize the
development of fields. The general geological data will thus have to be entered
in the numerical model of the field.
The model will be elaborated by integrating the results acquired during the
sedimentologic studies and put in shape by the geostatistical processing.
A first step will consist in achieving a 2-D model based, among others, on the
study of the images of the cliff.
The second step will consist in achieving a 3-D model deriving from the 2-D
model, the 3-D geostatistical modelling and the sedimentologic model.
A third step will consist in controlling the model according to real data from a
North Sea field.

STATE OF ADVANCEMENT :
****
Completed

RESULTS :
****
To solve the problem, heterogeneous reservoirs simulations have been set up :
1. a methodology for studying outcrops and geological data
2. a new method of simulations of random series : the method of trunketed
Gausian functions. The application of this method requires the respect of the
horizontal stationnarity. This 2D method tested on a gas storage site with the
parameters deriving from the study of a site in Yorkshire can only be

generalized if addtitional studies are carried out namely :
- consideration of the non stationnarity according to the horizontal
- consideration of the limits of the units and of their reference surface by
simultaneous simulations of lines (2D) or of surfaces (3D)
- acquisition and processing of new field data to set up database allowing to
calculate the geostatical parameters in view of calibration and/or as a
complement of the results obtained with the sole well data.
Some environments or changes of scale of the simulations (different magnitude
order of the heterogeneities) may lead to the elaboration of new additional
simulation methods.
This simulation method presents can be applied to a number of environments of
the fluvio-deltaic type caracterized by sub-parallel strata with perfectly
defined plane limits. The program allowing to run these simulations constitutes
a research tool. This tool should be industrialized to make its utilization
convivial.

REFERENCES :
****
PRELIMINARY RESULTS OF PERFORMED OPERATIONS HAVE BEEN THE OBJECT OF THE
FOLLOWING PUBLICATIONS :
1) "HETEROGENEITIES AND GEOMETRY OF FLUVIO-DELTAIC RESERVOIRS" BY G. MATHERON ET
AL.SPSEDIMENTARY BODIES IN A FLUVIO-DELTAIC RESERVOIR" BY C> RAVENNE ET AL.
2) "CONDITIONAL SIMULATION OF THE GEOMETRY OF FLUVIO-DELTAIC RESERVOIRS" BY G.
MATHERON ET AL. SPE PAPERS 16752 AND 16753 PRESENTED AT SPE ANNUAL TECHNICAL
CONFERENCE AND EXHIBITION, DALLAS SEPT. 1987.
3)
GEOMETRIE DE RESERVOIRS FLUVIO-DELTAIQUES DANS LES SERIES DU JURASSIQUE MOYEN DU
YORKSHIRE "BY R. ESCHARD ET AL. PRESENTED AT THE FIRST FRENCH SEDIMENTOLOGY
CONGRESS - PARIS NOVEMBER 1987.
4),5) TWO COMMUNITIES AT BSRG MEETING HELD AT LONDO

```
*****************************************************************************
* TITLE : DIRECT DETECTION OF HYDROCARBON          *      PROJECT NO       *
*         DEPOSITS USING ADVANCED SEISMIC          *                       *
*         TECHNIQUES                               *    TH./01072/86/UK/..  *
*                                                  *                       *
*****************************************************************************
* CONTRACTOR :                                     *    TELEPHONE NO        *
*    BRITISH GAS PLC                               *                       *
*                                                  *    01 736 3344 X 4140  *
*    LONDON RESEARCH STATION                       *                       *
*    MICHAEL ROAD                                  *                       *
*    UK - LONDON SW6 2AD                           *    TELEX NO            *
*                                                  *                       *
* TECHNICAL DIRECTOR :                             *    24670              *
*    DR. A. MELVIN                                 *                       *
*                                                  *                       *
*****************************************************************************
```

VERSION : 31/08/89

AIM OF THE PROJECT :
****
In the exploration for hydrocarbon deposits, seismic techniques are the methods
of first choice for the initial survey, both onshore and offshore. The aim of a
seismic survey is to delineate the sedimentary structures under the ground or
below the seabed and to use this delineation to identify potential structural
traps for hydrocarbons.
A seismic technique which is specifically sensitive to natural gas deposits and
to just lithology would be highly advantageous. The initial objective of the
British Gas project is to evaluate the feasibility of the developing such a
technique, which would be innovative and would facilitate the reliable location
of gas deposits. If the technique were adopted generally, a saving of some 80%
exploration drilling costs could be possible.

PROJECT DESCRIPTION :
****
Two complementary lines of investigation are being floowed, (1) the development
of a seismic source based on new principles, and (2) the use of interconversion
processes between shear waves and compressional waves which occur in sedimentary
lithology. The development of the new source must overcome the difficulty that
conventional sources which generate shear waves cannot be used near the surface
of the sea since shear waves cannot propagate through water.
Such sources must therefore be used at seabed level. Onshore, both vibrator and
explosive charge techniques are available for shear wave generation. A design
for a new seismic source based on controlled (supersonic) shock-wave generation
(without the use of explo9sives) has already been developed. This source can be
operated also for conventional developments as a high efficiency compressional
source of wide frequency bandwidth, but it has features in the mode of
generating shock waves which will allow it to generate shear waves of known
properties in sedimentary strata. The project as envisaged in this application
would involve both new source development and advanced signal processing.

STATE OF ADVANCEMENT :
****
Ongoing. The project is in the field trial phase where the seismic device has
been successfully tested. The results are being monitored and evaluated.

RESULTS :
****
The seismic device has now been fully-developed as a field prototype. The shock
wave valve design has now been fully-optimised and the source can be operated
routinely to give shock pressures of ca. 100 bar. Triggering of the device
repetitively is now possible by means of a flap valve pressure release on the
high pressure chamber of the device.
Based on the experience obtained with this 25 mm internal diameter prototype,
designs for 50 mm and 75 mm diameter sources have been generated for a planned
increase in acoustic power of an order of 10 in magnitude.
Tests in the laboratory have enabled us to characterise in detail the physical
performance of the seismic device and to study the attenuation of shock waves in
water.
Digital data collection and signal processing equipment has been developed in
the laboratory for field use with geophone and hydrophone detectors. Use has
been made of Fast Fourier Transform-based signal processing computer software to
process experimental seismograms on Compaq 386 microcomputer systems.
Field trials have taken place aimed at assessing the performance of the
prototype seismic source at the Purton borehole site near Swindon, Wiltshire.
This provices a simple sedimentary structure whre a one hundred metre depth of
Oxford clay lies on limestone with a well-defined reflector boundary. The four
boreholes on site are of approximately 100 metre depth and are cased.
A particular advantage is that both compressional and shear acoustic velocities
in the clay have been mapped in detail for various depths on the site by a
number of university geosciences groups.
In its field test form, the seismic source is mounted on a Land Rover four-wheel
drive vehicle and the recording instrumentation is carried in a separate van.
Field trial tests have been carried out with the source firing directly into the

ground. Hydrophones in the boreholes at various depths have been used to detect direct seismic compression waves and to measure wave velocities.
Geophones at offsets of up to 620 metres have provided information on direct surface waves, ground roll and reflections at the clay-limestone boundary for the source used in its compressional wave mode. Currently, shear wave studies are being undertaken in which geophones with acoustic responses in three dimensions are being used as detectors. This research is currently ongoing.

REFERENCES :
****

A UK PATENT FOR THE SEISMIC DEVICE HAS BEEN GRANTED AS GB 2165945A, PUBLISHED 23RD APRIL 1986. A UNITED STATES PATENT HAS ALSO BEEN GRANTED : PATENT NUMBER 4,667,766 DATED MAY 26, 1987.

```
*****************************************************************************
* TITLE : TWO DIMENSIONAL BALANCED SECTION        *       PROJECT NO       *
*         SOFTWARE                                 *                        *
*                                                  *   TH./01073/86/UK/..   *
*                                                  *                        *
*****************************************************************************
* CONTRACTOR :                                     *     TELEPHONE NO       *
*     MIDLAND VALLEY EXPLORATION LTD               *                        *
*                                                  *     041 332 2681       *
*     14 PARK CIRCUS                               *                        *
*     UK - GLASGOW G3 6AX                          *                        *
*                                                  *     TELEX NO           *
*                                                  *                        *
* TECHNICAL DIRECTOR :                             *     8950511            *
*     DR. A.D. GIBBS/DR.J.NICHOLSON                *                        *
*                                                  *                        *
*****************************************************************************
                                                       VERSION : 29/03/89
```

AIM OF THE PROJECT :
****
Section balancing is a powerful and simple but very time consuming method of
testing whether or not a seismic or geological map-controlled interpretation of
rock structure (on any scale, from prospect size to basin-wide) is geometrically
sensible. The aim of the project is to develop an interactive method which will
accept the interpreter's digitised first-attempt interpretations, perform checks
on bed lengths and areas, and allow serial restoration of movement on fault
surfaces, by any amount specified. It should calculate and substitute fault
shapes as an alternative to those initially input; accept any geometrical
configurations of beds and be applicable in all geological environments,
compressive or extensional or any mixture of styles; possess full graphics
editor facilities with add and delete options on-screen; and include a
decompaction capability to deal with the problem that rock sequences change
thickness as the sediment pile increases with time.

PROJECT DESCRIPTION :
****
New advances in balanced section construction and basin development have been
pioneered by MVE. These new advances facilitate the identification of new
exploration prospects and aid their interpretation. The construction of balanced
sections is now a vital step in validating seismically derived maps of prospects.
 The same techniques can be used to improve mapping of structurally complex
fields.
The planned program allows generation of new sections from the digitized input
by various operations, such as displacements on fault systems. Many sections can
be held in core at once, with the capability to select, view and operate upon
any one of them. Sections may be written to disk and later retrieved. They can
also be plotted.
The "active" operations which generate new sections from the input include the
simulation of displacement on a fault system. This operation can be used in two
ways: to remove displacements on faults which have offset marker horizons
(inverse modelling), or to simulate the progressive development of faults
(forward modelling). Another "active" operation is the construction of non-
planar fault profiles from the geometries of hanging wall rollovers.
The second class of operations are "passive", that is, they do not alter the
appearance of the section. There are two such operations: a calculation of total
bed length for each stratigraphic horizon, and a calculation for the area of a
specified formation or region of the section. Area calculations are used to
predict the estimated depth to detachment of controlling faults.
Program development and testing has proceeded rapidly and the scientific core of
the code is essentially complete. The program is graphics-intensive and the
initial version incorporates Tektronix software under licence and abstracted
from their Plot 10 STI; this form of the program has been written for a MicroVax
II, Tektronix 4107 or 4207 combination. Installations using the Tektronix
terminals have also been achieved on Nord machines, and several digitising and
plotter combinations have been employed. A GKS variant has been prepared and run
on a VAX workstation.Program development proceeded rapidly and the scientific
core ot the code is complete. The program is graphics-intensive and incorporates
Tektronix software under licence and abstracted from their Plot 10 STI; this
form of the program has been written for a MicroVax II, Tectronix 4107or 4207
combination. Installations using Tectronix terminals have also been achieved on
Nord and Data General machines, and various digitisingand plotter combinations
have been employedAn early GKS variant has been prepared and run on a VAX
workstation. GKS standard routines have been written to support different input
and ouput devices.

STATE OF ADVANCEMENT :
****
The program is now robust and is proving very effective in provididng a basis
for geologists and geophysicists to work together on interpretation problems.
Further improvements are planned and these centre on palinspastic reconstruction
method and mapping techniques. GKS implementation requires much further work,
partly because of operating system limitations outside our control; this
activity ia closely related to future workstation applications.

RESULTS :
****
Problems so far encountered have been in designing portability and compatibility
into the system.
Technical trials so far have confirmed that the program core is scientifically
sound.
Tests in a variety of basins have shown that BSP represents a significant level
of new capability in structure interpretation. Structure modelling can be put on
an objective basis and predictions can be tested systematically and very quickly.
 There is a great potential to develop the program.BSP is now used by three
major oil companies in exploration and development work, by two state geological
surveys, and is running on several academic sites.It is also on extended trials
with several more oil companies and promises to establish itself a significant
and distinctive interpretational software package.Users consider it to be a
specialist application but not one is requiring a high level of training. It is
the first program ot its type to allow routine balance exercises together with
full compaction/decompaction modelling capability.

REFERENCES :
****
BEACH,A.1987. A REGIONAL MODEL FOR LINKED TECTONICS IN EUROPE, P43-48IIN :BROOKS,
J. & GLENNIE K.(EDS)PETROLEUM GEOLOGY OF NORTH WEST EUROPE. GRAHAM AND TROTMAN
LONDON.
BEACH,A.,BIRD,T.AND GIBBS A.D. 1987.EXTENSIONAL TECTONICS AND CRUSTAL STRUCTURE:
DEEPP SEISMIC REFLECTION DATA FROM THE NORTHERN NORTH SEA VIKING GRABEN 467-476.
IN: COWARD,DEWEY AND HANCOCK (EDS),EXTENSIONAL TECTONICS,SP.PUB.28,GEOL.SOC.LOND.
BIRD< T.J.,BELLA.,GIBBS,A.D.& NICHOLSON,J. 1988.ASPECTS OF STRIKE-SLIP TECTONICS
IN THE INNER MORAY FIRTH BASIN,OFFSHORE SCOTLAND. NORSK GEOLOGISK TIDSSKRIFT,
VOL.67,NO 4PP.353-369.
GIBBS.A.D. 1987. LINKED TECTONICS OF THE NORTHERN NORTH SEA> IN: GEOL.SOC.LOND.
SPEC. PUBL. LOND. 640 PP.

```
***************************************************************************
* TITLE : RELATIVE PERMEABILITY MEASUREMENTS          *      PROJECT NO      *
*         UNDER SIMULATED RESERVOIR CONDITIONS        *                      *
*                                                     *  TH./01078/86/DE/..  *
*                                                     *                      *
***************************************************************************
* CONTRACTOR :                                        *    TELEPHONE NO      *
*     PREUSSAG AG                                     *                      *
*                                                     *    05176-17288       *
*     ARNDTSTRASSE 1                                  *                      *
*     DE - 3000 HANNOVER 1                            *                      *
*                                                     *    TELEX NO          *
*                                                     *                      *
* TECHNICAL DIRECTOR :                                *    92655             *
*     DR. R. SOBOTT                                   *                      *
*                                                     *                      *
***************************************************************************
                                                    VERSION : 06/09/89
```

AIM OF THE PROJECT :
****
The object of this project is to develop suitable methods for the determination
of oil/gas/water relative permeabilities under reservoir conditions in the
laboratory. The relative permeability data is of great importance in the
planning stage as well as the production development of secondary (waterflooding)
 and tertiary (chemical flooding, as drive, steam injection etc.) recovery
projects. The reservoir behaviour is simulated with a computer model which
depends heavily on reasonable input data including oil/water/gas relative
permeabilities. It is expected that measurements under simulated reservoir(pT-)
conditions yield more meaningful results and will improve the reservoir
engineering calculations. Of innovative character will be the installation of a
system that continuously monitors the saturation conditions in the core sample
during linear water/oil displacement experiments under reservoir conditions and
the development of a computer programme creating analytical

PROJECT DESCRIPTION :
****
The project comprises five stages:
1. Design and construction of a small-scale installation for linear flooding
experiments on sandstone plugs (3 cm diameter and 5 cm length)
2. Determination of oil/water relative permeabilities with the small flooding
installation. The suitable procedures will afterwards be employed on PREUSSAG's
high-pressure autoclave.
3. Reconstruction of the high-pressure autoclave for flooding experiments with
reservoir oil.
4. Measurement of oil/water and gas/water relative permeabilities under
reservoir conditions of an oil reservoir from the Osthannover area and a gas
reservoir from the Ostfriesland area.
5. Evaluation and interpretation of the collected data.
During the first stage of the project a small flooding installation is designed
and built in order to screen the suitable methods available for relative
permeability evaluation. This will save time and cost because only successfully
tested methods will be used for PREUSSAG's high-pressure autoclave.
The second stage includes the testing program for the small flooding
installation. Oil/water relative permeability will be evaluated from linear
flooding experiments and gas/water relative permeability measurements from
steady-stade flooding experiments. Hereby the average oil/water/gas saturations
in the core will be determined by electrical resistivity measurements.
The third stage will be the implementation of the selected methods on the high-
pressure autoclave. This requires the installation of pressure vessels for
reservoir oil and a separator at the autoclave exit.
In the fourth stage the high-pressure autoclave is put into operation for
oil/gas/water relative permeability measurements under reservoir conditions
simulating real reservoir parameters such as temperature, pore pressure,
overburden pressure, viscosity of fluids, etc...
Finally, the collected data is evaluated with respect to relative permeability
as a function of reservoir parameters. The results are presented in a commented
graphical and tabulated form. They will also be published in respective journals
and presented at appropriate conferences.

STATE OF ADVANCEMENT :
****
Stage 4. Oil/water relative permeability measurements under simulated reservoir
conditions have been comploeted. The gas/water relative permeability,
measurements are in preparation.

RESULTS :
****
Experimentation with the small-scale flooding apparatus resulted in the
selection of an unsteady state method (linear displacement) for the
determination of oil/water relative permeabilities and a steady state procedure
for the determination of gas/water relative permeabilities. The monitoring
system for the saturation conditions in the core sample which consists of two
plate electrodes at the core faces and four ring electrodes distributed along

the core sample produced good results for reservoir sandstones with low clay contents (3 %). The migration of a saturation discontinuity through the core sample could be observed and the impedance data evaluated with respect to average water saturation in the entire core as well as in the three segments between the ring electrodes. The measuring device for the determination of produced oil, water, and gas when flooding with recombined reservoir oil consists of a separator tank and a gas flowmeter. This device was used for checking the saturation data evaluated from the impedance measurements. Two experiments using dead crude oil have been performed uner pT-conditions of 200 bar/50 deg. with pore pressures ranging from 30 - 50 bar. Two further experiments were carried out with gas-recombined crude oil according to the reservoir conditions of an oilfield in the East Hanover province.

REFERENCES :
****
SOBOTT, R. (1988) : RELATIVE PERMEABILITATSMESSUNGEN UNTER LAGERSTATTENBEDINGUNGEN. STATUSREPORT 1988 GEOTECHNIK UND LAGERSTATTEN, HRSG. VON DER PBE, KFA JULICH, 647 - 658.

```
**********************************************************************
* TITLE : DOWNHOLE SEISMIC ENERGY SOURCE          *      PROJECT NO    *
*                                                 *                    *
*                                                 *   TH./01081/86/UK/.. *
*                                                 *                    *
**********************************************************************
* CONTRACTOR :                                    *   TELEPHONE NO     *
*    PRINCIPIA MECHANICA LTD                       *                    *
*                                                 *   01 831 6144      *
*    HIGH HOLBORN 233                              *                    *
*    UK - LONDON WC1V 7DJ.                         *                    *
*                                                 *   TELEX NO         *
*                                                 *                    *
* TECHNICAL DIRECTOR :                             *   267152           *
*    M. D.M. SIMPSON                               *                    *
*                                                 *                    *
**********************************************************************
```

VERSION : 29/11/89

AIM OF THE PROJECT :
****
The chief objective is to research a novel downhole seismic energy source and an
associated downhole detector sonde. These can be used in boreholes to generate
and receive controlled seismic signals. The source will be of the swept
frequency type, capable of generating signals up to and above the frequencies
conventionally achieved with surface based technologies. It shall be used for
cross-hole surveys, a technique which should enhance hydrocarbon reservoir
definition.

PROJECT DESCRIPTION :
****
The work is split into stages A to F as defined in the work programme. Briefly,
these are split into 4 groups :
1. A preliminary study phase which includes a review of existing techniques,
defining the desirable specification for such a device and proposing some
technical solutions. This first stage also includes identifying the software
needed to support such a source and adapting such software to the type of
surveys for which such a machine would be used. Stage 1 comprises phases A and D
of the work programme.
2. Technical design phase, in which the detailed engineering design of the
downhole source and the sensor investigation is undertaken, phases B and E of
the work programme. Phase B includes a shallow depth field test.
3. Construction phase of the seismic source and its associated control equipment.
4. Field trial and operational evaluation of the source.
Each phase will produce a report.

STATE OF ADVANCEMENT :
****
To date, phase A and D of the programme have been completed. A start on phase B
has been made, with the intention to produce a pre-prototype of the drive unit
of the source. A work plan for this stage has been prepared and a supplier of
the device identified.

RESULTS :
****
A review of the literature has shownb that the existing technologies are all
adaptations of marine devices. There is considerable industrial interest in such
a source. A pair of suitable concepts have been proposed, an electro-magnetic
tool and as an alternative, a servo-hydraulic source. The former is the
preferred option. A preliminary design has been prepared and this is still being
refined to improve its performance. A 10 cm diameter device is proposed, this
having the widest range of applications. Increasing the diameter rapidly
increases the force which may be generated. The design constraints are those of
producing sufficient magnetic flux in the confined space, solving the thermal
environment and source thermal load problems, getting sufficient primary
electrical power downhole using existing industrial hardware and providing a
sufficiently stiff clamping device for the borehole coupling.
The source requires, typically, about twice as much primary power downhole than
has been tried before. It is believed that we have obtained solutionsto these
problems.
To evaluate the design it is proposed to produce a pre-prototype drive unit,
perhaps using a simple hand-pumped hydraulic clamping syustem. The drive unit is
one of the key features of the tool. The pre-prototype will allow the
performance to be compared with the theoretical predictions and provide test
data to help resolve some of the theoretical problems uncovered in phase A, when
used in a calibrated and instrumented borehole.
A commercial seismographic processing package has been adapted for use in the
cross-hole and walk-away configurations and agreements with the owner obtained
to develop the system as required. A synthetic seismogram software package has
been written and implemented on an IBM-PC.
The pre-prototype field trial work plan has been defined and a contract with
Gearhart Tesel Ltd will fund laboratory trials of the drive unit at elevated
temperatures up to 150 deg by April 30 1990.

38

REFERENCES :
****
1. DESIGN DEFINITION DOCUMENT, DSS-PML-IS-001/88, ISSUE 1 AUG 4 1988
2. INVESTIGATION OF SEISMIC DATA PROCESSING TECHNIQUES, DSS-PML-RP-002/88, ISSUE 1, AUG 4 1988
3. THEORETICAL FORMULATION OF THE RADIATION IMPEDANCE OF A VERTICALLY VIBRATING DSS, DSS-PML-RP-003/88, ISSUE 1 AUG 4 1988
4. A REVIEW OF DOWNHOLE SOURCES, DSS-PML-RP-001/88, ISSUE 1, AUG 4 1988
5. PROGRESS REPORT ON RADIATION IMPEDANCE, DSS-PML-RP-008/88, ISSUE 1, AUG 4 1988
6. FINAL REPORT FOR PHASES A AND D, DSS-PML-RP-009/88, ISSUE 1, AUG 4 1988
7. ENGINEERING STUDY OF A DOWNHOLE SEISMIC SOURCE, DSS-PML-RP-010/88, ISSUE 1, AUG 4 1988
8. COMPUTATION OF THE RADIATION IMPEDANCE BY THE PRINCIPLE OF SUPERPOSITION, DSS-PML-RP-011/88

```
********************************************************************************
* TITLE : DEVELOPMENT OF A GEOMATHEMATICAL MODEL     *      PROJECT NO        *
*         APPLIED TO OIL RESERVOIRS                  *                        *
*                                                    *   TH./01083/86/PO/..   *
*                                                    *                        *
********************************************************************************
* CONTRACTOR :                                       *     TELEPHONE NO       *
*    PARTEX-COMPANHIA PORTUGUESA DE SERVICOS SA       *                        *
*                                                    *     73.50.13           *
*    AVENIDA 5 DE OUTUBRO, 160                        *                        *
*    PO - 1000 LISBOA                                 *                        *
*                                                    *     TELEX NO           *
*                                                    *                        *
* TECHNICAL DIRECTOR :                               *     14708              *
*    MR. A. DIOGO PINTO                               *                        *
*                                                    *                        *
********************************************************************************
                                                      VERSION : 15/01/90
```

AIM OF THE PROJECT :
****
The objective of the project is to build an integrated software package for
modelling oil reservoirs and describing the nature and disposition of the
heterogeneities that inevitably occur in petroliferous formations. This project
aims to develop a new approach based on a geomathematical model resulting from
the Theory of Regionalized Variables and from Kriging Technique, in order to
describe the spatial behaviour of reservoir properties, generate unbiased
profiles for use at different levels, define the error associated with the
estimation taking in account the position and spatial distribution of the
variable, detect structure trends and anisotropies, optimise of new wells
location, determine the best description of reservoir zonation.
The package will be flexible enough to permit the introduction of the practical
experience obtained by field geologists and reservoir engineers and designed to
be used by any petroleum technicien without computer science expertize.

PROJECT DESCRIPTION :
****
The plan of the project is divided in 3 phases and a final integrated stage :
1ST PHASE :
1. Screen review and assemble of the basic data regarding different oil
reservoirs;
2. Construction of the geomathematical model working out a set of computing
programs;
3. Initialization of the model, Kriging known values and checking the resulting
error of calculating various theoretical dispersion variances and comparing them
to the experimental values;
4. Analysis and evaluation of the results and impact on several areas.
a). variographic analysis and detection of the major structural characteristics
and anisotropies of the reservoirs, quantitative evaluation;
b). characterization and description of reservoir heterogeneities;
c). description of the reservoir properties spatial behaviour, generated
profiles and precision of the estimation;
d). contour mapping and comparision with classical techniques, evaluation of the
results;
e). degree of error associated with the estimation, real statistical meaning of
the values; impact on reservoir simulation outputs and reservoir estimation;
f). co-regionlization aiming to study inter-correlated variables simultaneously.
2ND PHASE :
1. Geological modelling using conditional simulation to provide a set of values
with the same variability as the studied phenomenon and coinciding with the
experimental values at the sample locations, and Minkovsky transforms for
characterizing the geometry of the reservoir;
2. Optimization of new wells locations based on the fact that the Kriging
variance does not depend on experimental values, but on the structure (variogram
model), on the geometric configuration of the domain to be estimated and on the
locations of the data points;
3. Subsurface reservoir model and impact on reserves calculation. Application of
the geomathematical model and Kriging technique to the study of the reservoir
geometry in order to reduce the uncertainty about it simulating the field
boundaries in a more accurate way.
3RD PHASE :
1. Automatic method of capturing and storing data. Design of a data-base with
specific characteristics of Petroleum information;
2. A new method for reservoir zonation based on the Correspondence analysis,
allowing the simultaneous treatment of quantitative and qualitative data;
3. Contribution of variography analysis and Mandelbrot's Theory of fractal
objects for the solution of the problems of scale in a reservoir; attempt to
relate them quantitatively.
LAST STAGE :
It consists of the integration of the results of all phases in a portable and
integrated package to be used by field engineers in order to solve practical oil
industry problems in exploration and production.

STATE OF ADVANCEMENT :

****
Phase I and II of the project are completed, the different software modules have been tested and validated through real case studies.
Some tasks of phase III as Reservoir Zonation by the application of Multivariate Statistical Techniques are already launched.
The next step of the project will be the integration and articulation of the different software modules into the final package.

RESULTS :
****
A continuous work has been done on package development framework with the organization of menus architecture and testing and integration of software modules namely related to variography and different estimation techniques. Special attention has been given to the refinement of the data base and graphic modules testing their interaction with the whole package. A careful attention has beendedicated to the different options available for the user and specific sub-menus to provide an easy manipulation of the different programmes have been implemented.
Reservoir description case studies have been performed and results discussed in order to assess the potentialities of geostatistical estimation procedures and their capability to convey geological information into quantitative models, useful for technical and economical purposes. A variety of estimation procedures, algorithms and programmes have been tested in a real case.
Results concerning the establishment of a consisten geostatistical methodology, developed on the grounds of transitive driging of indicator data and able to model the geometry of the relevant reservoir layers, have been presented and discussed.
Achievements in the field of Moddern Data Analysis techniques applied to variables spatially correlated have been expanded in order to enhance the Reservoir Quality Analysis. Improvements of horizontal and vertical zonation of oil reservoirs provide a better knowledge of the reservoirs internal architecture which is a key point for the planning of production and recovery operations. Finally more refined and detailed layering models can be built to be used in simulation studies.
In the last phase of the Project, methodologies have been developed and tested for prediction of permeability - based on Multiple Regression and Multidimensional Histograms - and for Simulation of the Shales Spatial distribution within the reservoirs - based on Morphological Kriging by conditioning the simulation directly to indicator data.

REFERENCES :
****
- MATHERON, G. "THE THEORY OF REGIONALISED VARIABLES AND ITS APPLICATIONS", C.M. M. FONTAINEBLEAU, 1971.
- JOURNEL, A. AND HUIJBREGTS, CH. "MINING GEOSTATISTICS", WILEY, 1981.
- COSTA E SILVA, A. "A NEW APPROACH TO THE CHARACTERIZATION OF RESERVOIR HETEROGENEITY BASED ON THE GEOMATHEMATICAL MODEL AND KRIGING TECHNIQUE" SPE 60TH TECHNICAL CONFERENCE, LAS VEGAS 1985.
- RIBEIRO, L. AND COSTA E SILVA, A. "A GEOMATHEMATICAL MODEL TESTED ON AN OIL RESERVOIR", APCOM, PENSILVANYA, 1986.

```
*****************************************************************************
* TITLE : GEOCHEMICAL EXPLORATION BY NITROGEN        *      PROJECT NO       *
*         ISOTOPE ANALYSIS OF UNDISTURBED SEA BED    *                       *
*         SAMPLES (GENIUS)                           *   TH./01084/86/DK/..   *
*                                                    *                       *
*****************************************************************************
* CONTRACTOR :                                       *      TELEPHONE NO      *
*    COWICONSULT CONSULTANT ENGINEERS                *                       *
*                                                    *    45 45 97 21 11      *
*    TEKNIKERBYEN 45                                 *                       *
*    DK - 2830 VIRUM                                 *                       *
*                                                    *      TELEX NO          *
*                                                    *                       *
* TECHNICAL DIRECTOR :                               *    37 280              *
*    MR. A. BJERRUM                                  *                       *
*                                                    *                       *
*****************************************************************************
                                                    VERSION : 28/08/89
```

AIM OF THE PROJECT :
****
DEVELOPMENT OF A NEW TECHNIQUE TO DETECT SUBSURFACE SOURCE ROCK BASEDDevelopment
of a new technique to detect subsurface source rock based on natural seepage of
hydrocarbons from the subsurface. The combination of seismic survey and
geochemical survey was proved to give important information for petroleum
exploration.
An easy and safe technique was developed to recover mundisturbed samples at
ambient pressure from the sea bed. The samples was analyzed for their content of
traces of hydrocarbons and nitrogen trapped in the soil. The purpose was to
determine the type and the age of source rock for hydrocarbon generation.

PROJECT DESCRIPTION :
****
Geochemical exploration for oil and gas is based on the fact that light gas
migrate from deep source rock to the surface. By taking undisturbed samples from
the sea bed and analyze the gas concentration it is possible to get an
indication of any subsurface source rock.
The method has to be used in combination with seismic interpretation and other
geological studies e.g. basin modelling as a useful additional source of
information.
The gas may be present either as free gas dissolved in the pore water or as gas
adsorbed on the grain surface. One of the problems by analyzing the free gas has
been risk of atmospheric contamination and escape of free gas from the sample
when the pressure is released.
To overcome these problems a new ambient pressure sampler has been developed and
tested. Laboratory equipment has been developed to extract the gas from the
sampler without atmospheric contamination.
A successful pilot test was carried out in West Baltic. 154 sea bed samples was
recovered from 64 locations. A comprehensive laboratory programme was carried
out including sediment analyses, gas chromatography (GC) and mass spectrometry
(MS).
The laboratory results were analyzed and compared with structured maps of the
survey area.
The project is divided into 5 phases comprising 12 activities :
PHASE 1 : CONCEPTUAL STUDIES
    - - Geochemical analyses, literature survey
    - Isotope measurement study
    - Core sampling requirements
PHASE 2 : DEVELOPMENT OF A CORE SAMPLER
    - Design of equipment
    - Construction and test
    - Marketing
PHASE 3 : PILOT TEST
    - Selection of test areas
    - Field work
    - Laboratory work
PHASE 4 : FINAL EVALUATION
    - Evaluation of GENIUS results
PHASE 5 : REPORTING AND PROJECT MANAGEMENT
    - Reporting
-   - Project Management.

STATE OF ADVANCEMENT :
****
Completed dec. 1988.

RESULTS :
****
The main conclusions of the project are :
    - An ambient pressure drop sampler (the Soderberg sampler) has been developed
and successfully tested.
    - Free gas and adsorbed gas from 154 samples was analyzed. A good correlation
between high gas anomalies and subsurface structure indicate that migration may
be correlated with the subsurface fault pattern.

- By gas chromatography and carbon isotope analysis it was possible to distinguish between biogenetic and thermogenetic gas.
- Attempt to correlate nitrogen gas to the subsurface structure failed probably due to atmospheric contamination of some of the samples.

REFERENCES :
****
GEOCHEMICAL EXPLORATION BY NITROGEN ISOTOPE ANALYYSIS OF UNDISTURBED SEA BED SAMPLES (GENIUS) : FINAL REPORT OCT. 1989.

```
*********************************************************************** ****
* TITLE : POSITION MEASURING SYSTEM FOR OFFSHORE      *      PROJECT NO      *
*         INSTALLATIONS                               *                      *
*                                                     *   TH./01086/86/DE/..  *
*                                                     *                      *
*********************************************************************** *****
* CONTRACTOR :                                        *     TELEPHONE NO     *
*    TELEFUNKEN SYSTEM TECHNIK                         *                      *
*                                                     *     040/8825-2216    *
*    BEHRINGSTRASSE 120                               *                      *
*    DE - 2000 HAMBURG 50                             *                      *
*                                                     *     TELEX NO         *
*                                                     *                      *
* TECHNICAL DIRECTOR :                                *     211925           *
*    MR. B. MICHAELSEN                                *                      *
*                                                     *                      *
***************************************************************************
                                                   VERSION : 13/07/89
```

AIM OF THE PROJECT :
****

Development and test of a position measuring system for great water depths.
Therefore inclination sensor systems in measuring cans are to be attached with
defined distances to the riser. From these inclination measurements a central
computer determines the displacement of the riser compared to the lower riser
end. Energy supply and connection of data lines to and from the sensor cans are
to be carried out as plugless inductive couplings. In addition to the position
displacement for supervision purpose the spatial curve of the whole riser is
determined.

PROJECT DESCRIPTION :
****

The main object of this R & D project is to build a prototype and to test the
riser inclination measuring system under real environmental and operational
conditions.
These tests are necessary, because this is the only way to examine :
- how the electrical behaviour of the conductor line change depending on diving
depth. For example, the capacity is depending on the permittivity of seawater,
which has to be taken into account for the data communication between the lower
and the central station.
- whether the configuration of sensor cans and the inductive data coupling stand
the test under real environmental conditions
The tests are organized in laboratory tests, shallow water tests with seawater
and pressure tests.
Additionally the following problems are to be investigated during this R & D
project :
- optimization of the design of sensor cans and transformers
- optimization of the fluxgate configuration (magnetic compass at lower end and
top station)
Handling of the system has to be clarified in addition
- especially, how the sensor cans are to be fastened to the riser. The mounting
of the system to the riser joints shall not substantially influence the riser
installation, it therefore has to be carried out simple and fast.
In addition to the design work the software has to be overworked and the test
programs have to be examined.

STATE OF ADVANCEMENT :
****

Hard and software development have been carried out. Laboratory, shallow water
and pressure tests have been performed successfully. Work continues on final
report.

RESULTS :
****

Components of the system are designed and fabricated. System is ready for
operation in water depths up to 2000 m with three sensor cans as tested.
Extension is possible up to water depths of appr. 3000 m with more than 20
substations, which can carry different sensors and control outputs (e.g. for
value control). Solutions for water depths up to appr. 6000 m are available.
Application of the system for different tasks not only riser inclination
measurement is possible due to flexible modification of the components and
software environment according to operator requirements.

REFERENCES :
****

- SCHROEDER, M.; KAEHLER, H.
POSITIONSMEBSYSTEM FUER OFFSHORE-EINRICHTUNGEN
- INTERMARITEC 80 - 407

c

```
***************************************************************************
* TITLE : DEVELOPMENT OF A FORMATION EVALUATION      *      PROJECT NO      *
*         TOOL FOR OILWELL BOREHOLE LOGGING          *                      *
*                                                    *   TH./01087/87/UK/..  *
*                                                    *                      *
***************************************************************************
* CONTRACTOR :                                       *    TELEPHONE NO       *
*    UNITED KINGDOM ATOMIC ENERGY AUTHORITY          *                      *
*                                                    *   235-24141 EXT.2901  *
*    NUCLEAR APPLICATIONS CENTRE                      *                      *
*    BUILDING 7 - HARWELL LABORATORY                 *                      *
*    UK-OXFORDSHIRE OX11 QRA                         *    TELEX NO           *
*                                                    *                      *
* TECHNICAL DIRECTOR :                               *    83135              *
*    MR V.J. WHEELER                                 *                      *
*                                                    *                      *
***************************************************************************
                                                         VERSION : 18/04/90
```

AIM OF THE PROJECT :
****
This project concerns the development of a novel type of formation evaluation
tool for use in oilwell borehole logging. It is aimed at improving the
capability for geochemical logging, including monitoring the position of the
oil/water interface in a producing well, and determination of the fluid
saturation level behind the well casing - these are two of the most important
parameters in production logging.
The design is based on a neutron source emitting a continuous flux of neutrons
into the rock formation surrounding the borehole, and the measurement of (i) the
coincidence of inelastically scattered fast neutrons with their associated
characteristic gamma-rays, and (ii) thermal neutron capture gamma-rays, to
provide information on rock and pore fluid composition.

PROJECT DESCRIPTION :
****
Technical approach
A neutron source emits a continuous flux of fast neutrons into the surrounding
environment, and these interact with the atomic nuclei of elements present in
the formation giving rise to gamma-rays of characteristic energies.
Spectrometric recording of these gamma-rays in well-defined energy "windows"
provides a means for determining elemental composition, and hence the
operational parameters of interest (rock type, fluid composition etc.).
Neutron interactions are of various types, but the two principal ones of
relevance are as follows :
- Fast neutrons are inelastically scattered by certain nuclei (eg. carbon,
oxygen) giving rise to gamma-rays which are in "coincidence" with the scattered
neutron.
- Following a sequence of collisions, neutrons are slowed down to so-called
"thermal" energies (<1 e V), and are then captured by nuclei, giving rise to
characteristic prompt gamma-rays. These too may recorded in the scintillation
detector, and such events can be used to give measurements of contens of Al, Si,
Ca, which are common elements in limestones and sandstones (typical formation
materials in oil reservoirs), and thus indicate the rock type.
The main novel feature of the proposed project is the design of a detector
system in which the scattered fast neutrons and their associated gamma-rays are
recorded in coincidence, thus providing time-dependent information that can
normally only be obtained from the use of a pulsed neutron generator. This means
that a cheaper, simpler and operationally more reliable neutron source (such as
a continuously-emitting radioisotope) can be used. By suitable electronic
processing of the coincident pulses, it should be possible to minimize
interference from borehole effects, and from unwanted reactions due to high-
energy neutrons.
The main problem to be overcome in designing a formation evaluation tool based
on the coincidence-counting principle is that only a relatively small number of
the total n, events will be recorded in coincidence, due to the small volume of
the detectors in the total volume interrogated, and this can lead to long data
collection times for adequate statistical accuracy. However, many of the events
recorded in conventional tools have to be discarded, and do not contribute to
the data-processing, and it is proposed to optimise source size, source-detector
geometry and the timing of coincidence events so that all of the events recorded
are useful.
We shall also briefly investigate the use of the new detector with a
conventional neutron generator, to determine whether it presents advantages in
spectrum processing with high energy neutrons (14 MeV) at higher fluxes than
those practicable with a radioisotope source.

STATE OF ADVANCEMENT :
****
Phase 1 (Initial Design Study) was completed on schedule (31 January 1989, and
Phase 2 (Experimental Assembly and Testing) was completed on 31 March 1990.

RESULTS :
****
The work programme described in section 2 (under the heading Phase 1) was

carried out, using simulated boreholes (open and cased) in limestone and
sandstone formations, with varying formation and borehole fluids (water and oil).
The concept of coincidence-counting to provide time-dependent data was
demonstrated successfully, using an 241 Am/Be neutron source sith a bismuth
germanate gamma-ray detector and a liquid scintillation fast neutron detector.
Some work was also carried out using a continuous flux of 14 MeV neutrons
(generated from an accelerator) to study the effects of the higher neutron
energy and flux. There was, as expected, an enhanced signal from carbon, and a
clear signal from oxygen (due to 16 O) at 6.13 MeV.
The main problem proved to be the low countrate associated with the poor
geometrical efficiency and the low neutron flux (-10 7 neutrons.s-1) from a
radioisotope source. This was particularly marked for cased holes.
Phase 2 of the project consisted of efforts to increase the countrate by
locating the neutron source in the middle of the neutron detector, and using the
signal from a neutron leaving the source to start the timing sequence, and the
arrival of a coincident gamma-ray to stop it. This has improved the efficiency
of the process considerably. A prototype electronics module for fast-timing
processes is now being designed, with a view to testing the complete
detectors/data-collection unit at elevated temperature, to investigate the
effect on detector response and circuit performance.

REFERENCES :
★★★★★★★★★★★★
THE FOLLOWING REPORTS HAVE BEEN PREPARED :
"DEVELOPMENT OF A FORMATION EVALUATION TOOL FOR OILWELL BOREHOLE LOGGING", B.W.
THOMAS AND V.J. WHEELER, AERE-G4768 (JULY 1988)
"APPLICATIONS OF SHORT-LIVED RADIOISOTOPE NEUTRON SOURCES IN OILWELL BOREHOLE
LOGGING", B.W; THOMAS AND V.J. WHEELER, AERE-G4732 (JULY 1988)
"AN EXPERIMENTAL ASSESSMENT OF COINCIDENCE COUNTING TECHNIQUES FOR OILWELL
BOREHOLE LOGGING", B.W. THOMAS, S.J. CONCHIE AND C.B. WARD, AERE-G5057 (MARCH
1989)
A PATENT APPLICATION (F87.26477 "BOREHOLE LOGGING-COINCIDENCE COUNTING", B.W.
THOMAS AND M.R. WORMALD) HAS BEEN FILED.
THE PROGRESS REPORT ON THE COMPLETION OF PHASE 2 WILL BE ISSUED SHORTLY.

```
****************************************************************************
* TITLE : INTERNAL STRUCTURE OF CLASTIC AND          *      PROJECT NO     *
*         CARBONATE RESERVOIR:STUDY OF               *                     *
*         HETEROGENEITIES ON REAL-LIFE CASES         *   TH./01088/87/IT/.. *
*                                                    *                     *
****************************************************************************
* CONTRACTOR :                                       *    TELEPHONE NO      *
*     AGIP S.P.A/SNEA (P) - C/O AGIP S.P.A.          *                     *
*                                                    *    (2) 520.5643     *
*     PIAZZA VANONI                                  *                     *
*     IT-20097 SAN DONATO MILANESE                   *                     *
*                                                    *      TELEX NO        *
*                                                    *                     *
* TECHNICAL DIRECTOR :                               *     310246          *
*     MR A. CARLINI                                  *                     *
*                                                    *                     *
****************************************************************************
```

VERSION : 30/03/89

AIM OF THE PROJECT :
****
Aim of the project is to develop a combination of synergetic techniques for the
recognition and definition of the internal structures of reservoirs using
advanced seismic methodologies and high-resolution logs analysis.

PROJECT DESCRIPTION :
****
The project comprehends two phases:
PHASE 1:
Its purpose is the development of innovative methodologies and technics and
improvement of the already tested tool performances.
The research will tackle two aspects: the seismic information analysis and the
fracture characterization by high resolution logs interpretation.
The first aspect's aim is to analyze the seismic datum respectively in prestack
domain and post-stack domain.
In pre-stack domain, the identification and study of the velocity and amplitude
anomalies, as function of the offset, will provide further physic information on
the reservoir rocks properties and their lateral variation.
In post-stack domain, using statistical methods, relationships between
significant seismic parameters and petrophysical properties, in given reservoir
intervals, will be searched.
In the second aspect, the research purpose is to study the type and distribution
of natural fracture in the reservoir, by high-resolution logs. The use of these
new tools requires a more detailed interpretation of the logs (BHTV-SHDT-FMS)
and the calibration of logs with the core data.
PHASE II:
Application to real cases.
The results of the technics and methodogogies, developed in the Phase I, will be
tested on suitable date and an application on real reservoir case will be tried :
 Cavone (AGIP), Yanga (SNEA(P)).

STATE OF ADVANCEMENT :
****
The Phase 1 of the project is still in progress with the pre-stack analyses and
high-resolution log analyses.
The phase 2 is continuing with the calibration of HR logs by analysing well
cores.

RESULTS :
****
PHASE I : Pre-stack analysis
Concerning the "recognition and inversion of stacking velocity anomalies" this
period of work has been devoted firstly to design the main configuration of the
software package composed of four programs; the software development and
implementation of each program is still in progress.
In particular for the detection of stacking velocities, automatic picking
algorithms of the coherence volume have been developed.
Concerning the "stratigraphic pre-stack analysis of seismic signals" a
methodology of AVO analysis has been studied and successfully tested to three
separate brightspots in the Po valley. Further theoretical studies of this
technique, to include evanescent waves phenomena in the reflectivity modeling,
have been also studied.
HR logs
The HR logs computer works are in progress with enphasis to the development of a
workstation software program (DIAMAGE) for pictures and HR logs interpretation.
Some basic functions as depth matching, visual interactive correlations and core
picture orientation have been already developed.
Concerning fracture interpretation DIAMAGE can provide a good tool. The
geologist can realize sedimentological or fracture study. The reservoir engineer
can detect high permeability layer for completion, and compare CPI results with
pictures or orientate the maximum stress and predict flooding anisotropy in
fractured reservoirs.
PHASE II : Calibration of the HR logs by analysing well cores
This phase of the project was aimed at developing a statistical analysis

methodology based on "cluster analysis" algorithm, allowing an accurate and more objective recognition of fracturation parameters in borehole, starting from the available log and core data.
The analysis was focused on travel time transit time and energy data of the various waveform components, given from sonic tools (SLC and SDT) and on high resolution logs (FMS) qualitative evaluation.
The fundamental phase of work was the log-facies and the HR logs characterization with core data.

```
******************************************************************************
* TITLE : RESERVOIR MULTIWELL GEOPHYSICS (PHASE    *      PROJECT NO        *
*         II)                                       *                        *
*                                                   *   TH./01091/87/FR/..  *
*                                                   *                        *
******************************************************************************
* CONTRACTOR :                                      *   TELEPHONE NO         *
*    GERTH                                          *                        *
*                                                   *   (1)47.52.69.27      *
*    4, AVENUE DE BOIS PREAU                        *                        *
*    FR-92502 RUEIL MALMAISON                       *                        *
*                                                   *   TELEX NO             *
*                                                   *                        *
* TECHNICAL DIRECTOR :                              *                        *
*    M. DELVAUX                                     *                        *
*                                                   *                        *
******************************************************************************
                                                       VERSION : 29/03/89
```

AIM OF THE PROJECT :
****
All geophysical methods envisaged within the framework of this project are based
on measurements made between two or more wells. Contract TH 01.50/85 "Reservoir
multiwell geophysics (phase 1)" proved the feasibility of two main types of
methods :
- methods based on acoustic wave propagation between wells
- methods based on electromagnetic signal measurement in a single well or
between two wells.
The present project entails experiments under real conditions in shallow
reservoirs.
The project is based on :
- new acquisition methods near and in shallow fields, at depths of around
several thousand meters
- processing techniques adapted to the specific characteristics of this type of
acquisition
- a methodology for integrated interpretation of all previously mentioned
processing and acquisition methods.
The scope and amplitude of future development and research will be governed by
the results obtained in this project.

PROJECT DESCRIPTION :
****
PHASE 1 : Inter-well seismic survey
Adaptation of prototypes : a new well source prototype, developed under contract
TH 01.067/86 for the bedsurface transposed seismic survey, will be tested under
inter-well seismic survey conditions.
Acquisition operations : acquisition campaigns, designed for the acquisition of
the volume of data required for the envisaged processing, will be completed, as
needed, by smaller scale operations, in order to test and characterize new
prototypes and their develop;ent.
Processing and interpretation :
The processing phase, which will follow acquisition operations, will be very
important, especially since the processing tools used at this point in the
project are not optimized.
PHASE 2 : Electromagnetic methods
Prototype finalization : Contract TH 01.50/85 made it possible to test a certain
number of sensors, to characterize them and to appraise their performance in a
drilling environment. Using these initial results, it will be necessary, in
order to carry out a true-size acquisition campaign on a shallow site, to
perform a certain number of tasks. These operations will begin after completion
of the feasibility study performed under contract TH 01.50/85 and after
competent subcontractors have been found.
Acauisition operations : These are the follow-up operations which will be
performed on the experimental sites used in the seismic survey of the phase 1
work programme. These operations will be prepared using previous models.
Processing and interpretation : research and implementation of the best adapted
facilities (algorithms); the efficiency of these facilities will be judged by
the practical results obtained during interpretation.

STATE OF ADVANCEMENT :
****
Phase1: Seismic between wells
The technical programme depended on the availability of an operational well
source, which is why it has been slightly delayed compared to the initial
planning and is presently under realisation for each item.
Phase 2:
Electromagnetic methods
The theoretical feasibility study has been completed. We are presently looking
for a motivated industrial partner to manufacture the experimental prototypes
mentioned earlier.

RESULTS :
****
Phase1 : Seismic between wells

Study of a device for the attenuation of tube waves. Experimental acquisition of data between wells presenting a spacing superior to 500 m. A data acquisition survey between four wells is underway. Continuation of processing and evaluation of a flexible migration/inversion software package.

Phase2 : Electromagnetic methods

Search for a partner has implied a certain publicity. B.P., Shell, the University of Koln and Schlumberger have recently showed some interest in the project. Negociations are underway.

```
****************************************************************************
*  TITLE : ACCESS (PHASE II)                    *      PROJECT NO        *
*                                               *                        *
*                                               *  TH./01092/87/FR/..    *
*                                               *                        *
****************************************************************************
*  CONTRACTOR :                                 *      TELEPHONE NO       *
*    GERTH                                       *                        *
*                                               *   1 47 52 61 39         *
*    4 AVENUE DE BOIS PREAU                      *                        *
*    F-92502 RUEIL MALMAISON                     *                        *
*                                               *    TELEX NO             *
*                                               *                        *
*  TECHNICAL DIRECTOR :                          *                        *
*    MR MILLE                                    *                        *
*                                               *                        *
****************************************************************************
```
                                          VERSION : 20/09/89

AIM OF THE PROJECT :
****
Petroleum exploration involves gathering data from numerous sources. These data
are highly heterogeneous and costly to gather.
Sophisticated interpretation methods are the key to successful exploration and
production entreprises. Those methods were developed on a case by case basis to
meet specific requirements.
Since they were developed at different times, they are not consistent from the
technological point of view.
The need for an integrated, sophisticated and up-to date system arose when
management accounted for the productivity gain to be expected from such an
integrated system.
This project covers most of the needs of an explorer and reservoir engineer. All
basic tasks concerning the reservoir evaluation are carried out (log evaluation,
mapping, estimates of hydrocarbons in place) as well as geophysical processing
and interpretation. A door is open to the reservoir engineering through a dual
pre and post processor.

PROJECT DESCRIPTION :
****
For practical purposes, the project was divided into two phases, each one being
subject of one contract with the EEC.
This note deals with Phase II of the project. During phase I (see TH/01047/85/FR)
 the standards tools (kernel), the basic applications dealing with geology and
geophysics were developed.
Phase II specifically deals with the integration of these applications as
developed during Phase I. The procedures in charge of ensuring the link between
applications will be built up during this phase.
Four main are involved during the second phase of the ACCESS project.
The seismic imagery. This complementary module integrates the geophysicist work
on high resolution graphic screen.
Two and three dimensions cross-sections and bloc diagrams will be made available
over which layering and hydrocarbons bearing zones will be made visible.
The tools necessary to interpolate between wells either a geological and/or a
seismic cross-section will be developed.
A full set of applications will be made available to interface those developed
during Phase I with reservoir geology and reservoir engineering.

RESULTS :
****
Commercialisation of the software under the name of INTEGRAL and finalization of
its development has been entrusted to PETROSYSTEMS, subsidiary of C.G.G.
(Compagnie generale de Giophysique). This company acquired the rights to grant
licence contracts to other enterprises or companies non member of GERTH having
conducted the project.
Today there exists no other software as integral and complete, that is
equivalent to INTEGRAL.

REFERENCES :
****
NO PUBLICATIONS NOR LECTURES WERE CARRIED OUT ON THE SUBJECT.

```
*********************************************************************************
* TITLE : DEVELOPMENT OF A WELL SITE SCREENING      *       PROJECT NO         *
*         METHOD FOR THE DETERMINATION OF           *                          *
*         HYDROCARBON SOURCE ROCK CHARACTERISTICS.  *    TH./01093/87/DE/..    *
*                                                   *                          *
*********************************************************************************
* CONTRACTOR :                                      *      TELEPHONE NO        *
*    TECHNICAL UNIVERSITY OF BERLIN                 *                          *
*                                                   *    30-314 72646 OR 696   *
*    ERNST REUTER PLATZ 1                           *                          *
*    D-1000 BERLIN 12                               *                          *
*    FRG                                            *      TELEX NO            *
*                                                   *                          *
* TECHNICAL DIRECTOR :                              *      184 262             *
*    PROF. DR. E. KLITZSCH                          *                          *
*                                                   *                          *
*********************************************************************************
                                                        VERSION : 13/06/89
```

AIM OF THE PROJECT :
****

Varying amounts of oil, gas, or both may be generated during burial of sediments
according to the type of kerogen and maturity (heating time and temperature).
Chemical analysis and pyrolysis techniques allow to distinguish the different
types of kerogen, their maturity and hydrocarbon potential. Elemental analysis,
however, is very timeconsuming and products of pyrolysis depending upon mineral
composition of the sediment and organic content, kerogen classification is often
unreliable. Using IR-spectroscopy an exact determination of the maturity and
hydrocarbon generation potential on kerogen concentrate is possible, as well as
a classification of bitumen and crude oil. The IR-spectroscopy can also provide
a quantitative determination of the mineral composition. The two innovations
project concern use of IR-spectroscopy and ultrasonic centrifugation and frigen
leaching in place of pyrolysis as a rapid well site screening technique.

PROJECT DESCRIPTION :
****

The IR procedure for analysis of kerogen isolated from its mineral matrix has
certain distinct advantages over the classical elemental analysis. It is
potentially simpler in its application, quicker and cheaper, and gives more
information. Compared to Rock-Eval analysis the results are more reliable. At
present the full analysis takes about 3 hours (incl. sample homogenisation,
bitumen extraction, TOC determination, kerogen isolation, evaluation and
interpretation).This compares with only 1 1/2 h for a Rock-Eval analysis (incl.
sample homogenisation, TOC and interpretation). By avoiding kerogen isolation
and making determinations and classifications on the homogenized original sample,
 it should be possible to reduce the IR analysis time to about that of a Rock-
Eval analysis. However, the IR results are anticipated to be more reliable and
to convey more information including the mineralogical characteristics of the
petroleum source rocks as well as a classification of bitumen and crude oil. The
proposed development of the method will adress the problem of calibrating the
organic amplitudes within the spectra of the samples without prior isolation of
the kerogen in order to reduce overall analysis time to a level competitive with
Rock-Eval. The following steps are necessary to calibrate "whole sample" IR
measurements :
1. quantitative determination of the mineral contents by IR spectroscopy and
   other methods (XRD, SEM, EMP)
2. computercontrolled stepwise compensation of the single minerals in the total
   spectra3. measurement of the organic bands in the compensared spectra that
are
   important for the calculation of kerogen type, maturity and hydrocarbon
   potential. For crude oil classification extensive comparative measurements
   (IR/standard procedures) are necessary.
Conventionally extraction of soluble organic matter (bitumen) is carried out by
a 48 h soxhlet extraction and saturated hydrocarbons are separated by some
chromatographic techniques (TLC, CC, or MPLC). It has been shown that addition
of frigen (CCl3F) to the bitume results into the frigen phase of mainly
saturated components which, while not identical to the saturates fraction, are
sufficiently similar for the determination of common chromatographic parameters
(CPI, Phytane/Pristane, bulk sterane distribution) (GANZ et al., in press). The
analysis time is typically reduced to a few minutes. In comparison to IR, only
limited further work is necessary to ensure consistency of parameters and ratios
with those determined on conventional saturates fractions. The intention of the
project is to reduce analysis time (possibly to as low as 30 min/homogenized
sample) by removing the kerogen isolation stage and refining procedures and
computational techniques to produce a comprehensive well site geochemical
screening method.

STATE OF ADVANCEMENT :
****

Ongoing. The quantitative determination of minerals and kerogens in
sedimentological samples a large number of varieties are performed for
calibration purposes.

RESULTS :

****
About 200 kerogens of different type and thermal maturation and more than 70
standards of pure minerals have been analysed and classified using IR-analysis
and conventional techniques. The new IR method is superior to all petroleum
source rock classification procedures so far. The accuracy of the method allows
more reliable detection of organic rich source rocks and indication of wether
they have produced oil or gas. Together with the mineralogical data obtained by
IR and the rapidly determined organic geochemical parameters (using frigen/GC)
statements about facies development and even reservoir qualities and migration
behaviour are possible. No other method furnishes that much information all at
once. It is therefore of the greatest interest to reduce analysis time in order
to not only make routine research in the laboratory possible but most
importantly to allow analysis continously on site during drilling. The main aim
of the project is to achieve a sample throughput similar to Rock-Eval type
pyrolysis techniques but at the same time to provide a much wider range of
information. In the event of this aim being achieved it is likely, in the nature
of geochemical interpretation considering as many parameters as possible, that
pyrolysis techniques would be retained and used alongside the new system. A
reduction in laboratory based routine work and an increase in the speed of
geochemistry based decisions would probably be the main outcome.
The major technical uncertainly is whether the accuracy at present achieved in
the IR method using isolated kerogens can be reproduced in the presence of the
mineral matrix, even after computer manipulation of the spectra. The crucial
point will be, how exact the quantitative mineral determination and the
following compensation of the spectra can be. By recent developments in the
procedure of sample preparation a standard deviation of less than 5 % was
achieved. According to this promising results we confidently expect that precise
determinations are possible at least in organic rich samples. The question
remains as to how successful the method will be on samples with minimal TOC
contents, in which the organic bands have only a very low intensity and
additional expansion of the spectrum is necessary.

REFERENCES :
****
GANZ, H. & KALKREUTH, W. (1987), FUEL, V.66, NO 5, 708-711
GANZ, H. KALKREUTH, W., ONER, F., M.J. & SMALL, J.S. (1987), IN : PETROLEUM
GEOLOGY OF NORTH WEST EUROPE (EDITED BY J. BROOKS AND K. GLENNIE), GRAHAM &
TROTMAN LTD, P.847-581.

```
*************************************************************************
* TITLE : DESIGN OF ACQUISITION PARAMETERS TO      *     PROJECT NO     *
*          SOLVE OPAQUE SEISMIC ZONES              *                    *
*                                                  *   TH./01097/87/ES/.. *
*                                                  *                    *
*************************************************************************
* CONTRACTOR :                                     *   TELEPHONE NO      *
*    REPSOL EXPLORACION S.A.                       *                    *
*                                                  *   (1)2.74.72.00     *
*    PEZ VOLADOR 2                                 *                    *
*    ES-28007 MADRID                               *                    *
*                                                  *   TELEX NO          *
*                                                  *                    *
* TECHNICAL DIRECTOR :                             *   49544             *
*    MR AGUSTIN ARRIETA MURILLO                    *                    *
*                                                  *                    *
*************************************************************************
                                                    VERSION : 24/02/88
```

AIM OF THE PROJECT :
****
The aim of this project is to study the viability of a general method to design
the parameters of acquisition, capable of solving the seismic zones containing
formations which are opaque when the conventional methods are used (such as
Olistostrome and Salt-diapirs). The method is based on the numerical simulation
of the travelling of the seismic waves through a physical-mathematical model of
the mentioned formations.

PROJECT DESCRIPTION :
****
The development of the project is divided in four stages :
STAGE 1.   : It includes the gathering, selection and analysis of the
                    available information.
                    It is subdivided into three parts :
STAGE 1.1.   : Selection of two areas containing a different type of opaque
                    formation each one. That is :
                 a) Olistostrome
                 b) Salt-tectonics.
STAGE 1.2   : Gathering of all the seismic and geological information
                    available.
                 a) Seismic lines
                 b) Well-logs
                 c) Surface geology
STAGE 1.3.   : Analysis of the available information.
STAGE 2.   : Build up of reliable geological models representing as well
                    as possible the structure of the studied formations.
STAGES 3/4   : These two stages are so closely related that it is difficult
                    to consider them separately. On the one hand, the
                    mathematical model should fit into the geological model. On
                    the other hand, it should be simple enough to be
                    handled with the existing methods for equation resolution.
                    These stages are subdivided into the following parts :
STAGES 3.1/4.1.   : Build up of the mathematical model for the formation being
                    studied and selection of the most suitable method for wave-
                    equation resolution among the existing methods such as ray
                    tracing, Kirchoff integral, Fourier discretization method,
                    Finite Differences, Finite Elements.
STAGE 4.2.   : Study of the available software for the solution chosen in 3.
                    1.
STAGE 4.3.   : Study of the viability of the development of a transmision
                    simulator for the travelling of the seismic waves through the
                    model built up in 3.1, using the method chosen in stage 4.1.

STATE OF ADVANCEMENT :
****
Zones to study have ;been selected.
Compilation of data is finished.
Analysis of data is finished.
Geological model has been generated.
Mathematical model generation and analysis of finite elements modelling method
have been started.

```
****************************************************************************
*  TITLE : DEVELOPMENT AND TEST OF A NEW PROBE FOR    *      PROJECT NO      *
*          IN SITU DEGASSING OF SOIL SAMPLES FOR      *                      *
*          ISOTOPE GEOCHEMICAL ANALYSIS.              *   TH./01103/87/DE/.. *
*                                                     *                      *
****************************************************************************
*  CONTRACTOR :                                       *    TELEPHONE NO      *
*     GCA, M. SCHMITT                                 *                      *
*                                                     *    5132 53579        *
*     WILHELMSTRASSE 36                               *                      *
*     D-3160 LEHRTE                                   *                      *
*     FRG                                             *    TELEX NO          *
*                                                     *                      *
*  TECHNICAL DIRECTOR :                               *    933521            *
*     M. SCHMITT                                      *                      *
*                                                     *                      *
****************************************************************************
                                                       VERSION : 29/03/89
```

AIM OF THE PROJECT :
****
Into light boiling destillates by application of the veba-combi-cracking (vcc)
technology in a 1 T/H pilot plant scale, The vcc process applies the features of
the former bergius-pier technology for the liquifaction of coals (high hydrogen
partial pressure, high reactor temperatures) using liquid phase hydrogenation
(lph) reactors for the primary conversion combined with gas phase (gph) reactors
for direct aftertreatment of the primary synthetic crude for quality improvment.
In the lph section a cheap one way additive is used, the gph reactors are filled
with standard refinery catalysts.

PROJECT DESCRIPTION :
****
The vcc process as developed by veba oel in cooperation with lurgi and intevep
is characterized by its extrmely high conversion efficiency of up to 95 wt. perc.
 based on the non-boiling portion of the residual oil. The process was developed
within the following phases since 1978 :
Phase I   : Basic bench scale reseach
Phase II   : design and construction of a pilot plant with a capacity of 1
t/hPhase III   : pilot plant operation

Phase IV   : design and construction of an commercial-scale plant for process
               demonstration.
Presently, the development is almost finished with the completion of the second
pilot plant operation phase (phase IIIA).
The development work within that phase covered the following fields of
experiments and theoretical research :
     - additive optimization
     - pressure reduction
     - evaluation of design data
     - evaluation of scale up factors
     - sreening of gas phase catalysts
     - long term tests.
Focal point of the development activities is the evaluation of design data for a
commercial vcc upgrader and optimization of the process for a given application.

STATE OF ADVANCEMENT :
****
Phase IIIA is finished, the test program was performed as scheduled, all
information needed for a commercial application of the vcc process was generated.
Commercial application itselve suffers today's very low oil prices.

RESULTS :
****
By application of the vcc technology several vacuum residues from conventional
and heavy crudes (arab heavy, bachaquero, Tiajuana, Morichal) and from a
visbreaking plant were almost completely converted (greater than 90 perc.) into
light distillates, some hydrocarbon gases and a small amount of hydrogenation re-
sidue which is suitable for the production of hydrogen via partial oxydation.
Feeding a vacuum bottom from a typical venezuelan crude having the following
analytical data.
Carbon   84,8 wtperc.
Hydrogen   10,4 wtperc.
Sulphur   3,3 wtperc.
Nitrogen   0,6 wtperc.
Vanadium   630 ppm
Nickel   75 ppm
The vcc process, 80 wtperc of a synthetic crude 15 produced which contains 27
perc. of naphta, 48 perc. of middle distillates and 25 perc. hydrogenated vqcuum
gasoil. Due to the application of the gas phase hydrogenation (the catalytic
fixed bed reactors directly combined to the primary conversion reactors), the
vcc syncrude having a relatively high hydrogen content is almost sulphur and
nitrogen free (less than 200 ppm each).
The vcc middle distillates are to be sold directly, the vcc naphta meets
reformer feed spezification and the vacuum gas oil is an excellent feedstock for

a ffc or a hydrocracker unit. As these results can be generalized for all the residues processed up to now it can be concluded that the vcc process is a well advanced technology to convert less valuable bottoms into light distillates thus making maximum use of the "bottom of the barrel".

REFERENCES :
****
US PATENT OFFICE NO. 4,350,051 SEPT. 21,1982, THOMPSON
US PATENT OFFICE NO. 3,539,299 NOV. 10,1970, THOMPSON
GOTH, M., DISS. 1983, GEOCHEMISCHE UNTERSUCHUNGEN AN BODENGASEM, TUBINGEN

```
************************************************************************************
* TITLE : DEVELOPMENT OF ADVANCED TECHNIQUES AND       *      PROJECT NO          *
*         INTERACTIVE MATHEMATICAL MODELS TO           *                          *
*         OPTIMIZE AQUIFERS PERFORMANCE AND FLUID      *   TH./01106/87/PO/..     *
*         INJECTION                                    *                          *
************************************************************************************
* CONTRACTOR :                                         *   TELEPHONE NO           *
*    PARTEX - COMPANHIA PORTUGUESA DE SERVICOS SA      *                          *
*                                                      *   73.50.13.              *
*    AVENIDA 5 DE OUTUBRO 160                           *                          *
*    PO-1000 LISBOA                                    *                          *
*                                                      *   TELEX NO               *
*                                                      *                          *
* TECHNICAL DIRECTOR :                                 *   14708                  *
*    ANTONIO DIOGO PINTO                               *                          *
*                                                      *                          *
************************************************************************************
                                                          VERSION : 20/02/90
```

AIM OF THE PROJECT :
****
The objective of the project is to develop a 3 - D interactive Mathematical
Model for use in continuous optimization of aquifers performance for full scale
water injection projects in oil reservoirs. The project searches technological
issues and new procedures in order to rationalize the full utilization of
aquifer potentialities in waterflood operations aiming at a better control of
fluids movement and displacement.

PROJECT DESCRIPTION :
****
Accorrding to the breakdown of the project it envolves three phases.
Phase I encompasses the screen, review and assembling of the basic data followed
by the Data Base organization, conception of the interactive mathematical model
and organization of the files for the well equipment selection.
Phase II encompasses the incorporation of several routines regarding aquifer
description procedures, Well Inflow Performance, Cross-Fow and Inter-Unit
Communication, Automatic Well Equipment Selection and the Graphics package.
Phase III encompasses the initialization and control of the model, testing
programme and validation procedures, interpretation of the results and
generalization of the model regarding the optimization of clustered systems and
influence on oil reservoirs.

STATE OF ADVANCEMENT :
****
Ongoing. Phase I is underway, some tasks are completed as the screen, review and
assembling of the basic data and the data files organization. The conception of
the model and Software development is underway as well.
Some tasks of phase II Namely the incorporation of Aquifer Description
Procedures and Well Inflow Performance routine are already launched.
Phase I is under completion, Phase II is underway and the tasds related will be
finalised in the next period.

RESULTS :
****
The project is ongoing and there were some developments involving areas related
to the conception and testing of the Interactive Mathematical Model for
Simulation of an aquifers system, design and incorporation of specific routines
regarding the Well Inflow Performance and Cross-flow modelling, and improvement
of the Data Base architecture.
New developments have been introduced or will, related to the selection of well
equipment and design and testing of different modules of the graphic package.
Furthermore new case studies have been performed for the initialization of the
model in order to test the general performance of the simulator and improve the
articulations between the different software modules.
The initialization of the model has been sucessful but the performance of the
simulator regarding its interaction with the data base and graphics package
needs to be improved and more complex case studies will be run parallel to the
development and testing of the software modules.
The initialization of the model has been sucessful but the performance of the
simulator regarding its interaction with the data base and graphics package
needs to be improved and more complex case studies will be run parallel to the
development and testing of the software modules.

REFERENCES :
****
- L.P. DAKE, "FUNDAMENTALS OF RESERVOIR ENGINEERING
, 1978
- D.W. PEACEMAN, "INTERPRETATION OF WELL-BLOCK PRESSURES IN NUMERICAL RESERVOIR
SIMULATION", SPE JOURNAL, 1977
- D.W. PEACEMAN, "INTERPRETATION OF WELL-BLOCK PRESSURES IN NUMERICAL RESERVOIR
SIMULATION WITH NON-SQUARE GRID BLOCKS AND ANISOTROPIC PERMEABILITY", SPE
JOURNAL, 1983
- H.L. STONE, "ITERATIVE SOLUTION OF IMPLICIT APPROXIMATIONS OF MULTI-
DIMENSIONAL PARTIAL DIFFERENTIAL EQUATIONS", SIAM 1968

- J.D. BREDEHOEFT AND G.F. PINDER, "DIGITAL ANALYSIS OF AREAL FLOW IN
MULTIAQUIFER GROUNDWATER SYSTEMS : A QUASI THREE DIMENSIONAL MODEL", WRR, 1970.

```
***************************************************************************
* TITLE : DEVELOPMENT OF AN IMPROVED MODEL FOR    *     PROJECT NO       *
*         PREDICTION OF THE PHASE BEHAVIOUR OF     *                      *
*         GAS CONDENSATE MIXTURES                  *   TH./01108/87/DK/.. *
*                                                  *                      *
***************************************************************************
* CONTRACTOR :                                     *   TELEPHONE NO       *
*     CALSEP A/S                                   *                      *
*                                                  *   45 42 87 66 46     *
*     LYNGBY HOVEDGADE 29                          *                      *
*     DK - 2800 LYNGBY                             *                      *
*                                                  *   TELEX NO           *
*                                                  *                      *
* TECHNICAL DIRECTOR :                             *   16600              *
*     MRS K.S. PEDERSEN                            *                      *
*                                                  *                      *
***************************************************************************
                                              VERSION : 21/05/90
```

AIM OF THE PROJECT :
****
The aim of the project is to make an improved model for phase equilibrium
calculations on gas condensate mixtures. The need arises because an increasing
part of the hydrocarbon production in the North Sea comes from gas condensate
fields. The phase behaviour of a gas condensate mixture is more complicated than
that of an oil mixture. In simulation studies on oil mixtures it is for many
propose acceptable to simulate the oil mixture as consisting of only two
components, "an oil component" and "a gas component". For gas condensate mixture
models which take into consideration the detailed composition are needed.

PROJECT DESCRIPTION :
****
The project will initially concentrate on the socalled cubic equations of state.
Based on the experience gained with these, more advanced models will be selected
and tested out. Part of the project will be on characterization of C7+ -
fractions. The project will be divided into the following parts: Part I:
Literature search, Part II: Performance of existing model, Part III: Development
of an improved model, Part IV: Documentation of the work.

STATE OF ADVANCEMENT :
****
Ongoing. PVT-data for 10 gas condensate mixtures and 6 oil mixtures from all
over the world have been collected. Three different cubic equations of state and
the group contribution equation of state, GC-EOS, have been tested for
simulation of the phase behaviour of these mixtures. The best results are
obtained with the SRK-equation modified as suggested by Peneloux.

RESULTS :
****
A procedure for predicting the phase behaviour of gas condensate mixtures with
good accuracy.

REFERENCES :
****
MOST IMPORTANT REFERENCES USED ARE:
SOAVE, G., CHEM.ENG. SCI. 27, 1972, 1197
PEDERSEN, K.S., THOMASSEN, P. AND FIEDENSLUND, AA., IND.ENG.CHEM.PROCESS DES.DEV.
24, 1985, 948
PEDERSEN, K.S., FREDENSLUND, AA AND THOMASSEN, P., "PROPERTIES OF OILS AND
NATURAL GASES", GULF PUBLISHING INC., HOUSTON, 1989
PEDERSEN, K.S., THOMASSEN, P. AND FREDENSLUND, AA., ADVANCES IN THERMODYNAMICS 7,
1989, 137.

```
*****************************************************************************
* TITLE : OPTIMIZATION OF SEISMIC ACQUISITION      *      PROJECT NO       *
*          METHODS WITH EMPHASIS ON SHALLOW         *                       *
*          APPLICATIONS.                            *   TH./01109/87/NL/..  *
*                                                   *                       *
*****************************************************************************
* CONTRACTOR :                                      *    TELEPHONE NO       *
*    TNO INSTITUTE OF APPLIED GEOSCIENCE            *                       *
*                                                   *   15.697185/697197    *
*    PO BOX 285                                     *                       *
*    NL-2600 AG DELFT                               *                       *
*                                                   *    TELEX NO           *
*                                                   *                       *
* TECHNICAL DIRECTOR :                              *    38071              *
*    J. RIDDER                                      *                       *
*                                                   *                       *
*****************************************************************************
                                              VERSION : 27/10/89
```

AIM OF THE PROJECT :
****
The main objective of the R&D programme is to optimize seismic reflection
acquisition methods for accurate delineation of shallow deposits in a macro
model and, by using this information, to improve the quality of deep seismic
reflection data.
Use will be made of full elastic data analyse and data processing techniques for
the determination of the macromodel of the shallow subsurface.
Much attention will be paid to facilities for quality control (acquisition
procedure and data).

PROJECT DESCRIPTION :
****
Deep seismic reflection surveys for hydrocarbons exploration and reservoir
delineation are severly hampered in areas with a complex shallow subsurface, due
to raybending. For the removal of the effect of raybending detailed information
about the structure and the seismic parameters of the shallow deposits is needed
(detailed macromodel). This information can be obtained by optimising high
resolution seismic reflections methods, for the determination of this detailed
macromodel. In numerous cases this accurate knowledge about shallow deposits is
indispensable for the interpretation and inversion of deep seismic data.
The important improvements that have to be realized for the efficient
determination of an accurate macromodel of the shallow subsurface are:
- development of a data acquisition system for efficient recording of two
component high resolution seismic data of high quality utilizing the proposed
data acquisition procedures.
- developments of techniques for the determination of a detailed macromodel of
the shallow subsurface applying for the inversion of techniques on two component
shot data.
The acquisition system main developments include extensive data quality control
facilities (incl. automatic position determination of remote units) implemented
on a PC based open central unit. Concerning the remote units, attention is paid
to the development of light weight cost effective units, by miniaturization of
the electronics and realizing the power supply over the field cable.
For the determination of macro models from the acquired data the following
processing steps must be adapted for pre-stack shot ordered data :
- decomposition of two component data set in P and S waves;
- estimation of coupled P/S macromodel;
- macromodel verification.
The developed techniques and instrumentation will be field tested.

STATE OF ADVANCEMENT :
****
Phase 1 and phase 2 : the definition of the data acquisition procedure and the
evaluation of existing technologies for possible use in the R&D programme is
finished. The development requirements were defined in a project proposal that
was granted in Septembre 1989. The actual development will start in January 1990.

RESULTS :
****
A data acquisition procedure and a data processing sequence for the estimation
of detailed macro models of the shallow subsurface has been defined. Development
needs were identified and the project plan for the R&D work.
In phases 1 and 2 of the project an evaluation of existing technology for
possible use in the R&D programme was carried out. The following aspects were
considered :
1) data acquisition procedure;
2) data acquisition systems;
3) sources and receivers.
The conclusions of this evaluation are summarized hereunder.
1. For determination of macromodels of the shallow subsurface useful information
is provided by P and S reflected waves. Both waves can be generated by burried
explosives. The data can be acquired by recording two components over a
sufficient long spread in order to obtain P-S conversions.
2. Data acquisition system that allow to perform high resolution seismic surveys

do not meet the requirements for :
- modern data acquisition procedures;
- high data quality;
- high productivity.
Major shortcomings of existing systems are :
- not sufficient quality control provided (data and system quality control);
- they are too bulky, heavy and expensive to be used for high resolution seismic surveys.
3. Existing sources and receivers satisfy the requirements set by the high resolution seismic method.

REFERENCES :
****
MEEKES J.A.C., SCHEFFERS B.C., RIDDER J., 1989
OPTIMIZATION OF HIGH RESOLUTION SEISMIC REFLECTION METHODS. PRESENTED AT THE 51ST ANNUAL MEETING AND TECHNICAL EXHIBITION OF THE EAEG IN BERLIN.

```
*****************************************************************************
* TITLE : ADVANCED PVT TECHNIQUES                  *      PROJECT NO      *
*                                                  *                      *
*                                                  *  TH./01110/87/UK/..  *
*                                                  *                      *
*****************************************************************************
* CONTRACTOR :                                     *  TELEPHONE NO        *
*    AEE WINFRITH                                   *                      *
*                                                  *   (0305)251888       *
*    UKAEA                                          *                      *
*    DORCHESTER                                     *                      *
*    UK-DORSET, DT2 8 DH                            *  TELEX NO            *
*                                                  *                      *
* TECHNICAL DIRECTOR :                             *   41231              *
*    DR N.A. BAILEY                                 *                      *
*                                                  *                      *
*****************************************************************************
```
VERSION : 04/09/89

AIM OF THE PROJECT :
****
This development programme, to be carried out at AEE Winfrith, is concerned with
the enhancement of PVT measurement techniques which can, eventually, be used in
an efficient and reliable fashion. Particular attention will be given to the
development of measurement techniques which can be incorporated into a PVT cell
or which require the removal of minimal quantities of fluids, while operating at
high levels of temperature and pressure. Attention would also be given to the
design of a mercury-free PVT cell which would incorporate the advanced
measurement techniques. Particular emphasis would be given to visual access to
the cell and the ability to gather up small quantities of a second phase for
accurate measurements. Such developments, if successful, will raise capabilities
well above existing levels.

PROJECT DESCRIPTION :
****
The work content of this one year study will be :
1. Critically review candidate measurement techniques from the point of view
   of accuracy and applicability to use with PVT cells at elevated pressure
   and temperature.
2. Select preferred measurement techniques for development.
3. Examine performance of selected techniques under laboratory conditions at
   lower temperature and pressure to establish viability of techniques.
4. Examine design requirements for a high temperature/high pressure mercury
   free PVT cell. Critical issues to include seals and materials.
5. Do conceptual design study of PVT cell incorporating preferred measurement
   techniques.
Priority will be given to the development of techniques for the measurement of
density and viscosity within PVT cells.
The programme will be divided into two six-months stages :
STAGE 1 (months 1-6) :
Candidate measurement techniques will be critically reviewed from the point of
view of accuracy and their applicability to use with PVT cells at elevated
pressure and temperature. The design requirements for the PVT cell would be
established and consideration given to materials issues which are likely to be
critical if the target temperatures and pressures are to be achieved. Preferred
measurement techniques would be selected for development. These activities would
involve detailed discussions with the users of PVT equipment in the oil industry
and service laboratories and equipment manufacturers.
STAGE 2 (months 7-12) :
The selected measurement techniques would be developed under less arduous
conditions of temperature and pressure while the design of a prototype PVT cell
for use at elevated pressure and temperature is carried out.

STATE OF ADVANCEMENT :
****
Ongoing. Development programme to start on 1 April 1988.
Work on this project started in October 1988. Stage 1 of the project is underway
with work being done to identify candidate techniques and test them in the
laboratory. Key features required for the cell have been identified and a
preliminary conceptual design prepared.

RESULTS :
****
Various techniques have been identified which could in principle, be used to
carry out the required measurements of density and viscosity.
Vibrating devices appear to be the most promising for both measurements raising
the possibility that both measurements might be made on a single device.
Preliminary tests have been carried out in the laboratorty on piezo crystal
devices and vibrating wires. Although the crystals showed a very clear
dependence in their resonant frequency on fluid density, various other aspects
of their performance suggested that they would not provide satisfactory
performance under cell conditions. The performance of a vibrating wire, from
preliminary experiments, look more promising. The effect of density on its
resonant frequency is readily observed and the damping of the vibration can be

recorded allowing the determination of fluid viscosity.
The key features required of a PVT cell if it is to accomodate such devices have been identified and it is clear that no presently available commercial cell meets tham all. A conceptual design for a new PVT cell has therefore been prepared and work will now proceed on the detailed evaluation of the design, identifying in particular those areas where development is needed.

```
*********************************************************************************
* TITLE : USE OF DRILLING NOISES TO GENERATE A        *       PROJECT NO       *
*         RECIPROCAL VSP                              *                         *
*                                                     *    TH./01113/87/IT/..   *
*                                                     *                         *
*********************************************************************************
* CONTRACTOR :                                        *     TELEPHONE NO        *
*     OSSERVATORIO GEOFISICO SPERIMENTALE DI TRIESTE  *                         *
*                                                     *     040-21401           *
*     VIALE ROMOLO GESSI 4                            *                         *
*     IT-34123 TRIESTE                                *                         *
*                                                     *     TELEX NO            *
*                                                     *                         *
* TECHNICAL DIRECTOR :                                *     460329              *
*     PROF. FABIO ROCCA                               *                         *
*                                                     *                         *
*********************************************************************************
                                                        VERSION : 23/08/89
```

AIM OF THE PROJECT :
****
It is well known that vertical seismic profiles (VSP) are useful tools for
predicting lithology ahead of the drill bit. The goal of the research is to
verify whether the seismic waves, both P ad S, generated by the bit during
drilling, could be utilized to generate a reciprocal VSP by properly disposed
receivers on the surface and in the subsurface.

PROJECT DESCRIPTION :
****
The first problem to solve is the identification of both the position and
signature of the source.
It is not true that the terrain is energized only at the position of the bit;
the drill touches the sides of the well and disturbances are generated. Moreover
the bit generates S waves as well as P waves during its rotation and hammering.
The undesired waves should be removed using their different frequency,
wavelength polarization using frequency filters, geophone groups and 3D
geophones.
It is necessary to stress that some of these waves might be useful, subsequently,
 for lithological purposes. In fact, S waves might very well enrich our data
base, provided that they are properly separated.
A graded experiment that removes, one by one, the uncertaines, would certainly
be useful since no modeling is currently able to simulate all the various, at
present unpredictable, effects on the bit.
Once the signature of the source is found, a crosscorrelation of that signature
with the signals recorded by the geophones will allow the determination of :
- differential propagation times between bit, geophones on the surface and
buried, and therefore seismic velocities
- bit location, using seismological techniques.
Finally once the location of the bit is determined, it will be necessary to
solve the problem of the prediction of the lithological characteristics in the
layers beyond the actual penetration.
If the upgoing wave only is recorded, filtering out all other non relevant waves,
 the autocorrelation of that signal, for positive times only, will provide a
signal that will contain, after the direct arrival, all the other reflections at
their proper times.

STATE OF ADVANCEMENT :
****
Ongoing.Following the programme sketched out for the first of the two phases in
which the research is divided, a seismic crew of O.G.S. made a seismic data
acquisition near the well named "Olzano 1", in Cremona province.
After data collection, all the information was sent to the Computing Center of
the Osservatorio Geofisico Sperimentale in Trieste, for processing.

RESULTS :
****
The possible results of the technique are numerous :
- identification of the position of the drill bit in space;
- identification of the position of the drill bit in the seismic time section;
- identification of the seismic velocities;
- determination of the VSP while drilling;
- prediction of lithological characteristics of the rocks ahead of the bit.

```
****************************************************************************
*  TITLE : INTEGRATED ELECTROMAGNETIC#EXPLORATION    *     PROJECT NO      *
*          SYSTEM FOR HYDROCARBONS                   *                     *
*                                                    *   TH./01116/88/DE/.. *
*                                                    *                     *
****************************************************************************
*  CONTRACTOR :                                      *    TELEPHONE NO      *
*     METRONIX MESSGERATE UND ELEKTRONIK GMBH        *                     *
*                                                    *    053-377007        *
*     PETZVALSTR. 36A                                *                     *
*     WEST-GERMANY                                   *                     *
*     DE - 3300 BRAUNSCHWEIG                          *    TELEX NO          *
*                                                    *                     *
*  TECHNICAL DIRECTOR :                              *    952311            *
*     MR. R. KARMANN                                 *                     *
*                                                    *                     *
****************************************************************************
                                             VERSION : 20/09/89
```

AIM OF THE PROJECT :
****
For the exploration of hydrocarbons mainly the Seismic method is used. It is
well known that with Seismics only unsufficient up to worthless results can be
achieve in areas with strong topography, coverage of basalts or complex geology.
Preceeding studies have shown that the electromagnetic exploration technique has
the potential to complete the seismics in these problem areas. To reach this
goal it is necessary to improve both, the instrumentation and the processing
techniques which are presently used.
The aim of the prject is to develop an extended electromagnetic exploration
technique which is based on automatic recording multistation configuration and
to proof the applicability in an area of complex geology.

PROJECT DESCRIPTION :
****
To improve the accuracy and productivity of the existing electromagnetic
exploration techniques, we have to follow, in principle, the same direction that
seismic methods have evolved in the past decades. From seismic methods it is
well known that a multireceiver-concept, combined with a related processing
concept, increases significantly the measurement accuracy, as well as the
resolution for the geological structure.
The layout of the planned multichannel, multireceiver-concept is as follows :
In the target area, N portable electromagnetic multichannel receivers, which are
running timesynchroneous with the reference station, will be used.
Synchronization between all receivers is accomplished by highly precise
synchronizable cloks. The reference station can be either a passive one
(receiver recording natural electromagnetic fields, or an active one
(alternating electrical dipole, generating electromagnetic fields in the
frequency band of interest).
Using Remote Reference techniques, the unmanned Reference Station should be
located in an area with approximately 1-D geological structure and low cultural
noise. The instrumentation of the passive Reference Station will be the same
type as the one at the receiver station. To each multichannel receiver, up to 8
eletrical field sensors, or 8 magnetic field sensors, or a mixture of both, can
be connected. When the N sites have been measured, the total N receiver
configuration moves to the adjacent sites, where the second set of sensors has
been implemented.
The basic idea behind this concept is, that due to the remote-reference
technique, all measurements in the target area can be processed in the same
manner, as if they were obtained at the same time in an M x N x 8 receiver array
(M is the number of N-station setups). In the example of M = 14 and N = 5 this
leads to an array of the size of S = 560.
If statistical independancy between the measurements is assumed, this leads to
an increase of accuracy of up to 24 compared with a single site measurement.
To test this new method in field applications it is planned to start with the
following hardware which has to be developed :
*   3 portable EM-multichannel stations
    (1 reference station, 2 moving stations)
*   5 sensors for each station (3 x H, 2 x E)
    and a high precision reference clock
*   2 portable data memories
*   1 computer and peripherals to be used
    at the base camp for data processing
The total project is divided into 3 phases :
In phase 1 the multireceiver instrument is developed. For testing the
multireceiver configuration, in addition to the first instrument, two additional
sets of instruments have to be built. The development of the basic processing
software enables the checkout of the instrument prototype in a field test.
Phase 2 focuses on the development and field testing of the processing software.

STATE OF ADVANCEMENT :
****
Ongoing with the implementation of the advanced electromagnetic processing
system and the development of the extended EM interpretation software, the
phases 2 and 3 of the project are worked out. Up to now, the prototype and one

electromagnetic exploration receiver have been built up and tested. The third is in production.

RESULTS :
****
The development of the geophysical exploration instrument is nearly finished. The complete system will contain three multichannel EM-receivers, magnetic field sensors and electric field probes, and a high precision clock.
A field experiment was carried out to test and improve the equipment. The result shows clearly that the hardware comes up to the expectations.

REFERENCES :
****
BAHR, K. INTERPRETATION OF THE MAGNETOTELLURIC IMPEDANCE TENSOR REGIONAL INDUCTION AND LOCAL TELLURIC DISTORTION, J. GEOPHYS., 1987.

```
*******************************************************************************
* TITLE : POLYSEIS                                    *       PROJECT NO      *
*                                                     *                       *
*                                                     *   TH./01118/88/FR/..  *
*                                                     *                       *
*******************************************************************************
* CONTRACTOR :                                        *    TELEPHONE NO       *
*    GERTH                                            *                       *
*                                                     *    1 47 52 69 27      *
*    AVENUE DU BOIS PREAU 4                           *                       *
*    F - 92502 RUEIL MALMAISON                        *                       *
*                                                     *    TELEX NO           *
*                                                     *                       *
* TECHNICAL DIRECTOR :                                *    FAX 1 47 52 69 27  *
*    MR RIALAN                                        *                       *
*                                                     *                       *
*******************************************************************************
                                                         VERSION : 05/06/90
```

AIM OF THE PROJECT :
****
The aim of the product is to develop a 3D seismic data-acquisition system by
radio transmission, which cumulates the advantage of existing systems while
avoiding their disadvantages.

PROJECT DESCRIPTION :
****
The project consists in creating a system capable of transmitting, under
conditions close to real time, seismic data for 1000 seismic channels on several
radio frequencies semisequentially. To achieve this radio transmission, a
digital modulation technique will be used to transmit a great density of digital
data in a reduced radio-wave spectrum by implementing low power compatible with
national regulations. DCUs (data concentrator unit) capable of receiving
modulations from about 100 stations will e linked to the CRU either by UHF radio-
wave linkage or by a single optical cable, or else by a combination of these two
types of linkages, as function of the natural of the ground. The flexibility of
the system will also be increased by the possibility of preprocessing the
seismic information such as the stacking of shots in acquisition stations (RTUs).
 The possibility of remote controlling from the CRU of variable transmission
velocities suited to different regulations will be a third type of flexibility
of the system. A new analog-to-digital conversion concept will be adopted to be
able to transfrom seismic data into a digital form when they are aquired in situ.
 The different parts of the system will be designed by progressively building,
in the laboratory, four mock-ups of RTU telemetering acquisition stations, one
mock-up of the DCU and one CRU, DCU. Fields tests will then be undertaken with
the aim of being representative of different transmission possibilities between
the CRU, DCU and telemetering stations for a POLYSEIS system simulating 1000
seismic channels. These tests will require the development, as part of the
POLYSEIS project, of the following pieces of equipment : 60 prototypes of RTUs,
4 DCUs, 2 CRUs.
The project involves 3 phases :
Phase 1 : Feasibility of the POLYSEIS system
Phase 2 : Development of prototype subsets of the POLYSEIS system
Phase 3 : Field tests of the POLYSEIS system.

STATE OF ADVANCEMENT :
****
Phase 1 : Feasibility
This phase was completed by the end of 1989.

RESULTS :
****
Results of phase 1 are satisfactory. The main targets of the project appear to
be feasible :
- real-time acquisition of a great number of seismic traces
- the flexibility of the equipment according to different field configurations :
shallow water, mountains, tropical forests...

```
*****************************************************************************
* TITLE : GEOSCOPE                                    *      PROJECT NO      *
*                                                     *                      *
*                                                     *   TH./01119/88/FR/.. *
*                                                     *                      *
*****************************************************************************
* CONTRACTOR :                                        *    TELEPHONE NO      *
*    GERTH#COMPAGNIE GENERALE DE GEOPHYSICQUE         *                      *
*                                                     *   33(1)47.52.61.39   *
*    PRAKLA-SEISMOS                                   *                      *
*                                                     *                      *
*                                                     *    TELEX NO          *
*                                                     *                      *
* TECHNICAL DIRECTOR :                                *   33(1)47.52.69.27   *
*    MR FRECHU                                        *                      *
*                                                     *                      *
*****************************************************************************
                                                       VERSION : 15/02/90
```

AIM OF THE PROJECT :
****
Oil exploration today is faced with an increased complexity due to relative
limitations of the existing softwares used to process the seismic data in order
to discover the oil traps currently searched for. These limitations can be
reduced by suppressing strong hypotheses that are commonly used (common depth
point) and introducing, at the processing stage, geological information. The
objective of the GEOSCOPE project is then to develop a new software which will
integrate the latest evolutions in seismic processing based on more general
hypotheses (tomography, prestack depth migration, inversion, section balancing)
in conjunction with an interpretive approach of problems. All this will be based
on recent improvements of the computer technology (computing power, graphic
capabilities and large mass storage accessible volumes), which are to be taken
into account to make the project feasible.

PROJECT DESCRIPTION :
****
The project is divided into 3 phases.
PHASE 1 : consists in the analysis of the whole project from different point of
view. The different geophysical products which shall be developed have to be
specified on bases that integrate new features and can be accessed by all
participants to the project. The tools that shall be used by the different
companies to develop the software have to be standardized to reduce the costs at
the development level and also at the maintenance level. Moreover, an important
constraint will be to make the software portable in many possible environments,
at least the one used by the different participants to the project. With the
same objectives, common tools have to be specified to store and access the data
as well as to develop the graphic interfaces.From the objectives, an estimation
of the necessary resources shall be done and finally a planning shall be
designed. If at the end of this phase, the project was found to be non feasible
or too expensive or had to be stopped for other reasons, the other phases shall
not be undertaken.
PHASE 2 correspond to the developpment of the two dimensional aspects of the
project. The different modules that would be selected by the different
participants (prestack depth migration, tomography, reflectivity or impedance
estimation, section balancing and possibly some others) would be worked out at
the research stage first and then developed. In thid phase, even if the products
are conceived in 2D, the model should be designed in 3D and managed in 3D. An
access to the data in 2D should be made available for the applications of the
second phase but the organization of the data in 3D should be made accessible to
the end user.
PHASE 3 would be the extension of these research works to the three dimensional
case. This should be done with the same techniques as the one that have been
developed in the second phase, wwith possibly some extensions specific to the 3D
case at least to manaage the data, visualize it and input interpretated data
from it (we may have at this stage 4D data, for we intend to work with prestack
data in most of the previous techniques).

STATE OF ADVANCEMENT :
****
Ongoing

RESULTS :
****
Interactive environments for seismic processing are being tested where the
software from Stanford University called "SEPlib" is mostly involved.
Three dimensional depth modelers are also being tested. The consortium named
"GOCAD" seems well suited to the needs of the project.
The User Interface topic is being covered. C + + as a programming language, X-
Windows for window management seem promising. Develpment of a graphic library
has been made.

```
*****************************************************************************
* TITLE : ACOUSTIC MEASUREMENT ON RESERVOIR ROCK     *     PROJECT NO       *
*                                                     *                      *
*                                                     *   TH./01120/88/NL/.. *
*                                                     *                      *
*****************************************************************************
* CONTRACTOR :                                        *   TELEPHONE NO       *
*     DELFT UNIVERSITY OF TECHNOLOGY                  *                      *
*                                                     *   (015) 782310       *
*     BUREAU HOGESCHOOL                               *                      *
*     COLLEGE VAN BESTUUR                             *                      *
*     P.O. BOX 5                                      *   TELEX NO           *
*                                                     *                      *
* TECHNICAL DIRECTOR :                                *   38151              *
*     MR. J.P. VAN BAAREN                             *                      *
*                                                     *                      *
*****************************************************************************
                                               VERSION : 16/11/89
```

AIM OF THE PROJECT :
****
The research objective is to obtain the reservoir rock properties via "sonic log" measurements in the borehole. The project aims at the development of an improved interpretation technique for acoustic recordings. Acoustic measurements will be performed under atmospheric and under high pressure/temperature conditions in the laboratory on both artificial and natural (outcrop) rock samples.
The improved interpretation procedure should make it possible to determine permeability, lithology, rock strength and types of porosity from the acoustic measurement in the borehole. The laboratory results will be applied to acoustic field recordings.

PROJECT DESCRIPTION :
****
The procedure aims at the investigation of the effect on acoustic response of the change in one reservoir property while all other properties remain constant. A general interpretation technique will be derived from this.
a. Production artificial rocks
Production of artificial rock in the laboratory and changing the following properties: lithology, volume of cement binding the grains, grainsize, porosity and sorting.
b. Taking outcrop samples
Take "outcrop" samples in the field and vary the following properties within limits: lithology, cement, grainsize, porosity, and sorting. The samples have to be taken in quarries. On small samples the reservoir properties will be measured after which the bigger blocks can be selected.
c. Modification atm. system
Modify the atmospheric measurement system. The objective is to increase the flexibility in measuring configurations and varying geometries of the rock.
d. Atm. measurements
Measure acoustic properties on the samples at atmospheric conditions. The measurements can be performed on a flat surface and a borehole cylindrical surface with varying distances.
e. Modification high P/T system
Modification of the high pressure, temperature reactor to make it possible to measure displacements in horizontal and vertical directions. Making the heating mechanismen more effective and less damaging for the samples. Adjusting the transducers configuration.
f. High P/T measurements
Measure acoustic properties on the samples at high pressure and temperature. The samples will have a maximum height of 0.6 m and diameter of 0.4 m. The temperature and pressure range should represent the in-situ circumstances down to 4000 m depth.
g. Evaluation procedure
Development of an interpretation technique to obtain the reservoir properties from the acoustic recording in the laboratory using experiment and theory.
h. Field applications
Apply the developed evaluation procedure to the field logs and check the results with the core measurement results available for those wells. The acoustic field recording of several wells are in our possession.

STATE OF ADVANCEMENT :
****
Ongoing.

RESULTS :
****
A. Production artificial rocks
Artificial rocks has been made to investigate the accuracy of the porosity determination. Report will be published in the second half year of 1989.
B. Taking outcrop samples
During this period the first steps in the procedure for taking outcrop samples were started. Interviews were made with geologists, societies and companies and a literature survey has been started.

C. Modify atm, system
A rotation table needed for the transmission measurements has been constructed and is tested.
D. Atm. measurements
A start has been made to perform transmission measurements to investigate the influence of the fluid viscosity on the recorded micro-seismograms.
E. Modification high P/T system
The drawings for an easier operation method to load the large reservoir sample in the borehole simulator are finished. Construction of the new system is started.
G. Evaluation procedure
The material constraints and physical simplifications used in the Biot theory have been studied.
H. Field applications
Core porosities obtained from cores taken in the field using the mercury method have to be corrected, because this method introduces a systematic error.

REFERENCES :
****
- ABSTRACT : ACOUSTIC MEASUREMENTS ON ARTIFICIAL ROCK. BY J.P. VAN BAAREN, H.K.J. HELLER AND R. VISSER, DELFT UNIVERSITY OF TECHNOLOGY.
- GEERITS, T.W. AND J.P. VAN BAAREN, BIOT THEORY, MATERIAL CONSTRAINTS AND PHYSICAL SIMPLIFICATIONS, DELFT UNIVERSITY OF TECHNOLOGY, MAY 1989.
- VAN BAAREN J.P., REPORT ON THE OUTCROPS OF THE CARBONIFEROUS AACHEN COAL DISTRICT, DELFT UNIVERSITY OF TECHNOLOGY, JUNE 89.

```
*********************************************************************
* TITLE : INVERSION OF LAND SEISMIC DATA          *   PROJECT NO    *
* *                                               *                 *
* *                                               *  TH./01121/88/NL/.. *
* *                                               *                 *
*********************************************************************
* CONTRACTOR :                                    *  TELEPHONE NO   *
*    DELFT UNIVERSITY TECHNOLOGY                  *                 *
* *                                               *  015 781328     *
*    COLLEGE VAN BESTUUR PO BOX 5 - NL 2600 AA DELFT  *             *
* *                                               *                 *
* *                                               *  TELEX NO       *
* TECHNICAL DIRECTOR :                            *  38151          *
*    PROF.A.M.ZIOLKOWSKI                          *                 *
* *                                               *                 *
*********************************************************************
                                                   VERSION : 26/07/89
```

AIM OF THE PROJECT :
****
The objective is to develop and test existing inversion theory to obtain a
robust theory to invert seismic data with the minimum possible number of
constraints on the Earth model. The theory will be extended to cope with elastic
effects and arbitrarily dipping, locally plane, reflectors. The test will
consist of (a) acquiring seismic reflection data on a line over a logged well,
(b) inverting the data to obtain densities and velocities, (c) putting the
theory at risk by comparing the inverted results with the data from the well.
Inversion is normally done either by iterative forward modelling, changing all
parameters of the Earth simultaneously, or by assuming a model of the Earth in
which the velocities are already known and the reflections are assumed to be
primaries only. Our approach is different in that the Earth model is determined
from the top down in a recursive layer stripping scheme.

PROJECT DESCRIPTION :
****
We shall do the following:
1- Develop a new analytic forward model for an elastic Earth with locally plane,
dipping reflectors. For a discussion of the theory developed on this subject see
Ziolkowski et al. (1988). We are left with the following subjects which should
improve and extend the method developed thus far:
(a) Extension to a three dimensional Earth with arbitrarily dipping locally
plane reflectors. The idea is to combine the reflectivity method and the theory
for the plane wave response of a 3D Earth with plane dipping reflectors as
described by Diebold (1987) and Richards and Witte (1987). This extension will
remove the former constraint that the Earth model should have horizontal plane
layers.
(b) Extension to an elastic Earth. We will use the forward theories of DuCloux
(1986) and Frasier (1970). So far the Earth model we have used is acoustic. This
extension will make it possible to cope with phenomena such as P to S
conversionsm which become important at large offsets and with large dip.
2- Develop the corresponding inversion scheme. For a discussion of the theory
developed on this subject see Ziolkowski et al. (1988). We are left with the
following subjects which should improve and extend the method developed thus far:
(a) Removal of the effect of the free surface. We shall look in the (o-x) and (o-
p) domain to remove these effects using a wave theoretical approach. The
multiples generated in the top layer can be removed in principle if the velocity
of the top layer is nown.known.
(b) Extraction of the wavelet. There are two alternative approaches. One is
based on the Critical Reflection Theorem (Fokkema and Ziolkowski, 1987); the
other is based on the scaling law approach (Ziolkowski et al., 1980) and will
require three shots of different quantities of dynamite at each shot point.
(c) Extraction of the density and velocity from primary impedance profiles.
After inversion we are left with a reflection coefficient series for each plane
wave component of the data. Using the pre-critical plane wave components we can
calculate from these series the interval velocities and densities in the layered
sequence. We anticipate that the velocities will be easier to determine than the
densities.(d) Development of a robust t-p tranformation for point source data in
a three dimensional structure. The inversion we intend to apply requires the
data in the t-p domain, so if we want to include dip in 3D structures we should
be able to transform the original x-y-t data to that domain. The theoretical
foundation for this transformation has been provided by Brysk and McCowan (1986).
3- Develop and test software for inverting real seismic data using these schemes.
4- Record a 2D seismic data set that satisfies minimum sampling criteria for our
inversion scheme, and for which the source signature can be measured or
estimated accurately.
5- Apply the processing scheme to our recorded 2D data set and to a 3D data set
from an oil field in the Netherlands..

STATE OF ADVANCEMENT :
****
Ongoing. We have simultaneously investigated a number of points described in
section 2. We have formulated a scheme for simultaneous dip correction and layer
stripping inversion (2.1.a). We have investigated the influence of elastic

effects on the final results calculated with the acoustic inversion scheme (2.1. b). We have deceloped a scheme to remove the effects of the free surface (2.2.a). We are working on a robust scheme to extract the dynamite wavelet (2.2.b).

RESULTS :
****

We developed a scheme to calculate the full plane wave response for a three dimensional acoustic Earth, where layer interfaces are plane but may have arbitrary dip and strike. In addition to primary reflections, multiple reflections within the layers may be taken into account up to any specified order. The scheme is similar to the reflectivity method for a plane horizontally layered acoustic Earth. The proposed forward modelling scheme is recursive in two ways:first the response is calculated layer by layer starting at the bottom; secondly, the calculation of each order of multiple reflection uses the results from the previous, lower order, of multiple reflection. In this way the redundancy in the scheme is minimized, thus making the scheme fast. However, the scheme requires the simultaneous calculation of the response of all possible plane waves, and gives in return all possible reflected plane waves up to a given order of multiple reflection within one layer; this plane wave response is calculated at every layer interface for every incident wave. This is in contrast to the reflectivity method which considers only one horizontal slowness at a time for all layers.

We have formulated a scheme for simultaneous correction of dip and layer stripping inversion. In this scheme the dip of the first macro layer interface is determined together with the velocity in the first layer. With these data it is possible to apply our modified layer stripping inversion up to and including the first interface, removing all internal multiples related to this interface. Then the dip of the second macro interface is determined. This dip is again used in the inversion to propagate downwards to the second interface. This scheme will be applied repeated until the last interface has been inverted. The result will consist of a dip-corrected primary reflection coefficient series for each p-value.

Research on the influence of elastic effects showed that these effects have little influence on the inversion results calculated with the acoustic inversion scheme, for small angles of incidence (<10 degrees) and when the dip of the interfaces is neglected (Holstege 1989). Therefore, we can not expect to get any reliable information on the elastic parameters of the Earth from the pre critical p-trqces when the dip of the interfaces is small. This conclusion justifies our decision to tackle first the problem of dip, before we investigate the possibility to invert also for the elastic parameters of the Earth. We have included in our inversion scheme a method to remove the effects of the free surface. This method requires only knowledge of the source and receiver geometry and the velocity in the top layer. In this scheme we compensate for the source and receiver ghost and remove the surface related multiples. The results of this scheme were presented in the workshop on inversion at the EAEG meeting in Berlin, and in a paper at the same meeting.

REFERENCES :
****

KOSTER, J.K. 1988. TRUE AMPLITUDE FORWARD MODELLING OF 3D STRUCTURES IN THE t-P DOMAIN, INCLUDING INTER-BED MULTIPLES. PAPER PRESENTED AT THE 1988 SEG MEETING IN LOS ANGELES.
ZIOLKOWSKI,A.M.,ET ALII,1989. INVERSION OF SEISMIC DATA, PAPER PRESENTED AT THE 1989 EAEG MEETING IN BERLIN.
HOLSTEGE,G.C.J., 1989. THE ACOUSTIC ASSUMPTION AND THE CONVOLUTIONAL MODEL OF THE STACKED TRACE,M.SC.THEIS,DELFT UNIVERSITY OF TECHNOLOGY, FACULTY OF MINING AND PETROLEUM ENGINEERING, SECTION APPLIED GEOPHYSICS, JULY 1989.

```
*****************************************************************************
* TITLE : ACQUISITION METHOD FOR 3D SEISMIC      *      PROJECT NO         *
*          SURVEYS WITH A PSEUDO-CASUAL          *                         *
*          DISTRIBUTION OF RECEIVERS AND SOURCES *   TH./01122/88/IT/..    *
*          IN HIGHLY POPULATED ARE               *                         *
*****************************************************************************
* CONTRACTOR :                                   *   TELEPHONE NO          *
*    AGIP SPA                                     *                         *
*                                                *   520 5819              *
*    AGIP/OPSI VIA FABIANI 3                      *                         *
*    I-20097 SAN DONATO - MILANO                  *                         *
*                                                *   TELEX NO              *
*                                                *                         *
* TECHNICAL DIRECTOR :                            *   310246               *
*    ING.P.V. RAVERA                              *                         *
*                                                *                         *
*****************************************************************************
```

VERSION : 07/12/89

AIM OF THE PROJECT :
****
The principal aim of the project consists of the study and carrying out of a 3d
seismic survey operatively achievable in areas of difficult access, and the
promotion of an appreciable reduction in exploration expenses. Indeed, it it's
true that 3d seismic can yield important results in hydrocarbon exploration,
it's also true that normally employed techniques, which presuppose regularly
spread layouts, heavily limit its carrying out in areas either highly urbanized
or characterized by critical environmental conditions.
In order to overcome this inconvenience an acauisition technique, based on a
pseudo-casual distribution of receivers and sources, is evaluated.
Of course, besides the alternative acauisition scheme, the project provides for
the preparation of an opportune sequence of processing and interpretation of the
collected data.

PROJECT DESCRIPTION :
****
The realization of the project will be divided in four fundamental phases.
- First phase (foreseen length of 6 months).
It includes the feasibility study, the planning and the operative choosings.
Definition of a new methodology which, basing itself on the scouting data and on
the geological features of the targets, will allow the evaluation of the best
distribution of sources and receivers.
Feasibility study, based on the already existing technology, aimed at the
individuation of the instrumentation needed to optimize the application of the
new methodology.
Once defined the theoretical aspects, we'll proceed to their sperimentation
beginning with a detailed scouting of the choosen area first and then
elaborating the survey project.
- Second phase (foreseen length of 8 months) : acquisition of data.
We shall proceed to the finalization of the project, in agreement with the subcontractor, starting off with the actual acquisition.
- Third phase (foreseen length of 10 months) : data processing.
specification of the best processing seauence, taking into account the extra-
difficulties introduced by the random distribution.
- Fourth phase (foreseen length of 3 months) : interpretation and evaluation of
data.
The last phase envisages the interpretation of he obtained data and the
verification of results to be pursued zith the aid of modeling.

STATE OF ADVANCEMENT :
****
THE PROJECT WAS STARTED ON APRIL THE 1ST 1989.
IT IS BEING IMPLEMENTED BY STUDYING THE AREAS AND GEOPHYSICAL CONSTRAINTS; A
COMPUTERIZED SIMULATION OF THE ACQUISITION LAY-OUT IS GOING TO BE CARRIED OUT.

RESULTS :
****
The study and application of the described project should allow the achievement
of the following results, which are considered fundamental in hydrocarbon
exploration conducted in uninvestigated areas :
- perfecting the projectig criteria, on a mathematical-statistical base, so as
to guarantee maximum reliability of the final result and easy implementation;
- definition of different practcal realizations and evaluation of their limits
with respect to the areas and equipments available;
- definition of the basic configuration and specifications of a recording system
to meet the application requirements of this type of survey;
- improvement of the data processing software for the random grouping envisages
in the acquisition phase.
This software could be utilized or extended to large scale so that the
methodology could be tested also in other surveys;
- definition of standards for the data classification in order to allow a more
flexible exchange of data among geophysical and exploration companies.

```
*******************************************************************************
* TITLE : CONCEPTS FOR OFFSHORE EXPLORATION IN    *      PROJECT NO        *
*         PACK-ICE AREAS                          *                        *
*                                                 *   TH./01127/88/DK/..   *
*                                                 *                        *
*******************************************************************************
* CONTRACTOR :                                    *   TELEPHONE NO         *
*    COWI CONSULT                                 *                        *
*                                                 *   45 45 97 21 11       *
*    45 TEKNIKERBYEN - DK 2830 VIRUM              *                        *
*                                                 *                        *
*                                                 *   TELEX NO             *
*                                                 *                        *
* TECHNICAL DIRECTOR :                            *   37280                *
*    MR ANDERS BJERRUM                            *                        *
*                                                 *                        *
*******************************************************************************
                                                     VERSION : 15/01/90
```

AIM OF THE PROJECT :
The objective of the PACK-ICE project is to investigate methods for seismic data
acquisition in Arctic waters with dense ice concentrations. In particular the
area offshore East Greenland in the region 73 deg.C - 80 deg.C has been studied.
Presently, only limited seismic data acquisition and no drilling activities have
mbeen carried out in the area or in other offshore pack-ice areas with similar
conditions.

PROJECT DESCRIPTION :
Conventional system
When towing a streamer and source in dense ice the system has to be modified to
prevent damage or an alternative system shall be developed. The PACK-ICE team
has found it important to base the development as close as possible on known
industry practice since the existing seismic methods provide a well proven
technology. This means that any consequences of changes from the known systems
shall be carefully evaluated. The evaluations include :
- Operational Safety
- Data Quality
- Environmental Impact, and
- Costs.
Studies
The studies undertaken in Phase 1 (1989) include :
- Environmental conditions (ice, weather, water)
- Marine operations (icebreaking, towing positioning, reconnaissance)
- Seismic Systems (source, streamer, depth control)
- Risks (environmental impact, emergencies).
Phase 2
Some preliminary conclusions can be made from Phase 1 but several problems are
still unsolved and should be covered in Phase 2 planned for in 1990.
Ice and Weather Conditions
When planning a survey great variations in ice and weather conditions even in a
few days should be anticipated. Optimal ice and weather conditions are found in
August and September but with great variations from one year to the other. A
summer sea ice model for ice forecasting is being developed. The single year ice
is mostly less than 21 mp thick and is mixed up with 10-30 m or thicker
multiyear ice brought south by the East Greenland Current.
Seismic Concepts
Four concepts for Seismic Survey are studied :
- Icebreaker towable systems
- Ice floe based systems
- A free flating digi-buoy system DIGISEIS , and
- Submarine Seismic.
The two most promising concepts seem to be icebreaker towable systems and
submarine seismic. The other systems may be feasible as infill systems for use
in areas where access by an icebreaker is impossible.
Cost evaluation and risk analysis are to be carried out for the most promising
concepts.

STATE OF ADVANCEMENT :
Phase 1 - Conceptual study, jan 1989 - March 1990
Phase 2 - Detailed Concept Development, April 1990 - April 1991
Phase 3 - Detailed Design Construction and Testing
          Planned to start in 1991.

REFERENCES :
A GEOLOGICAL CONDITIONS, NORTHEAST GREENLAND SHELF (COWICONSULLT, FEBRUARY 1989,
PACK-ICE, WO 2.1-0101)
TYPICAL PROFILES OF WATER TEMPERATURE AND SALINITY ON THE NORTH EAST GREENLAND
SHELF (DANISH HYDRAULIC INSTITUTE, FEBRUARY 1989, WO 2.1-3206)
ENVIRONMENTAL IMPACT OF SEISMIC EXPLORATION IN THE NEGS AREA (COWICONSULT, APRIL
1989, WO 2.1-0112)

HIGHLIGHTING PROBLEMS RELATED TO THE PACK-ICE - INNOVATIVE SEISMIC SYSTEMS (COWICONSULT, FEBRUARY 1989)

AMBIENT NOISE IN ARCTIC WATERS (ODEGAARD AND DANNESKIOLD-SAMSOE, FEBRUARY 1989, WO 2.1-3301)

MARINE SEISMIC SOURCES (NORTH SEA GEOPHYSICAL, FEBRUARY 1989, WO 2.1-3504)

```
****************************************************************************
*  TITLE : DEVELOPMENT OF A TUNABLE FREQUENCY         *       PROJECT NO      *
*          SEISMIC SOURCE FOR VSP IN THE              *                       *
*          TOMOGRAPHIC IMAGING OF HYDROCARBON         *     TH./01128/89/UK/.. *
*          RESERVOIRS.                                *                       *
****************************************************************************
*  CONTRACTOR :                                       *   TELEPHONE NO        *
*     BRITISH GAS PLC                                 *                       *
*                                                     *     01-736 3344       *
*     LONDON RESEARCH STATION                         *                       *
*     MICHAEL ROAD                                    *                       *
*     UK - LONDON SW6 2AD                             *     TELEX NO          *
*                                                     *                       *
*  TECHNICAL DIRECTOR :                               *     24670             *
*     DR A. MELVIN                                    *                       *
*                                                     *                       *
****************************************************************************
```
VERSION : 14/12/89

AIM OF THE PROJECT :

In recent years, Vertical Seismic Profiling (VSP) has been used extensively for the improved delineation of oil and gas reservoirs to determine their size and shape and the optimum siting for production well drilling. A review of the current position by the Society of Exploration geologists came to the conclusion that the major problem to be solved in VSP was the provision of a reliable and effective downhole seismic source. The aim of this project is to develop a new type of VSP source which can be tuned to single frequencies in the range 10-10 kHz and, hence, have a pre-selected spatial resolution varying from whole-reservoir mapping to local detailed examination of strata.

PROJECT DESCRIPTION :

The source under development is based on a physical principle which is entirely new for seismic sources. The source makes use of some of the hardware developed for an earlier seismic project supported by DG-XVII but operates on an entirely different principle. The device operates by firing a supersonic shock through a gas into a vertical column of liquid. A train of waves then develops in the liquid at a single frequency determined by the nature of the liquid, the height of the liquid column and the source dimensions. The pressure in the liquid at the end wall varies from zero to ca. 100 bar at the pre-determined frequency in the range 10-10 kHz. A flexible diaphragm at the end of the source tube acts as the means of transmitting acoustic energy into the environment. The project will involve development of the device principle into a practical field VSP source.
The first phase of the project (starting in April 1990) will involve laboratory development and optimisation of the design of the tunable seismic source. Since gas-liquid dynamics form an important aprt of the principle of operation of the device, some effort on gas dynamic modelling using appropriate computer simulation programs will be required and the results used to make sure that adequate experimental tests are carried out. A major part of the development will be the optimisation of the diaphragm pressure transmitter design.
Laboratory tests of the performance of the device will be carried out on prepared rock specimens, both dry and saturated with water and/or hydrocarbon condensate. These tests will be at the high frequency end of the device range, where the spatial resolution is at its highest.
The second phase of the work will involve field instrumentation tests on the Fulham works in preparation for tests off-site. Shallow boreholes will be drilled (less than 10 m in depth) and the emphasis will be on getting the field instrumentation ad logistics right, rather than providing geological information.
The third phase of the project will involve testing the source for VSP at established borehole sites in the United Kingdom. Much of this phase of the work will be done at the Purton borehole site which is geologically well-characterised. Frequency-dependent attenuation studies will be carried out in the range 10-10 kHz for both compression and shear waves.
The fourth phase of the project will involve profile mapping on sets of industrial boreholes drilled for hydrocarbon reservoir evaluation and aquifer evaluation. The outcome will be the evaluation of the device as a VSP source for tomographic imaging of a reservoir, using both compression and shear wave measurements and attenuation studies at well-defined frequencies.

STATE OF ADVANCEMENT :

Ongoing

RESULTS :

Preliminary laboratory results have been obtained which have allowed us to establish the working principle of the device. No results available will now.

REFERENCES :

A UK PATENT APPLICATION HAS BEEN MADE IN 1989 TO COVER THE DESIGN OF THE DEVICE AS A VSP SOURCE.

D

```
***************************************************************************
* TITLE : SISTRE : AN ADVANCED 3D TOOL TO IMAGE    *      PROJECT NO      *
*         COMPLEX STRUCTURES.                       *                      *
*                                                   *   TH./01129/89/FR/.. *
*                                                   *                      *
***************************************************************************
* CONTRACTOR :                                      *    TELEPHONE NO      *
*    GERTH                                          *                      *
*                                                   *    47 52 61 39       *
*                                                   *                      *
*                                                   *                      *
*                                                   *    TELEX NO          *
*                                                   *                      *
* TECHNICAL DIRECTOR :                              *                      *
*    MR. J.J. RAOULT                                *                      *
*                                                   *                      *
***************************************************************************
                                              VERSION : 12/03/90
```

AIM OF THE PROJECT :
****
On the geophysical side, to :
- create a new generation of time to depth conversion tool based on 3D ray
tracing, inversion techniques and a depth model editor.
- Provide the user with a common environment for all the tools developed in the
project.
On the computer sciences side, to :
- Derive a methodology to carry out suchg a project,
- Make the programming environment and methodology portable,
- Propose them as a new standard.

PROJECT DESCRIPTION :
****
The SISTRE project has been subdivided in four topics :
Phase 1 : Environment
Phase 2 : 3D Depth modeler
Phase 3 : Macro-model estimation/3D ray tracing
Phase 4 : Macro-model estimation/3D rtravel time inversion.
The last three technical topics are further subdivided in modules which contain
isolated functionalities and algorithms. These modules require data access to
and retrival from common interface to guarantee maximum communication between
modules and optimal user flexibility for the manipulation, retrival and storage
of data. The environment topic has been divided in isolated modules to warrant a
structured and transparent software package.

STATE OF ADVANCEMENT :
****
Ongoing

```
*********************************************************************************
* TITLE : CHARACTERISATION OF ROCK FORMATIONS FOR     *      PROJECT NO        *
*         THE IMPROVED CALIBRATION OF NUCLEAR         *                        *
*         LOGGING TOOLS                               *    TH./01132/89/UK/..  *
*                                                     *                        *
*********************************************************************************
* CONTRACTOR :                                        *      TELEPHONE NO      *
*    WINFRITH PETROLEUM TECHNOLOGY                    *                        *
*                                                     *     0305-251888        *
*    DORCHESTER                                       *                        *
*    DORSET                                           *                        *
*    UK-DT2 8DH                                       *     TELEX NO           *
*                                                     *                        *
* TECHNICAL DIRECTOR :                                *     41231              *
*    MR J. LOCKE                                      *                        *
*                                                     *                        *
*********************************************************************************
                                                         VERSION : 14/03/90
```

AIM OF THE PROJECT :
****
The primary objective of the project is to develop a range of calibration rock
formations for nuclear logging tools which will serve as a basis to satisfy the
requirements of the oil industry into the next century. These requirements are :
 (i) the development of the test formations which will be needed for Primary
Standards to calibrate Wireline and MWD tools;
 (ii) the development of a representative range of borehole environments which
can be used to check the various corrections made to tool responses.
The dey innovative features of this proposal are (i) the use of nuclear
techniques to characterise the test formations combined with (ii) the use of
Monte Carlo code Mc BEND to predict absolute responses of neutron tools from a
knowledge of the absolute neutron source strengths. This will enable the code to
be used not only for the calculation of environmental corrections but also to
extrapolate to field conditions with the aid of nuclear core analysis.

PROJECT DESCRIPTION :
****
Five specific tasks are envisaged :
Task 1 : Open-Hole Porosity Scale
For porosity calibration in the three classical reservoir rocks, sandstone,
limestone and dolomite, nine formations will be contained in five tanks
saturated with fresh water. Three porosity values will be provided in each
formation.
Task 2 : MWD Calibration Pits
The MWD facilities will comprise two fresh water tanks. One will contain two
formations of sandstone with different porosities; the other two formations of
limestone with different porosities. In two additional tanks sandstone
formations will be saturated with a high-salinity and a low-salinity fluid
respectively.
Task 3 : Cased-Hole Calibration Apparatus
The above open-hole formations will also serve to calibrate pulsed-neutron tools
and a range of suitable casings with annular cement regions will be provided.
Task 4 : Testing the Formations
The formation tests will be carried out with an agreed suite of logging tools as
soon as they have been installed and saturated. The results will complete the
characterisation data set for each tank including comparison of computer
predictions with measurement for each tool in each formation.
Task 5 : Computer Program Development
The Monte Carlo code McBEND used for these calculations will be specially for
this work with the aim of simplyfing the data input//output modules and
introducing an automatic procedure for acceleration.
During the fourth year of the project, the test pits will be available to
sponsors so that a thorough evaluation can lead to their acceptance as Industry
Standards.

STATE OF ADVANCEMENT :
****
ONGOING

```
*****************************************************************************
* TITLE : NEW TECHNOLOGIES DEVELOPMENT IN THE          *     PROJECT NO     *
*          STUDY OF PETROLEUM BASINS THERMICITY        *                    *
*                                                      *   TH./01133/89/FR/..*
*                                                      *                    *
*****************************************************************************
* CONTRACTOR :                                         *   TELEPHONE NO     *
*    CREGU                                             *                    *
*                                                      *   83.41.21.86      *
*    BP 23                                             *                    *
*    FR - 54501 VANDOEUVRE-LES-NANCY CEDEX             *                    *
*                                                      *   TELEX NO         *
*                                                      *                    *
* TECHNICAL DIRECTOR :                                 *   960934           *
*    MR. M. PAGEL                                      *                    *
*                                                      *                    *
*****************************************************************************
                                                         VERSION : 20/12/89
```

AIM OF THE PROJECT :
****
The aim of this project is to elaborate new reliable techniques for exploration
of complex geological areas like the North Sea, from data obtained by the study
of hydrocarbon bearing fluid inclusions and the comparative study of two
geothermometers (Organic matter and Fission Tracks).
This program is characterized by a new approach of petroleum basin thermicity
study, the achievement of which is bounded by several rules :
- the combined use of direct investigation tools
- the choice of thermal markers, the application of which is compatible with the
temperature range of oil window
- the experimental calibration of these markers
This is an original and alternative way to obtain more precise time-temperature
data to be introduced in computer programs designed for the genesis and
migration of hydrocarbons.

PROJECT DESCRIPTION :
****
The program is divided into three stages.
Stage 1 : Physical modelling
Both simple hydrocarbon water mixtures (n-alkane, cycloalkanes, aromatics...)
and crude oils will be trapped in minerals (halite, sylvite) in order to check
the analytical techniques and to stimulate the natural geological systems. A
pyrolysis technique allows to perform experimental simulations of organic matter
maturation and fission tracks annealing under high pressures and high
temperatures in a confined medium. A new device to determine routinely the
length of fission track will be elaborated.
Stage 2 : Analysis
Different microprobe (microthermometry, UV and IR microspectrometry, Raman,
laser-mass spectrometry...) and global techniques (N.M.R and gas chromatography
will be applied in order to assess the chemistry of the trapped hydrocarbon and
water mixtures in the natural fluid inclusions.
Stage 3 : All the new or improved devices and kinetic parameters obtained for
organic matter and fission tracks analyses will be useful to provide new
constraints for time-temperature data introduced in computer programs for basin
modelling. This will be applied to a petroleum reservoir.
The study of hydrocarbon bearing fluid inclusions in common minerals of
reservoirs and the numerical simulation of petroleum genesis and migration
prediction from the geothermometers, will result in the production of a fast,
reliable and inexpensive prospecting guide.

STATE OF ADVANCEMENT :
****
Ongoing. Hydrocarbon inclusions have been synthetized. The new device to
determine the routinely the length of fission tracks is in construction.
Pyrolysis experiments in a confined system have been performed at high pressures
and temperatures. Choice of NMR spectrometer and definition of optimal
analytical conditions have been realized.

RESULTS :
****
Alkane and benzene inclusions have been synthetized in sylvite crystals below
100 deg. C at atmospheric pressure. This has allowed to make a precise
calibration curve for microthermometric measurements and to show that the
quantification of the infrared results by CH2/CH3 ratio measurements is limited
by an analytical problem. Preliminary results on the fission tracks annealing
are available. Time-temperature couples for pyrolisis experiments have been
calculated.

REFERENCES :
****
PIRONON J. - HYDROCARBON FLUID INCLUSION SYNTHESIS AT LOW TEMPERATURE. AMERICAN
MINERALOGIST (UNDER PRESS).
PIRONON J. AND BARRES O. - SEMI-QUANTITATIVE FT-IR MICROANALYSIS LIMITS;
EVIDENCE FROM SYNTHETIC HYDROCARBON FLUID INCLUSIONS. GEOCHIM. COSMOCHIM. ACTA

(UNDER PRESS).
LANDAIS P., MICHELS R., POTY B. AND MONTHIOUX M. (1989) : PYROLYSIS OF ORGANIC
MATTER IN COLD-SEAL PRESSURE AUTOCLAVES. EXPERIMENTAL APPROACH AND APPLICATIONS
J. ANAL. APPL. PYROL, 16, 103-115.

```
****************************************************************************
* TITLE : FRACTURED RESERVOIRS - DEVELOPMENT OF          *    PROJECT NO      *
*          NEW MATHERMATICAL MODELS TO IMPROVE THE       *                    *
*          CONFIDENCE IN SIMULATION INPUT DATA.          *  TH./01136/89/PO/.. *
*                                                        *                    *
****************************************************************************
* CONTRACTOR :                                           *   TELEPHONE NO     *
*    PARTEX - COMPANHIA PORTUGUESA DE SERVICOS SA         *                    *
*                                                        *   735013           *
*    AV. 5 DE OUTUBRO 160                                 *                    *
*    PO - 1000 LISBOA                                     *                    *
*                                                        *   TELEX NO         *
*                                                        *                    *
* TECHNICAL DIRECTOR :                                   *   14708            *
*    A. DIOGO PINTO                                       *                    *
*                                                        *                    *
****************************************************************************
                                                    VERSION : 12/03/90
```

AIM OF THE PROJECT :
****
To produce a software package for modelling Naturally Fractured Reservoirs,
incorporating all the improvements achieved by a new conceptual and sthocastic
model, able to produce input data compatible with the major reservoir simulators.
A new sthocastic model will be developed, able to describe the fracture pattern
within the reservoir and the interaction between matrix blocks and fractures.
The use of Multivariate Statistical Analysis will be lead to a simultaneous
treatment of quantitative and qualitative variables allowing to establish
correlations between the fractures density and lithology, rock types, etc.. in
order to identify the fractured zones of the reservoir. The combined use of
these techniques and geostatistics will lead to a correct combination of data
from different scales and sources.

PROJECT DESCRIPTION :
****
Namely in Naturally Fractured Reservoirs, the simulation results are quite often
assumed to be more accurate than they really are and the predicted outcome of
the simulators are not consistent with actual reservoir performance.
This is so because the input data are not consistent with the overall reservoir
architecture, due to the absence of a conceptual model able to describe,
independently from scales, the fracture pattern, treating fractures descriptors
as random vaiables used to generate Fractures spatial distribution for input in
simulation of Dual-Porosity Systems.
It is felt that the degree of fracturing within the reservoir is a critical
factor affecting the flow of fluids and the ultimate recovery.
In this new synergetic approach close cooperate geology, petrophysics,
geophysics, reservoir engineering, mathematics and statistics. The use of
multivariate data analysis and geostatistics will allow a simultaneous treatment
of qualitative and quantitative data.
Furthermore, the exploration of a new technique like the fractals in order to
tackle the geometric and scale problems, can open new directions of research.
Finally, a software package will be developed, incorporating all these new
improvements, and able to produce input data compatible with the major reservoir
simulators.
This project is split in the following main stages :
a) Structural geology analysis and approach to the scale transfer problem;
b) Integrated geological and reservoir engineering approach for sthocastic
modelling of fractures;
c) Use of sthocastic fractures model for reservoir description purposes;
d) Quantification of reservoir heterogeneity;
e) Package Development and Impact on Production Planning.

STATE OF ADVANCEMENT :
****
Ongoingre

RESULTS :
****
The final outcome of this project will be an integrated software package for
modelling naturally fractured oil reservoirs, incorporating all the improvements
achieved by this new model and able to produce input data compatible with the
major Reservoir Simulators available in the market.
This new software will contribute to :
. Guidelines for ecploration, development and production of Fractured Reservoirs;
. Prediction of reservoir characteristics of fractured zones;
. Quantification of fracture storage capacity and production conductivity;
. Impact of fractures pattern in the ultimate oil recovery; consequences for
residual oil distribution;

REFERENCES :
****
ABBOTT, D., 1985 - "FRACTURE ZONE STATISTICS" - EOS, VOL. 66, NR 18.
BARENBLATT, G., ZHELTOV, Y. AND KOCHINA, I. - JOURNAL APPLIED MATH AND MECHANICS,
  NR 5 1286, 1960.

CHILES, J.P. - "MODELISATION GEOSTATISTIQUE DE RESEAUX DE FRACTURES" - PROCEEDINGS OF INTERNATIONAL GEOSTATISTICS CONGRESS - AVIGNON, SEPTEMBER 1988.
DOWNEY, M.W., 1988 - "FAULTING AND HYDROCARBON ENTRAPMENT" - A.A.P.G. BULL., VOL. 72/2, 180.

DRILLING

```
************************************************************************
* TITLE : DEVELOPMENT OF A DEEP WATER DRILLING    *      PROJECT NO     *
*         UNIT.                                   *                     *
*                                                 *    TH./02017/83/NL/.. *
*                                                 *                     *
************************************************************************
* CONTRACTOR :                                    *      TELEPHONE NO    *
*    V.O.F. DESDEC  DEEP SEMI DESIGN CONSULTANTS  *                     *
*                                                 *    010-4260426       *
*    'S - GRAVELANDSEWEG 557                      *                     *
*    POSTBUS 687                                  *                     *
*    NL - 3100NAR SCHIEDAM                        *    TELEX NO          *
*                                                 *                     *
* TECHNICAL DIRECTOR :                            *    25628 MSC NL      *
*    MR. G.J. SCHEPMAN                            *                     *
*                                                 *                     *
************************************************************************
                                             VERSION : 22/09/89
```

AIM OF THE PROJECT :
****
Aim of the project is the design of a drilling vessel for severe environment and
deep water.

PROJECT DESCRIPTION :
****
The basic objective is to design a vessel which combines the advantages of a
drill-ship and a semi-submersible. This resulted in the DSS-10.000. This is a
dynamic positioned column stabilized vessel, characterized by the two floaters,
four stability columns, one large central column with an enclosed moonpool and a
box-type upper hull.
Overall length   : 109,5 m
Overall width   : 64 m
Operating draft   : 26,3 m
Operating displacement : 65,000 tons
Accommodation   : 114 persons
Variable deckload   : 4,700 mtons
Total variable deckload : 10,000 mtons
Drilling depth   : 25,000 ft
Water depth   :10,000 ft
The center column will be used for the storage of the entire marine riser and
creates the space for an enclosed moonpool and a BOP assembling and testing area.
 Riser joints (85 ft. length) are stored vertically in carousels. Patents have
been applied for.

STATE OF ADVANCEMENT :
****
Basic design package has been finished. Model testing has been completed and
analyzed. Approval has been obtained from DNV and ABS. Technical evaluation with
oil company engineering departments has been finished. Building cost has been
determined. Adaptations for specific requirements are studied, depending on
actual questions of interested parties.

RESULTS :
****
The design of the DSS-1000 has been completed and documented in drawings,
reports and specifications. Patents have been applied for a specific system on
board. The 5 column principle stands as a feasible and valuable feature for deep
water drilling. The design package is at sufficient level to enable immediate
start of a building project if and when conditions permit.
The design will be valuable for drilling in deep, rough water such as north-west
of the U.K. or in Norwegian waters. It may serve as well as a base case for
different applications, e.g. floating production offshore. No immediate follow
up is expected in view of the relatively low level of the crude oil price, which
makes exploration of harsh areas presently unattractive.

REFERENCES :
****
ARTICLES IN SEVERAL TRADE JOURNALS, SUCH AS "SCHIP EN WERF", SEPTEMBER 1984.
PAPERS WERE PRESENTED AT :
- DOT OCTOBER 1985
- RINA SYMPOSIUM ON DEVELOPMENTS IN DEEPER WATERS, OCTOBER 1986
- 3RD CONFERENCE ON "THE WAY FORWARD FOR FLOATING PRODUCTION SYSTEMS, LONDON,
DECEMBER 1987
- CEC SYMPOSIUM, LUXEMBOURG, MARCH 1988.

```
************************************************************************
* TITLE : APPLIANCE FOR AUTOMATIC CONTROL OF          *     PROJECT NO      *
*         HOISTING OPERATION IN A DRILLING MAST        *                     *
*         OF DERRICK.                                  *   TH./02018/84/FR/..  *
*                                                      *                     *
************************************************************************
* CONTRACTOR :                                         *    TELEPHONE NO     *
*     P.S.O.                                           *                     *
*                                                      *   1 47.49.15.52     *
*     17-19 RUE DES GRANDES TERRES                     *                     *
*     FR - 92500 RUEIL-MALMAISON                       *                     *
*                                                      *    TELEX NO         *
*                                                      *                     *
* TECHNICAL DIRECTOR :                                 *    203310           *
*     MR. G. GAZEL-ANTHOINE                            *                     *
*                                                      *                     *
************************************************************************
                                              VERSION : 25/10/89
```

AIM OF THE PROJECT :
****
The aim of the project is to set an apparatus allowing an accurate stop of the
drill string at a predetermined position without waste of time and with full
safety precautions.

PROJECT DESCRIPTION :
****
The overall development involves :
- a sensor, which detects rotation or displacement of an element of the
cinematic chain;
- a sensor detects tool joints passage to permit correction of divergence
between actual and calculated position;
- a tension load sensor gives a signal proportional to the suspended loac.
Finally a control unit acquires and treats the data and allows displaying of the
information.
The project has been carried out with the following phases :
1- completion of a prototype
2- industrial preparation and preliminary simulation
3- comprehensive operational tests.

STATE OF ADVANCEMENT :
****
The system has been manufactured and tested successfully.
Long term validation tests concerning reliability are going on in operational
configuration.

RESULTS :
****
) The system applied to drawworks control works :
- with a 1,5 cm stopping accuracy for long travels
- and with a 5 cm accuracy for short travels (less than 2 m)
It controls maximum trip in and out speeds with a 10 % accuracy.
2) The system used for suspended weight control allows to drill in automatic
mode.
3) The computer has been completed by a sequential automatic mode which controls
other tools used in tripping (slips, iron roughneck, elevators, pipe handling
system)
This has allowed to achieve tripping in full automatic mode with computation of
intermediate stopping points and starting up of suitable tools.
For the moment, the cycle duration is about 3 mn with a trip in speed limited to
2 m/s.

```
*********************************************************************************
* TITLE : DEVELOPMENT OF A SUBSEA WIRELESS SYSTEM.      *     PROJECT NO       *
*                                                       *                      *
*                                                       *   TH./02019/84/UK/..  *
*                                                       *                      *
*********************************************************************************
* CONTRACTOR :                                          *    TELEPHONE NO      *
*    ADVANCED PRODUCTION TECHNOLOGY LTD.                *                      *
*                                                       *   01 748 4600        *
*    TRAFALGAR HOUSE                                    *                      *
*    HAMMERSMITH INTERNATIONAL CENTRE                   *                      *
*    UK - LONDON W6 8DM                                 *    TELEX NO          *
*                                                       *                      *
* TECHNICAL DIRECTOR :                                  *    262227            *
*                                                       *                      *
*                                                       *                      *
*********************************************************************************
```

VERSION : 12/12/89

AIM OF THE PROJECT :
****
To demonstrate the feasibility of preforming wireline service from a Monohull
Vessel thus drastically reducing the cost of such operations.

PROJECT DESCRIPTION :
****
Wireline interventions are an accepted method of performing a wide variety of
downhole maintenance, inspection and monitoring taks. These taks are normally
performed either from the Platform Celler Deck or from the Deck of a Semi
Submersible Work-Over Vessel in the case of subsea completions. In both cases a
riser from seabed to deck (either permanent riser or temporarily installed) is
used as the means of access to the borehole.
this project targeted the ability of a high-technology Field Support Vessel to
be used in place of the much more expensive Semi-Submersible Work-Over Vessel.
To achieve the project objectives a design study was undertaken which defined
the necessary parameters that would be required by a system.
Operational avaiilibility; the range of tasks that could be required; the motion
response characteristics and many other factors were considered.
A Basic Offshore Trial Unit was engineered and built during 1985. This basic
unit was aimed at an early demonstration that the system could be successfully
operated in the North Sea. The system was deployed on the British Argyll in june
1986 and successfully performed a trial operation on a Duncan well under the
operatorship of Hamilton Brothers Oil + Gas Ltd.
The system comprised a Derrick and Heave compensated load and wireline units
together with a subsea lubricator. The whole system was mounted on a Dynamically
Positioned; Roll-Stabilised; Twin-Moonpol support vessel the BRITISH ARGYLL
which is owned and operated by APT's parent company British Underwater
Engineering Ltd.
Since successful demonstration of the basic unit the project has progressed. Its
current status is that a commercially viable unit is now being constructed which
embodies the lessons from the Basic Offshore Trial. In particular APT are
totally redesigning the Subsea Lubricator element so as to provide access for
both slick line and electric braided cable applications. This technology
continues to be supported by oil company operators. In particular the support of
Amerada Hess is to be acknowledged in the forthcoming build phase.

STATE OF ADVANCEMENT :
****
Abandoned after engineering of the commercial unit.

RESULTS :
****
The Basic Offshore Trial unit demonstrated the feasibility of this mode of
operation.
Numerous improvements were identified as a results of the trial. These
particularly concerned the subsea lubricator which displayed several undesirable
characteristics of a troublesome and time-consuming nature. A more robust
Derrick and compensation unit was also indicated to be desirable when operating
over more complex xmas trees.
Project has been abandoned.

REFERENCES :
****
- "THE WHOLE OF THE FIELD SUPPORT VESSEL IN SUBSEA FIELD DEVELOPMENTS" - BAXTER
AND EDE, MARGINAL + DEEPWATER OILFIELD DEVELOPMENT CONFERENCE, LONDON APRIL 86.
- THE BRITISH ARGYLL DSV WIRELINING SYSTEM - HUBER (HAMILTON BROS), OFFSHORE
EUROPE EXHIBITION 1985.
- "MARGINAL FIELDS" SPECIAL FEATURE - OCEAN INDUSTRY MAGAZINE, NOVEMBER 86.

```
*****************************************************************************
* TITLE : STUDY OF REALIZATION OF A SYSTEM          *        PROJECT NO     *
*         ENABLING THE DISPLAY OF DRILLING          *                       *
*         PARAMETERS AND THE OPTIMIZATION OF        *    TH./02020/84/FR/..  *
*         COSTS.                                    *                       *
*****************************************************************************
* CONTRACTOR :                                      *      TELEPHONE NO     *
*      SYMINEX                                       *                       *
*                                                   *      91 739003        *
*      BOULEVARD DE L'OCEAN 2                        *                       *
*      FR - 13275 MARSEILLE CEDEX 9                  *                       *
*                                                   *      TELEX NO         *
*                                                   *                       *
* TECHNICAL DIRECTOR :                              *      400563           *
*      C. CRIADO                                     *                       *
*                                                   *                       *
*****************************************************************************
                                                      VERSION : 06/11/89
```

AIM OF THE PROJECT :
****
To develop oil drilling rigs equipment which helps the operator to take
decisions during drilling operations by informing him in a reliable manner and
in real time about the tendancies of the present or of past situations. This is
achieved by means of theoretical computations and forecasted statistical reports
to guarantee a reliable display and recording of all the parameters necessary to
enable the rational exploitation of the rig.

PROJECT DESCRIPTION :
****
The project will start from the "VISUFORA" system, developed by SYMINEX to
acquire and graphically display drilling and deviation parameters and to record
them.
This system will be developed along 2 directions :
- improvement and extension of existing programmes using only information from
the well being drilled.
- exploitation of the records obtained from the wells previously drilled on the
field.

STATE OF ADVANCEMENT :
****
Completed.

RESULTS :
****
Stage 1 : Development of new programs
Drilling parameter computations and their digital and analogical displays have
been completed. Round tripping detection is possible automatically and the stock
control of drill pipes is operational. Regarding casing, the casing phase
parameters and stock control of the casing can be monitored. Definition of the
cementing programme is achieved. Drilling cost software has been implemented.
Additional software (hydraulics, alarms, tables edition; editions (print-out)
have been included.
Blow out controle module has been completed offering quick help by computing the
necessary parameters for the monitoring of hydraulic pressure changes.
Acquisition :
Up to 40 sensors and transduccers can be accepted by the standard acquisition
unit.
All signals are filtered and checked for validity.
Calculation :
Over 100 derived parameters are continuously calculated from the input signals
every 250 ms.
Display :
Real-time anmimated displays, tables and charts of the drilling parameters are
shown on large colour CRT specially housed for industrial and classed zones.
Displays and printout are simply selected form function keys.
Stage 2 : Programs for exploitation for statistical analysis of the results on
previous wells of the same field is operational. Directional well path analysis
software is completed.
Data storage :
A record of up to 20 days drilling activities for 30 channels can be stored on
the 67 Mbyte tape cartridge. Results of the last 24 hours operations are
immediately available for consultation on the CRT and for printing out the daily
report.
Print out :
Any display on request and well logs plotted against time or depth on the graph
plotter.

```
***********************************************************************
*  TITLE : NEW LINING TECHNOLOGIES FOR DRILLING    *    PROJECT NO     *
*          AND PRODUCTION EQUIPMENT.               *                   *
*                                                  *   TH./02021/84/FR/.. *
*                                                  *                   *
***********************************************************************
*  CONTRACTOR :                                    *   TELEPHONE NO     *
*     COATING DEVELOPPEMENT S.A.                   *                   *
*                                                  *   75 41.40.38      *
*     74 rue des Acieries                          *                   *
*     FR-42000 SAINT ETIENNE                       *                   *
*                                                  *   TELEX NO         *
*                                                  *                   *
*  TECHNICAL DIRECTOR :                            *   345079           *
*     MR NISIO                                     *                   *
*                                                  *                   *
***********************************************************************
                                                      VERSION : 29/03/89
```

AIM OF THE PROJECT :
****
This project consits in developing a method allowing to deposit on drilling and
production equipment, a product presenting mechanical characteristics
(resistance to abrasion and shocks) superior to those of tungstene carbide and
slightly inferior to those of the diamond, a synthetic polycristalline, of
equivalent cost or even less expensive than tungstene carbide.

PROJECT DESCRIPTION :
****
The project consists in achieving a pilot unit allowing to process a great
number of elements, within conditions representative of an industrial
manufacture.
The number of parts to be treated and their diversity require an important
number of tests within various domains involved in the oil industry : drilling
(progress test, wear test) and production (erosion and abrasion test, corrosion
tests)
Parallel to the pilot assembling and test surveys, the project also includes a
fundamental study covering the physical and chemical analysis of the
constituents of the lining, the texture of the deposit, and the interaction
between the deposit and the different substrata studied.
The project comprises 4 phases:
PHASE 1 - Fundamental study:
Systematic study of deposit parameters (temperature, pressure, gaz nature,
flowrates, proportions of mixtures), by characterising the structural properties
of the products (nature of phases, interaction with substratum, deposit, balance
diagram), by determining the mechanical properties and the correlations between
the structural properties and the mechanical properties.
PHASE 2 - fabrication of pilot unit:
These tasks will consits, depending on the database deriving from the
fundamental study, in determining the adequate instrumentation, in computing the
thermal capacities of the kilns and in optimising working parameters so to
elaborate the equipment which will be tested later on a real site.
PHASE 3 - Manufacture of prototypes :
The pilot unit allows application of hard lining on real size prototypes
specially made for the cell and in-situ tests: drill bits, stabilizers, tool-
joints, swiveling equipment, nozzles, cutting tools, cutting plates.
PHASE 4 - On-site tests
Prototype elements are first tested on cells or test benches, so to minimise the
risk of interrupting a drill by presenting an equipment, whose reliability would
not have been previously tested.

STATE OF ADVANCEMENT :
****
Ongoing

RESULTS :
****
PHASE 1-Fundamental study
Different underlayers have been investigated as diffusion barrier. The CVD
titanium carbide seems to be suitable.
PHASE 2-Construction of a pilot unit
The plasma assisted CVD reactor is ready for use now. Deposition studies at low
temperature are beginning now.
PHASE 3-Manufacture of prototypes
Depositions are realized on tungsten carbide sintered cutting tolls.
PHASE 4-In-cell and on-site test
The diffusion barrier for tungsten carbide cutting tools and the hard coating
have been realized at the same time, with a strong adhesion.

REFERENCES :
****
A PATENT HAS BEEN APPLIED FOR BY DIAMANT BOART IN FRANCE ON 27 JUNE 1984 UNDER
THE NUMBER 84-10290
A SECOND PATENT HAS BEEN APPLIED FOR IN FRANCE BY COATING DEVELOPPEMENT ON
NOVEMBER 14TH 1986 UNDER THE NUMBER 86-16288

```
****************************************************************************
* TITLE : OPTIMIZATION OF DRILLING OPERATIONS      *       PROJECT NO      *
*                                                  *                       *
*                                                  *   TH./02024/84/FR/..   *
*                                                  *                       *
****************************************************************************
* CONTRACTOR :                                     *     TELEPHONE NO      *
*    GERTH                                         *                       *
*                                                  *     1 47 52 61 39      *
*                                                  *                       *
*    AVENUE DE BOIS PREAU 4                        *                       *
*    FR - 92502 RUEIL-MALMAISON                    *                       *
*                                                  *     TELEX NO          *
*                                                  *                       *
* TECHNICAL DIRECTOR :                             *     203050            *
*    MR TRAONMILIN                                 *                       *
*                                                  *                       *
****************************************************************************
```
                                                        VERSION : 05/09/89

AIM OF THE PROJECT :
****
The object of this project is to develop new methods allowing to improve the
immediate drilling procedures, in view of a bettrer control of drilling
operations through new measurements while drilling (MWD) methods and subsequent
reduction of its costs.

PROJECT DESCRIPTION :
****
The main trend of the project is to exploit and integrate, in real time, the
information linked to the major physical phenomena which intervene in the
optimization of drilling procedures. These phenomena concern mainly the
following three aspects :
1. Directional behaviour of drill strings
2. Resistance of walls through friction analysis
3. Behaviour of drill bits while cutting formation.
The present project consists in preparing, not only the in-well experiments to
be run during drilling operations, but also tests on the appropriate test bench.

STATE OF ADVANCEMENT :
****
Completed within the scope of this contract by 31 December 1986 and pursued
within the scope of contract TH 02.028/85 with further tests in wells and on
benches.

RESULTS :
****
The project is divided in 4 phases :
1. Phenomenological studies : bibliographical study on each item,
2. Physical modelling : design of a test bench for down scale modelling of
bottom hole assembly,
3. Numerical modelling in three dimensional manner of the directional behaviour
of the bottom hole assembly,
4. Methodology for test programmes and data acquisition systems.
PHENOMENOLOGICAL STUDIES
The major part of the work consisted in bibliographical studies relative to
above mentioned aspects. These studies allowed to specify the aspects which were
to be the object of a more thorough study, namely :
- the importance of friction in the behaviour of drill strings
- the utilisation of MWD tools to limit wedging against the walls and the
detection of abnormal pressure areas
- the mechanical aspect of the tool progress at the cutting front, particurlarly
modelling of the relation between the weight on the bit and its torque so to
obtain characterization of the bit or of the rock and information on the state
of wear of the bit while drilling
- the importance of hydraulic phenomena linked to :
- the flowrate and the mud pressure
- the characteristics of the drilled rock versus filtration
- the geometry of the drilling tool.
NUMERICAL MODELLING
The composition of drill strings and the choice of parameters during directional
drilling are generally based on acquired know-how. The purpose is to create
computing aids to measure the distorsions and the contact force of the drill
strings. Studies thus consisted in choosing mathematical modelling methods for
the drill strings and coding them to enable computer calculation. These models
predict the behaviour of the inclination and the azimuth of drill strings.
PHYSICAL MODELLING
The purpose was to prepare bench tests so to complete the in-well tests
scheduled during the next test phase. The studies covered both the detailed
design and engineering of a drill string simulation bench at reduced scale and
the possibility of modelling drilling turbines compatible with the simulation
algorithms of the drill strings. It appears necessary to measure the linear
weight and the inertia of turbine. Negociations were carried out with various
suppliers of turbines for the adaptation of their benches and experimentation
procedures have been set-up.

TEST METHODOLOGY
The purpose was to design the appropriate methodology for each set of
measurements in well and studies allowed to specify :
- the tests programme
- the nature of parameters to be measured
- the frequency of recordings
- the equipment
- choice of wells
- choice of data acquisition methods.
Two preliminary sets have been run to adapt methodology to operational realities
: quality of the measurements, stresses on normal drilling. These tests were run
on wells L4A4 in Holland then on HAA4423 and TM25 in Indonesia. During these
tests the data acquisition method has been adapted to suit the field conditions.
Data analysis software packages have also been elaborated.

```
****************************************************************************
* TITLE : COMPLETION OF HORIZONTAL DRAINS.          *     PROJECT NO      *
*                                                   *                     *
*                                                   *  TH./02025/84/FR/..  *
*                                                   *                     *
****************************************************************************
* CONTRACTOR :                                      *   TELEPHONE NO      *
*    GERTH                                          *                     *
*                                                   *   1 47 52 61 39     *
*    AVENUE DE BOIS PREAU 4                          *                     *
*    FR - 92500 RUEIL-MALMAISON                      *                     *
*                                                   *   TELEX NO          *
*                                                   *                     *
* TECHNICAL DIRECTOR :                              *   203050            *
*    MR. SPREUX                                     *                     *
*                                                   *                     *
****************************************************************************
                                               VERSION : 11/03/88
```

AIM OF THE PROJECT :
****
   The object of this project is to determine the techniques, equipment and
procedures specific to horizontal wells regarding both the acquisition and
interpretation of field data and the means of achieving the appropriate well
completion. Considering the field heterogeneties encountered by the drain, its
optimized exploitation may require the implementation of a selective completion
providing the possibility of chosing the production areas.

PROJECT DESCRIPTION :
****
The project has been divided in two main phases, a study or engineering phase
and a completion phase.
The object of the engineering phase is to determine the means regarding the
acquisition and interpretation of field data around a well. These are
characterized by two main elements, production measurements or logs and
interpretation methods.
Regarding the first element, an attempt will be made to obtain a number of
measurements and their interpretation while considering the particular feature
of the horizontal flow. As for the second element, an attempt will be made,
first, to schematize the horizontal variations of the geological characteristics
of the reservoir and to interprete and model the well tests and production with
or without interface.
The completion phase is more technological and its object consists in making a
selective completion.
Depending on the nature of the reservoir, the liner can be either cemented and
slotted or separated in several preslotted sections by means of special
equipment. Both systems constitute the first two studies of this phase. In order
to improve production conditions, an attempt will also be made to find the means
of simulating and treating such a drain. Last of all, and in as much as possible,
 all these results will be finalized or confirmed by real in-well tests.

STATE OF ADVANCEMENT :
****
The project is well underway and results are positive regarding the completion
technology inspite of the lack of real in-well experience. Studies within the
engineering phase have made it possible to adapt a set of analytical and
numerical models allowing, in a first step, to assess the opportunity of
exploiting a reservoir by horizontal drains, and in a second step, to interprete
the data collected during the development and exploitation of a field.

RESULTS :
****
a) ENGINEERING OF HORIZONTAL DRAINS
MEASUREMENTS
- Determination of the section of a horizontal drain. Tests run on a test bench
showed that a measuring equipment of the SHDT type provided diameters by defect
within an error rate of 8%, which is rather important and needs to be taken into
account when restituting the shape and the volume of the drain.
- The study on sonds transport means in a well fitted with its tubing (where
methods using SIMPHOR type strings cannot be applied) has led to the choice of
two methods "PUMP OUR STINGER" and "COILED TUBING CONVEYED". Both were tested
with success.
- A test bench has allowed to test the measurements made by bottomhole
flowmeters under horizontal diphasic flow conditions. It revealed that to be
properly interpreted, these measures required additional data on parameters such
as the water content of the fluid and the nature of the flow.
INTERPRETATION METHODS
These studies should provide a better comprehension and quantification of the
evolution of the well and reservoir parameters while producing with horizontal
drains.
- The interpretation of well tests provides us with data on the vertical and
horizontal permeabilities near the well and on the residual skin. Analytic and
numerical models have been adapted to the specific case of horizontal wells.
These models have to be confirmed on actual data.

- A study on the evolution of the interfaces provides today easy analytic methods for the preliminary studies of specific cases and numerical models have also been studied in detail for the examination of real cases. Regarding the reservoir, this allows to predict the production before and after the breakthrough, the breakthrough duration, the critical flowrate, etc.
- Regarding the knowledge of the lateral evolution of the geological characteristics of the reservoirs, studies enabled a preliminary evaluation of the influence of the horizontality on the measures made by logging tools and provided specifications allowing to recalibrate the logging sequences with the cores.

b) COMPLETION

Cementing : to perform a hole-layer interface taking into account the variations of the characteristics of the reservoir, the liner has been cemented then perforated according to chosen zones. An important number of laboratory studies and tests showed the possibility of achieving a correct liner cementation while modifying the formulation of the slurries, the rheological stresses with the spacer, the positioning scheme.

Another possibility of sectioning the drain is to use external packers. Here again, experience has proved the feasibility. The problem was to reduce the risks of damaging the packers during their descent and to find an efficient positioning procedure. Regarding the well equipment, studies have also been carried out in view to complete with appropriate internal production equipment the selectivity made by the liner. Very rare are the possibilities really applicable.

REFERENCES :
****
- "ECOULEMENT DES FLUIDES AUTOUR D'UN PUITS HORIZONTAL-ESSAIS DE PUITS"
SPE-ROME-FEBRUARY 1985
- "HORIZONTAL WELLS PRODUCTION TECHNIQUES IN HETEROGENEOUS RESERVOIRS"
SPE-BAHRAIN-MARCH 1985
- "HORIZONTAL WELLS PRODUCTION AFTER FIVE YEARS"
SPE-LAS VEGAS-SEPTEMBER 1985
- "PRESSURE ANALYSIS FOR HORIZONTAL WELLS"
SPE-LAS VEGAS-SEPTEMBER 1985
- "NEW PRODUCTION LOGGING TECHNIQUES FOR HORIZONTAL WELLS"
SPE-LAS VEGAS-SEPTEMBER 1985
- "ANALYTIC 2D MODELS OF WATER CRESTING BEFORE BREAKTHROUGH FOR HORIZONTAL WELLS"
SPE-NEW ORLEANS-OCTOBER 1986
- "SOME PRACTICAL FORMULAS TO PREDICT HORIZONTAL WELL BEHAVIOUR"
SPE-NEW ORLEANS-OCTOBER 1986
- "MUD AND CEMENT FOR HORIZONTAL WELLS"
SPE-NEW ORLEANS-OCTOBER 1986

```
***********************************************************************
* TITLE : OPTIMIZATION OF DRILLING OPERATIONS      *     PROJECT NO    *
*         (PHASE II)                               *                   *
*                                                  *   TH./02028/85/FR/..  *
*                                                  *                   *
***********************************************************************
* CONTRACTOR :                                     *   TELEPHONE NO    *
*     GERTH                                        *                   *
*                                                  *   1 47.52.61.39   *
*     4, AVENUE DE BOIS PREAU                      *                   *
*     FR - 92502 RUEIL-MALMAISON                   *                   *
*                                                  *   TELEX NO        *
*                                                  *                   *
* TECHNICAL DIRECTOR :                             *   203 050         *
*     MR. TRAONMILIN                               *                   *
*                                                  *                   *
***********************************************************************
                                                   VERSION : 01/07/87
```

AIM OF THE PROJECT :
The aim of the project is to develop new methods to improve the conduct of
drilling operations, while using the physical data related to the drilling
process. The topics selected to conduct this research are directly related to
the making of new hole. This activity is given priority here, as it uses up to
40% of the time spent on a well. The other basis for this research is in
assuming that progress relies on an improved knowledge of downhole physics,
until now mostly approached through statistical models. The necessary basis for
the expected improvements is the acquisition of models. The areas concerned are
for this project:
- the behaviour of the borehole walls as seen through the transmission
  losses of torque and weight on bit from surface to the drill bit,
- the directional behaviour of drill strings,
- the behaviour of the drill bit cutting face while drilling.

PROJECT DESCRIPTION :
The main activities of the project are:
- well data acquisition,
- test bench data acquisition,
- interpretation of data,
- integration of results in the conduct of drilling operations.
These steps have been prepared by the project nr TH 2024/84 during which
bibliographies, physical and numerical modelling and preliminary tests were
performed.
These activities are carried out in a specific manner for each of the topics
mentioned, except for the well data acquisition which is common to all of them.
WELL DATA ACQUISITION
Drilling data are acquired essentially by specialized teams using data recorders
brought to and used on the well site under their direct supervision. Both
downhole and surface data are recorded. A complex treatment is afterwards
required to put all the needed data in files adapted to the interpretation work
pertaining to the other activities.
TEST BENCH DATA ACQUISITION
Such data are acquired in a laboratory environment much less subject to
environmental and process noises.
The test benches involved are:
- a drilling bench for bits up to 6" (152 mm) in diameter able to drill 50 mm
  long rock samples,
- a reduced scale bottom hole assembly (BHA) model, representing about 50 m
  length of BHA,
- turbine test benches for static or dynamic measurements.
INTERPRETATION OF DATA
The comparison between well and/or bench data and the output of the theoretical
models attempting to simulate downhole physics is made systematically. The
models are thereafter completed or modified to fit better with the processes
observed.
INTEGRATION OF RESULTS IN THE CONDUCT OF DRILLING OPERATIONS
The integration of the results in the daily drilling procedure requires various
approaches depending on the topic studied. It will often require the use of
computing means and their adaptation to the relatively rough environment of the
drilling site. The results will either be integrated in automation processes, or
at a lower frequency of occurence, participate in decision making and equipment
selection.

STATE OF ADVANCEMENT :
Completed : most of the test bench and well data acquisition has been performed.
The interpretation of the acquired data has been done. The integration of the
results in the drilling procedures has been studied.

RESULTS :
The main results are:

94

- the discovery of practical laws for the transmission of torque and weight
  on bit,
- computer codes for the simulation in three dimensions, statically and
  dynamically, of the bottom hole assemblies directional and vibratory
  behaviour,
- a better knowledge of the torque-weight relationships at the bit cutting face.

REFERENCES :
****
- PETROLE ET TECHNIQUE NR 322 - JANVIER 1986
  OPTIMISATION DE LA CONDUITE DU FORAGE
- SPE 15466 (SOCIETY OF PETROLEUM ENGINEERS MEETING - NEW ORLEANS - OCT 86)
  ORPHEE 3D : STATIC AND DYNAMIC TRIDIMENSIONAL BHA COMPUTER MODELS
- OTC 5510 (OCT 87)
  FIELD DATA ANALYSIS OF WEIGHT AND TORQUE TRANSMISSION TO THE DRILL BIT
  (PLANNED)
- 12TH WORLD PETROLEUM CONGRESS - 1987 (TOPIC 10 - BLOCK IV - PAPER 3)
  OPTIMISATION DU FORAGE : CAMPAGNES DE MESURES ET MODELISATION

```
************************************************************************
* TITLE : DRILLING AUTOMATION (PART 2)          *     PROJECT NO      *
*                                               *                     *
*                                               *  TH./02029/85/FR/.. *
*                                               *                     *
************************************************************************
* CONTRACTOR :                                  *    TELEPHONE NO     *
*    GERTH                                       *                     *
*                                               *   1 47.52.61.39     *
*    4 AVENUE DE BOIS PREAU                      *                     *
*    FR - 92502 RUEIL-MALMAISON                  *                     *
*                                               *    TELEX NO          *
*                                               *                     *
* TECHNICAL DIRECTOR :                           *    203 050          *
*    MR. R. LE ROC'H                             *                     *
*                                               *                     *
************************************************************************
```

VERSION : 01/11/89

AIM OF THE PROJECT :
****
The final purpose of this project is the full automation of the drilling
operation. The aim of Part 2 is to obtain a complete automation of screwing and
racking devices in the derrick. These devices have already been automated
individually within the scope of project TH 02024/84 Drilling automation Part 1.
The last phase of this project is to complete automation of the power swivel
previously developed within part 1 thanks to the handling equipment depending
directly on the power swivel allowing overdrill and automatic connection and
disconnection operations at any height inside the derrick. The new design of
this second generation Power Swivel handling equipment is modular, so future
improvements could be made separately for each module and also the drilling
process will not be stopped because of troubles of only one module.

PROJECT DESCRIPTION :
****
1. Improve the automation of hooking systems namely by increasing the handling
capacity of the lower racking machine for the triples up to 9 tons so to handle
the drill collars.
2. Realise a single automation linking the following operations :
 - screwing-unscrewing (BIL-K)
 - transfer of BIL-K to bottom racking
 - racking on lower position
 - racking upper position.
3. Improve handling beyond the drilling floor area to replace pneumatic
wirelines and winches presently used by safer systems based on jacks, winches
and robots.
4. Manufacture of an actuator for the brake band
5. Realisation of power swivel automation equipment, namely :
 - connection module (or tool joint breaker): this module is able to make
   the connection between the power swivel and the drill stem at any height
   inside the derrick, and also to dismantle or to connect lower
   parts of the power swivel shaft for easy maintenance
   (i.e. : Kelly cocks, Saver sub).
 - handling module : this module improves handling operations during
   tripping phases by tilting, to convenient position, the elevator
   which catches the pipe standing out the racking board,
 - threading compensation module : this module avoids the damage of the
   tool joint thread during connection of power swivel to drill stem.
   Without this module the total weight of the power swivel (19 tons)
   should be supported by the power swivel saver
   sub pin threads and by the upper tool joint box threads of the drill stem.

STATE OF ADVANCEMENT :
****
Completed

RESULTS :
****
Phase 1 - Start-up and manual operating tests completed.
Phase 2 - Automation intended for the synchronization of machines have been
supplied and assembled on the GELOS site in a pressurized panel. Synchronization
softwares have been written, tested in laboratories and at real scale.
Phase 3 - Software intended for the comprehension of PSO commands by the BLM
machines has been written and tested.
Phase 4 - After manual start-up of racking machine, all the BLM machines will be
ready for connection to the PSO computer for the handling integrated
robotisation.
Phase 5 - Completed.

REFERENCES :
****
ONE ELECTRICAL POWER SWIVEL FOR BRITISH PETROLEUM (MILLER PROJECT) TO BE
DELIVERED EARLY 1990.

```
****************************************************************************
* TITLE : AUTOMATIC DOWNHOLE NAVIGATION CONTROL      *      PROJECT NO      *
*                                                    *                      *
*                                                    *   TH./02037/86/DE/..  *
*                                                    *                      *
****************************************************************************
* CONTRACTOR :                                       *   TELEPHONE NO       *
*    EASTMAN CHRISTENSEN GMBH                         *                      *
*                                                    *   05141 203-1        *
*    POSTFACH 309                                    *                      *
*    CHRISTENSSTRASSE 1                              *                      *
*    DE - 3100 CELLE                                 *   TELEX NO           *
*                                                    *                      *
* TECHNICAL DIRECTOR :                               *   925 149            *
*    DR.-ING VOLKER KRUEGER                          *                      *
*                                                    *                      *
****************************************************************************
                                                     VERSION : 01/03/90
```

AIM OF THE PROJECT :
****
Controlled directional drilling is steadily increasing. During the past few years "steerable systems" to be controlled from the surface have been developed and were successfully introduced by Eastman Christensen and others. These steerable or navigational systems allow directional or straight hole drilling without tripping the string out of the hole.
With increasing depth and inclination of the borehole, the adjustment of the navigation tools becomes more difficult.Therefore, it is the aim of this project to develop an automatic system to directly adjust the navigation tool downhole. The positioning device will receive the control commands from a measuring device. A downhole closed control loop holds the preset direction at a fixed value by controlling the tool face by means of a regulating system.

PROJECT DESCRIPTION :
****
By assistance of LITERATURE AND PATENT RESEARCHES, and under consideration of the technical know-how on string components and drilling techniques as presently available, possible solutions and concepts have to be summarized and evaluated. As a result of the SYSTEM STUDY decisions will be made as to the design of the automatic tool face control.
The ADJUSTING DEVICE and the MEASURING AND CONTROL SYSTEM form the main two components of the downhole tool which allows to automatically control the course of a borehole. The adjusting device may rotate in one or the other direction as long as the sensor system monitors an angle deviating from the desired value. As an alternative principle, the adjusting device may receive the deviation signal and process the required angle without continuous feedback from the transducer. The measuring system for determining the angle position of the tool consists basically of a "tool face indicator" which is presently under field testing. The currently installed electronics may be modified in such a way that the rotary position for activating the adjusting device is measured once and transmitted to the adjusting device for further processing. This principle is valid if the adjusting mechanism works on an internal closed loop basis. As a second alternative, the tool face indicator may act as a rotary angle sensor continuously measuring and comparing the actual and desired values thus forming a sensor for the control loop. A decision as to whether to proceed with either solution will not be made before having evaluated possible contrary factors and having finally decided upon tool requirements. Finally, the COMPLETE TOOL has to be laid out and manufactured. The combination consisting of adjusting device and measuring and control system will be lab and field tested during the second half of the project time. Field testing will be performed in connection with the SYSTEM'S SURFACE COMPONENTS. which are necessary to monitor the dwnhole performance of the closed loop device.

STATE OF ADVANCEMENT :
****
Ongoing progress is being reported. The first subassemblies are under longtime reliability tests. A vertical drilling system as a first development step has been expanded by centering elements to avoid rotation of the sensitive sensors. To maintain the maximum knowledge of ex-house development the system study and patent literature evaluation has been continued. The development of shock proof and high temperature electronics is under work.

RESULTS :
****
Out of the general systems study two major but different principles have been extracted for further follow up. One principle uses a so-called steerable motor as the main component. It's tool face will be automatically adjusted by a rotation of the motor housing in order to drill into the preplanned direction. The other main alternative - called the adjustable or steerable stabilizer - features movable stabilizer ribs which automatically counteract a deviation from the planned course. As measuring and control systems can be built in closer to the drill bit for the latter alternative, thus reacting eacting earlier to any potential deviation, this system has been selected for prototyping first. For this system several design alternatives have been drafted.

Three candidates remained, of these the most promising and versatile control
system is now being built for first laboratory and field testing. Along with the
design of the mechanical hardware, several subassemblies and usable components
were intensively lab tested with respect to reliable functioning and wear
resistance. Detailed drawing of one system subsequently has been completed and
manufacturing has been started. Out of the pre-tested components, alternatives
can be immediately chosen if malfunctions of the components occur which are now
under manufacturing.

For sensing deviations and for the control of the hydraulic circuits, redundant
alternatives have also been laid out and are being followed up in order to
reduce the development risk. For this control system for example, available
electronics have been used as a basis. They are being modified for the different
requirements such as smaller size and control circuits in addition to the
existing measuring features. Also, the development of an even more sophisticated
device, operating with more sensors and more control features, has been started.
In order to guarantee at least the minimum functions, work is also done for a
simple analog measuring and control circuit.

In cooperation with universities the modelling of drilling assemblies has begun
to simulate their behaviour and, especially, compute the various modes of
vibration and also directional tendencies.

This work is being done for both the identification of critical drilling
conditions, which could either damage components or would influence the accuracy
of the sensing system, as well as the determination of the directional drilling
behaviour of the system.

As an aid to bothy system's alternatives, the development of an expandable
stabilizer has been started. For this tool, which shall improve the overall
steering capacity of automatically operated systems, a hydraulic/mechanical
design was completed. But is has now been decided to also operate this device
with electronic means to reduce the number of mechanical parts and to also
utilize the control systems which are under development for other devices.
System's study has been continued as well as the patent and literature search.

REFERENCES :
****
NO PUBLICATIONS OR LECTURES PREPARED UP UNTIL NOW.

```
***********************************************************************
* TITLE : OPTIMIZED DRILLING TO REDUCE COSTS       *    PROJECT NO    *
*         (PHASE 1)                                *                  *
*                                                  *  TH./02040/87/FR/.. *
*                                                  *                  *
***********************************************************************
* CONTRACTOR :                                     *   TELEPHONE NO   *
*    GERTH                                          *                  *
*                                                  *  1 47526139      *
*    4 AV DE BOIS PREAU - F-92502 RUEIL MALMAISON. *                  *
*                                                  *                  *
*                                                  *   TELEX NO       *
*                                                  *                  *
* TECHNICAL DIRECTOR :                             *                  *
*    MR DOREL                                       *                  *
*                                                  *                  *
***********************************************************************
                                                      VERSION : 30/06/89
```

AIM OF THE PROJECT :
****
The purpose of drilling a borehole is to acquire better knowledge of geological
formations crossed and to ensure production under the best performance and
safety conditions, taking time and cost into account. A major part of this gain
can be obtained by full control, at all times. of essential drilling parameters
(bed and surface), so as to enable their interpretation, correlation and, from
this, optimization of all drilling operations. These operations could then be
automated, to a varying degree depending on the local contraints, using
softwares or even artificial intelligence.
In order to pinpoint the various activities envisaged in this project, the work
program has been broken down into three main technological areas :
- DRILLING MEASUREMENTS.
- DRILLING HYDRAULICS.
- TRAJECTORY CONTROL.

PROJECT DESCRIPTION :
****
a) DRILLING MEASUREMENT
- The evaluation of the current state of the art for mud-logging techniques and
measurement while drilling (MWD) will be given in reports relating to the
services offered by this new generation of measurement facilities proposed by
the contractors.
- Elaboration of graphic measurement softwares, written processing and automatic
comparison with data from other sources, mainly the wireline.
- Feasibility study of measurement operating methods, aimed at characterizing
formation fluids through gas index measurement.
Feasibility study of a parameter calculation method for formation (permeability,
porosity, drillability, pore pressure). The initial step will be a critical
study of the products offered by the contractors in this field.
- Feasibility study for the drafting of a PDC-and three-cone-rock bit wear law
that is reliable and covers a wide field of applications. This study will be
based on a limited study of bed wear and a detailed phenomenological analysis.
b) DRILLING HYDRAULICS
The studies, development works and test setup and on-site experiments under real
operatintg conditions should lead to :
- the improvement of tools and bottomhole equipment and optimization of their
use. The hydraulic flow studies and tests carried out on test setups will make
it possible to ask equipment manufacturers to modify their products and increase
their performances, efficiency and service life. As regards drilling tools,
special attention will be given to tools with synthetic-diamond blades (PDC).
- operational procedures and methods making it possible to make, on site, the
best selection of hydraulic parameters, in function of conditions encountered,
and to define the specifications best adapted to the drilling fluid.
- decision-making softwares enabling the site manager to modify and control
hydraulic and mechanical drilling parameters and drilling fluid rheological
characteristics, as a function of the measurements made while drilling and the
condition of the well.
c) CONTROL OF TRAJECTORY
- An analysis and definition of requirements will be provided for trajectory
selection.
It is already known that a "friction" software should be developed. If this
analysis shows that there is a need for other softwares, their feasibility will
be evaluated. The general analysis of the "friction" software and the special
software will be carried out during phase 1 and up to completion time (probably
without validation during this phase).
- As regards trajectory selection, the following will be provided :
. a general analysis of a software simulating the behaviour of a packing with a
deflection tool (bent stub or offset stabilization) and, if possible, the
elaboration of the software (probably without validation during phase 1),
. the statistical analysis of the azimuth behaviour of the bottomhole assembly,
depending on the formation dip,
. results on digital model validation using the microdrilling test bench.

STATE OF ADVANCEMENT :

****
During Phase I entitled "Methodologies" object of the present contract, the proposed objectives are considered as attained. Some items need to be pursued during Phase II, object of contract TH 02050/88 as to the additional experimentations, experimental development and/or the finalization of softwares. Nevertheless, the results and conclusions of this first phase have enabled GERTH members to carry out the second phase of this important project.

RESULTS :
****
a) Drilling measurement
Detailed studies of systems ALSGEO (GEOSERVICE), GEMDAS (EXLOG) AND ADVISOR (ANADRILL) revealed the unsufficiency of some of the systems as to the representation of data in real time (penetration speed, gas shows...), the little progress of sensors (flow of mud inlet and outlet) and the variable performances of degassers. On the other hand, these systems offer the possibility of replaying the logs, the creation of databases, a follow-up in real time, the connection with work stations and the integration of mud-logging measurements and of MWD measurements.
A report on MWD state of art shows that MWD service companies are generally capable of running measures of good quality comparable to those made by wireline. Thanks to the development of mud-logging data softwares it is now possible in connection with new systems to automatically process mud-logging and MWD data depending on the depth and soon on the time.
Interpretation of quantitative and qualitative measurements is blocked by the poor efficiency of degassers. Methods for evaluating the formation pressure and porosity are faced with two problems : the unaccuracy of drillability and bit wear formula and the consideration of lithology in computing models.
On site tests for setting up bit wear criteria during drilling were run on a single well and under difficult operating conditions. These tests have not allowed to validate the interpretation method. Operating conditions were reviewed and will be implemented in the future.
b) Drilling hydraulics
A computing link (meshing of bit and borehole, post processor for presenting results) has been set up for studying hydraulics at bit and stabilizer level. This link has been validated for a simple geometry tool thanks to tests carried out on the two envisaged benches. These tests showed that modelling could be simplified.
Concerning the cleaning of the well and use of hydraulic power, priority was given to the computing of pressure drops in the drilling fluid circuit and the computing of annular pressures during operations. The validity of the numerical program without hypothesis on the rheological model has been checked in the case of wells drilled with water-based fluids, or a shallow depths with oil-based fluids. Calibration and validation of the software for oil based fluids at important depth will be performed on a test prepared on a real well and which will be run during phase II.
The study of hydraulics at the drill front has implied the development of a software simulating flows in the bit area, with the help of a reservoir flow model. The influence of the pressure distribution in the sub-jacent block on the bit penetration speed will be determined thanks to experiments run on the drilling bench.
c) Control of trajectory
Requirements were analysed in the form of a survey. This survey shows the need for a friction computing software and a previsional trajectory calculation software.

```
****************************************************************************
*  TITLE : MEASUREMENT AND INTERPRETATION OF        *      PROJECT NO      *
*          DRILLING VIBRATIONS                      *                      *
*                                                   *   TH./02041/87/UK/..  *
*                                                   *                      *
****************************************************************************
*  CONTRACTOR :                                     *    TELEPHONE NO      *
*     SCHLUMBERGER CAMBRIDGE RESEARCH               *                      *
*                                                   *   44-(223)325200     *
*     P.O. BOX 153                                  *                      *
*     UK-CAMBRIDGE CB3 OHG                          *                      *
*                                                   *    TELEX NO          *
*                                                   *                      *
*  TECHNICAL DIRECTOR :                             *    818872            *
*     MR B. PELTIER                                 *                      *
*                                                   *                      *
****************************************************************************
```

VERSION : 01/12/89

AIM OF THE PROJECT :
****

It is proposed to develop a set of equipment and software which can be installed
on an operating oil well drilling rig, which will measure, record and interpret
those drilling-induced vibrations which are transmitted to the surface.
Laboratory experiments have shown that drill-bit vibrations can be interpreted
to give information about the degree of wear of the teeth of a three-cone bit,
and to detect incipient or complete failure of the cone bearings. It has also
been shown that the vibration signature is modified by the quality of cleaning
of the hole bottom, and by the nature of the formations being penetrated.
It is the intention now to apply this knowledge to the development of field-
worthy hardware and interpretation methods.

PROJECT DESCRIPTION :
****

It is believed that surface vibrations contain information about the state of
the drilling equipment, and about the nature of the formations being penetrated.
Specifically, laboratory experiments have shown that it is possible to detect
the state of wear of the drill-bit teeth, and to detect incipient or complete
failure of the cone bearings. If the same interpretations can be carried out in
the field, it will on the one hand become possible to leave the drill-bit at
work until it has reached a predetermined state of wear, and on the other, it
will be possible to detect failure of the bit bearings before cone loss occurs.
In addition, it has been seen that the vibration signature of the bit is
modified by the quality of cleaning of the hole bottom : if these observations
can be developed into a valid diagnostic method, it will be possible to optimise
the hydraulic parameters so as to avoid both inadequate and excessive flow.
Laboratory and field experiments have also shown that the drilling vibrations
are modified by the nature of the rock being penetrated. Correct and timely
detection of a change in lithology is important for the selection of points at
which to set casing, to detect the penetration of the bit into reservoir rocks,
and in the avoidance of the damaging effects of drilling into over-pressured
zones without adequate precautions (kicks and blow-outs).
The detailed phases of development are as follows :
1. To model the bit behaviour, so as to have an understanding of the nature and
meaning of the initial signal
2. To validate this model by simple tests under atmospheric conditions, but
using a real bit drilling real rock.
3. To verify that similar signals are generated by the bit when it operates in a
downhole environment, and to understand them.
4. To model and understand the transmission of these signals from the bit to the
surface through the drill-string.
5. By comparing simultaneous downhole and surface vibration records, to validate
the drill-string model
6. To develop the necessary surface measuring and recording equipment.
7. To develop and apply the correct signal-processing techniques in order to
extract meaningful information from the vibration signals.
8. To interpret these signals in terms of changes in bit status or operating
conditions, or changes in lithology.

STATE OF ADVANCEMENT :
****

Theoretical models of the bit and drillstring have been developed; Laboratory
experiments have studied bit vibrations and the effects of wear and lithology.
Field data have been collected and compared with theoretical and experimental
results. Development of signal processing and interpretation algorithms for
surface vibrations continues.

RESULTS :
****

1) Numerical modelling of the bit :
A numerical model of a roller cone bit has been used to predict the vibration
spectra produced when rotating the bit at a uniform speed.
2) Validation of bit model :
The bit model has been validated with laboratory experiments under atmospheric

conditions.
3) Signals generated by bit in downhole environment :
Bit vibrations have been recorded in a large number of full scale drilling
machine experiments covering a range of downhole pressure and lithology
conditions at elevated temperature. Bit signal is modified by lithology
variations. Field data suggests that downhole vibrations are dominated by
drillstring effects.
4) Modelling of drillstring transfer function :
Numerical models of the drillstring for axial and torsional modes of vibration
have been developed and extended to include the effects of propagation loss.
5) Validation of drillstring models :
Predictions from the numerical models have been compared to surface vibration
data collected in field experiments. Major spectral features in the data agree
well with the models if the correct downhole boundary conditions are applied.
6) Surface measurement equipment :
Surface instrumentation to record drilling vibrations has been developed and
tested in a number of field experiments.
7) Development of signal processing techniques :
Signal processing algorithms are being developed to extract meaningful data from
surface vibrations. Prototype algorithms have been applied to laboratory and
field data.
8) Interpretation :
Interpretation of processed data in terms of lithology, bit condition and
drillstring/borehole interaction continues.

REFERENCES :
****

COOPER G.A., LESAGE M., SHEPPARD M. AND WAND P. : "THE INTERPRETATION OF TRICONE
DRILL BIT VIBRATIONS FOR BIT WEAR AND ROCK TYPE" 1987 RETC PROCEEDINGS, VOLUME 1.
PRESENTED AT RAPID EXCAVATION AND TUNNELING CONFERENCE, NEW ORLEANS, JUNE 1987.

```
*********************************************************************************
* TITLE : MUD ANALYSIS AND CONTROL SYSTEM FOR      *      PROJECT NO        *
*         DRILLING                                 *                        *
*                                                  *    TH./02042/87/UK/..  *
*                                                  *                        *
*********************************************************************************
* CONTRACTOR :                                     *      TELEPHONE NO       *
*    SCHLUMBERGER CAMBRIDGE RESEARCH               *                        *
*                                                  *    44-(223)325200      *
*    PO BOX 153                                    *                        *
*    UK-CAMBRIDGE CB3 OHG                          *                        *
*                                                  *      TELEX NO          *
*                                                  *                        *
* TECHNICAL DIRECTOR :                             *      818872            *
*    MR. H.J. MANNING                              *                        *
*                                                  *                        *
*********************************************************************************
                                                      VERSION : 01/06/89
```

AIM OF THE PROJECT :
****
The project seeks to develop a comprehensive mud analysis and management system
for use on the rig during drilling. For safe and effecient drilling, mud
composition must be precisely controlled to achieve design specifications for
density (for hole stability), rheology (for hole cleaning) and fluid loss (to
avoid drill string sticking); and to minimise reactivity towards chemically
active shales. The key innovatory idea in this proposal is to provide a new
suite of measurements on the rig for real-time monitoring of mud composition
during drilling. This capability is to be combined with new software tools
describing mud circulation and interaction with the wellbore, downhole mud
chemistry and mud conditioning to generate a set of new logs.

PROJECT DESCRIPTION :
****
The project comprises four parts :
1. MUD ANALYSIS AND GENERATION OF A LOG
The development at SCR of an analytical method for ionic substances in mud
filtrate is complete; this can be applied to mud solids and cuttings, allowing a
complete characterisation of the mud. In the case of shales the cuttings cation
exchange capacity (an indicator of clay type and hence reactivity) can be
determined. The development of analytical methods has now been extended to
include polymeric and surfactant mud additives.
2. MODEL OF MUD CIRCULATION AND INTERACTIONS AND INTERPRETATION OF THE LOG
Time-varying data on the return mud cannot be used directly to derive
information about downhole processes because of the distortion of signals
produced by flow in the annulus. Four major effects need to be allowed for :
a) Changes due to addition of drilled and eroded rock, influxes or deliberate
additions of mud chemicals;
b) fluid loss;
c) chemical reactions between mud filtrate, mud solids and rock;
d) dispersion (mixing) caused by flow up the annulus.
It is therefore proposed to develop a numerical mud circulation model
incorporating these processes; this will form a key component of an interpreted
chemical log.
3. WELLBORE SIMULATOR EXPERIMENTAL PROGRAMME ON MUD/SHALE INTERACTIONS.
The SCR Wellbore Simulator will be deployed at intervals throughout the project
to validate methods and models by direct experiment under controlled conditions.
Three particular features of the Simulator deserve emphasis. First, muds are
circulated in the Simulator at representative borehole pressures, temperatures
and velocities through an annulus. Second, mud can be drawn off to provide
laboratory samples for precision analysis. Third, shale rock samples of 60 cm
diameter are used, so that considerable depths of penetration/erosion may be
studied. The Simulator is equipped with ultrasonic instrumentation to allow
changes in hole diameter through erosion or swelling to be measured continuously
while circulating.
The large-scale Wellbore Simulator experiments are supported by laboratory
investigations of shale/mud reactivity in a high pressure diffusion cell. Its
main purpose will be to provide data on the reaction between stressed,
chemically active shales and mud fluids, especially on the role of cations in
modifying the rate and extent of hydrational swelling.

STATE OF ADVANCEMENT :
****
The project is in the laboratory phase. The design specification for a field mud
analysis and control system for ions has now been completed. Analytical methods
for polymers are still under development.

RESULTS :
****
Spectroscopic methods have been developed for the identification and
quantification of polymeric components in mud filtrate. The techniques for
measuring molecular weight distributions are established and have been applied
extensively to cellulosic drilling fluid components.
The system for ionic analysis of muds has been refined : fully automated sample

preparation techniques have been developed, and the speed/accuracy of the analysis improved. A field mud analysis and control system has been fully specified, comprising :
a. mud sampling
b. sample preparation
c. chemical analysis
d. data collection and interpretation.
Further field tracer experiments have been analysed to quantify the degree of hydrodynamic dispersion which occurs to mud as it moves up the annulus, and the extent of ion exchange processes in the wellbore.
The ability of the SCR chemical reaction simulator to model composition changes in bentonitie-based muds, as conditions and key components vary, has been demonstrated.
Experiments on mud-shale interactions have been carried out to relate shale swelling to water activity differences. The ability of potassium ions to inhibit the rate and extent of swelling has been quantified.

REFERENCES :
****
1. THE ION CONTENT AND MINERALOGY OF A NORTH SEA CRETACEOUS SHALE FORMATION :
T.G.J. JONES, T.L. HUGHES AND P. TOMKINS, CLAY MINERALS 24, 393-410 (1989)
2. THE CHEMICAL COMPOSITION OF CMC AND ITS RELATIONSHIP TO THE RHEOLOGY AND FLUID LOSS OF DRILLING FLUIDS.
T.G.J. JONES, T.L. HUGHES AND O.H. HOUWEN, IADC/SPE 20000, ACCEPTED FOR PUBLICATION, FEB. 1990.

```
****************************************************************************
* TITLE : ENHANCED OIL RECOVERY - MODIFICATION        *     PROJECT NO      *
*         AND CONTINUATION OF THE ARTICULATED         *                     *
*         DRILLSTRING.                                *  TH./02047/87/DE/..  *
*                                                     *                     *
****************************************************************************
* CONTRACTOR :                                        *    TELEPHONE NO      *
*   PREUSSAG AG#ERDOL UND ERDGAS                      *                     *
*                                                     *    05176/171        *
*   POB 1220                                          *                     *
*   D-3155 EDEMISSEN                                  *                     *
*                                                     *    TELEX NO          *
*                                                     *                     *
* TECHNICAL DIRECTOR :                                *    92655            *
*   DIPL. ING. BERNHARD PREVEDEL                      *                     *
*                                                     *                     *
****************************************************************************
                                                         VERSION : 17/08/89
```

AIM OF THE PROJECT :
****

"The project is based and shall continue the results of the development of an
'articulated drill string' from the former research and development project 'new
deep hole drilling technique II' by Preussag Erdoel une Erdgas. It's
continuation has the objectives to design and build prototype drilling and
completion tools in order to achieve by that means a horizontal section of
several hyndred meters inside a hydrocarbon layer via a short radius inclination-
buildup section with highest tolerances in inclination and direction.

PROJECT DESCRIPTION :
****

The development consists of seven project stages related to three topics
(subjects) :
1. Articulated drill string (ADS)
      STAGE 1 : state - of - the - art report
      STAGE 2 : Modification of the existing drill string as well as the mud
                motor plus adaption of flexible drill pipes
      STAGE 3 : adaption of the flexible drill string, a two way EM MWD
                system and bits.
2. Flexible drill string
      STAGE 4 : state - of - the - art report
      STAGE 5 : laboratory test, configuration and adaption work
3. Completion and production methods
      STAGE 6 : state - of - the - art report and theoretical solution of
                problems
      STAGE 7 : configuration work, testing, modifications and reports.

STATE OF ADVANCEMENT :
****

"The next generation drilling tool derivated from the articulated drillstring in
the articulated downhole motor (ADM), a prototype of which in 4,75" toolsize has
been build and successfully fieldtested".

RESULTS :
****

"The general resumee of the ADM-fieldtest Lehrte 38 was the confirmation, that
the single version of the 4,75" articulated downhole-motor (ADM) is supplying
with 10 kW mechanical power output enough energy for max. 6,25" holesize in
order to drill a similar short radius well economically in the future.
YThe feature to combine individual motor sections across articulations was
therefore and due to high vibration environment declined.
A solution for holesize up to 8,5" will be a 6,75" ADM of the same design as the
4,75" version, which shall be assembled and possibly fieldtested in the second
half of 1989.
One objective concerning short radius drilling operation for the future will be
the technique to drill several horizontal sidetracks inside the reservoir out of
one existing well. Taking into account this fact Preussag developed a concept
for a multiple settable and retrieveable Packer-Whipstock system. This tool will
be set prior to kicking off the first horizontal sidetrack in the existing well
at kickoff depth, remains for the entire drilling period downhole and is
retrieved and reset afterwards in a higher wellsection for the following
sidetrack.
This procedure was also tested for the first time in Lehrte 38 and shall be
further fieldtested with a complete Packer-Whipstock assembly for a period of a
normal drilling operation in a well in Italy later 1989.
With respect to the poor performance of 3,5" Aluminium Drillpipe for short
radius drilling, Preussage developed a concept for high grade steel Compressive
Drillpipe. This pipe reprents ferromagnetic of non-magnetic alloys and are
physically identical to normal drillpipe, with the exception of having 4,75"
wear-knots distributed over the pipebody in order to create a higher apparent
stiffness against pipe buckling in a 6" hole in compressed mode, not loosing the
capability of it's flexibility to trip over a 30 m radius hole profile".

REFERENCES :

```
************
WITHIN THE DEVELOPMENT OF THE ADS, THREE PUBLICATIONS HAVE BEEN MADE:
1. A. STEENBOCK     NEUE DRAINHOLEBOHRTECHNIK
                    ERDOL-ERDGAS, 101. IG., 3, MARZ 1985
2. B. PREVEDEL      NEW TECHNIQUES IN HORIZONTAL AND DRAINHOLE
                    DRILLING OPTIMIZATION : LEHRTE 41
                    LATERAL DRILLING PROJECT, SPE 15694
3. A. STEENBOCK     ALTERNATIVE DRAINHOLE DRILLING SYSTEM -
   D. MULLER-LINK   THEORETICAL BACKGROUND AND PRACTICAL EXPERIENCE
                    SPE 16564
```

```
****************************************************************************
*  TITLE : SUBSEA WIRELINE UNIT                      *      PROJECT NO      *
*                                                    *                      *
*                                                    *   TH./02048/87/DK/.. *
*                                                    *                      *
****************************************************************************
*  CONTRACTOR :                                      *    TELEPHONE NO      *
*     LIC ENGINEERING                                *                      *
*                                                    *    45 1 62 16 42     *
*     EHLERSVEJ 24                                   *                      *
*     DK-2900 HELLERUP                               *                      *
*                                                    *    TELEX NO          *
*                                                    *                      *
*  TECHNICAL DIRECTOR :                              *    21437             *
*     HENRIK NEDERGAARD                              *                      *
*                                                    *                      *
****************************************************************************
```

VERSION : 14/09/89

AIM OF THE PROJECT :
****
The objective of the project is to develop a Wireline Unit for servicing and
maintenance operations of subsea wellheads. The wireline unit shall be operated
from smaller sized surface support vessels and the system will reduce the need
for standard workover riser systems operated from heavy drilling rigs.
A system supported by small vessels will not only lead to cost savings as a
result of a considerable reduction in mobilization/demobilization time, but as
more vessels are available in this category wireline operations can be planned
on shorter notice thus reducing the production loss period.

PROJECT DESCRIPTION :
****
It is the intention of this project to develop and test a subsea wireline unit.
In doing this we will break away from the present trend of routing the wirelines
from the ship through a stuffing box to the tubing. Instead the system will
consist of the following components :
1) A subsea wireline unit which contains winches and control system for wireline
operations. This unit will be landed on the subsea christmas tree (after the
tree cap has been removed). The unit is fully capable of sustaining the well
pressures.
2) From this position wireline operations in the tubing can take place under
full well pressure by remote control from the support vessel via an umbilical.
Once connected to the top of the x-mas tree and made ready for operation,
wireline services can be performed unaffected by surface weather conditions.
As the wireline winch is a part of the Subsea Wireline Unit, relevant wireline
tools/equipment required for the actual downhole work to be performed, have to
be hooked-up/connected to the wireline itsself and loaded into the Subsea
Wireline Unit prior to launching the complete system.
A prototype of the unit will be designed, procured and assembled by the project
group. Fabrication of certain items will be subcontracted to relevant shops and
suppliers based on tendering. The testing will be made by the project group at
selected test sites.
The project can be broken down in the following phases :
1) Preliminary layout and engineering. This part comprises layout and functional
requirements to the system. System definitions.
2) Detailed design. Detailed drawings and specifications. Long lead items will
be ordered and procured.
3) Fabrication. Fabrication of individual units will be made according to tender.
 Subcontractors are envisaged for steel works and electronics.
4) Preliminary verification and adjustments. Testing of unit. Assembly and
initial testing will be made by personnel of the group. Testing of performance
will be made at site of subcontractor.
5) Final verification testing. Made by subcontractor with supervising personnel
from project group.

STATE OF ADVANCEMENT :
****
Phase 1 of the project is completed. Phase 2, Detailed Engineering, is initiated
Planning of the fabrication of the prototype of the subsea wireline unit has
also been initiated.

RESULTS :
****
A report covering the work in Phase 1 was released in December 1988. This report
describes the different stages in the development of the design concept from the
first to the final chosen scheme. The conclusion of the work is that the most
economical and simple solution is a subsea winch placed on the lubricator
directly exposed to sea water. The seal between winch and lubricator is a
stuffing box. The novelty of the approach is the remotely controlled winch
placed on top of the lubricator.

```
****************************************************************************
* TITLE : A SUBSEA COILED TUBING SYSTEM        *      PROJECT NO       *
*                                              *                       *
*                                              * TH./02049/87/UK/..    *
*                                              *                       *
****************************************************************************
* CONTRACTOR :                                 *    TELEPHONE NO       *
*    WHARTON WILLIAMS                           *                       *
*                                              *    0224 722877         *
*    STONEYWOOD IND PARK                        *                       *
*    DYCE                                       *                       *
*    ABERDEEN AB2 ODF                           *    TELEX NO           *
*                                              *                       *
* TECHNICAL DIRECTOR :                          *    73368              *
*    MR. C. HEADWORTH                           *                       *
*                                              *                       *
****************************************************************************
                                              VERSION : 03/04/90
```

AIM OF THE PROJECT :
Coiled tubing work on subsea completions is currently achieved using semi-submersible vessels with a rigid marine riser attached to the wellhead assembly. The coiled tubing is injected through a pressure containment device at the top of the riser against well pressure down into the well.
The project objective is to design and develop equipment to locate the tubing injector directly onto the wellhead assembly thus obviating the need for a rigid marine riser and large surface support vessel.
The system will comprise of a handling equipment for a positive deployment and controlled landing on the wellhead assembly, the coiled tubing system with marinised subsea components and a hydraulic control package.

PROJECT DESCRIPTION :
The project will comprise of five major activity areas which are as follows :
Modification To Existing Coiled Tubing : After initial system design the components to be used subsea will be modified for underwater use. This primarily involves marinisation of the tubing injector assembly which will be deployed from the support vessel and landed on the subsea xmas tree. The workscope involves definition of an operations specification, environmental loading analysis, marinisation of the surface equipment design and procurement of the injector followed by an assembly and testing phase.
Deployment System : A Handling system which will include lifting, umbilical and guidewire winches for subsea and topside equipment manipulation. Positive handling (using heave compensation) on the surface vessel and soft landing of the subsea equipment on the wellhead are prime objectives. Constant tension guidewire winches will be used to accommodate vessel motions and an deploying the various control and monitoring umbilicals for the subsea equipment. The handling system will include a main lifting winch whose wire will be motion compensated for deployment and recovery of the subsea coiled tubing equipment. For the deployment system components the development phases will include initial design, finalisation of the engineering package, procurement, assembly and testing.
ROV Support Tooling : The system will be designed to be remotely operated with subsea tasks undertaken by construction class ROV's. Some tooling design will be necessary for certain work areas.
Control System : Hydraulic control of xmas tree functions, coiled tubing equipment and the handling system will be carried out from a surface control cabin. The system will include control umbilicals and hydraulic power packs. The workscope includes design, specification, procurement and testing of the umbilicals plus the surface control panels.
System Trials : Trials will be carried out, on land, to function test critical system components. These dry trials will involve mobilisation of the equipment to the trials site where intergration testing of the subsea equipment followed by hook up, function and load testing of the entire equipment package will be performed.

STATE OF ADVANCEMENT :
Completed

RESULTS :
All activities were successfully completed.
An operation using the handling frame and active heave compensation system has been successfully carried out offshore but the coiled tubing system, although called for in the customer's well servicing programme, has not been used.
The system was mobilised with the intention of backflowing a water injection well by injecting nitrogen followed by flushing of flowlines on two subsea wells.
 Favourable hole conditions in the subsea water injection well however, meant that use of the coiled tubing was unnecessary so the system remains for the present time, untried offshore.

REFERENCES :

E

****
COBB, C.C., HEADWORTH C.S., 'A SUBSEA REELED TUBING SERVICE UNIT", OFFSHORE
EUROPE CONFERENCE, SEPTEMBER 1989, SPE 19277.
COWAN P, HEADWORTH C, "SERVICING SUBSEA WELLS WITH SUBSEA REELED TUBING"
UNDERWTAER CONFERENCE, MARCH 1990.

```
**********************************************************************
* TITLE : OPTIMIZED DRILLING TO REDUCE COSTS        *    PROJECT NO    *
*          (PHASE II)                               *                 *
*                                                   *  TH./02050/88/FR/.. *
*                                                   *                 *
**********************************************************************
* CONTRACTOR :                                      *    TELEPHONE NO   *
*     GERTH                                         *                 *
*                                                   *  33(1)47526139   *
*     4 AV. DE BOIS PREAU                           *                 *
*     FR-92502 RUEIL-MALMAISON                      *                 *
*                                                   *    TELEX NO       *
*                                                   *                 *
* TECHNICAL DIRECTOR :                              *                 *
*     MR DOREL                                      *                 *
*                                                   *                 *
**********************************************************************
                                            VERSION : 07/11/89
```

AIM OF THE PROJECT :
****

The purpose of drilling for oil is twofold : first of all, to define, within the
best safety and performance conditions, and taking into account time and cost,
the best adapted technical facilities for the production of hydrocarbons, and
also to acquire the best possible knowledge of the formation that is drilled.
An appreciable reduction in cost may be attained by constant mastery (via data
acquisition, interpretation and modelling) of the essential drilling parameters
(surface and strata), and by using this data to optimize drilling operations. It
is then possible to practically automize these operations for optimal control of
the drilling rig developed and designed by EUROFOR.

PROJECT DESCRIPTION :
****

The project concentrates on the following predominant aspects :
a) Drilling measurement :
- approximation of gaseous indexes. Measurement representativity will be
improved by the use of a new standardization formula and the definition of
degasser efficiency,
- pose pressure evaluation by introducing correction or lithologic factors,
deduced from MWD well logs, in the calculation module, as well as drill bit wear
laws under the given drilling fluid rheological conditions,
- the study of gas plugs at rod connection time as an overpressure indicator by
modelling of the phenomenon,
- the definition, through site measurement campaigns, of the drill bit wear laws
- the estimation of the stratum relative permeability
b) Drilling hydraulics :
- stabilizer and drill bit hydraulics : validation and optimization of the flow
calculation chain for simple well patterns and its extension to complex patterns.
 To seek validated optimized patterns on banches and on the site,
- well cleaning and the use of hydraulic power :
  . validation and utilization, during specific tests or on site, of an annular
pressure determination model and of a model for upraising of well cuttings in
inclined wells
  . continuation of well erosion studies and research on prevention methods.
- hydraulics at working face : validation of the numerical model developed in
Phase I (contract TH 02.040/87)
c) Trajectory control :
- trajectory selection :
  . study of friction and development of a calculation software,
  . development of a well profile calculation module (left curve),
  . characterization of seam inclination and structure
- trajectory calculation : this study of the physics of directional drilling
will lead to improved prediction of drill string behavious by :
  . developing numerical models,
  . readjusting numerical models using a complete range of tests on a test bench,
on actual wells under controlled experimental conditions and on actual wells
under operational conditions,
  . simulating the role of the stratum (geological characteristics, inclination,
stress conditions) on drill string behaviour, by bench tests and by statistical
analysis of actual borings.
- trajectory control : the development works will consist in establishing the
anti-collision criteria to be introduced in uncertainty ellipse calculation.
Project synthesis will consis of :
- the drafting of the operational procedures and operator manuals for the
softwares developed,
- the definition methodology for the optimum trajectory of a well,
- the possible integration, depending on its feasibility, of the drill
bits/stratum interactions and the drillability models in a model called the
"feed" model.

STATE OF ADVANCEMENT :
****

Nearly all the studies are underway and their achievement is planned for next
year. The study of gas plugs at rod connection time has been reported during the

current period, after the results obtained on the studies of the pore pressure and of the overburden pressure.

RESULTS :
****
Drilling measurement
A measurement campaign has been done with a new equipment for the measurement of the gaseous indexis. The results show a good improvement of the degasser efficiency, but in some cases, the data interpretation remains critical.
Some experiments on drilling bench to define drill bit wear are achieved during June 1989. A site measurement campaign is decided.
Two series of the resistivity logging with a new MWD tool giving data on resistivity, gamma-ray and neutronic density, were realized in North Sea.
Drilling hydrauliczs
After the development of the fow calculation chain at tool stabilizer level during the Phase I, the study is continuing. The analysis of the results is made. Specific tests on site are achieved to study the parameters governing the upraising of well cuttings in inclined wells. The results will be used to validate the model.
A comparison between the results obtained during the Phase II on the experimental well of the Rogaland Institute (Norway) and the results calculated with various models of the literature will be made.
Trajectory control
The study of friction and the development of the model (FRISSON) are achieved.
A new model on the determination of the well pattern (COTRAO) was undertaken. The software to predict the behaviour of a drillstring with one or more bent sub has been developed, a sensitivity study was made on the angle of the bent sub and on its place in the drill string. A comparison with real data on site has been made.
The study of the role of the inclination of the stratum on drill string behaviour was made by experiments on the drilling bench.
The statistical analysis of the azimuth behaviour during the drilling phases 17 1/2 on North Alwyn field conducts to an estimation of a "lead angle".

REFERENCES :
****
FOR SOME OF THESE STUDIES, THERE WERE WRITTEN REPORTS.

```
********************************************************************************
* TITLE : COMPUTERIZED MANAGEMENT SYSTEM FOR      *      PROJECT NO          *
*         MOBILE DRILLING RIGS - COMMAND          *                          *
*                                                 *   TH./02053/88/DK/..     *
*                                                 *                          *
********************************************************************************
* CONTRACTOR :                                    *      TELEPHONE NO        *
*    THE COMMAND CONSORTIUM                        *                          *
*                                                 *    31 410022             *
*    BASLEV                                        *                          *
*    155 ROEDOVRE CENTRUM                          *                          *
*    DK-2610 ROEDOVRE                              *      TELEX NO            *
*                                                 *                          *
* TECHNICAL DIRECTOR :                            *                          *
*    MR U. GOTTSCHALK                              *                          *
*                                                 *                          *
********************************************************************************
```

VERSION : 14/03/90

AIM OF THE PROJECT :
****
The aim of the project is to develop an integrated modular computer concept
comprising of supervision, control, planning and follow-up and reporting on
- Drilling operations
- Drilling programpme status and planning
- Completation operations
- Logistics
- Resources (equipment, materials, labour force)
- Safety systems
- Ballast system

PROJECT DESCRIPTION :
****
The total system will comprise the following basic modules with possibilities
for future extensions.
- An operations database
- Modular multiprocessor microcomputer or microcomputer in local network
- Modular and flexible process interface to existing or system furnished
sensoring devices
- Centralized man machine communications interface
- Application software
This project is divided into 5 phases consisting of
* PHASE 1 - System Design Specifications
In this part of the project the overall requirements to the system is defined.
*PHASE 2 - System Development
In this part of the project the data structure, the data base system and the
system commands are developed and implemented.
PHASE 3 - Data Acquisition and Transmission
This part of the project includes selection/development of hardware.
PHASE 4 - Application Software
The development of application programmes and menues alerting procedures etc.
are included. It should further be possible for the users to define their own
applications.
PHASE 5 - Implementation and Testing
This part of the project includes installation for testing and final development.

STATE OF ADVANCEMENT :
****
The project is in the predesign phase (Phase 1). A prestudy has been completed
and a demo-computer model developed.

RESULTS :
****
The primary results achieved by the system will be :
1) To transmit step by step activities required in the drilling program to the
rig personnel, via screen message or hard copy printouts.
2) To monitor the feedback from individual activities in the drilling process
and to compare this information to profile requirements simulated in the
drilling program and the database.
3) To inform and to log such corrections and decisions which were made to
satisfy the system. All activities wire be recorded in terms of time and
sequence and reports issued as required.
The benefits derived from the system will reduce administrative load on
personnel, thus interuptions of drilling operations are given parameters.
Unified, automated documentation of all processes, given improved planning and
follow-up supporting increased drilling productivity.
A prototype of the system comprising a range of the modules will be installed on
a drilling rig in a late phase of the project. The specific purpose of the
prototype is to establish confidence in the integrated system, operational
experience and direct input for improvements and extensions.

REFERENCES :
****
"REMOTE CONTROL AND MONITORING SYSTEM FOR THE FUTURE". CONCEPTUAL PROPOSAL

112

PRESENTED AT A SEMINAR IN OCTOBER 1986 IN COPENHAGEN - J. JENSEN AND E. DYHR.
COMMAND SYSTEM PRESTUDY - SUBMITTED TO THE DANISH GOVERNMENT AGENCY "INDUSTRI-OG
HANDELSSTYRELSEN" OCTOBER 1988 - THE COMMAND CONSORTIUM.

```
**************************************************************************
* TITLE : LESCODRILL - LOW COST DRILLING SYSTEM      *    PROJECT NO     *
*                                                    *                   *
*                                                    *  TH./02056/89/BE/..*
*                                                    *                   *
**************************************************************************
* CONTRACTOR :                                       *    TELEPHONE NO   *
*    DIAMANT BOART STRATABIT                         *                   *
*                                                    *    348 36 99      *
*    AV.DU PONT DE LUTTRE 74                         *                   *
*    B - 1190 BRUXELLES                              *                   *
*                                                    *    TELEX NO       *
*                                                    *                   *
* TECHNICAL DIRECTOR :                               *    26253          *
*    ROBERT DELWICHE                                 *                   *
*                                                    *                   *
**************************************************************************
                                            VERSION : 14/12/89
```

AIM OF THE PROJECT :
To develop a system to drill slimhole wells to 1500 m, including design and
construction of the rig and specialized accessories, and down hole equipment,
such as rods, safety system, computerized data acquisition, etc...
The basic equipment comes from the mineral exploration industry and will be
adapted to the petroleum industry.

PROJECT DESCRIPTION :
The project is to be conducted in four phases.$Phase 1 - Basic
Engineering
Definition of techniques and procedures that are to be applied to slimhole
drilling including :Down hole equipment :
- rods and casings (optimisation of annular)
- safety equipment
- tools
- cementation technology
Surface equipment :
- safety equipment (BOP)
- data acauisition system
- mud system
- generalized automatisation of the rig
Investigation of specialized equipment such as :
- logging tool
- testing and production equipment
- survey tool
- automatisation of core analysis
Phase 2 - Prototype definition
Field test on existing rig to define specialized tooling and drill pipe versus
hole size.
Choice of technical options and definition of final rig and accessories
specifications.
Phase 3 - Prototype realisation
Construction of the prototype rig capable to drill in safe conditions to 1500 m
in sedimentary formations.
Phase 4 - Tests on experimental well
The new rig is to be tested in a well known petroleum environment.

STATE OF ADVANCEMENT :
The works started on 1st of August 1989.
The phases 1 and 2 are at this moment ongoing, the first one being half way.
About phase 2, we are now modifying certain elements of the existing mineral
drilling machine, and building the workshop around this rig.

REFERENCES :
A FIRST LECTURE HAS BEEN GIVEN TO BP ON THE GOAL AND THE AIM OF THE PROJECT TO
GET THEIR INPUT, THE FUTURE LECTURE WILL BE GIVEN TO ELF AND TOTAL.

PRODUCTION SYSTEMS

```
***********************************************************************
*  TITLE : NEW CONCEPT FOR FIXED OFFSHORE        *      PROJECT NO      *
*          PLATFORMS.                            *                      *
*                                                *  TH./03132/83/NL/..  *
*                                                *                      *
***********************************************************************
*  CONTRACTOR :                                  *     TELEPHONE NO      *
*     CONSULTING ENGINEERS H. VETH BV            *                      *
*                                                *     078 131 944       *
*     POSTBUS 274                                *                      *
*     NL - 3300 AG DORDRECHT                     *                      *
*                                                *     TELEX NO          *
*                                                *                      *
*  TECHNICAL DIRECTOR :                          *     29452             *
*     MR. IR. R. VAN DE WAAL                     *                      *
*                                                *                      *
***********************************************************************
                                                  VERSION : 06/02/90
```

AIM OF THE PROJECT :
****
The development of a monotower production platform of steel. Preliminary study
shows a cost advantage of a steel monotower of 20 percent compared with
conventional jacket structures. Moreover the innerside of the column is suitable
for use of oil-storage of 500.000 barrels oil. This project must prove in more
detail the advantages of this concept and must beacon its boundaries.

PROJECT DESCRIPTION :
****
In close cooperation with one or more oil companies case-studies have to be done
for some locations. Therefore next items must be examined:
- foundation;
- structural analysis;
- process requirements;
- transport and installation;
- construction method;
- drifting ice problems;
- IMR aspects.

STATE OF ADVANCEMENT :
****
After finishing the basic designs, studies have been done after special
applications of the F.S.P. The application of the F.S.P. in artic regions is
very promising.

RESULTS :
****
Feasibility study has been finished. Static and dynamic calculations done.
Application study for North Sea, Artic and Brazilian conditions cost
calculations done. Main advantages are cost saving of 20 to 40 percent in
fabrication cost saving in IMR. Simple fabrication and assembly methods.
Horizontal or vertical assembly and transport. Risers, conductors and wellheads
protected inside monoleg. Capture of underwater oil spill. Considerable reserve
strength after damage. Favourable dynamic and fatigue behaviour. Quality
assurance given by Lloyds Register of shipping.

```
***************************************************************************
* TITLE : USE OF HYDRAULICALLY DRIVEN SUBSEA       *      PROJECT NO      *
*         PUMPS FOR ARTIFICIAL LIFT AND REMOTE     *                      *
*         FIELD PRODUCTION.                        *   TH./03136/83/UK/..  *
*                                                  *                      *
***************************************************************************
* CONTRACTOR :                                     *     TELEPHONE NO     *
*    BRITOIL PLC                                   *                      *
*                                                  *    01 409 25 25      *
*    ST VINCENT STREET 150                         *                      *
*    UK - GLASGOW G2 5LJ                           *                      *
*                                                  *       TELEX NO       *
*                                                  *                      *
* TECHNICAL DIRECTOR :                             *     881 20 71        *
*    MR. J. ANDERSON                               *                      *
*                                                  *                      *
***************************************************************************
                                           VERSION : 18/03/86
```

AIM OF THE PROJECT :
****
To appraise the detailed engineering and economic aspects of alternative
downhole and subsea pumping systems both proven and novel, including an
assessment of proven systems based on operator experience in the North Sea where
natural drive is weak or absent.

PROJECT DESCRIPTION :
****
The study envisaged the study of the historical record of the problems
experienced in the artificial lift pumping systems as applied to North Sea
reservoirs. A review of the philosophies adopted by the various alternatives and
the resulting consequences which affect the downhole completion details, topside
power supply and interface connections with satellite wells. Sufficient detailed
engineering to be carried out to give a firm basis for costing a complete
integrated system to allow a cost/benefit analysis to be made.

STATE OF ADVANCEMENT :
****
Because of Britoil's involvement in a North Sea Field where a hydraulically
driven downhole pump was being tested, the study was delayed to enable an
assessment to be made on its performance which could have altered the direction
and necessity for the study. The rapid advance in subsea pump technology in the
intervening period negated the major part of the study and it was decided to
limit the study to considering the reliability of the electrically driven
submersible pumps on Beatrice.

```
**************************************************************************
* TITLE : THE DEVELOPMENT OF A DIVERLESS AND          *     PROJECT NO      *
*         GUIDELINELESS SUB-SEA CHRISTMAS TREE TO     *                     *
*         BE USED IN DEEP WATER.                      *   TH./03138/83/IT/.. *
*                                                     *                     *
**************************************************************************
* CONTRACTOR :                                        *    TELEPHONE NO     *
*    AGIP SPA                                          *                     *
*                                                     *     02/5201         *
*    C.P.120 69                                       *                     *
*    IT - 20120 MILANO                                *                     *
*                                                     *    TELEX NO         *
*                                                     *                     *
* TECHNICAL DIRECTOR :                                *   310 246 - ENI     *
*    MR. P. TASSINI-TEIN                              *                     *
*                                                     *                     *
**************************************************************************
                                                     VERSION : 01/09/89
```

AIM OF THE PROJECT :
****
Scope of this project is the development of a new generation of subsea system,
specifically conceived for hydrocarbon production in deep (200-600 m w.d.) and
very deep (beyond 600 m) waters, including a dedicated maintenance device.

PROJECT DESCRIPTION :
****
The object of the research is the evaluation of the feasibility of sub-sea
production systems which do not require divers("DIVERLESS" systems) or
guidelines ("GUIDELINELESS" systems) for installation, operation and maintenance.
 Therefore, the concepts of maintenance studied in previous projects will be
transferred into a purpose designed operating system and the design a subsea
christmas tree suitable for use in deep and very deep waters (including flowline
connection system and maintenance device) will be produced.
The design of the maintenance device will be made simultaneously with the system
to be serviced, in order ot obtain an intervention tool integrated in the system,
 with obvious advantages from the operating point of view, and consequently
improved reliability.
The prototype construction and long-term tests (scope of the phase II and III of
the project) will allow the evaluation of the system's functioning and
reliability.

STATE OF ADVANCEMENT :
****
The detailed design of the systems is complete.

RESULTS :
****
Main results of the project are the specifications of the production system, the
maintenance system and the operative procedures.
PRODUCTION SYSTEM
The modular production system has the following features. A rectangular platform,
 called "flowline frame", is located on the drilling well-head. The X-mas tree
is lodged on the central part of the flowline frame right upon the well-head.
The flowline connection system is located on one end of the flowline frame and
the control system and umbilical connection on the opposite side. In order to
perform pigging operations from shore, two flowlines and two service lines will
be layed.
Three landing areas are prepared for the maintenance vehicle : the first one on
the flowline connection area, the second one on the opposite side on the control
system area and the third one upon the X-mas tree.
Positioned on the first one, the subsea vehicle, can install the flowline
connectors and perform mechanical override of the X-mas tree valves that, in
order to simplify the operations, are located on the same front side.
MAINTENANCE SYSTEM
The philosophy of the maintenance operations is to bring to surgface the
defective componenets in order to repair or substitute and reinstall them on the
production system. This has been obtained modularizing the production system and
interfacing it to dedicated work modules of the maintenance system. The
maintenance system consists of the following modules;
- A non floating main vehicle hanged on a strenght cable, called Master Vehicle.
- Six work modules connected in turn to the MV in order to perform specific
functions.
- The surface handling system and control system.

```
****************************************************************************
* TITLE : CONTRUCTION AND ASSEMBLY TECHNIQUES FOR    *      PROJECT NO      *
*         STEEL PLATFORMS IN DEEP WATER.             *                      *
*                                                    *   TH./03140/83/IT/.. *
*                                                    *                      *
****************************************************************************
* CONTRACTOR :                                       *     TELEPHONE NO     *
*     TECNOMARE SPA                                  *                      *
*                                                    *     041 796711       *
*     SAN MARCO 3584                                 *                      *
*     IT - VENEZIA                                   *                      *
*                                                    *     TELEX NO         *
*                                                    *                      *
* TECHNICAL DIRECTOR :                               *     410484           *
*     MR. L. BEGHETTO                                *                      *
*                                                    *                      *
****************************************************************************
                                              VERSION : 06/07/89
```

AIM OF THE PROJECT :
****
Purpose of the present research project is the sudy of both the basic problems
and the procedures concerning the construction, assembling, transportation
phases of the fixed offshore platform in 350 m W.D.

PROJECT DESCRIPTION :
****
The project is subdivised into four phases with the following tasks :
- analysis of basic platform components and calculation programs;
- evaluation of alternative tripod configuration and their construction and
marine operations implication to select the most promising concepts;
- definition of the selected concepts and definition of construction, assembly
and installation procedures;
- development of a computer program for the definition of construction schedule
and costs.
The project was completed within November 1988.

STATE OF ADVANCEMENT :
****
Completed

RESULTS :
****
The results are in accordance with the aim of the project and are now available
to third parties.
The tripod configuration developed has the following characteristics :
- it is the only tripod concept developed that can be directly assembed at the
field site without need of a sheltered deep water site;
- can be installed over a pre-drilled template.
Following main results have been achieved :
- development of a computer procedure for the dynamic linear analysis of a multi
body floating system;
- procedure for the design of sandwich type and shell stiffned cylinder under
external pressure and axial loads;
- evaluation of fatigue behaviour and evaluation of residual life with Fracture
Mechanic Analysis;
- development of a computer procedure for the stochastic fatigue analysis of the
main components of offshore structures;
- development of a computer program for the evaluation of construction schedule
and costs for offshore structures;
- naval basin model tests of the main critical stages of the tripod marine
operations.

```
**********************************************************************
* TITLE : TENSION LEG PLATFORM WITH HIGH PAYLOAD     *      PROJECT NO      *
*         FOR A NATURAL GAS LIQUEFACTION PLANT IN    *                      *
*         WATER-DEPTHS OF 500 TO 1,000 METERS.       *   TH./03150/84/DE/.. *
*                                                    *                      *
**********************************************************************
* CONTRACTOR :                                       *    TELEPHONE NO      *
*    SALZGITTER AG                                   *                      *
*                                                    *   030/88.42.97-15    *
*    POSTFACH 15 06 27                               *                      *
*    DE - 1000 BERLIN 15                             *                      *
*                                                    *    TELEX NO          *
*                                                    *                      *
* TECHNICAL DIRECTOR :                               *    308 611           *
*    DR-ING. PIETSCH                                 *                      *
*                                                    *                      *
**********************************************************************
                                                       VERSION : 23/01/90
```

AIM OF THE PROJECT :
****
The project concentrates on the development of a tension leg platform for high
deck loads as a supporting structure for a natural gas liquefaction plant in
water up to 1,000 m deep, in order to economically exploit natural gas
reservoirs in deep-water locations.

PROJECT DESCRIPTION :
****
The development work examines from a technical standpoint the possibility of
using a TLP with a buffer store as a support structure for a natural gas
liquefaction plant in water up to 1,000 m deep. The overall system comprises the
foundation components, tensioning system, the buoyancy body, the jack-up
platform and the transfer system for the product.

STATE OF ADVANCEMENT :
****
Ongoing the project is in the design and construction phase.

RESULTS :
****
Within this research project, technical solutions have been developed for the
exploitation of offshore natural gas deposits in greater depths of water.
The entire system consists of the foundation components (pile templates), the
buoyancy body, the flex-jointed tension leg anchors, the coupling devices and
the adjusting mechanisms, as well as of the jack-up platform which is deposited
with its supporting columns (pillars) upon the buoyancy body once this has been
braced.
  The desired reduction in expenditure has lead to a far-reaching integration of
functions which were hitherto assigned to separate components of the system :
- the external storage of the liquefied natural gas should be integrated into
the production platform, respectively into the buoyancy body;
- the indirect offloading of the liquefied natural gas, for example, via an
articulated tower, should take place directly from the platform to the LNG
tanker.
The savings in capital investment which can be achieved, will be, according to
present-day findings, of about 25 % permitting a corresponding reduction of
fixed costs resulting from interest payments. Additionally, savings of variable
costs will also be considerable, as the relatively high expenses for maintenance,
 repairs and staffing of the external temporary storage facilities and the
articulated tower will be economized.
In keeping with the purpose of the R and D project to construct a tension-leg
platform for high deck loads as supporting system for a natural gas liquefaction
plant including LNG intermediate storage and direct loading facility, two
variants have been developed.
The difference between them consists mainly in the location of the storage tanks
for the LNG :
- storage facility in the platform
- storage facility in the buoyancy body.
Both of the proposed TLP designs offer a storage capacity of approx. 43000 cubic
meters of LNG.
The net load for both of the solutions proposed, comes to about 36000 tons i.e.
about 40% of the total (gross) weight of the tension leg platform. The storage
capacity is in direct proportion to the transfer tankers of 50000 m3. Three
tankers will be used in shuttle service for offshore loading, so as to interrupt
LNG production only during prolonged periods of bad weather (docking possible
over 90 % and loading over 95% of the year).

```
*************************************************************************
* TITLE : COMPLIANT MONOTOWER FOR OIL/GAS        *      PROJECT NO       *
*         PRODUCTION IN DEEP WATER MARGINAL      *                       *
*         FIELDS.                                *   TH./03151/84/IT/..   *
*                                                *                       *
*************************************************************************
* CONTRACTOR :                                   *     TELEPHONE NO       *
*    TECNOMARE SPA                               *                       *
*                                                *     041 796 711        *
*    SAN MARCO 2091                              *                       *
*    IT - 30124 VENEZIA                          *                       *
*                                                *     TELEX NO           *
*                                                *                       *
* TECHNICAL DIRECTOR :                           *     410484             *
*    MR. G. MINETTO                              *                       *
*                                                *                       *
*************************************************************************
                                                  VERSION : 01/08/89
```

AIM OF THE PROJECT :
****
Development of systems which should render economically attractive the
exploitation of numerous oil and gas marginal fields in deep water up to 1,000 m.
w.d.

PROJECT DESCRIPTION :
****
Study, design and model testing of new concepts of structures, the COMPLIANT
MONOTOWER, proposed for oil and gas production from marginal fields in very deep
waters up to 1,000 m.w.d.
The basic concept is a very slender and flexibile structure, with complete
structural continuity from the seabed to the sea surface. Such structural
continuity allow a certain number of conductors to run along the structures as
in a conventional fixed platform and the possibility to put the wellheads on the
deck surface, with strong benefit of the operation maintenance process.
The project is subdivided into two phases :
PHASE 1 includes : . a general technical-economic system analysis; the study of
the structural configurations together with the development of the procedures of
analysis and computer program; the preliminary design of the monotower and
definition of the installation/operation procedure.
PHASE 2 includes : . final configurations definitions and analysis; execution of
model tests and comparison of theoretical and experimental results; final
definition of the installation/operation procedures; issue of final design
documentation.

STATE OF ADVANCEMENT :
****
Completed. BASIC DESIGN PHASE.

RESULTS :
****
The results obtained during the development of the compliant monotower research
project allow to assert its global feasibility for the exploitation of deep
water hydrocarbon fields.
In particular we can assert the feasibility of the following three systems
configuration :
- an axial piled compliant tower for exploitation of hydrocarbon fields in water
depth from 270./.320 m to 500./.550 m with conventional completion and operating
deck weight up to 25000 t.
- a flexible tower for exploitation of hydrocarbon fields in w.d. ranging from
500./.550 m up to 800 m with conventional completion and operating deck weight
up to 25000 t.
- a monopile for exploitation of gas fields in water depth from 270./.300 m up
to 800 m with underwater wellhead and operating deck weight up to 25000 t.
From an economic point of view the compliant monotowers appear very attractive
due to their low structural weight.
In particular the costs of the installed structure compared with other
alternatives developed for similar scenarios (as conventional jackets and TLPs)
evidentiated the competitiveness of the compliant monotower in the overall range
of technical applicability of the concept with the maximum advantage in the
range of 350./.650 m w.d.
During the project development the following main critical components have been
identified and defined at a feasibility level : Reins, grouted connections,
transition module and foundation template.
For the actual applicability of the concepts detailed studies will be required
in the following areas :
- the qualification of the high or very high strength steel needed in the reins,
for the configurations which require axial piles;
- the new configuration of the grouted connections of reins or columns which
allows a more efficient transfer of very high axial loads.
Transition module and the new configuration of the grouted connections are
patent pending.

REFERENCES :

122

****
NICOLUSSI F., NISTA A., PELLIZZARI L. - A COMPREHENSIVE STUDY AND ANALYSIS OF A COMPLIANT MONOTOWER FOR VERY DEEP WATER. PROCEEDING OF THE ICES-BB INTERNATIONAL CONFERENCE ON COMPUTATIONAL ENGINEERING SCIENCE. ATALANTA, GEORGIA, USA - APRIL 1988.
2/ NICOLUSSI F., NISTA A., VANORE S. - COMPLIANT STEEL TOWER FOR DEEP WATER PLATFORMS. PROCEEDING OF THE DOT-89 5TH INTERNATIONAL CONFERENCE AND EXHIBITION 16-17-18 OCTOBER 1989 - MARBELLA (SPAIN).

```
*******************************************************************************
* TITLE : OIL FIELD TESTING OF A NEW COMPACT    *      PROJECT NO         *
*         SEPARATOR OF 25,000 BBL/DAY CAPACITY.  *                        *
*                                                *   TH./03153/84/FR/..    *
*                                                *                        *
*******************************************************************************
* CONTRACTOR :                                   *      TELEPHONE NO       *
*    BERTIN & CIE                                *                        *
*                                                *      34.81.85.00        *
*    B.P. 3                                      *                        *
*    FR - 78373 PLAISIR CEDEX                    *                        *
*                                                *      TELEX NO           *
*                                                *                        *
* TECHNICAL DIRECTOR :                           *      696231             *
*    MM. J.Y. DEYSSON/M. REYBILLET               *                        *
*                                                *                        *
*******************************************************************************
                                                     VERSION : 03/03/87
```

AIM OF THE PROJECT :
****
Test a new separator concept for off-shore oil production, incorporating novel
features for improved gas, oil and water separation, resulting in reduced weight
for both vessel and structure.

PROJECT DESCRIPTION :
****
The current project is the test of a 25 000 BBL/day separator under true oil
field operating conditions. The program is comprised of the following phases :
1. Preparation of a 25 000 BBL/day compact separator for the tests. Design,
procurement and setting of the necessary control and safety devices.
2. Shipment of the skid and erection of same on the Obagi Site.
3. Test of performance under various flow rate and GOR configurations : oil in
gas, gas in oil, water in oil, oil in water.

STATE OF ADVANCEMENT :
****
Completed. The whole program has been completed on March 1986 with good results.
In order to make some operational improvements and to carry out a long duration
testing of the separator on an off-shore platform, a community technological
development (Hydrocarbons) project was registered at the EEC (DG 17) in April
1986.

RESULTS :
****
The testing of the BERTIN three-phase compact separator on the OBAGI oilfield
(ELF NIGERIA) occured according to the programme laid down. The tests made it
possible to set the figure for the separator practical operating capacity in its
present state at 15,000 bbl/d. This figure corresponds to that of separators
currently used in off-shore production.
Under such conditions, the gain in volume and in ground surface area in relation
to a conventional horizontal separator with equivalent performance is very
considerable.
The BERTIN separator has a tank volume of 11 m3 as against 26 m3 for an
equivalent classical horizontal separator and a ground surface area of 2 m2 as
against 18 m2.
Naturally, these advantages are slightly attenuated if one considers the entire
skidmounted separator rather than the tank, but they are still highly
appreciable when considered from the point of view of a policy of price
reduction involving a reduction of the load at the head of the structures.
Also, all the improvements proposed should make it possible to increase the
separator capacity from 15 to 20 or 25,000 bbl/d (which is the practical
capacity limit found on a platform) and, by a long-duration campaign on an offshore production platform, to prepare the marketing of the apparatus.
Even now, it is already possible to anticipate the following off-shore
applications for the BERTIN three-phase compact separator :
. 1st stage HP separator
. 2nd stage MP separator
. test separator

```
******************************************************************************
* TITLE : RECOVERY FROM VERY SMALL FIELDS. STUDY       *    PROJECT NO      *
*                                                      *                    *
*                                                      *  TH./03154/84/UK/.. *
*                                                      *                    *
******************************************************************************
* CONTRACTOR :                                         *   TELEPHONE NO     *
*    BRITOIL PLC                                       *                    *
*                                                      *   041 2254754      *
*    301 ST VINCENT STREET                             *                    *
*    UK-GLASGOW G2 5DD                                 *                    *
*                                                      *   TELEX NO         *
*                                                      *                    *
* TECHNICAL DIRECTOR :                                 *   777633           *
*    MR. A.M.THOMPSON                                  *                    *
*                                                      *                    *
******************************************************************************
                                             VERSION : 27/09/89
```

AIM OF THE PROJECT :
****
The study objective was to consider methods by which a small reservoir of 8/10,000 barrels per day throughput and having a 5 year life could be developed at minimum capital expenditure.

PROJECT DESCRIPTION :
****
A study to explore the hypothesis that the use of a buoy/tanker system could offer significant advantages for the development of small reservoirs and could reduce the time span between project start and first oil export. The project comprises two phases, the first studying a number of design options including a Britoil developed Flexible Riser and Mooring System (FRAMS). This system embodies a flexible riser conveying well fluid and injection water, and envisages using standard flexible flowlines. A simple deck-mounted connection unit on the vessel bow mates with the flexible riser. The second phase of the work will enhance the selected basic design concept to achieve a reduction in the technical complexity of the novel untried components. A parametric study on field size and throughput is also included.

STATE OF ADVANCEMENT :
****
The study is now complete and has indicated that small fields (of about 10m bbl recoverable reserves) could be developed economically.

RESULTS :
****
The project has shown that the Flexible Riser and Mooring System (FRAMLS) is an economical method of developing a 10 m bbl oil reserve in the Central North Sea. The subsea buoy and 60000 DWT tanker, suitably converted for production purposes, have suffcient uptime to be economically attractive. In severe weather the production tanker disconnects from the FRAMS mooring and either leaves the area for discharge at a European port, or waits on suitable weather to re-connect and restart production. The simple swivel-less connection offers a cost-effective means of providing fluid paths for both well fluid and water injection.

```
*********************************************************************
*  TITLE : CONCRETE PLATFORMS FOR DEEP WATER        *    PROJECT NO     *
*                                                   *                   *
*                                                   *  TH./03159/84/UK/..*
*                                                   *                   *
*********************************************************************
*  CONTRACTOR :                                     *   TELEPHONE NO    *
*     MAC ALPINE OFFSHORE LTD                       *                   *
*                                                   *   0442 233444     *
*     ST ALBANS ROAD                                *                   *
*     HEMEL HEMPSTEAD                               *                   *
*     UK-HP2 4ATA                                   *   TELEX NO        *
*                                                   *                   *
*  TECHNICAL DIRECTOR :                             *   825955          *
*     MR. M.J. COLLARD                              *                   *
*                                                   *                   *
*********************************************************************
                                                      VERSION : 01/09/89
```

AIM OF THE PROJECT :
****
To investigate the technical and economic potential for the use of concrete
hulls for floating platforms in water depths typically 400 m extending to 1000
meters.

PROJECT DESCRIPTION :
****
The main technical considerations in designing concrete hulls for tension leg
platforms were examined and the feasibility of, construction in the UK was
investigated and confirmed. An outline design and cost estimate were developed
for a concrete TLP suitable for a 27000 tonne payload is a water depth of 400
metres located West of Shetland.
The hull configurations examined had four columns and four pontoons, both
circular and rectangular section pontoons being considered. Design requirements
for the hull included practical considerations of construction, deck mating and
installation, floating stability, structural strength and installation, floating
stability, structural strength and serviceability. For the tethers upper and
lower tension limits suitable as preliminary design criteria were derived. These
are critical to be feasibility of the combined hull tether system and
relationships between the main variables were established.
Analyses of wave-induced tether tension were performed on a range of trial
configurations for which hull dimensions, mean density and draught were varied.
Other sources of tension variation, calculated or estimated were added to the
results to give maximum and minimum tension estimates. From a parametric study
of the data, optimisation procedures were developed enabling configuration s
complying with the design criteria to be readily identified. The data were
updated by incorporating improved estimates of hull density enabling a
particular configuration to be selected for structural analysis.
A survey was made of the basic requirements for a concrete Catenary Moored
Floater (CMF). For the two primary hull types, viz. gravity stabilised and
column stabilised, relationships were established between hull dimensions and
draught. These enabled configurations ssatisfying particular stability criterial
to be identified. An interim assessment was made and the scope for further work
outlined.

STATE OF ADVANCEMENT :
****
Completed

RESULTS :
****
The project demonstrated that the relatively large mass of concrete hull is no
disadvantage in cost and payload capacity, given appropriate design and
sufficient draught. Thick walls lead to a simple structure enabling external
pressure to be resisted without the need for stiffeners.
A fatigue assessment of critical areas at the pontoon node intersection showed
that cumulative fatigue damage was an order of magnitude less than the normally
accepted maximum permissible damage for an offshore concrete structure.
A survey of potential UK construction sites showed that a choice exists of
several sites suitable for constructing TLP hulls. Two stage construction would
be required, the first stage being built in a dry dock followed after float-out
by completion of hull construction afloat on inshore moorings.

```
*******************************************************************************
* TITLE : A PROGRAMME FOR DEVELOPMENT OF            *      PROJECT NO         *
*         STANDARDIZED PRODUCTION FACILITIES FOR    *                         *
*         THE EXPLOITATION OF MARGINAL              *   TH./03160/84/UK/..     *
*         HYDROCARBON RESERVES                      *                         *
*******************************************************************************
* CONTRACTOR :                                      *   TELEPHONE NO          *
*    WIMPEY OFFSHORE                                *                         *
*                                                   *   01 748 2000           *
*    27 HAMMERSMITH GROVE                            *                         *
*    UK -                                            *                         *
*                                                   *   TELEX NO              *
*                                                   *                         *
* TECHNICAL DIRECTOR :                              *   25666                 *
*    DR. I.E. TEBBETT                               *                         *
*                                                   *                         *
*******************************************************************************
                                            VERSION : 25/02/87
```

AIM OF THE PROJECT :
****
The original aim of the project was to evaluate the potential for standardised
production facilities in European waters; to develop a basic design, procurement
and fabrication premise for a standard production facilityu; to prepare a
preliminary design package and establish the order of costs for its design,
fabrication and installation and to further develop the preliminary package into
a fully optimised and standardised design package in market-ready form.

PROJECT DESCRIPTION :
****
Each of the above aims was to be represented by a separate phase of the project.
Phase 1, 2 and 3 were each structured to include a "decision point' that would
govern whether or not the project should proceed to the succeeding phase.
Towards the end of Phase 2, the aim of the study was revised due to market and
technology developments. The work was ultimately performed in four phases with
marketing activity common throughout :
Phase 1 - Data Gathering and Classification
Phase 2 - Design Premise
Phase 3 - Preliminary Design
Phase 4 - Conceptual Sector Studies.
The use of a standardised production facility (SPF) is proposed as a viable
option for development of offshore gas fields in relatively shallow waters.
Savings in both cost and schedule are achievable, although some inevitable
mismatching will limit these advantages. Some fourteen fields (twelve UK, two
Dutch), amongst a total of sixty prospective gas fields in the Southern North
Sea Basin, have been identified as nominally suitable for application of the
standardised facility.
These fields have assessed recoverable reserves of between 100 and 400 billion
cubic feet of dry gas, and are located in water depths in the range of 25 to 50
meters; the "target" ranges for the SPF. It is believed that up to twenty-three
additional fields (thirteen in the UK Sector) could be judged suitable when
their reserve sizes are known. The objective was not merely to produce a
standardised facility, but to standardise at all levels, such that cost
advantages would be enhanced further after the second and subsequent
installations.

STATE OF ADVANCEMENT :
****
Design

RESULTS :
****
The use of a standardised production facility (SPF) is proposed as a viable
option for development of offshore gas fields in relatively shallow waters.
The most promising hydrocarbon resource type is "gas-only", ie. where the
liquid/gas ratio is less than 10 barrels/million SCF, and where on-hydrocarbon
components in the wellstream tend to be minimal.
In terms of processing capacity, it is concluded that design throughputs should
be sized within a range suitable for development of the smpall gas fields,
having total reserves of between 0.1 and 0.4 TCT of gas.
By far the majority of field prospects in the above range are located in the
Southern Basin of the North Sea, where water depths fall predominantly into the
range of 25-50 metres.
This suggests that such a depth range might be appropriate to standardisation of
a complete topsides/substructure facility. The prospects for anything more than
partial standardisation of substructures, however, has been shown to be limited.
Whilst the well-developed infrastructure of the Southern North Sea Basin will
tend to lower the economic viability threshold in this area, there should be no
difficulty in applying standardised facilities at any European offshore location
falling within the selected field size and processing capacity. It is merely
that the threshold will vary from location to location, and from sector to
sector.
Neglecting any reduction in lead time to production achievable through
standardisation, initial indications are that CAPEX savings of up to 13% are
possible, accompanied by OPEX savings of up to 10%.

```
************************************************************************
* TITLE : DEVELOPMENT OF A COST EFFECTIVE SUBSEA     *     PROJECT NO     *
*         PRODUCTION SYSTEM FOR MARGINAL AND DEEP    *                    *
*         WATER HYDROCARBON FIELDS.                  *   TH./03161/84/UK/.. *
*                                                    *                    *
************************************************************************
* CONTRACTOR :                                       *    TELEPHONE NO    *
*    SHEERWEY TECHNOLOGY GROUP LTD                    *                    *
*                                                    *    0932 75 2742    *
*    SIEMENS HOUSE                                   *                    *
*    WINDMILL ROAD                                   *                    *
*    UK SUNBURY-ON-THAMES,MIDDLESEX TW16 7HS         *    TELEX NO        *
*                                                    *                    *
* TECHNICAL DIRECTOR :                               *    8951091         *
*    MR.P. TOWERS-PERKINS                            *                    *
*                                                    *                    *
************************************************************************
                                              VERSION : 23/10/89
```

AIM OF THE PROJECT :
****
To develop computer based simulation methods to analyse various field
development options in order to produce optimum subsea production systems.

PROJECT DESCRIPTION :
****
A new approach is required to design and assess alternate production systems
such that safety, performance and economics are integral part of overall system
development. Such an approach would provide a more complete appreciation of the
"risks" involved and will also permit design optimisation to be conducted. This
should embrace all practical engineering aspects pertaining to technical
categories of field development and, to be effective, must be formalised and
"computerised". The work programme is designed to provide the facility to meet
this need. The essential components of this facility are :
* A computerised database which contains basic data on reliability, repair,
maintenance, operating procedures, weather, repair vessels and costs.
* A method of generating a computerised representation of the various subsystems.
* A means of generating an analysis model based upon subsystem representation
and the database.
* Rapid analysis of the subsystem performance over field life taking into
account failure, repair and maintenance of all operating modes which effect
production and safety.
* "Designer Friendly" interface for data input, program control and information
output.

STATE OF ADVANCEMENT :
****
All activities of the project are complete as of 30 september 1989.

RESULTS :
****
Initial Results from the program give excellent agreement with alternative
method of calculation. The failure, repair and availability models have all been
tested with in-house data and provide sensible results which have been verified
as far as possible against hand calculation.

```
***************************************************************************
* TITLE : HYDROCARBON PRODUCTION-SYSTEMS IN ARTIC     *     PROJECT NO     *
*         WATER.                                      *                     *
*                                                     *   TH./03162/84/IT/.. *
*                                                     *                     *
***************************************************************************
* CONTRACTOR :                                        *   TELEPHONE NO      *
*    TECNOMARE SPA                                    *                     *
*                                                     *   041 796 711       *
*    SAN MARCO 2091                                   *                     *
*    IT - 30124 VENEZIA                               *                     *
*------------------------------------------------------*   TELEX NO         *
*                                                     *                     *
* TECHNICAL DIRECTOR :                                *   410484            *
*    MR. G. MINETTO                                   *                     *
*                                                     *                     *
***************************************************************************
                                            VERSION : 01/08/89
```

AIM OF THE PROJECT :
****
This research project is aimed at studying the feasibility of hydrocarbon
production systems for Arctic waters and their preliminary design.

PROJECT DESCRIPTION :
****
The project is divided in 2 phases :
- The first include : A) Analysis of the basic problems posed by Artic, such as
the definition of the environmental data, the ice/structure and ground/structure
interactions; B) Feasibility studies of some production systems suitable for the
most interesting areas.
- The second phase includes preliminary design of some production systems
identified in feasibility studies.

STATE OF ADVANCEMENT :
****
PHASE 1 has been completed.
Phase 2 is in progress : preliminary design of a fixed steel gravity platform
for Okhotsk Sea in shallow water is at an advanced stage; preliminary design of
a floating production system for South East Canada in 150-250 m w.d. is at the
beginning.

RESULTS :
****
Main results can be summarized as follows:
Acquisition of environmental conditions or Arctic and Sub-Arctic offshore areas.
Acquisition of different formulation in evaluating the loads due to the static
or almost static and dynamic ice-structure interaction.
Identification of possible problems on platforms in earthquake conditions and
methodologies to be used on sismic analysis in arctic conditions
The most promising Arctic offshore exploitation scenarios have been identified
and design premises for feasibilities studies have been performed
In feasibility studies the following three different production systems have
been identified feasible :
- RAP (Removable Arctic Platform) for Arctic seas (Beaufort and Russian Barents
sea) characterized by water depths in the range of 25-80 metres and severe ice
conditions
- TSG (Tecnomare Steel Gravity) platform for Sub-Arctic areas (Bering and
Okhotsk seas) characterized by water depths in the range of 20-60 meters with
severe sismic and moderate ice conditions.
- FPS (Floating Production Systems) for Sub-Arctic seas (South-East Canada and
Barents seas) characterized by iceberg and severe environmental conditions.
- Design premises for preliminary design of a fixed steel gravity platform for
Okhotsk Sea and floating production system for South East Canada have been
performed.
- A procedure for Ice Management have been identified.
- A seismic analysis procedure have been developed.

```
************************************************************************
* TITLE : POSEIDON (PHASE 1)                    *     PROJECT NO      *
*                                               *                     *
*                                               *  TH./03164/84/FR/.. *
*                                               *                     *
************************************************************************
* CONTRACTOR :                                  *    TELEPHONE NO     *
*     GERTH                                     *                     *
*                                               *   1 47 52 61 39     *
*     AVENUE DE BOIS PREAU 4                     *                     *
*     FR - 92502 RUEIL-MALMAISON                *                     *
*                                               *    TELEX NO         *
*                                               *                     *
* TECHNICAL DIRECTOR :                          *   203050            *
*     MR A. CASTELA/MR. B. DARDE                *                     *
*                                               *                     *
************************************************************************
                                                 VERSION : 30/06/87
```

AIM OF THE PROJECT :
****
The aim of this project is to develop new techniques allowing a reduction of the
costs associated with offshore hydrocarbon production. In the near future, the
trend will go towards development of smaller hydrocarbon deposits situated under
an ever increasing water depth. In order to stay cost effective, the "all
subsea" scheme appears very promising : it will permit already proven subsea
production stations to be implemented at a longer distance from the process
facilities (situated onshore or on an available platform). This entails
basically mastering of two phase pumping, subsea motors, diverless subsea
production, two phase flow transportation techniques.

PROJECT DESCRIPTION :
****
The "all subsea" production scheme is based on two phase transport technology
and is characterized by :
- Elimination of the petroleum effluent separation units, thus also eliminating
platforms and all surface installations at the production site.
- Installation of subsea equipment of modular design highly flexible operation
and ease of maintenance.
- Use of a pumping system and a polyphasic pipeline.
- Reduction to a minimum of the on-site processing installations, which will be
strictly limited to the requirements of the polyphasic pumping and
transportation.
The present project will essentially consist of research and development in the
most critical areas of the concept.
Accordingly, the project is divided into two main activities :
MULTIPHASE PUMPING
Basically this activity consists in the study of the basic components of the
polyphasic production pumping unit, together with production of models and
testing of these models in a test bench.
- Homogenizing-regulation cell : regulates the flow at the inlet of the pump,
  and also acts as a choke, when needed.
- Compression cells : compresses the effluent.
- Separation-recycling cell : isolates a portion of the liquid phase,
  enabling lubrication of the bearings with gas-free liquid,
  and also recycling the liquid to the homogenizer, in order to lower the gas
content, when needed.
- Mechanical components; this activity aims selecting the proper materials for
the industrial pump.
SUBSEA STATION
- Subsea wellhead compatible with a down hole pump
- Subsea matable electrical connector : design and full scale development of a 1
MW prototype.
- Subsea motors for multiphase pumps : design and development of a full scale 1
MW subsea motor.
- Laying of heavy equipment on sea bottom.
- Connexion between the export pipeline and the subsea production station.
- Subsea pig launcher.

STATE OF ADVANCEMENT :
****
Completed

RESULTS :
****
ITEM 1 :
Multiphase pumping
All four subactivities are in a design phase. They should be completed during
1987 (see "Poseidon - Phase II).
The main results concern the compression cells. Several pump prototypes were
developed and extensively tested on a test bench. The performance of those
prototypes demonstrated the capability of a larger size pump to cope with
industrial requirements, in the future
A low pressure prototype of the homogenizing unit has been tested, and improved.

Results are positive. A high pressure prototype is currently under development.
The recycling cell prototype is still in an early development stage.
Improvements are still required, as performances are too low. This prototype is
of the centrifugal type.
A mechanical components bench test is under development as well as a short pump
prototype, for test purpose.
ITEM 2 : Subsea station
Subsea wellhead compatible with a down hole pump : after preliminary
investigations, this project was discarded because it was not deemed justified,
being too speculative : the reliability of a down hole pump is still low, when
considering the duration and cost of a replacement after failure, below a subsea
wellhead.
ITEM 3 : Subsea matable electrical connector
The full scale, 1 MW, prototype has been designed, developed and tested in a
pressure caisson. It was connected and disconnected 16 times in sea water,
under pressure, without any loss of insulation.
ITEM 4 : Subsea motors
A full scale subsea prototype has been developed. It is capable of delivering 1
MW at 8000 rpm. It is presently undergoing long term subsea trials out of the
scope of the present contract (contract TH 03.172/85).
ITEM 5 : Pig launching station
Detailed engineering studies are available. No major problem was encountered.
ITEM 6 : Pipeline connection
A basic study was performed, which established that two methods can be envisaged
for this study : either welding connection or a mechanical connection. The last
one should be more readily available, for deep water application.
ITEM 7 : Heavy equipment laying
A feasibility study is available. It identifies the problems areas, studies them
and suggests solutions. Heavy lift vessels, now available, reduces even further
the problem.

REFERENCES :
****
TECHNICAL MEETING : "ENVIRONMENTAL EXTREMES"
SPONSORED BY CESTA - FEBRUARY 25TH, 1986-MARSEILLE, "ADVANCED TECHNIQUES FOR A
HOSTILE ENVIRONMENT" BY B. DARDE

```
*****************************************************************************
* TITLE : NEW TREATMENT PROCESS FOR OIL AND WATER     *    PROJECT NO      *
*         EMULSION ON OFFSHORE PLATFORMS              *                    *
*                                                     *   TH./03165/84/FR/..*
*                                                     *                    *
*****************************************************************************
* CONTRACTOR :                                        *   TELEPHONE NO      *
*    GERTH                                            *                    *
*                                                     *   1 47 52 61 39     *
*    AVENUE DE BOIS PREAU 4                           *                    *
*    FR - 92502 RUEIL-MALMAISON                       *                    *
*                                                     *   TELEX NO          *
*                                                     *                    *
* TECHNICAL DIRECTOR :                                *   203050            *
*    MR. J.C. GAY & MR. C. SCHRANZ                    *                    *
*                                                     *                    *
*****************************************************************************
                                                         VERSION : 01/02/88
```

AIM OF THE PROJECT :
****
The quantity of water associated to the crude generally represents a significant
part of the well effluents. This water must be separated as early as possible
from the commercial part of the output. Specifications of the maximum water and
salt concentration in the crude are stipulated by the refiners. In addition, the
waste waters are subjected to environmental contraints covering their
hydrocarbon concentrations.
Since the efficiency of primary separation process is not sufficient, heavy and
cumbersome equipments are generally set up in order to achieve the required
specifications.
The objectives of this study are both :
- to find out means of reducing the volumes and the fineness of the oil and
water emulsions to be treated on offshore platforms,
- to develop smaller and lighter equipments for the settling of these emulsions.

PROJECT DESCRIPTION :
****
Three types of equipment capable of bringing about significant gains, were
initialy selected :
a) HYPO-EMULSIFYING CHOKES AND VALVES
The formulas defining the droplet size distribution of a liquid dispersed in
another liquid by turbulent dissipation show that it is possible to increase the
diameter of the droplets dispersed :
- by reducing the pressure drop, which is achieved by dividing this into several
stages instead of a single stage,
- by increasing the energy dissipation volume, which corresponds to increasing
the frictional surface areas that generate the turbulence,
- by increasing the concentration of the phase dispersed, resulting in partial
local separation and in subsequently applying the pressure drop to each of the
roughly separated phases.
Models applying the above processes :
- a multi-port stage valve,
- a rotating flow diode,
- a long pipe type valve.
were developed for parallel testing with a single diaphragm representing valves
that already exist, thus providing a reference for comparison with other systems.
b) DE-OILING CYCLONE
Second a de-oiling cyclone (emulsion of oil in water) capable of accommodating
to wide variations in flow (1 to 4) and oil concentrations whilst at the same
time reducing in a proportion of 1 to 10 the overall dimensions and weights of
existing equipments fulfilling the same functions.
This device, working on a segregation principle, will multiply the difference in
specific gravity between the oil and the water phases by the huge centrifugal
acceleration developed in a vortex flow.
On the basis of a mathematical model initialy set up, a plexiglass scale of a
conventional cyclone was built first.
After extensive testing of this model a new concept of hydrocyclone has emerged
and a modified scale model was developed. Finally an industrial prototype was
built and has been tested in an offshore field.
c) ELECTROSTATIC DEHYDRATOR/DESALTER
Design improvements and increases in efficiency of the electrostatic dehydrator
was found possible, which should result in significant reduction in the size and
volume of the existing equipments. the initial analysis of the problem has
focussed on the following axis of research :
- the design and the development of a sampling/blistering procedure enabling to
clot the droplet size distribution of the emulsion for future characterization,
- the analysis by means of radioacive tracers of the flow of the different
phases in refinery desalter,
- the pilot testing of the new developments available on the market like the
three electrode technology (Bielectric from Petrolite) or of the horizontal flow
type (HTI from Hydrotech).

STATE OF ADVANCEMENT :
****

a) Hypo-emulsifying chokes and values. Laboratory work is completed but industrial development is abandoned due to lack of commercial interest.
b) De-oiling cyclone. Industrial prototype field test being successful the industrialization will be implemented.
c) Electrostatic dehydrator/desalter. A new technology scale pilot is underwent trials on a test rig. A reduction of the settling time by a factor of 5 can already be expected, compared to a conventional electrostatic desalter.

RESULTS :
****

a) HYPO-EMULSIFYING CHOKES AND VALVES

The three types of valves were initially tested on low concentration oil in water emulsions. The efficiency of each type of valve was quantified and the validity of the theoretical experiments initially set up was confirmed. These results were checked in the case of oil in brine emulsions, higher concentration emulsions an finally in the presence of gas. The relationship between the parameters being defined a conceptual design of two types of valves/chokes was completed. Although no potential problem was identified in the technical development of these equipments, the difficulties anticipated in the marketing approach of the customers in conjunction with the uncertainties of the oil industry linked to the low oil price level, have made necessary to cancel this project.

b) DE-OILING CYCLONE

Testing of the conventional type scale model has enabled first to optimize the general set up of the cyclone. Then an adjustable injection head was defined in order to improve the flexibility in flow rate. The efficiency of the cyclone was then improved furthermore by the addition of internal devices increasing the hydraulic stability of the stream. Two patent applications were filed. The cyclone performances were then comparable to those of a tilted plate separator but a much smaller weight and volume. However the results were much lower than what was anticipated from the theoretical calculations.Therefore the laboratory studies were resumed in order identify the origin of this problem. Based on this study a new concept was defined leading to a different type of hydrocyclone. An industrial prototype has been built and tested on an offshore field. Excellent results were obtained but some minor improvements are still required in order to reduced the pressure drop requirement. Further scale model testing are currently outstanding.

c) ELECTROSTATIC DEHYDRATOR/DESALTER

The sampling/blistering procedure studies lead to the development of a method which has been patented. However, this system can only be used on laboratory fluids and attempts to extend it to the field conditions have failed.
Testing of newly available process have shown some possibilities of reduction in the sizes of the equipments. However the work on these pilots was stopped when studies on a new concept of equipment have revealed that larger progress could be achieved. A scale pilot is being tested resulting in significant improvement in performance and residence time. In the future it will be possible to prove that the weight and volume could be reduced yet further by optimizing this new technology.

REFERENCES :
****

A PAPER ENTITLED "OFFSHORE OILY WATER TREATMENT : NEW DEVELOPMENT IN HYDROCYCLONES" WAS PREPARED AND PRESENTED AT OFFSHORE EUROPE 87 IN ABERDEEN (GB), 11 SEPTEMBER 1987.

```
******************************************************************************
* TITLE : CONAT-DEEPWATER PRODUCTION TOWER.          *      PROJECT NO      *
*                                                    *                      *
*                                                    *   TH./03167/84/DE/.. *
*                                                    *                      *
******************************************************************************
* CONTRACTOR :                                       *    TELEPHONE NO      *
*    ABTEILUNG ZENTRALE TECHNIK#BILFINGER + BERGER   *                      *
*                                                    *   040 229 23 124     *
*    CARL-REISS-PLATZ 1-5                             *                      *
*    D-6800 MANNHEIM 1                                *                      *
*                                                    *    TELEX NO          *
*                                                    *                      *
* TECHNICAL DIRECTOR :                               *    21 11 86          *
*    MR PETER WAGNER                                 *                      *
*                                                    *                      *
******************************************************************************
                                                   VERSION : 06/02/90
```

AIM OF THE PROJECT :
****
Development of an articulated tower used in waterdepths between 300 m and 450 m
in the North Sea as an alternative for fixed platforms or as satellite platforms
to them. A parameters study performed shows the deckweight that can be carried
by towers in the waterdepth range and the deflections these towers undergo. A
structure that is optimum with regard to operational conditions and the design
of which is guided by practice, safety and cost was subject to the R&D work. A
component of special importance of the R&D work was the design of a joint for
the articulation which is able to transfer high loads from the column into the
base and which is, at the same time, maintenance free and resistant to wear and
tear. Construction methods and configuration of the tower have been developed
and worked on to assure a vertical construction of the tower and a partial
installation of the topsides close to shore in sheltered areas.

PROJECT DESCRIPTION :
****
The main components of the substructure are the bottle shaped column uprighted
by its buoyancy and attached by a ball joint to the base structure.
The concrete column has a circular cross section and is due to its shape in
biaxial compression without the requirement of prestressing. In case that higher
deckloads have to be installed inshore, the column has to be surrounded by a
multicell buoyancy body, closed at its upper and lower ends by domes. This
buoyancy body is monolithically concreted together with the column and serves
for floating stability during tow out. Several as risers are installed inside
the shaft.
The joint of articulation is of the ball - and socket-type made of stainless
steel and under permanent compression without any tension member. The horizontal
and vertical forces affecting the column are transferred from the upper
spherical shell to the lower sperical shell into the foundations. The moments
about the vertical axix are transferred by the torsion lock into the base. All
slide faces are plated with a metallic matrix comparable with a tin-lead-bronce
alloy in which solid lubricants are embedded.
The material has low friction coefficient and has excellent self lubricant
features which are further improved by an oil which will be injected into the
sliding faces. All parts subject to wear are designed with a sufficient safety
margin so that during 50 years lifetime no replacement is necessary. The inside
of the joint is separated from the inside of the column by a gastight and
pressure resistant bulkhead. The ball joint is permanently filled with
compressed air balancing the outer water pressure. No sealings for the operation
of the joint are required. Access into the joint is provided. The ball joint is
surrounded by a spherical air bell which keeps the ball joint under "dry"
conditions and prevents the permanent contact of sand-contaminated seawater with
the outer surface of the joint.
The concrete base is designed as an arrangement of vertical cylinders
surrounding a central thick-walled cylinder to which the ball joint is fastened.
Each of these cylinders is closed at its lower and upper ends by spherical domes.
 The base structure is resting on a skirt of 16,00 m long sheel piles
penetrating the subsoil. The base is equipped with installations for
waterballast, input of compressed air and vent pipes. Whether or not the base
has storage capacity depends on the field conditions.
All production - and export risers are installed inside the column. The
productiontrees are installed on the deck.

STATE OF ADVANCEMENT :
****
Completed are the conceptual design of an articulated production tower for a gas
field in 335 m waterdepth a deck load of 18.000 to. Topsides installed inshore
amount to 15.000 to. Alternatively an articulated tower for the same field has
been investigated but in this case with offshore installation of topsides. A
further development was an articulated tower for a field in 400 m waterdepth and
the creation of storage capacity within the structure.

RESULTS :
****

134

Comprehensive investigations have shown that the system has a high degree of reliability and integrity.
The articulated production tower has an Eigenperiod of To = 89 sec. The conservative calculation of the cumulative frequency occurence of inclinations during the lifetime shows that inclinations more than 1 deg will occur less than 10 %, inclinations more than 2 deg less than 0,05 % of the time. Thus the expected downtime of the operation of the system due to inclination is relatively low. An analysis of earthquake response based on specified frequency content and peak ground acceleration for the probable event with a return period of 100 years as well as for the extreme event with a return period of 10.000 years have been performed with the result that the PLS code and the ULS code provisions can be observed. The effects of a subsea gas blow out affecting the motion behaviour of the column have been investigated. The loss of buoyancy leads to a reduced restoring coefficient which does not endanger the integrity of the structure. Due to its articulated connection the structure is insensitive to non uniform settlements. Extensive fatigue calculations show good fatigue resistance. The hydrostatic pressure acting on the conical section of the bottle-shaped column together with the circumferential hydrostatic pressure are producing a biaxial compression state in the column's shaft which is favourable with respect to fatigue resistance.
In the ball joint no high local stress variations are to be expected under all SLS/ULS loading conditions. Ther permanent vertical force acting on the ball joints results in small variation according to wave loading. Therefore the ball joint is in a state of stress that is also favourable with respect to fatigue resistance. By the evaluation of long term statistical data on wave loading a lifetime for the production risers of 10.000 years is predicted being 400 times as much as the planned operational lifetime. The design of conductors and guiding pipes inside the column provides a double barrier to the inside of the column.
The althernative design of a production tower in 335 m waterdepth with offshore installed topsides shows a saving of about 30 % of the concrete masses of the column versus a production tower where the system has to be towed out with prior installed topsides. The design of a production tower in 400 m waterdepth has a very improved motion behaviour versus the 335 m version.

REFERENCES :
****
- " DESIGN AND SYSTEMATIC CONCEPT DEVELOPMENT OF ARTICULATED OFFSHORE TOWERS"
OFFSHORE BRASIL 1978
- "TEST CONAT. LARGE-SCALE TEST IN THE VICINITY OF THE RESEARCH PLATFORM
'NORDSEE'" EUROPE LONDON 1980
- "CONAT - THE ARTICULATED OFFSHORE SYSTEM"
  DEEP OFFSHORE TECHNOLOGY CONFERENCE 1981
- ARTICULATED PRODUCTION TOWER FOR DEEPWATER
  OTC 1984
- INTERNATIONAL WORKSHOP ON CONCRETE FOR OFFSHORE STRUCTURES, ST. JOHN'S 1986.

```
*****************************************************************************
* TITLE : EXTRA-DEVELOPMENT OF EXPERT SYSTEM      *      PROJECT NO       *
*         TECHNOLOGY FOR RISER ANALYSIS AND       *                       *
*         DESIGN                                  *   TH./03171/84/IR/..  *
*                                                 *                       *
*****************************************************************************
* CONTRACTOR :                                    *   TELEPHONE NO        *
*     STELLAR INT. LTD                            *                       *
*                                                 *   353 91 66455        *
*     BUTTERMILK WALK 3                           *                       *
*     IR-GALWAY                                   *                       *
*                                                 *   TELEX NO            *
*                                                 *                       *
* TECHNICAL DIRECTOR :                            *   50094               *
*     MR J. CONROY                                *                       *
*                                                 *                       *
*****************************************************************************
                                               VERSION : 31/10/89
```

AIM OF THE PROJECT :
****
The objective of the EXTRA project is the synthesis of expert systems technology
and advanced riser analysis and design techniques into an integrated marketable
software package.
A major innovation is the possibility of accumulating engineering knowledge and
company procedures into a productive design tool for a variety of commonly used
offshore systems such as marine risers, loading towers and flexible risers.

PROJECT DESCRIPTION :
****
The project includes an intelligent pre-processor that assists in setting up the
analysis of a riser system, refined solution techniques for steel and flexible
risers and an intelligent post-processor that incorporates design checks and
code certification.
Using the pre-programmed expert knowledge, a user of EXTRA is guided and advised
through all the modelling and analytical steps associated with the finite
element analysis of riser systems. Special rules and requirements based on a
particular company's experiences and background may be incorporated into the
analysis and results post-processing including some report preparation. The
EXTRA system is very useful for training purposes for new or re-assigned staff
since it performs as an expert tutor.

STATE OF ADVANCEMENT :
****
The EXTRA package has been implemented using the OPS5 production system language
to control the Expert System and to callout to the various analysis packages via
local BASIC, FORTRAN and system language modules.

RESULTS :
****
The package acts as an expert assistant in the design of exploration and
production, rigid and flexible risers and articulated loading columns and towers.
A preprocessor aids the user in the definition of the model in a form suitable
for the analysis routines; a program strategy controller decides the most
apprppriate analysis techniques to be followed. Finally, a postprocessor
performs design code checks and certification, before advising the user of the
suitability of his model. Graphical output is available to the user at all
stages to provide a clear, understandable representation of results.
Results include the satisfactory design of two API test case risers, namely API-
500-21.5-2-R and API-1500-21.5-2-R, using frequency domain and time domain
analyses respectively. The 2-D flexible riser section is also operational and
satisfactory results have been obtained from the design of a number of flexible
riser production systems.
A problem encountered is that of costs and programming effort in interfacing
between knowledge and analysis engineering systems in the software and hardware
context. Present-day software tools and computers are designed for one or the
other, but no integrated approach for communication between the two different
streams is currently available. Accelerated use of expert systems by industry in
general is generating new operating systems that may overcome this problem.

REFERENCES :
****
D. BOYLE, J.F. MC NAMARA, R. O' SULLIVAN, OFFSHORE MECHANICS AND ARTIC
ENGINEERING, SIXTH INTERNATIONAL SYMPOSIUM, HOUSTON, TEXAS, 1-6 MARCH, 1987.
J.F. MC NAMARA,R. O' SULLIVAN, "EXPERT SYSTEM BLACKBOARD CONCEPT APPLIED TO
MECHANICAL ANALYSIS AND DESIGN", PROCEEDINGS OF THE 5TH IRISH DURABILITY AND
FRACTURE CONFERENCE, CORK, RTC CORK, OCTOBER 1-2, 1987, ISBN-087849-573-8, PP.
97-113.

```
***********************************************************************
* TITLE : POSEIDON (PHASE II)                    *     PROJECT NO     *
*                                                *                    *
*                                                *  TH./03172/85/FR/.. *
*                                                *                    *
***********************************************************************
* CONTRACTOR :                                   *    TELEPHONE NO    *
*    GERTH                                        *                    *
*                                                *   1 47 49 02 14    *
*    4 AVENUE DE BOIS PREAU                       *                    *
*    FR - 92502 RUEIL-MALMAISON                  *                    *
*                                                *    TELEX NO        *
*                                                *                    *
* TECHNICAL DIRECTOR :                           *    203 050         *
*    MR A.CASTELLA (INST.FRANCAIS DU PETROLE)    *                    *
*                                                *                    *
***********************************************************************
                                          VERSION : 31/10/89
```

AIM OF THE PROJECT :
****
The aim of the Poseidon project is to develop new techniques allowing a
reduction of the costs associated with offshore hydrocarbon production. Poseidon
(phase I) project focussed on the design studies of the main subcomponents of
the "all subsea" scheme. Poseidon (phase II) project is basically dedicated to
the trials of the most critical amongst those subcomponents, namely the
multiphasic pumping system and the subsea drive system.

PROJECT DESCRIPTION :
****
The present Poseidon Phase II project comprises :
1) Development of a reduced scale, complete and marinized multiphase pumping
system, derived from the results obtained during the Poseidon Phase I project.
This development is divided into four subsequent phases:
- evaluation of the results acquired from the Poseidon Phase I project,
- basic engineering and definition of the architecture of the system,
- detailed engineering of the reduced scale prototype,
- fabrication and bench tests of the prototype.
2) Long term subsea trials of the full scale drive assembly developed within
Poseidon Phase I project.
The main activities are:
- fabrication of the test platform,
- integration of the drive assembly into the test platform,
- shore tests,
- deep water tests
- test analysis and equipment appraisal.
3) Theoretical study of the electrical network supplying power to a future
subsea station through the means of long distance subsea cables.

STATE OF ADVANCEMENT :
****
- Item 1 : multiphase pumping
  System = Ongoing
- Item 2 : subsea station and wellhead = completed.

RESULTS :
****
ITEM 1 : MULTIPHASE PUMPING SYSTEM
ONGOING
Simulation of the pumping system behaviour (completed) : this activity allows to
define and size up the prototype pumping system architecture.
Prototype pumping system : the fabrication has been completed; the test loop is
upgraded for maximum flowrate of 25 000 bbl/d and 90 bar maximum pressure. The
first phase of bench tests is completed (hydraulic characterization). Second
phase (endurance) is ongoing and planned to be completed in January 90.
Hydraulic testing of the industrial prototype has shown an excellent behaviour
from pure liquid to pure gas at the pump inlet, and the ability to handle severe
slugging conditions. Endurance testing on the bench followed by long term field
testing are the next essential steps to be completed before envisaging a first
industrial application, onshore of offshore.
ITEM 2 : SUBSEA STATION AND WELLHEAD
COMPLETED
The subsea testing of the drive assembly has been shortened by an insulation
failure of the stator windings of the motor.
Expert appraisals are completed. New studies have been launched, the aim being
to have a complete system ready by the end of 1989. First applications are
contemplated in the North Sea.
The shore trials of the full scale motor have been successfully completed. The
drive system totalled in excess of 3000 subsea running hours without major
problem, early in December, when a short circuit occured in the stator windings.
However all other electric, mechanical and electronic components behaved
perfectly during this test as attested by the expert appraisals. Some
improvement studies based on the results from subsea testing and appraisals have
been launched.

Basic engineering of the remote system of a future subsea station using Data transmission by optic fibre is completed. Theoretical studies have been performed, concerning the electrical behaviour of the long distance power supply system. Both steady state and transient states (start up, short circuits...) have been studied. This activity will be completed when the electronic variable frequency converter power supply case is covered.

REFERENCES :
****
DOT 1987 - MONTE CARLO: DEVELOPMENT AND TESTS OF THE POSEIDON SUBSEA MOTOR.
OTC 1988 _ HOUSTON: ONE MEGAWATT SUBSEA MATEABLE ELECTRIC CONNECTOR KEY TO MULTIPHASE PUMP DRIVE ASSEMBLY - NOW FIELD PROVEN (B. DARDE - TOTAL CFP).
OTC 1988 - HOUSTON: DEVELOPMENT OF A TWO-PHASE OIL PUMPING SYSTEM FOR EVACUATING SUBSEA PRODUCTION WITHOUT PROCESSING OVER LONG DISTANCE POSEIDON PROJECT (M. ARNAUDEAU - IFP). DOT 1989 - MARBELLA :POSEIDON : THE SUBSEA MULTIPHASE BOOSTING TECHNOLOGY, ON THE WAY TOWARDS INDUSTRIAL APPLICATIONS (A.FAFAILLE-TOTAL CFP).

138

```
****************************************************************************
* TITLE : PLATINE (PHASE 1)                    *    PROJECT NO         *
*                                              *                       *
*                                              *    TH./03173/85/FR/.. *
*                                              *                       *
****************************************************************************
* CONTRACTOR :                                 *    TELEPHONE NO       *
*    GERTH                                     *                       *
*                                              *    1 47.52.61.39      *
*    AVENUE DE BOIS PREAU 4                    *                       *
*    FR - 92502 RUEIL-MALMAISON               *                       *
*                                              *    TELEX NO           *
*                                              *                       *
* TECHNICAL DIRECTOR :                         *    560 804            *
*    MR. P. LEFEVRE                            *                       *
*                                              *                       *
****************************************************************************
                                              VERSION : 07/11/89
```

AIM OF THE PROJECT :
****
The aim of the PLATINE project is to specify a central production platform,
unmanned,teleoperated from a control room located onshore, with a yearly (or
half-yearly) visit for inspection and overhauling.
The intermediate aims consist of working out and testing pilot units related to
sub-systems and testing new equipments.
The project has been divided into 4 phases : PLATINE - Phase 1 (1986-1988);
PLATINE - Phase 2 (1987-1989); PLATINE - Phase 3 (1988-1990).89); PLATINE -
Phase 4 (1989-1991).

PROJECT DESCRIPTION :
****
This project consists of several items :
- methodology studies;
- test and adaptation of equipment;
- development or adaptation of maintenance and production softwares.
The adopted policy will aim at simplifying the general architecture and
examining the existing processes or methods critically. Use of new techniques or
materials will be systematically assessed.
The facilities have been divided into 7 main sub-systems : 4 production sub-
systems (gas-lift, sea water injection, separation-treatment, well pumping) and
3 sub-systems non specific to oil industry (energy/rotating machinery-
safety/reliability-automation sytem).
- The gas-lift-sub-system study has 3 parts : conception and construction of a 2-
phase flowmeter - automatic start-up of a well in high pressure gas-lift - real
time optimization of gas-lift injection.
- The sea water injection sub-system study aims at selecting sea water treatment
processes before injection and at specifing the best means of lifting and
pumping.
- The separation-treatment sub-system study aims at specifing a separation-
treatment chain compact, optimal and adapted to the project.
- A study related to the various means of well pumping will enable us to select
the (or the two) best means and to specify the automation works to be done.
- An on-condition plan will be worked out ; the main lines of the future on-
condition system (including probably an expert system) will be defined.
- A comprehensive statistical investigation of rotating machines will be carried
out. Besides, the development of new technologies in this field will be followed
and directed in connection with the manufacturers ; the main planned actions are
related to peripheral/centrifugal compressors, variable speed engines, magnetic
bearings, and gas packing.
- The automation system includes sensors, making their measurements reliable,
communication system and expert systems.
    The reliability of an automation system is mainly dependent on the
reliability of sensors. The actual sensors will be tested then adapted, if
needed for offshore oil conditions; their measurements will be made reliable by
means of coherence treating numerical system. Prototype systems of optic fiber
transmission will be tested; the development of optic sensors by manufacturers
will be followed and guided.
- The safety/reliability module includes regulation studies, reliability studies
related to equipment or systems safety equipment and platform surveillance (with
possibly a future robot development).

STATE OF ADVANCEMENT :
****
The methodology studies performed have allowed to define work programs, to carry
out sensors testing and to initiate a condition based maintenance program.

RESULTS :
****
Two modules are being studied : gas-lift and sea water injection. Besides, for
the other modules, investigations have been made and future works guided.
- Gas-lift : A prototype 2-phase flowmeter has been built and tested; after
modification results appeared not sufficient and this application has been
cancelled.

The automatic start-up has been defined with unique injection point. The gas-lift optimization will be carried out in testing the wells only when well behaviour has changed. Monitoring will be through a permanent bottomhole pressure sensor or, in a degraded mode, by permanent knowledge of wellhead temperature and/or wellhead casing pressure.
- Well pumping : Various means of well pumping have been evaluated resulting in the selection of the long-stroke pumping set.
- Sea water injection : The principles have been assessed and processes have been selected : filtration, deoxygenation, chloration and bacterial treatment. Reliable and accurate enough control analysers have been searched for (filtration quality, bacteria content, O2, Cl2 content...).
- Rotating machines : A comprehensive investigation has been made into (NEACP) subsidiaries, oil companies and other industries. It pointed out that a lot of existing equipment or facilities have failure rates inconsistent with PLATINE. Besides, the development of new technologies by manufacturers has been followed and guided (peripheral/centrifugal compressors, magnetic bearings, gas packings and variable speed engines).
- Maintenance : A list of curative or preventive works and trouble shooting on existing platforms has been drawn up. Optimal electric scheme has been studied (comparing for example offshore generation to onshore generation with cable transportation).
A condition based maintenance program has been carried out. It is the basis for an expert system applied to turbines and compressors but applicable to any kind of rotating machine.
An on-line event generator has been worked out; it consists of a common kernel and a specific module dedicated to electro-compressors. An off-line expert system for compressor maintenance (fault diagnosis) has been developed.
- Sensors : a large number of process sensors (gas and liquid flowmeters, level sensors, BSW meters) and pressure and temperature bottomhole sensors have been selected and purchased; sensors have been tested; the results have lead to the definition of modifications to be applied to these sensors in order to meet the PLATINE requirements.
- Miscellaneous : A preliminary study of the interest of a surveillance robot has been made.

F

```
************************************************************************
* TITLE : WATER PRODUCTION IN OFFSHORE WELLS          *    PROJECT NO      *
*                                                     *                    *
*                                                     *  TH./03174/85/FR/.. *
*                                                     *                    *
************************************************************************
* CONTRACTOR :                                        *    TELEPHONE NO    *
*    GERTH                                            *                    *
*                                                     *   1 47.52.61.39    *
*    AVENUE DE BOIS PREAU 4                           *                    *
*    FR - 92502 RUEIL-MALMAISON                       *                    *
*                                                     *    TELEX NO        *
*                                                     *                    *
* TECHNICAL DIRECTOR :                                *    203 050         *
*    MR. KOHLER                                       *                    *
*                                                     *                    *
************************************************************************
                                              VERSION : 31/03/89
```

AIM OF THE PROJECT :
****
The aim of the project is to develop in the laboratory processes based on water
soluble polymers in order to reduce in a selective manner the water production
in production wells without affecting the oil or gas production. Application on
a high temperature production well preferably in offshore conditions.

PROJECT DESCRIPTION :
****
The injection of water soluble polymers in producing wells is one of the most
performing processes for reducing water production. The process consists in
adsorbing in the vicinity of the producing wells a given amount of polymer
capable of considerably reducing the water production without affecting the oil
or gas production.
As water production from reservoirs presents different characteristics both, of
reservoir rock composition and permeability and also of produced brine salinity
and temperature, it is proposed to search for the appropriate solutions for
various environmental conditions and to apply them on producing wells. Indeed,
existing processes, based mainly on crosslinked polymer systems are considered
to reduce in a great extent the oil or gas production and are applied
indifferently on all cases of excess water production independently of reservoir
characteristics.

STATE OF ADVANCEMENT :
****
Completed laboratory work on phase 1.1 "Water production from high temperature
wells" by selecting a process based on a non-ionic polysaccharide.
Completed laboratory work on phase 1.2 "Medium salinity and temperature water
production".
Completed laboratory work on phase 1.3 "High salinity and temperature water
production".
Preparation, design and implementation in December 1988 of the field test
corresponding to phase 2.

RESULTS :
****
Phase 1.1
- Two original processes, one based on a hydrolyzed polyacrylamide, the other on
a non-ionic polysaccharide were compared in order to solve a real field case.
Polysaccharide type polymers were found to resist better than polyacrylamides to
elevated temperatures and to mechanical degradation. Numerical simulation showed
that it is preferable to inject the polymer near the oil water contact zone
rather than through existing perforations at the top of the oil zone.
Phase 1.2
- Field handling and injectivity of a non-ionic polysaccharide were tested both
on field cores in the laboratory and on an abandoned well in order to chose the
best conditions to perform a real field case. Results of the fiel injectivity
test are presented.
Feasibility studies were performed in the laboratory in order to select the most
performing process for a field operated by ELF-AQUITAINE. Two processes
developed by the partners and two competitive processes developed by service
companies were compared. Finally our process 2 was selected for the field
applications. Numerical simulations were performed and pointed out to the
treatment of a well producing oil having a lower viscosity than the previously
selected one.
Phase 1.3
- Feasibility studies of process 2 treatment for a high temperature, high
salinity oil producing well in offshore conditions.
Phase 2
- Preparation, design and implementation of a process 2 treatment on BARBIER 16,
an oil producing well operated by ELF AQUITAINE in offshore conditions.

REFERENCES :
****
SPE PAPER NR 16274 "THE ROLE OF ADSORPTION IN POLYMER PROPAGATION THROUGH

RESERVOIR ROCKS", SAN ANTONIO, TEXAS, FEBRUARY 1987.
PATENT FIELD FRANCE 87/03224, MARCH 6, 1987 - "PROCESS FOR REDUCING SELECTIVELY
WATER INFLOW IN OIL OR GAS PRODUCTION WELLS.
COMMUNICATION AT THE 4TH EUROPEAN SYMPOSIUM ON EOR, HAMBBURG, OCTOBER 27TH, 29,
1987 - "MODIFICATION OF WATER/OIL AND WATER/GAS RELATIVE PERMEABILITIES AFTER
POLYMER TREATMENT OF OIL OR GAS WELLS".
TECHNICAL AUDIT, NOOVEMBER 3RD, 1987.
COMMUNICATION AT THE 3RD SYMPOSIUM ON NEW TECHNOLOGIES FOR THE EXPLORATION AND
EXPLOITATION OF OIL AND GAS RESOURCES, LUXEMBURG, MARCH 22-24, 1988 -
"EVALUATION OF WATER SHUT-OFF TREAMENTS FOR SEVERAL .

```
************************************************************************
* TITLE : DIPHASIC COMPRESSION                    *     PROJECT NO     *
*                                                 *                    *
*                                                 *  TH./03175/85/FR/..*
*                                                 *                    *
************************************************************************
* CONTRACTOR :                                    *   TELEPHONE NO     *
*    GERTH                                         *                    *
*                                                 *   1 47 52 61 39    *
*    4, AVENUE DE BOIS PREAU                       *                    *
*    FR - 92500 RUEIL-MALMAISON                    *                    *
*                                                 *   TELEX NO         *
*                                                 *                    *
* TECHNICAL DIRECTOR :                            *   203 050          *
*    MR. P. DOUINEAU                               *                    *
*                                                 *                    *
************************************************************************
```
                                            VERSION : 18/04/88

AIM OF THE PROJECT :
****
The aim of this project is to develop a new production system able to boost the
stream of gas wells which are producing some liquids in addition to the gas.

PROJECT DESCRIPTION :
****
The production system is based on the use of a new equipment : a diphasic
compressor capable to boost the gas as well as the liquid. The selected type of
compressor is a screw compressor.
The project applies to a programme which purpose is mainly to design, execute
and test at field conditions a screw compressor which accepts gas containing a
liquid fraction.
Slugging will be also a matter of concern during the project as slug could
seriously damage the machine : a slug breaker will also be tested.
Attention is also paid to hydrates problems, which formation condition has to be
predicted as accurately as possible.
As regards liquid content, the project considers gas containing 2 percent
(volume) at suction conditions.

STATE OF ADVANCEMENT :
****
A gas field was selected to be the one on which the studies were to be
concentrated and where the new equipment could have been tested.
Operating conditions were estimated and preliminary design of the machine
performed.
A bibliography study was made on hydrates formation condition in diphasic
streams.

```
********************************************************************************
* TITLE : NEW RANGE OF COMPLETION EQUIPMENT       *      PROJECT NO      *
*                                                 *                      *
*                                                 *  TH./03176/85/FR/..  *
*                                                 *                      *
********************************************************************************
* CONTRACTOR :                                    *     TELEPHONE NO     *
*    GERTH                                         *                      *
*                                                 *   1 47.52.61.39      *
*    4 AVENUE DE BOIS PREAU                        *                      *
*    FR - 92502 RUEIL-MALMAISON                    *                      *
*                                                 *     TELEX NO         *
*                                                 *                      *
* TECHNICAL DIRECTOR :                            *     203 050          *
*    MR ALAIN BASSE                                *                      *
*                                                 *                      *
********************************************************************************
                                              VERSION : 31/07/89
```

AIM OF THE PROJECT :
****
This project consists of studying a new line of completion equipment to meet
four main objectives :
- safety implementation when bringing wells into production
- increment of production capacity
- long exploitation life of equipment, despite hostile environments (corrosion
and abrasion by effluent)
- reduction of exploitation costs by using different production modes.
presently the line of products on the market is all US brand and of old design,
while in this project, a new technology is used, the originality of which is
bound to the polyfunctional features of its different components, in particular
latching and sealing systems. The latter will be similar for each equipment
within this new line of completion equipment.

PROJECT DESCRIPTION :
****
The main completion equipment aimed at in this project are the following :
- Safety valve associated with a concentric gas lift system
- Subsurface controlled subsurface safety valve (SCSSV) wireline retrievable
- Safety valve for rod pumping wells
- Annular safety packers
- Circulating valve
- Tubing retrievable safety valve.
The project will consist of integrating polyfunctional systems in each equipment,
to meet safety, reliability and simplicity of use in the field of requirements.
The project development will consist in the 3 running stages :
- Engineering study
- Prototype manufacturing
- Cell and field testing
Prototypes of this equipment are made for the purpose of destructive tests and
to obtain API certificate (14A specifications).

STATE OF ADVANCEMENT :
****
Ongoing

RESULTS :
****
The present status of the various equipment of the project is as follows :
- Safety valve associated with a concentric gas-lift system : second design
completed. Prototype presently on long duration field testing.
-Design of the system to be run with a coil tubing under process.
-Design of q "stinger" to be installed in the Xmas tree and to be retracted in
order to close the two Xmas tree master valves, under process.
-Surface controlled Subsurface Safety Valve (SCSSV) wireline retrievable -
prototypes of nominal size 3 1/2 and 4 1/2 manufactured. Long duration field
testing under process and API qualification passed (14A Specification). Diameter
2 7/8 and 5 1/2 have still to be studied.
-Safety valve for rod pumping wells : study completed.
- Annular safety packers : inflate packers tested. One packer on long duration
test with Gaz de France.
-Circulating valve : study showed that to obtain a good and a long term
reliability of the equipment no elastomer has to be used to ensure the dynamic
sealing of the system.
-Tubing retrievable safety valve : same remark as for the circulating valve.
Safety valve associated with a concentric gas-lift system
The new latching device (hydraulically activated) successfully tested both in
Total Handil field and in Serept Ashtart field, has proved its long term
reliability.
A new latching device has been studied in order to install the valve in CAMCO BP-
6 landing nipple (nominal size 3 1/2).
The prototype of the latch for XL Otis nipple has been successfully tested in
workshop. But after discussions with wireline specialists it appears that it is
better to leave the dogs floating and to latch them hydraulically when they set

ın landing nipple profile.
Surface controlled subsurface safety valve (SCSSV wireline retrievable)
API tests (Spec.14A) were passed successfully for nominal size 3 1/2 and 4 1/2
nominal size valve and has in South West Research Institute in San Antonio (Texas)
. Nevertheless, sandy test (Class II) was missed for 4 1/2 nominal size valve
and has to be taken again.
3 1/2 and 4 1/2 size are presently under long duration test in Total Indonesia
(gas field) and in Elf Angola.
Safety valve for rod pumping wells
Good results were obtained with a valve which can be installed at 3000 ft depth
with no pressure inside the tubing.
Annular safety packers
Tests were made with inflatable packer in our workshop. Equipment is presently
on long duration field testing with Gaz de France.
Circulating valve
It is now proved that a study has to be undertaken with metal/metal sealing
system in order to keep a lonng term reliability (as long as the well life).
Tubing retrievable safety valve
Same remark as for the circulating valve.

REFERENCES :
****
SIX PATENTS HAVE BEEN APPLIED FOR IN FRANCE ON JULY 29, 1986, UNDER THE
FOLLOWING NUMBERS : 8611417 THROUGH 8611422, FOR THE WIRELINE RETRIEVABLE
SURFACE CONTROLLED SUBSURFACE SAFETY VALVE. USA AND CANADA EXTENSIONS.

```
****************************************************************************
* TITLE : IN-WELL LONG DRIVE OFFSHORE PUMPING      *      PROJECT NO       *
*         UNIT                                      *                       *
*                                                   *   TH./03178/85/FR/..  *
*                                                   *                       *
****************************************************************************
* CONTRACTOR :                                      *   TELEPHONE NO        *
*     MAPE                                          *                       *
*                                                   *   40433411            *
*     BOULEVARD DU MARECHAL JUIN, 14                *                       *
*     FR - 44100 NANTES                             *                       *
*                                                   *   TELEX NO            *
*                                                   *                       *
* TECHNICAL DIRECTOR :                              *   .711436             *
*     MR. CHARDONNEAU                               *                       *
*                                                   *                       *
****************************************************************************
                                                   VERSION : 30/06/89
```

AIM OF THE PROJECT :
****
The object of this project is to develop a thorough pumping equipment for drill
strings from the bottomhole pump to the surface unit, specially adapted to
offshore pumping, but also suitable for onshore pumping.
Therefore, the surface will have an integrated jack at the wellhead in such a
manner that its dimensions are minimized compared to other types of existing
beam units. This technology will allow :
- grouping of wells with very near cross axes (offshore platforms or onshore
  cluster);
- easy intervention on a well whithout having to interrupt the nearby well(s)
  in the case of clustered wells;
- considerable masking of pumping unit compared to other types of units in
  the case of protected offshore or onshore sites;
- improvement, compared to conventional units, of the pumping drive (long
  drive) without handicaping the driving part (beam unit reducer), and thus,
  a more economical and reliable adaptation to the production of viscous
  and/or gas containing crudes.

PROJECT DESCRIPTION :
****
1. Study and realisation of a "minimum" surface unit prototype aimed at
   controlling the validity of the basic principle of this unit. This prototype
   consits in :
     - a hydraulic jack temporarily fitted, for practical reasons, directly
       on the standard wellhead, thus higher than the wellhead.
     - an electrically driven hydraulic unit mounted on skid laved on the floor
       independly of the jack
     - a unit/jack connection formed by flexible and hydraulic piping.
   Its main characteristics are the following :
   - Maximum pumping feel = 25 600 Lbs
   - Three possible drive lengths = 100", 125", 144"
   - Six rythms also distributed up to 9 cp/mn maxi, for the 144" drive length
   - Power = 100 HP
   Hence, its name : V 256.144
   This prototype has no means of balancing the dead weight of the pumping
   drill string.
2. On-site measurements of prototype pumping parameters and development of
   computation method
     - of the pumping efforts at the jack rod
     - of the required engine power to pump a given flowrate at a given depth,
       without having to balance the deadweight of the pumping drill string.
3. Study and fabrication of a second prototype unit fitted with a hydraulic
   jack directly above the wellhead but with a hydraulic unit that recovers
   the deadweight energy of the pumping drill string during its lowering.
4. Study and construction of bottomhole pumps to api requirements, but with
   dismountable barrel of important length (not within api standards)
   On-site tests of different prototypes.
5. Study and construction of bottomhole pump, long drive with piston beyond
   API standards.
   Tests on site.
6. Study and construction of sucker-rods 1 1/4 at API standards, in special
   steel with dispersoids
     - Previous realisation of prototypes of existing dimensions (3/4" - 7/8"
       - 1")
     - Fatigue tests on test bench, comparison with rods commercially known
     - Realisation of 1 1/4" prototypes
     - Fatigue tests on test bench under loading conditions similar to previous
       fatigue tests. Commercial rods cannot be used as this dimension does not
       exist
7. Study and construction of sucker rods in composite materials
     - Construction of prototypes
     - Dynamic tests on fatigue bench, comparison with best quality American rods
       (FIBERFLEX)
     - Fiability test of complete drill string on significant site.

8. Integration of above subassemblies and preparation of tests for a complete
   system on the site

STATE OF ADVANCEMENT :
****
1 & 2 : Completed
2.3 : Prototype tested at the site in December 1988.
2.5 : Prototype pump THC 6 5/8" x 5 3/4" x 40 ft under continuous service in
Congo since 19 january 1988.
3 : Pending
4 : Partially completed.
5 : Study completed - ramp under construction awaiting for test site decision.
6.1 & 6.2 : completed.
6.3 & 6.4 : will not be made (cannot be financially justified).
7.1 & 7.2 : underway.
8 : temporarily delayed.
9 : pending.

RESULTS :
****
2.5 : On-site fatigue test of prototype interrupted end 1988 independent of pump
due to well incident. Retrieval and observation of pump. Satisfactory pump
condition except for the barrel chrome plating which goes away by whole strips.
2.7.1 : Satisfactory fabrication and fatigue tests of prototype rods of
diameters of 19 mm (3/4") and 25 mm (1").
2.7.2 : Comparative f dynamic test on fatigue bench with US FIBERFLEX rods.
Satisfactory for 3/4" rods.

REFERENCES :
****
MAPE DOCUMENT UNDER PREPARATION FOR THE JACK UNIT.

```
************************************************************************
* TITLE : CONCEPT FOR SUBSEA SYSTEMS IN SHALLOW    *    PROJECT NO    *
*         WATER                                     *                  *
*                                                   *  TH./03179/85/DK/..  *
*                                                   *                  *
************************************************************************
* CONTRACTOR :                                      *    TELEPHONE NO   *
*    AALBORG VAERFT A/S                             *                  *
*                                                   *   45 98 163333   *
*    P.O. BOX 661                                   *                  *
*    DK - 9100 AALBORG                              *                  *
*                                                   *    TELEX NO       *
*                                                   *                  *
* TECHNICAL DIRECTOR :                              *    69705          *
*    MESSRS JOERGENSEN (AALBORG VAERFT A/S)/OLSON   *                  *
*                                                   *                  *
************************************************************************
                                             VERSION : 06/10/89
```

AIM OF THE PROJECT :
****
The aim of the project is to develop a number of subsea systems which can be
installed in the North Sea, the Baltic Sea or in surrounding waters for the
production of hydrocarbons from economical marginal fields. The systems shall be
regarded as alternatives to light weight platforms which may not be attractive
of economical and environmental reasons in the deeper areas and in areas with
heavy ship traffic and large ice forces.

PROJECT DESCRIPTION :
****
A large number of known fields have not been developed due to their marginal
economics. The main objective of this project is to investigate the technical
and economical feasibility of marginal field development concepts based on
subsea systems.
Marginal fields can in general be defined as fields, whose recoverable reserves
are minimal and cannot be developed economically utilizing conventional
structures and processing facilities.
The definition applies in principle to two different categories of fields :
a. A peripheral marginal field allowing the produced oil/gas to be linked to an
existing facility for processing or tied into existing pipelines for
transportation to said facility.
b. A geographical marginal field remotely located from existing facilities or
pipelines and believed to hold insufficient reserves to make conventional
production methods economical.
In order to improve the economics of hydrocarbon production from marginal fields
in the more shallow parts of the North Sea, the Baltic Sea and surrounding
waters and in order to decrease the field size limit for economically viable
field developments a review of possible development options will be carried out.
This results in the identification of a number of concepts for the subsea
systems which can compete favourably with more traditional solutions.
Further it is to be expected that removal of platforms will be made compulsory
in the future when an oil/gas field is abandoned. This will add to the cost of
these solutions, where as subsea systems will not have this disadvantage.
Therefore, it is proposed to perform design verification for a number of
alternatives. Emphasis will be placed on features requiring development or
redesign beyond existing technology in order to meet the specific application.

STATE OF ADVANCEMENT :
****
The project was divided into 30 activities, each covering a well defined number
of objectives to be studied.

RESULTS :
****
The working groups made up from the contractors have reported on the following
design results achieved.
- Environmental DTA, conditioning, criterias, and loads.
- Well data for oil and gas wells.
- Assessment and comparison of risks for platform and subsea system, ice, ship
impact, fishing gerr, etc.
- Design cases and standards.
- Economic and technical review.
- Template, wellheads, and manifold.
- Protection, geometry and structure.
- Two phase flow, oil and gas condensate pipelines.
- Pipelines and control lines, overall review.
- Installation, drilling, and completion incl. tie-back.
- Tie-in, diver assisted and remote controlled.
- Proces facilities, offshore and shore based.
- Control systems, valves and actuators.
- Power supply and signal transmission.
- Operation of systems.
- Maintenace, methods, equipment, and personel.
- Risk analysis, failure mode and effect analysis.

- Preliminary economic analysis.
Each of the referenced topies are documented in seperate design reports.

REFERENCES :
****
TECHNICAL MEETINGS :
- SEMINAR ON SUBSEA SYSTEMS : DONG HOERSHOLM. DECEMBER 1986.
DANSK INGENIORFORENING 1987
DANMARKS TEKNISKE HOEJSKOLE 1987
EEC 1987.

```
****************************************************************************
* TITLE : DEVELOPMENT OF VARIABLE DRAUGHT SEMI-       *     PROJECT NO      *
*         SUBMERSIBLE CONCEPTS                         *                     *
*                                                      *  TH./03180/85/UK/.. *
*                                                      *                     *
****************************************************************************
* CONTRACTOR :                                         *    TELEPHONE NO     *
*    WIMPEY OFFSHORE ENGINEERS AND CONSTRUCTORS LTD    *                     *
*                                                      *    748 2000         *
*    27 HAMMERSMITH GROVE                              *                     *
*    UK-LONDON W6                                      *                     *
*                                                      *    TELEX NO         *
*                                                      *                     *
* TECHNICAL DIRECTOR :                                 *    25666            *
*    DR J.R. WILLIAMS                                  *                     *
*                                                      *                     *
****************************************************************************
                                              VERSION : 01/08/89
```

AIM OF THE PROJECT :
****
The aim of the project is to develop a standard design for a new type of
floating platform which is suitable for providing a number of functions (such as
drilling, workover and production) in a range of water depths including deep
water. The platform is a variable draught semi-submersible STAbilised PLAtform
(STAPLA) and the results from the marriage of two existing technologies, namely
semi-submersibles and jack-up platforms. By incorporating a variable draught
facility, it is intended to combine the restricted motion response
characteristics normally associated with purpose built deep draught vessels and
the inshore maintenance/modification capabilities of conventional semi-
submersible designs. The benefits of STAPLA, compared to similar alternative
floating systems, are being investigated and evaluated in terms of: improved
motion behaviour; wider operating capabilities and reduced downtime; and lower
capital and operating costs.

PROJECT DESCRIPTION :
****
The work is divided into three phases with two 'break points' at the ends of
Phases I and II when progress towards meeting the project aims will be assessed
and the content of subsequent phases may be modified/updated to incorporate any
new findings.
Phase I - Feasibility studies and definition of design premise
Phase I of the work consisted of the following tasks :
- Defining the design premise. This included establishing design information on
field specifications, environmental criteria, topsides facilities, structural
configuration (desk, hull, legs and jacking mechanism), design codes and marine
operations.
- Carrying out hydronamic model testing to demonstrate the anticipated improved
motion response characteristics at an early stage in the project.
- Setting up and using stability and vessel response analysis software to carry
out a limited parametric study.
- Analysing critical structural components, including the potoon
raising/lowering mechanism.
Phase II - Parametric studies and component refinement
The overall objective of Phase II is to develop a particular STAPLA design to a
level acceptable to certification Authorities for concept approval.
Phase II will be split into two :
(1) Phase IIA : This consists of two main areas of activity :
- detailed development of the STAPLA concept
- parallel development of the systems on which a STAPLA unit will rely.
The first area will consist of analytical parametric studies. Other tasks that
will be undertaken are stability analyses. The second area will include the
development of active and/or passive positioning systems, rigid and flexible
riser systems, subsea equipment required for use with a STAPLA unit and
potential export systems specifically applicable for use with a STAPLA unit.
(2) Phase IIB : This will be established from the conclusions of Phase I and may,
 if necessary, consist of structural tests on components which are unique to the
STAPLA concept due to its variable draught.
Phase III - Conceptual Design and Marketing :
Phase III will entail development and finalisation of the design to the level
required for conceptual engineering purposes.
The detailed planning of a marketing strategy will be part of Phase III with
back-up provided from comparative technical and economical appraisals between
STAPLA and alternative floating platforms.

STATE OF ADVANCEMENT :
****
Main project activities commenced on 1 September 1986 and the preliminary
design/testing tasks of Phase I (as described above) are now complete. Phase IIA
has commenced and is ongoing.

RESULTS :
****
From a review of existing floating productions systems, initial topsides design

criteria have been established to enable model testing to be carried out and to provide a basis for discussions with operators, certifying authorities, etc. The typical weight and space profile produced is for a facility capable of producing 100,000 bopd (or gas equivalent) in a hostile environment. To be self sufficient within supply boat cycles, the platform topsides includes separation of crude oil, water injection, gas lift, generation, utilities, living quarters for 120 people with helicopter access, flexible riser platform, diving back-up and seawater/fire water lift pumps. The deck area is approximately 90 metres by 90 metres and the total topsides operating weight of the initial STAPLA vessel is approximately 20,000 tonnes (including deck steelwork).

Environmental criteria have been established for a number of potential locations identified in European waters for the deployment of STAPLA (Nothern north sea, Haltenbanken, Tromso Patch, West of Shetland Porcupine Basin, Celtic Sea and the Mediterranean), hydrodynamic model tests have been carried out at 1:100 scale to establish a preferred vessel configuration.

The testing programme fully demonstrated the substantial reduction in motion response that can be obtained on having the pontoons located at a deep draught. However, while Design 1 has excellent motion response characteristics, the steel content is high. The following tasks were to quantify the steel content of Design 1 and to correlate analytical motion response with the model test results. The results of these two activities enabled the Design 1 to be developed in order to achieve a compromise between motion and steel content. This design development led to Design 2.

It was concluded that both STAPLA designs provide motion responses which are considerable better than conventional semi-submersibles and compare fabourably with those of a tension leg platform (TLP). The motion responses of STAPLA Designs 1 and 2 are 20 % and 40 % respectively, of greater flexibility, of those existing floating production facilities.

A compromise hull configuration has been established which provides the best balance of cost/weight low motion responses whilst using existing technology. This has particularly been achieved by a refuction in operating draught from 80 m to 50 m.

Mooring system will consist of a length of heavy chain, partially on the seabed, linked to the vessel by means of a wire (or synthetic) rope.

A review of riser systems has identified a bouyant rigid riser with flexible lines linked to the vessel as being the present preferred riser configuration.

As part of Phase IIA a complete reassessment of the required payload has been made with reference to existing vessels.

Steel weights have been estimated for columns, pontoons and deck from parametric equations and by comparison with existing vessels.

REFERENCES :
****
WILLIAMS JR, AND DAVISON J J.
"DEVELOPMENT OF A NEW CONCEPT FLOATING PRODUCTION". JOINT OFFSHORE GROUP INTERNATIONAL SYMPOSIUM ON NEW OIL AND GAS TECHNOLOGIES, LUXEMBOURG, 1988.
WALKER S AND WILLIAMS J R,
"STAPLA - A VARIABLE GEOMTRY SEMI-SUBMERSIBLE FOR FLOATING PRODUCTION". JOINT OFFSHORE GROUP INTERNATIONAL CONFERENCE ON FLOATING PRODUCTION SYSTEMSS FOR THE 1990'S RINA, NOVEMBER 1988.

```
****************************************************************************
* TITLE : SUBSEA WELLHEAD SEPARATION SYSTEM          *    PROJECT NO      *
*                                                    *                    *
*                                                    *  TH./03182/85/UK/.. *
*                                                    *                    *
****************************************************************************
* CONTRACTOR :                                       *   TELEPHONE NO     *
*    BRITISH OFFSHORE ENGINEERING TECHNOLOGY LTD     *                    *
*                                                    *   01-828 9797      *
*    18TH FLOOR, PORTLAND HOUSE                      *                    *
*    STAG PLACE                                      *                    *
*    UK - LONDON SW1E 5BH                            *   TELEX NO         *
*                                                    *                    *
* TECHNICAL DIRECTOR :                               *   8950676          *
*    B W SONGHURST                                   *                    *
*                                                    *                    *
****************************************************************************
```

VERSION : 05/02/90

AIM OF THE PROJECT :
****
To design and develop a subsea separator for offshore oil fields.
To fabricate a Pilot Unit to demonstrate the technology.
To operate the Pilot Unit under realistic conditions to demonstrate its
viability.

PROJECT DESCRIPTION :
****
The project will be carried out in two phases.
The first phase produced a conceptual design for a unit suitable for a range of
small N.W.E.C.S. fields with typically 10,000 bbls/day production and a field
life of nominally 5 years. The second phase will be the design, building and
fabrication of a Pilot Unit and the demonstration under realistic operating
conditions of this unit.
The initial concept was a three stage unit which was subsequently developed
into a two stage unit to separate oil, water and gas. For cases where flaring is
acceptable, a floating flare design has been developed. The separated fluids are
pumped to a tanker or nearby platform.

STATE OF ADVANCEMENT :
****
The conceptual study has been completed and the Pilot Unit commissioned and
trials have been carried out.

RESULTS :
****
The first phase has shown that the concept is viable and economically attractive
compared with other methods of small field development. The second phase has
shown that the unit can be operated to produce required separated products of
gas, oil and water.

REFERENCES :
****
PAPERS HAVE BEEN PUBLISHED BY THE FOLLOWING : OTC 1989 CONFERENCE
                                              SPE OFFSHORE EUROPE
CONFERENCE
PAPERS WILL BE PRESENTED AT OTC 1990 ON OPERATIONAL EXPERIENCE.

```
**********************************************************************
* TITLE : FLOATING PRODUCTION SYSTEM FOR THE       *     PROJECT NO      *
*         EXPLOITATION OF DEEP WATER                *                     *
*         MEDITERRANEAN OIL FIELDS                  *   TH./03183/85/IT/.. *
*                                                   *                     *
**********************************************************************
* CONTRACTOR :                                      *   TELEPHONE NO      *
*     AGIP SPA                                       *                     *
*                                                   *   02 5205969        *
*     AGIP-TEIN                                      *                     *
*     P.O. BOX 12069                                 *                     *
*     IT - 20120 MILANO                              *   TELEX NO          *
*                                                   *                     *
* TECHNICAL DIRECTOR :                              *   310 246           *
*     MR. P. TASSINI                                 *                     *
*                                                   *                     *
**********************************************************************
```

VERSION : 06/09/89

AIM OF THE PROJECT :
****
The scope of the research is the development of a floating production system
suitable for the exploitation of hydrocarbon fields in very deep waters in the
Mediterranean Sea. The system to be developed in the research consists of a
Tension Leg type platform and the relevant production riser.
The following are believed to be the main innovative aspects of the project:
- firstly, this project must be considered as the basis of a possible industrial
application of the Tension Leg solution in very deep waters (830 m, Aquila field)
.
- secondly, the installation procedures foreseen are innovative for what it
concerns assembling and deployment of tethers (it is this the only one detailed
study of a TLP with welded anchoring lines) and for all the technical and
operative aspects influenced by the high water depth.

PROJECT DESCRIPTION :
****
The present contract follows 3 previous one (TH.03121/82) related to stage 1 of
this research project. The present phase (Stage 2) aims at testing critical
components and procedures, as well as at re-examing and perfecting the system
design in the light of results obtained during Stage 1 and at preparing design
procedures and computer programs required for the engineering phase.
The following components and systems will be tried and tested :
Automatic Welding Machine (WM) and Non Destructive Equipment (NDE); for the
development of welding and fabrication inspection procedures, and for a first
complete check of their practical performance.
Tether Laying Experimental Plant for reproducing the overall laying sequence;
this will include, in addition to the above mentioned WM and NDE, equipment for
pipe handling, positioning and centering which will allow for a full scale
testing of the tether lowering operations.
- Lower anchor connector (on scale model); to verify its correct working and in-
service behaviour.
- Stress joint; to test the material and the welding process.
- Corrosion and cathodic protection deep water experimental station; to study
the effects of deep marine environment on corrosion phenomena.
- Foundation piles; to verify the behaviour tension piles under cyclic loads.
- Completion system; to verify the capability and behaviour of component(s).
The critical analysis of the design will be directed mainly towards: hull
configuration, tethers, installation procedures, risers and completion system.
Results of basin tests, of particular studies and analysis presently in progress
and precertification information will form the basis of the design review.
The design of certain components and aspects of particular importance will be
greatly improved. Among these:
- most important structural modes, with special emphasis on fatigue analysis
- risers
- behaviour of tethers
- temporary mooring system
- installation operations
Special computer procedures will be prepared for Tension Leg type structures,
mainly with regard to the overall calculation of the platform, the detailed
structural analysis of hull, tethers and risers.
Design procedures indicating basic design criteria and methods, including
procedures for weight control, will be elaborated as well as standards for
certain structural components.

STATE OF ADVANCEMENT :
****
Ongoing project. The critical revision and design optimization of platform and
mooring system configuration was completed. Computer codes for global motion
analysis and for platform and tethers/risers structural analysis were developed
and integrated.
The front end engineering of the prototypes and plants to be tested was
completed. Activities in progress are : Preparation of design and calculation
procedures for TLP engineering. Detailed design and fabication of the prototypes
to be tested.

RESULTS :
****
OPTIMIZATION OF PLATFORM AND MOORING SYSTEM CONFIGURATION :
The hull and deck configuration was studied and optimized through extensive sensitivity hydrodynamic studies.
COMPLETION SYSTEM AND RISERS :
The topside equipment layout was revised and integrated with the new deck structural configuration.
The riser components and stress joint configuration were analysed both by the functional and structural point of view.
The safety level of the whole completion system was increased by including, at the mud line level, a newly developed component : the Underwater Safety Block (USB).
CALCULATION PROCEDURES :
A complete set of calculation procedures was developed for TLP global motion analysis both in service and during installation, for tether and riser dynamic analysis, for hull and deck and for riser and tethers structural analysis of the significant parameters.
EXPERIMENTAL TESTS ON CRITICAL COMPONENTS/PROCEDURES
GTAW welding machine prototype with three torches and NDE system for assessing weld quality and will be installed and operated within a Tether Laying Experiment Plant (TLEP) which is in the design stage.
Basic design of mobile cathodic Protection Tester is completed. Field tests with Veritec in progress to study influence of high water on different steel grades. Joint industry project with NGI on pile behaviour.

154

```
*****************************************************************************
* TITLE : SEA TESTS ON RISERS FOR FLOATING        *      PROJECT NO        *
*          PLATFORMS                               *                        *
*                                                  *   TH./03184/85/IT/..   *
*                                                  *                        *
*****************************************************************************
* CONTRACTOR :                                     *    TELEPHONE NO        *
*     AGIP SPA                                     *                        *
*                                                  *    02 520 59 69        *
*     AGIP-TEIN                                    *                        *
*     P.O. BOX 12069                               *                        *
*     IT - 20120 MILANO                            *   TELEX NO             *
*                                                  *                        *
* TECHNICAL DIRECTOR :                             *    310 246             *
*     MR. P. TASSINI                               *                        *
*                                                  *                        *
*****************************************************************************
```
                                                        VERSION : 06/11/89

AIM OF THE PROJECT :
The scope of this research project is to organise and carry out sea tests on
instrumented risers, in order to acquire experimental data on the dynamic
behaviour of operating risers suitablefor verifying the capability of the
available theoretical calculation procedures of compliant stuctures.
oreover, aim of the project is to carry out a direct experimental verification
of the conditions which cause hydroelastic vibrations induced by vortex shedding,
 both in the presence and absence of a second riser and the consequences which
such phenomena have on the stresses and the displacements of the riser.

PROJECT DESCRIPTION :
This project constitutes phase one of a two phase project. The phase one is
dedicated to design of the experimental set up; the phase two, which will follow
this one, will be dedicated to the building of the test rig, to the sea tests
and to data analysis.
The basic objective is to install two instrumented risers on an already existing
structure (a fixed platform) in water depth of about 75 metres. Such a structure
will act as a support to the equipment required to carry out the tests.
The test rig will consist of :
- support structure
- motion simulation trolley supporting a rotating table
- tensioner
The instrumentation and data acquisition system will consist of three parts :
- instruments to measure risers' dynamic behaviour
- instruments to measure environmental conditions
- equipment for data acquisition and storage
The measurement of the risers' dynamic behaviour will be achieved by means of a
sufficient number of instrumented pipe sections placed on each riser.
The signals from all the sensors will be collected by a data acquisition system
based on a computer.
The data acquired during the tests will be suitable processed and compared with
theoretical simulations.
On the basis of this comparison, verification of the theoretical calculation
procedures will follow, in order to establish wether or not the available
procedures are adequate and to produce modifications of the existing calculation
procedures or to develop new ones, where appropriate.

STATE OF ADVANCEMENT :
Completed Project. The Detailed design of the system has been completed. The
detailed specifications for the vendor of the equipment have been issued.

RESULTS :
The project will develop and validate the calculation procedures needed to
design production risers for floating systems suitable for exploitation of deep
water hydrocarbon fields located in the European contiental shelf.
Development will refer mainly to :
- Instrumentations systems
- Oaata acquisition and analysis procedures
- Calculation procedures
The main result of this project is the complete design of a sea test facility to
perform full scale tests on risers for floating production systems.

REFERENCES :
AGIP MOVES ON DEEP WATER FACILITY DESIGNS.
OCEAN INDUSTRY, APRIL 1986 PP 156-158.
"SEA TESTS ON RISERS FOR FLOATING PLATFORMS PROCEEDINGS" JOF THE 3RD HYDROCARBON
SYMPOSIUM.
LUXEMBURG - MARCH 1988
AND 4TH DEEP OFFSHORE TECHNOLOGY CONFERENCE - MONTECARLO- OCTOBER 1987.

```
*****************************************************************************
* TITLE : TENSION LEG PLATFORM INVESTIGATION FOR      *     PROJECT NO      *
*         WATER DEPTH RANGING BETWEEN 200 AND         *                     *
*         1200 M.                                     *   TH./03185/85/HE/.. *
*                                                     *                     *
*****************************************************************************
* CONTRACTOR :                                        *   TELEPHONE NO      *
*    ALFAPI                                           *                     *
*                                                     *   01-6532579        *
*    304 MESSOGHION AVE.                              *                     *
*    HE-15562 ATHENS                                  *                     *
*                                                     *   TELEX NO          *
*                                                     *                     *
* TECHNICAL DIRECTOR :                                *   223296            *
*    MESSRS ANGELOPOULOS/PAPANIKAS                    *                     *
*                                                     *                     *
*****************************************************************************
                                                  VERSION : 25/10/89
```

AIM OF THE PROJECT :
****
To investigate in detail the application of a Tension Leg Platform (TLP) for the
Mediterranean Sea and, in particular, for the Greek area. Specifically, one type
of TLP for depths of 200 to 1,200 m is being considered. The execution of this
study from theoretical point of view is innovative and the results will be very
important for the application of such constructions . in Greece as well as in
other relative areas.

PROJECT DESCRIPTION :
****
The application of a TLP type platform is associated with risks due to its
dynamic sinking and the environmental conditions in general. This study includes
the prediction of TLP behaviour in the time domain with interactions between
hydrodynamic and structural design, including non-linearities in a wide range of
the TLP applications. The estimation of the techno-economical feasibility of the
production, erection and multi-usage of a TLP in a peripheral European area is
also being considered. The project is divided in the following main stages:
1. Environmental and design specifications
2. Calculation of the behaviour of the TLP
3. Foundation concepts
4. Application in South Kavalla region
5. Economic analysis according to the Greek industry capabilities
6. Project management.

STATE OF ADVANCEMENT :
****
Completed

RESULTS :
****
Final result of the project is the general techno-economical feasibility of the
production, erection and multi-usage of a high technology system as a TLP for
the Mediterranean Sea and in particular for the Greek area.
A preliminary design of two platforms having different payloads and for water
depths up to 500 and 1200 m respectively, was carried out based on the
environmental conditions of the greek area of South Kavalla, which differ
greatly with those of other seas where such systems have already been installed
(i.e. North Sea).
- Each unit is a semisubmersible of double symmetric type and consists of four
cylindrical columns connected by four rectangular pontoons. An integrated deck,
housing process equipment, auxiliaries and quarters, offers rigidity and good
structural continuity to the overall structure.
- The anchoring system of the vertical type, proved as the best one, is composed
of cables.
-The gravity type foundation was selected as the more effective solution.
Main emphasis in the study was given to the precise simulation of platform's
hydrodynamic behaviour, including a large number of non-linearities annd only a
time-domain approach was feasible. The analysis was performed by a highly degree
ALFAPI developed computational system using the F.E.M. (Finite Element Method).
The analysis showed that due to the constraints introduced by the tethering
system, the platforms show an excellent hydrodynamic performance while resonant
phenomena are to a large extend avoided.
As a consequence the study proved that the TLP concept is a solution for the
Mediterranean sea, specially for marginal fields, due to its low cost of
installation, high hydrodynamic performance and relatively easy way of
transportation.

REFERENCES :
****
THE PROGRESS OF THE WHOLE PROJECT IS DOCUMENTATED BY INTERNAL REPORTS AND IS
PRESENTED BY INTERIM REPORTS TO EEC. PRESENTATION OF RESULTS AND DISSEMINATION
OF KNOWLEDHE IS PERFORMED BY SPECIAL LECTURES AND PUBLICATION OF PAPERS IN
SCIENTIFIC PERIODICALS.US

156

```
***********************************************************************
* TITLE : DEVELOPMENT OF A SEMISUBMERGED,          *     PROJECT NO     *
*         TENSIONED CONCRETE CASING FOR THE        *                    *
*         OFFSHORE PRODUCTION OF LIQUID AND        *  TH./03186/85/DE/.. *
*         LIQUEFIED HYDROCARBONS                   *                    *
***********************************************************************
* CONTRACTOR :                                     *  TELEPHONE NO       *
*   SALZGITTER AG                                  *                    *
*                                                  *  030 88 42 97 26    *
*   ABTEILUNG FORSCHUNG UND ENTWICKLUNG            *                    *
*   POSTFACH 15 06 27                              *                    *
*   DE - 1000 BERLIN 15                            *  TELEX NO           *
*                                                  *                    *
* TECHNICAL DIRECTOR :                             *  185 655            *
*   DR-ING H.J.WESSEL                              *                    *
*                                                  *                    *
***********************************************************************
                                              VERSION : 20/09/89
```

AIM OF THE PROJECT :
****
From 1978 to 1984, the project group made up of HDW, LGA, Zueblin and Salzgitter
AG developed two production systems for use in deep-water areas which are
particularly suited to produce natural gas from offshore sites which have to be
termed marginal. The support structures for the necessary process facilities
comprise either a tension leg platform (TLP) or a concrete casting situated on
the sea-bed.
The aim of the project is to take the fundamental components of these two
research and development projects as a basis on which to draw up a new concept.
The idea behind this new project is to reduce investment costs to such a degree
that, even in the face of decreasing prices for energy, it will be possible to
produce oil and gas from marginal offshore fields economically. One particular
advantage of the project described here is that the system can also be used in
areas where ice can be expected.

PROJECT DESCRIPTION :
****
The purpose of the technical development under study is to design a production
system for small deposits in deep-sea areas. In this connection, priority is
given to keeping the investment and system costs as low as possible, so as to
make the "marginal" fields in the northern part of the North Sea accessible to
the European consumer, also at the present level of energy prices.
The main items of the project:
- Development of a semisubmerged tensioned concrete casing, which serves
simultaneously as a buoyancy body, a production casing and a storage casing as
well as the development of the foundation components (gravity foundation, piled
foundation) suited to the particular buoyancy conditions and the various
conditions on and in the sea-bed.
- Construction of a tensioning system to anchor the concrete casing, to the
foundation body and with regard to particular buoyancy conditions.
- Adaptation of the process plant with regard to the determined casing to the
foundation body with regard to particular buoyancy conditions.
- Design of an integrated intermediate store for different products for
intermittent tanker transport.
- Development of a system for transferring the separated hydrocarbons from the
submerged casing to the tanker.
- Adaptation of the developed riser technique to the submerged production
casing.
- Modification of the overall system for application in arctic offshore areas
(behaviour in driftice, pack ice, etc.).
Determination of the capital investment, estimation of the operating costs and
final considerations of economic efficiency.
- Performance of model tests.

STATE OF ADVANCEMENT :
****
The project is in the final phase during which the concluding report is composed.

RESULTS :
****
The area chosen for installation is the sea off the Norwegian coast to the north
of the 62nd degree of latitude. The facility will be designed merely to
accommodate the equipment required to process the oil and gas. Drilling
facilities will not be installed. It is assumed that the underwater field has
already been developed and that the processing facility can be connected up to
an existing piping system. The quality characteristics of the auxiliary and
ancillary systems needed for the self-sufficient running of the plant have been
specified.
Due to the depressed market situation processing facilities for methane
liquefaction are not considered. It is, however, planned to re-inject this
constituent with associated gas to stimulate oil production. A liquefaction
plant could be considered in a separate concrete casing if demand should require
it.
In order to achieve the largest degree of flexibility for installation of the

processing equipment, it is considered to accommodate the facility in one integrated room.

To use the production system in deep water, an intermediate store on the sea bed would result in large cross-sectional dimensions in order to withstand the high hydrostatic pressure which in turn would call for high investment costs.

When placing the intermediate store in the tensioned concrete casing, special significance is placed on the ballast system, for it would have to compensate variations in buoyancy arising when filling and emptying the intermediate store. In such a case, the hydrodynamic characteristics of the overall structure should alter as little as possible.

In contrast to the static requirements which require resistance to high pressure, the production procedure calls for easy-to-assemble elements, i.e. with flat surfaces. A compromise must be found between both requirements.

The following most important characteristics values were determined:
- effective cubic capacity
- buoyancy volume
- intermediate storage volume
- ballast space volume
- immersion depth in a floating mode.

Parallel to the casing design, fundamental studies have been conducted for installation of tensioning and coupling systems of the tension legs, in the concrete casing.

Flex joints (30.000 kW), which connect the tension legs with the tensioning arrangement, anchored in the concrete casing have been studied.

Investigations about the intermediate store to be integrated into the overall system consider the following parameters:
- production rate
- strategy for tanker operation taking into consideration adequate stand-by capacity
- buoyancy differences between tanks filled with water and those filled with products.

```
*******************************************************************************
* TITLE : DEVELOPMENT OF REUSABLE CONCRETE          *      PROJECT NO        *
*         PLATFORMS FOR MARGINAL FIELDS IN THE      *                        *
*         NORTH SEA                                 *   TH./03187/85/DK/..    *
*                                                   *                        *
*******************************************************************************
* CONTRACTOR :                                      *   TELEPHONE NO         *
*     CHRISTIANI & NIELSEN A/S                      *                        *
*                                                   *   45 114 12 33         *
*     VESTER FARIMAGSGADE 41                        *                        *
*     DK - 1501 COPENHAGEN                          *                        *
*                                                   *   TELEX NO             *
*                                                   *                        *
* TECHNICAL DIRECTOR :                              *   22336                *
*     MR. T. MORDHORST                              *                        *
*                                                   *                        *
*******************************************************************************
                                                       VERSION : 05/02/90
```

AIM OF THE PROJECT :
****
On the basis of the tender project for a direct founded concrete platform on the
marginal Rolf field in the Danish sector of the North Sea it is proposed to
develop methods and to carry out project details in such a way that the platform
may be reused on another marginal field after the first one has been depleted.
The need for a reusable off-shore platform has been furthered by the development
in the North Sea for exploitation of earlier found hydrocarbon reservoirs which
at the time of exploration were not considered to be commercially exploitable.

PROJECT DESCRIPTION :
****
The proposed concrete platform may be refloated and moved to another location
provided that the following activities can be carried out:
- Disconnecting pipe systems etc.
- Refloating of platform.
- Inspection/cleaning of skirt and skirt departments.
- Inspection/cleaning of pipe systems etc. for ballasting/injection.
- Transport to another location.
- Sinking of platform.
- Injection of skirt departments.
All the activities mentioned are actually based upon proven techniques but may
impede on another. However, it is assumed that working procedures may be
developed and that suitable materials may be selected, which allow for the above
order of operations. As an example it may be mentioned that an injection system
which has been used for cement mortar is not reusable right away. Likewise,
comprehensive studies are required to ensure that the material to be used for
filling the void between the platform bottom and the sea bottom can be removed
in an economical way. Cement mortar is presumably not suitable and therefore it
may be required to find another material.
In connection with the refloating of the platform, it is necessary to analyse
again and possibly change the proposed sinking procedure to make it fully
reversible with due regard to an assumed 2-10 years' interval between the two
operations.
Likewise, the original platform design will have to be analysed for possible new
loading cases.
The above studies and investigations shall result in the description of
procedures, technical calculations and drawings.

STATE OF ADVANCEMENT :
****
Completed

RESULTS :
****
The study has shown that for certain geotechnical conditions likely to be found
to a reasonable extent in the EEC-part of the North Sea, a reinforced concrete
monotower platform of the gravity type may be installed at water depths from
approximately 40 m up to 70 m.
The platform, which initially is designed for environmental loads governing at
the high water level has subsequently been checked for positioning at lower
levels with 10 m interval applying the corresponding modifications of the
environmental loads.
The horizontal extension of the hexagonal caisson is determined in order to
allow the major jack-up drilling rigs operating in the North Sea to be placed in
a working position over the platform.
The platform is made removable by applying a short skirt, which enable the
platforms to be pulled free of the sea bed, when water pressure is introduced in
the different skirt departments. The platform occupies 3,117 m2 and contains
approx. 17,000 m3 of reinforced concrete. The size is determined partly by the
stability at the different water depths which are intended to be covered within
the range of application partly by the hability of being able to float without
application of expensive and unhandy floating tanks. In spite of the
considerable amount of concrete required, the platform is considred competitive,
due to

1) the minimizing of the offshore works proper and
2) the possibility of reusing, which may limit the cost for future application
to the cost of a part of the marine works only.
The platform can advantageously be constructed at water depths as low as
approximately 16 m and towed to the ballasting site through shallow water.
Due to its flexibility it is specially suited for the marginal oil fields in the
North Sea.
The cost of the first installation is estimated to be DKK 240 millions, however,
subsequent installations are much less. In case more than one platform will be
required at the same time, savings can be obtained by applying the same dry dock,
slip form and fitting out quay for the different platforms.
Even though international standards for the removal and retention of obsolete
platforms has not yet been agreed upon, there is no doubt, that the worldwide
industry faces a multibillion dollar expenditure as it grapples with the
disposal of a large amount of installations as offshore fields stop producing.
Considering these prospects, the removable and reusable platform has a great
advantage, always being scheduled for reinstallation in another offshore field.

```
***************************************************************************
* TITLE : DEVELOPMENT AND APPLICATION OF         *      PROJECT NO        *
*         COMPOSITE STRUCTURAL SYSTEM FOR         *                        *
*         OFFSHORE PLATFORMS.                     *   TH./03189/85/UK/..   *
*                                                 *                        *
***************************************************************************
* CONTRACTOR :                                    *     TELEPHONE NO       *
*    TAYLOR WOODROW CONSTRUCTION LTD              *                        *
*                                                 *     01 578 2366        *
*    TAYWOOD HOUSE, 345 RUISLIP ROAD              *                        *
*    SOUTHALL                                     *                        *
*    UK - MIDDELSEX UB1 2QX                       *     TELEX NO           *
*                                                 *                        *
* TECHNICAL DIRECTOR :                            *     24428              *
*    MR. J.R. SMITH                               *                        *
*                                                 *                        *
***************************************************************************
                                                   VERSION : 01/08/89
```

AIM OF THE PROJECT :
****
The principal objective of the investigation is to develop the application of
composite steel/concrete sandwich construction for offshore engineering in order
to exploit the potential benefits of faster construction, more efficient use of
material and greater resistance to local loads.

PROJECT DESCRIPTION :
****
The programme will develop two reference designs of platforom structure, one for
an Artic environment and the other for a more benign environment. A design
method for composite construction will be developed and the validity checked by
a programme of confirmatory structural testing.Certification of the design
method will be sought.

STATE OF ADVANCEMENT :
****
Programme curtailed owing to potential use of the technology being delayed as a
result of developments in arctic regions.

RESULTS :
****
State of the art study and initial modeltesting.

```
*******************************************************************************
* TITLE : EXAMINATION OF FLEXIBLE OFFSHORE PIPE      *     PROJECT NO        *
*         SYSTEMS                                     *                       *
*                                                     *   TH./03190/85/DE/..  *
*                                                     *                       *
*******************************************************************************
* CONTRACTOR :                                        *    TELEPHONE NO       *
*     PAG-O-FLEX                                       *                       *
*                                                     *   0211 6505 378       *
*     IN DEN DIKEN 16                                 *                       *
*     DE - 4000 DUESSELDORF 30                        *                       *
*                                                     *    TELEX NO           *
*                                                     *                       *
* TECHNICAL DIRECTOR :                                *    8588857            *
*     DR. M. PEUKER                                   *                       *
*                                                     *                       *
*******************************************************************************
                                                      VERSION : 29/03/89
```

AIM OF THE PROJECT :
****
Within this project the applicability of flexible pipes for riser systems in the offshore industry will be proven. Therefore the behaviour of the pipes is investigated under various static and dynamic load conditions. Furthermore S/N-curves for flexible pipes which are not yet available will be determined and by further developed computer programs it will be possible to calculate lifetime of risers.

PROJECT DESCRIPTION :
****
This project is a joint industry programme with the LMT/Aachen and six oil companies. For a standard flexible pipe for subsea application (6in/6000 psi) the static data (burst pressure, stiffness data, external pressure resistance) as well as the dynamic data (tensible and bending fatigue data with different load amplitudes) is calculated in order to evaluate fatigue failure criteria. The fatigue test data is used to develop S/N curves for estimatiom of lifetime for flexible riser systems by computer analysis. For the evaluation of fatigue failure criteria the main components of a flexible pipe, i.e. the rubber material and the reinforcement steel cord wire, are investigated in dynamic tests. Furthermore the adaptability of the rubber material as liner material is checked in chemical tests with various fluid media.
Upon completion of the dynamic tests of the standard flexible pipe, alternative designs (varying inner diameter, pressure rate and liner type) are also investigated in static and dynamic tests.

STATE OF ADVANCEMENT :
****
The project was completed in December 1987

RESULTS :
****
The theoretical static qualities of the standard flexible pipe have been confirmed by test results. The expected number of cycles in the dynamic tests has been exceeded and the tests have been terminated at an unexpectedly high number of cycles without any obvious defect of the pipes. The compound tests of the elastomer material and the steel cord are completed and show good performance. The expected behaviour of the different elastomer materials in the chemical tests has been demonstrated.

REFERENCES :
****
"STATIC AND DYNAMIC PROPERTIES OF A 6" 6000 PSI FLEXIBLE PIPE FOR THE OFFSHORE MARKET"
V. PERZBORN & J. HYSKY, 3RD HYDROCARBON SYMPOSIUM, LUXEMBOURG, 1988.

```
*********************************************************************
* TITLE : ROTATING TURRET ASSEMBLY FOR OIL       *    PROJECT NO    *
*         RECEIVING, HANDLING AND SHIP MOORING    *                 *
*                                                 *                 *
*                                                 *  TH./03191/85/UK/..  *
*                                                 *                 *
*********************************************************************
* CONTRACTOR :                                    *  TELEPHONE NO    *
*    GEC ALSTHOM MECHANICAL HANDLING LIMITED      *                 *
*                                                 *  0533 750750     *
*    CAMBRIDGE ROAD                               *                 *
*    WHETSTONE                                    *                 *
*    LEICESTER                                    *  TELEX NO        *
*                                                 *                 *
* TECHNICAL DIRECTOR :                            *  347344          *
*    MR.J. MCPHIE                                 *                 *
*                                                 *                 *
*********************************************************************
                                            VERSION : 06/02/90
```

AIM OF THE PROJECT :
****

To provide an engineered design of a turret system which can be incorporated
into floating production system for the exploitation of marginal oil fields.
The turret will allow production to continue despite changing wave and wind
direction by allowing the ship to rotate around its centre line, the turret is
moored to the seabed.
The turret will contain much of the necessary production equipment and will
incorporate a multipath system for the production, injection, inspection and
test lines.

PROJECT DESCRIPTION :
****

A rotating turret assembly for oil receiving, handling and ship mooring and
designed for a floating production vessel used for the exploitation of offshore
oil reserves incorporates a number of innovative features.
The turret concept is covered under British Patent Specification No. 1447413 in
the name of GEC Mechanical Handling Ltd. and the design has already been
progressed through the preliminary stages by this company in collaboration with
two other major U.K. organisations, Foster Wheeler Petroleum Development Ltd.
and YARD Ltd.
Based on this conceptual work a turret design will now be established to meet
the current specification of requirements and design criteria.
The areas of technological innovation to be developed during the project include
the following:
a) Turret structure.
b) Support bearings and drive system.
c) Mooring line handling equipment.
d) Riser handling equipment.
e) Multipath fluid transfer system.
In each of the above areas of the overall system the work will be carried
forward to the preparation of detailed assembly drawings.
Interface systems will also be worked on as required to develop a complete
turret assembly.

STATE OF ADVANCEMENT :
****
Work is now complete.

RESULTS :
****
Detail development work has been carried out on the multipath fluid transfer
system, mooring line handling equipment, riser handling equipment and on turret
production facility.
Dialogue has been maintained with specialist equipment manufacturers and their
recommendations incorporated where appropriate into the design.
Arrangement drawings have been prepared for each of these systems based on
turret configurations for the COMPASS 50/205 vessel.
During the course of developing the multipath hose wrapping system, emphasis has
been placed on minimising the overall dimensions to limit windage and top weight,
 but with due consideration to operational efficiency and maintenance access and
also to ensure that the demands of the system are compatible with current hose
performance.
A feature of the system is the provision of a standby single path high pressure
swivel to swivel to facilitate uninterrupted transfer of fluids during hose
replacement.
The system designed can accommodate up to 7 hose wrapped services and the
configuration permits relative ship/turret rotation up to +/-410 degrees.
In response to specifc mooring data prepared by YARD Ltd. a mooring line
handling system has been designed suitable for mooring COMPASS vessel in water
depths up to 340m.
The equipment comprises propietary and special purpose mooring machinery
arranged on the turret to handle and stow 9 mooring lines of chain and wire rope.
A mooring sequence diagram has been prepared to illustrate the handling
procedure.

A practical equipment layout of valves, chokes, pipework and manifolds on the turret manifold deck has been established.
A feasibility study into the marketing prospects of the Rotating Turret System has been prepared.
From the technical and market investigations no optimum solution has wolved which would justify the development of a turrer design in the abstract.

```
****************************************************************************
*  TITLE : DEVELOPMENT PROJECT CONCERNING          *      PROJECT NO        *
*          EXPLORATION AND PRODUCTION OF MARGINAL   *                        *
*          HYDROCARBON DEPOSITS                     *  TH./03192/85/DK/..    *
*                                                   *                        *
****************************************************************************
*  CONTRACTOR :                                     *      TELEPHONE NO      *
*     DANSK OLIE & GASPRODUKTION A/S                *                        *
*                                                   *      02 57 20 44       *
*     SLOTSMARKEN 16                                *                        *
*     DK - 2970 HOERSHOLM                           *                        *
*                                                   *      TELEX NO          *
*                                                   *                        *
*  TECHNICAL DIRECTOR :                             *      21259             *
*     MR. LARS H. GAD                               *                        *
*                                                   *                        *
****************************************************************************
                                           VERSION : 08/09/89
```

AIM OF THE PROJECT :
****
To encourage exploration and commercial production at marginal hydrocarbon
deposits associated with gases of low calorific value containing impurities at
marcaptans and hydrogensulphide. By development of a mobile well test/production
unit with a new low cost gas disposal/utilization process.

PROJECT DESCRIPTION :
****
PHASE I, FEASIBILITY STUDY
Comparison of possible processes for disposal of low grade sulphur contamined
gases. Selection of the most economical and technical feasible process. Basic
design. Economical evaluation of commercial feasibility.
 PHASE II, DESIGN AND CONSTRUCTION OF A MOBILE DEMONSTRATION UNIT
Design and construction of a mobile demonstration unit, transportation to site,
and site construction/installation.
PHASE III, TESTING AND DEMONSTRATION
   MAIN ACIVITIES
III.a Well opening, preparation and tie-in unit
III.b Testing of unit
III.c Preparation of final report.
Reopening of an excisting exploration well at a marginal low grade gas deposit.
1 years test of technical feasibility.

STATE OF ADVANCEMENT :
****
The feasibility study is completed. A low cost process, which catalytical and
thermal incineration is found to be economical/technical feasible for disposal
of low grade gases which cannot burn naturally (flame).
Phase II and III will not be performed.

RESULTS :
****
A new process which combines catalytical and thermal incineration seems to be
technical and economical feasible for disposal of low grade associated gases,
which cannot be burned in a normal flame.
The study concluded that gases with high inert content and imprities of sulphus
compounds associated with production of oil and gas condensate can be disposed
safety. The technically feasible process is to be made rather flexible towards
change in gas composition and flow rates, by introducing interchangeable buruer
nozzles. It is a simple and reliable process unit, having few moving parts, a
low weight and small dimension. The unit will make it economicaly attractive to
produce oil reservoirs which contain associated gases of high inert content.

REFERENCES :
****
DEVELOPMENT PROJECT CONCERNING EXPLORATION AND PRODUCTION OF MARGINAL
HYDROCARBON DEPOSITS. FEASIBILITY STUDY REPORT, NOV. 1986.

```
****************************************************************************
* TITLE : POLYMERIC ADDITIVES FOR FRACFLUIDS        *        PROJECT NO     *
*                                                   *                       *
*                                                   *    TH./03193/86/DE/..  *
*                                                   *                       *
****************************************************************************
* CONTRACTOR :                                      *     TELEPHONE NO       *
*     CASSELLA AKTIENGESELLSCHAFT                    *                       *
*                                                   *     069 41092114       *
*     HANAUER LANDSTRASSE 526                        *                       *
*     DE - 6000 FRANKFURT/MAIN 61                    *                       *
*                                                   *     TELEX NO           *
*                                                   *                       *
* TECHNICAL DIRECTOR :                              *     411 208            *
*     DR. ENGELHARDT                                *                       *
*                                                   *                       *
****************************************************************************
                                                     VERSION : 06/09/89
```

AIM OF THE PROJECT :
****
To improve hydraulic fracturing technology it is important to create new frac
fluid polymers with increased thermostability for use in a variety of formations.
The scope of this project is the synthesis of new versatile polymers for frac
fluid application.

PROJECT DESCRIPTION :
****
A number of suitable monomer compounds were selected and model homopolymers
synthesized. The thermostability of aqueous solutions of these homopolymers was
tested under various conditions.
According to the results of these measurements model-copolymers were synthesized.
The thermostability and rheologic behaviour of aqueous solutions were tested.
The synthesis of co-polymers of different chemical composition and structure was
carried out according to different polymerisation processes and initiator
systems which were found to be suitable.
The polymers were characterized by measuring: molecular weight, residual monomer
content, molecular weight distribution, rheological data at different
temperatures and crosslinking behaviour to viscoelastic fluids.
Shear stability, pumpability and break-out under formation conditions were also
tested.

STATE OF ADVANCEMENT :
****
Until now the project proceeded according to schedule.

RESULTS :
****
Polymers were synthesized which showed good performance within the temperature
range up to 150 deg. C as far as the rheological data were concerned.
The crosslinking reaction of aqueous polymersolutions was carried out by the
addition of metal cations yielding macromolecular chelate-complexes, which show
desirable viskoelastic properties.
The degradation of the gelled fluids under formation conditions can be performed
by breaker systems. New breaker systems are under evaluation.
Continuing the synthesis of new polymers based on the information received so
far and testing the properties our present systems will be further optimized.
A more detailed study of our systems under simulated formation conditions will
be done at the University of Clausthal.
The studies carried out at the University of Clausthal, showed that fracturing
fluids based on the synthesized new watersoluble polymers, performed, performed
by far better compared to the present technical state as far as thermostability
and tolerance to high saline brines are concerned.
However some formation damage was observed, which could possibly be avoided by
changing the molecular weight of the polymer. Development into this directions
is going on.
New oilbased fracturing fluids were developed, based on novel polymer-tenside
systems. The rheological properties at different conditions are under evaluation.
Patent applications are presently filed.
More detailed information will be given in the next report.

166

```
*****************************************************************************
* TITLE : FLOATING PRODUCTION SYSTEM DEVELOPMENT,        *     PROJECT NO     *
*          DYNAMICALLY POSITIONED TANKER                 *                    *
*                                                        *  TH./03194/86/UK/.. *
*                                                        *                    *
*****************************************************************************
* CONTRACTOR :                                           *   TELEPHONE NO     *
*    BP EXPLORATION                                      *                    *
*                                                        *   01 920 8000      *
*    BRITANNIC HOUSE                                     *                    *
*    MOOR LANE                                           *                    *
*    UK - LONDON EC2Y 9BR                                *   TELEX NO         *
*                                                        *                    *
* TECHNICAL DIRECTOR :                                   *   888811           *
*    MR CJ EDGERTON                                      *                    *
*                                                        *                    *
*****************************************************************************
```

VERSION : 04/09/89

AIM OF THE PROJECT :
****
A new system based on a dynamically positioned (DP) tanker vessel, progressing
from previous work on the SWOPS concept is to be developed to be used for
extended well testing and for production and delivery of oil from marginal
fields.
The primary aims are to increase the efficiency and to reduce the capital and
operating costs of oil extraction in the North Sea, and to extend the concept
for worldwide application.

PROJECT DESCRIPTION :
****
The system involves a converted tanker connected via flexible risers to one or
more subsea wellheads, or to a riser base and flowline manifold system, handling
systems for deployment of risers, a riser swivel unit, process equipment and a
novel DPP thruster system concept for maintening the vessel on station.
The programme of work for the project includes development work on the new DP
thruster system with model simulation, tank testing and control system design;
development of the technology of multiple, flexible riser systems and methods of
riser suspension; and design studies of devices and systems to accommodate
rotation of the riser and swivel-less capabilities particular to this concept.
It will be necessary to verify the operational behaviour of the process units
such as separators, water separation, injection and lift systems, as subjected
to the motions of the vessel. Development work will also be required to ensure
cost reduction through reconsideration of space and weight of process and
related plant and economic utilisation of other types of production system. A
thorough examination of the design will be made to reduce the costs of all the
major system components.
The feasibility of the DP Tanker Scheme has been studied by BP using a purpose
designed computer simulation. Using this tool, a novel arrangement of thruster
units was conceived which promised lower power requirements than the
conventional designs. Model ship tank testing and further simulation work is
necessary to confirm and extend the concept.
Design work during 1985 developed the conventional DP tanker concept for
extended well testing (EWT). More recently, it became apparent that early
production could be achieved and hence extended production, using the DP Tanker.
The advantages over single point mooring systems are essentially due to the
mobility of the vessel from one oilfield to another, the system's independence
in environmental survival conditions and its particular suitability for deep
water oilfields.
The project was executed in three phases. Phase 1, which was completed in 1986,
included DP simulation and ship model testing, modular process design studies,
and the design of riser handling equipment. In 1987 phase 2A covered conceptual
work on the heliflex swivel concept (which was patented), and preliminary riser
analysis work. Phase 2B, in 1988, comprised completion of the riser analysis and
heliflex swivel work, and also two new studies covering flare design and an
offshore loading option.

STATE OF ADVANCEMENT :
****
The results are now available for applications.

RESULTS :
****
This project has resulted in the outline design for a low cost `sailaway'
dynamically positioned (DP) floating production system. The DP production tanker
is a feasible concept. The cost of the vessel was kept down in two main ways.
Firstly, by converting an existing tanker rather than commissioning a newbuilt
vessel. Process units would be installed on the deck, and designed for a minimum
of structural modification and hook-up. Secondly, DP power would be concentrated
at the ship's bow, with the stern being permitted to weathervane about a bow
mounted flexible riser.
The feasibility of this bow-centred DP, as opposed to the midships-centred
system used by conventional DP vessels, was tested by simulation studies and by
ship model testing. Power savings of the order of 30 % were demonstrated.

The process design work covered several modules, on the basis that relevant modules could be selected and a detailed design completed once a field application were identified. These included modules for separation, metering, water injection, and gas injection/gas lift. Weight and cost estimates were prepared on the basis of a 20 MBD production rate with entrained gas up to 1000 SCF/BBL.

Three studies were undertaken on the riser system. The first concerned a retractible cantilevered riser handling unit, to support the flexible riser safely over the ship's bow. This incorporated a conventional multipath swivel. A parallel study concerned the development of BP's heliflex swivel concept. This device, based on coiled flexible pipes, could replace the toroidal swivel and lead to increased reliability. A third study concerned the configuration and maximum water depth potential of the risers themselves. After examining the configuration options this work showed that flexible riser systems could be engineered for the DP production tanker in water depths up to at least 600 metres.

The cost of purchasing a suitable trading tanker, and converting it into a DP production tanker, would be of the order of L50 m.

REFERENCES :
****
"THE DEVELOPMENT OF A BOW DP PRODUCTION TANKER SYSTEM"
- PAPER PRESENTED AT THE EUROPEAN COMMISSION'S THIRD HYDROCARBONS SYMPOSIUM - LUXEMBOURG MARCH 1988.

```
****************************************************************************
*  TITLE : ON-FIELD MEASURES OF RESIDUAL OIL         *     PROJECT NO      *
*          SATURATIONS                               *                     *
*                                                    *   TH./03195/86/FR/.. *
*                                                    *                     *
****************************************************************************
*  CONTRACTOR :                                      *    TELEPHONE NO     *
*     AGELFI C/O GERTH                               *                     *
*                                                    *    1 47.52.61.39    *
*     4, AVENUE DE BOIS PREAU                         *                     *
*     FR - 92502 RUEIL MALMAISON                     *                     *
*                                                    *    TELEX NO         *
*                                                    *                     *
*  TECHNICAL DIRECTOR :                              *    203 050          *
*     MR LUIGI TERZI                                 *                     *
*                                                    *                     *
****************************************************************************
```

VERSION : 30/06/89

AIM OF THE PROJECT :
****
The project consists in developing methods of determining residual oil
saturation in water-flooded reservoirs. Several methods will be tested and
compared in different application cases: logs or tracer experiments.

PROJECT DESCRIPTION :
****
The project is to be developed in two main phases:
PHASE 1 - preliminary studies, laboratory activities and development of computer
programs.
The research will start with an extensive survey of technical literature on the
different methods employed for evaluating the oil saturation in situ. At the
same time an analysis of the reservoirs operated by SNEA(P) and AGIP will be
carried out to find out the wells allowing experiments under the best operating
conditions. The aim of the lab activity will be the development of an
experimental methodology for selecting and characterizing the most suitable
tracer for each field where the SOR measurements are to be carried out. The
modelling activity is intented to develop computer program and to adapt existing
software packages for the interpretation of logs or tracer experiments.
PHASE 2 - preparation, execution and interpretation of the field test.
In the selected well sites, facilities will be installed.
Logs or log-inject-logs will be the first technique to be experimented.
By using the surface facilities installed at the well location, the well tracing
tests will be completed.
The software packages developed in the first phase will be used for the
quantitative log analysis, the modelling and the interpretation of the tracing
tests.

STATE OF ADVANCEMENT :
****
Phase 1 :
The lab tests required to choose the best tracers for the SWTT in Torrente Tona
Field (Italy), and for the interwell experiment at Chuelles (France) were
completed.
Phase 2 :
The operations on Cortemaggiore field were carried out on two wells. The
interpretation of the preliminary field test run in the other site was completed.

RESULTS :
****
First phase
- The tracers for SWTTs on the first selected sites were chosen and their
partition coefficients under reservoir conditions were determined
- It was shown that the hydrolysis kinetics is strongly affected by the pH
- The radioactive tracers for the log-inject-log (gamma) were selected and
tested in the laboratory
- An extensive selection of radioactive tracers for the interwell tracer test is
still in progress. Tracers which will satisfy safety and partition coefficients
requirements seem difficult to find.
- Numerical models designed to match tracer tests were tested.
Second phase :
The workover on Cortemaggiore 56 was suspended because of the bad mechanical
condition of the well casing. However, the field activity was resumed on
Cortemaggiore 57 and SWTT and LIL tests were completed. The interpretation of
field data is under way.
On the other site (Chuelles) the final experiment of interwell tracer was
performed. The sampling and analysis of the fluids produced are not yet
completed. Some experiments of LIL were also carried out in an offshore well
located in Cameroun. The log interpretation is still under way.
Overseas branches of both companies have been inquired on the possibility of
finding new sites for field tests.

```
************************************************************************* *******
* TITLE : DATA TRANSMISSION WITHOUT CABLE FOR      *       PROJECT NO       *
*         PRODUCING WELLS                          *                        *
*                                                  *   TH./03197/86/FR/..   *
*                                                  *                        *
************************************************************************* *********
* CONTRACTOR :                                     *       TELEPHONE NO     *
*    SYMINEX                                        *                        *
*                                                  *     91.73.90.03        *
*    3, BLD DE L'OCEAN                              *                        *
*    FR - 13275 MARSEILLE CEDEX 09                  *                        *
*                                                  *       TELEX NO         *
*                                                  *                        *
* TECHNICAL DIRECTOR :                             *       400 563          *
*    MR. A. BAUDRY                                 *                        *
*                                                  *                        *
*********************************************************************************
                                                    VERSION : 07/11/89
```

AIM OF THE PROJECT :
****
This project concerns the study and realization of an acquisition unit for
bottom well measurements having only short time lag between the measurements and
the transmission of the data by a cableless system. The knowledge of the
pressure and temperature at the bottom of producing wells is fundamental for a
better appreciation of the pay zone and optimization of production.
Disadvantages of actual systems are related to the problem encountered by cable
transmission or the lag between data acquisition and interpretation.
The proposed system ensures data transmission by using small elements, carried
up by the fluid and in which the desired information is stored.
These memories are recovered at the surface and read by an appropriate computer.
This apparatus aims at eliminating the need of cables as well as reducing costs
by utilising well-know gas-lift technology.

PROJECT DESCRIPTION :
****
The project is to be studied in several parts :
- Bottom unit : the acquisition unit is wireline set in a side pocket mandrel
installed at the bottom of the production tubing (traditional operation for gas-
lift valves).
It includes :
* Pressure and temperature sensors
* Microprocessor based electronics
* Data storage memories and ejection module
* Power supply
- Surface unit : The recovery of the memories when they reach the surface is
made by a memory trap set on the well head prior to the choke.
  It does not need to close the well.
- Data read out unit : Data storage memories are read on the surface by means of
an appropiate interface and dedicated soft on a portable computer, data being
stored on a magnetic support as well as printed.

STATE OF ADVANCEMENT :
****
Hybrid technology has been developped for the memories chip : data and energy
transfer by inductive field is successfully achieved. The final mechanics for
gaz lift and wire-line tools is under development after prototype trials.

RESULTS :
****
The different modules of the equipment have been developped
. Bottom equipment : electronic prototype of the management system have been
achieved.
Low consumption CMOS components have been used offering a maximum consumption of
2,50 mA.
This electronics enables :
- data acquisition of the measured parameters (temperature, pressure of bore
hole) in the annexe memory of the CPU
- the interface between the CPU data storage and emitting module to the
ejectable memory chip
- the management of the memory ejection
- real calendar time storage with the data
- maximum temperature environmment : 160 deg. C.
- maximum pressure environment : 500 bars.
. Software enables the CPU ROM programmation with the number of acquisition to
be made, the frequency of ejection (every hours, every days, every week,...)
. Memory chips has several Kilo-Octets of memory offering more than 1 month 12
bits data storage autonomy on one chip
. Ejection mechanics have been studied and designed.
. Surface reading electronics and mechanics have been completed.

```
*******************************************************************
* TITLE : FLOODED MEMBER DETECTOR            *    PROJECT NO      *
*                                            *                    *
*                                            *   TH./03198/86/FR/.. *
*                                            *                    *
*******************************************************************
* CONTRACTOR :                               *   TELEPHONE NO     *
*    SYMINEX                                  *                    *
*                                            *   91.73.90.03      *
*    2, BLD DE L'OCEAN                        *                    *
*    FR - 13275 MARSEILLE CEDEX 09            *                    *
*                                            *   TELEX NO         *
*                                            *                    *
* TECHNICAL DIRECTOR :                        *   400 563          *
*    MR. C. CRIADO                            *                    *
*                                            *                    *
*******************************************************************
```

                                            VERSION : 07/11/89

AIM OF THE PROJECT :
****
The objective of this project is to design a Flooded Member Detector whose
characteristics are aimed at reducing as far as possible the cost of inspection
campaigns on offshore jackets.
The maintenance and control of these structures are relatively important part of
the production costs. The design of this new instrument includes the following
features :
- Operation by non-NDT trained diver
- Requires no umbilical connection to the surface
- Automatic interpretation by a micro computer (no specialist being required for
its use and interpretation).

PROJECT DESCRIPTION :
****
The priorities for the project were to produce an instrument.
This project has been conducted in order to respect the basic concepts which :
- is simple to use
- has no cable to the surface
- requires no operator interpretation
The study has been separated into several phases :
- Ergonomy
- Measurement principle and laboratory testing
- Data storage
- Interpretation
The ergonomy has been studied in conjunction with divers in order to obtain a
final product which is easy to handle underwater (submerged weight, shape,
dimensions, etc..).
The measurement principle is based on the detection of reflected ultra-sonic
impulses by means of an analogue circuit and a transducer (study of frequency,
type of transducer, associated electronics).
The data are compiled by a microprocessor and stored in RAM CMOS memories
(electronics and shoft).
They are read by a dedicated program on a microcomputer once the diver is back
on surface (choice of microcomputer, soft, etc...).

STATE OF ADVANCEMENT :
****
After the first phases of functionnal analysis and design, the realisation of a
first prototype detector both hardware and software has been successfully
performed. Some final software developments utilities and test trials have to be
completed.

RESULTS :
****
The final product should be tested under real conditions on offshore structures
before next summer (1987).
Problems encountered were due to :
- Use of the apparatus by non NDT trained divers.
  * Shape and weight, several solutions were submitted to divers.
  * Positioning on members (size to be checked range from 8 to 60 inches
diameter).
  A special front has been developed using magnets to amintain the instrument
in position.
- The choice of transducer frequency and electronic design :
  * The analogue circuit (very high amplification of the reflected signals is
needed).
- The miniaturization of all the electronic component (data storage, power
supply, microprocessor, analogic circuit).
- The choice of microcomputer (a new product from EPSON with printer has been
selected).
All the equipment may be carried in a briefcase.
After some difficulties in analog ultrasonic electronics adapted to the divise
transducers tested and problem of positioning on members, a first prototype has
been successfully achieved.

The different aims have been reached :
- it can be used by non specialist divers, interpretation of the results is automatic by computer, stored in the detector and then transfered on the surface reporting microcomputer.
- there is no umbilical connection to the surface offering a complete autonomy for the diver. Also clear and simple informations by means of leds are offered to the diver : "measurement in progress", "flooding detected", "unvalid measurement", "battery discharged".
- positioning on members with magnets enabling members inspections from 8 to 60 inches diameterbracons.
- miniaturisation of the electronics (shape/weight optimisation) and low consumption components (for greater autonomy) has been achieved. All analog circuits, digital circuits, datastorage, power supply, and microprocessor electronics has been integrated in a small light and easy to handle "ultra-sonic gun".
- the surface equipment has been integrated in a portable brief case containing the reporting microcomputer, a thermic printer, a battery charger, and the read-out cable.

```
**************************************************************************
* TITLE : CLUSTOIL (PHASE 1)                          *    PROJECT NO     *
*                                                     *                   *
*                                                     *  TH./03200/86/FR/..*
*                                                     *                   *
**************************************************************************
* CONTRACTOR :                                        *  TELEPHONE NO     *
*    GERTH                                            *                   *
*                                                     *  1 47.52.61.39    *
*    4, AVENUE DE BOIS PREAU                          *                   *
*    FR - 92502 RUEIL-MALMAISON                       *                   *
*                                                     *  TELEX NO         *
*                                                     *                   *
* TECHNICAL DIRECTOR :                                *  203 050          *
*    MR. J. PHILIPPOT                                 *                   *
*                                                     *                   *
**************************************************************************
```

VERSION : 31/03/88

AIM OF THE PROJECT :
****
The project is aimed at studying subsea stations for oil production. The study
is based on the following fundamental assumptions : modular design, diverless
installation and maintenance, handling operations from light surface supports.
In addition, the concepts will use the results of previous studies and should
lead to a significant cost reduction, both in investments and in development and
maintenance costs.

PROJECT DESCRIPTION :
****
The whole study includes three phases :
- A feasibility phase
- A test phase for components and subassemblies
- An overall test phase on the offshore pilot.
The present project only applies to the first phase and concerns two difrent
subsea stations.
- The first one, called CLUSTOIL, will be designed for a North Sea application
with the following fundamental assumptions : cluster of twelve wells, total
production : 30000 m3/d, water depth : 350 meters, distance from treatment site :
 about 10 km, production life time : 15-20 years.
- The second one, called SAPHIR, will be designed for a Gulf-of-Guinea
application with the main following requirements : cluster of six boosted wells,
total production : 2000 m3/d, water depth : 160 meters, distance to onshore
treatment site : 8 km, soft foundation, production life time : 12 years.
For both stations, the feasibility phase will include two stages :
1. Definition of subsea station design
The design will be specified according to exploitation conditions, production
techniques, reliability of components and maintenance means.
2. Detailed study of subsea station
Once the layout is specified, the different modules and their links and
connections will be the subject of detailed studies.
Modules and related connections studies will be carried out while keeping in
mind the following objectives :
- acquisition of maximum reliability for the overall system
- optimization of dimensions and weight of different modules to facilitate
handling operations.
The main subassemblies are : template, production wellhead, manifold,
telecontrol system, production and electro-hydraulic connections.

STATE OF ADVANCEMENT :
****
The studies focused on the Subsea station designed for a Gulf-of-Guinea
application : SAPHIR. The general station design has been specified and the
detail engineering of some modules has been carried out. The design of the
station intended for a North Sea application (CLUSTOIL) has been completed.

RESULTS :
****
SAPHIR
The station is a six-well cluster. X-trees are installed around an octagonal
guide base supported by a central conductor pipe. Each X-tree is divided into
two parts : the safety block tree and the production block tree. The sensitive
equipments such as pressure transducers, remote-controlled choke and the most
prompted valves are housed in the production block tree. In case of failure,
this block can be entirely removed and replaced by a new one using a dedicated
vessel. During this operation, the safety block tree insures well safety.
A common inert manifold gathers the fluids from the six wells and allows
automatic connections between the X-trees and flowlines.
The subsea station is remotely controlled from the surface control room. Coded
messages are sent to the central control module through an underwater cable.
These messages activate the relevant pod installed on X-tree and activate the
hydraulic pressure system driven from the surface through a hydraulic bundle.
Three lines are tied to the cluster, the 10" production line, the 6" test line
and the 4" service line.

The installation of the whole station is carried out from the drilling rig through its moon pool.
Feasibility of the design, such as defined in 1986 has been confirmed by two detailed studies: Manifold study and Flowline connection study.
The functional analysis of the production block has been completed.
Feasibility of the design as defined in 1986 has already been controlled by two detailed studies : Manifold Study and Flowline connection Study.
The functional analysis of the production tree followed by a study on the reliability-availability-safety, has made it possible to confirm the choice made in the general architecture of SAPHIR in 1986, that is:
The production tree is divided in two blocks : the safety block and the production block. The production block comprises all the sensitive components and the telecontrol Pod. The safety block assures the safety of the well during the replacement of a production block.
CLUSTOIL:
The station is a twelve-well cluster. The solutions used for the SAPHIR design are fitted and re-used for the CLUSTOIL design.
The critical point of the design is the connection of the important diameter production manifold.
The connection of the 18" production flowline on the station is achieved thanks to a module comprising 2 mechanical connectors and a flexible loop.
SPECIFIC STUDIES
- During exploitation mode, the Pod receives from a platform electro-hydraulic energy and electric signals through a subsea umbilical.
During intervention, energy and signals are generated on a floating support placed above the station and conveyed to the Pod through a special umbilical. At each operation the subsea umbilical is automatically isolated and the Pod can no longer be controlled from the platform.

```
*******************************************************************************
* TITLE : DEEPWATER SUBSEA PRODUCTION SYSTEM AND     *      PROJECT NO      *
*         MAINTENANCE DEVICE - PHASE 2 - STAGE A     *                      *
*                                                    *   TH./03201/86/IT/.. *
*                                                    *                      *
*******************************************************************************
* CONTRACTOR :                                       *    TELEPHONE NO      *
*    AGIP SPA                                        *                      *
*                                                    *    02 5201           *
*    C.P. 12069                                      *                      *
*    IT - 20120 MILANO                               *                      *
*                                                    *    TELEX NO          *
*                                                    *                      *
* TECHNICAL DIRECTOR :                               *    310 246           *
*    MR. P. TASSINI - TEIN                           *                      *
*                                                    *                      *
*******************************************************************************
                                               VERSION : 04/09/89
```

AIM OF THE PROJECT :
****
Scope of this project is the development of a new generation of subsea systems
specifically conceived for hydrocarbon production in deep (200 - 600 m w.d.) and
very deep beyond 600 m) waters including a dedicated maintenance device.
Therefore diverless and guidelineless Techniques for installation, maintenannce
and work over are basic design criteria.

PROJECT DESCRIPTION :
****
The design philosophy is based on a Master Vehicle (MV) with a certain number of
work Modules (WM). The different modules of the system are designed with the
scope to follow a defined procedure in each configurations of :
- First installation, - Normal maintenance, - Repair activity, - Work over
activity,
- Desassembling and recovery.
The MV consists of a non buyoand vehicle that provides to each WM, positioning,
power supply, control, communication and monitoring. The MV carries the main
plants used by the WM's i.e Energy Conversion Plant, Accpistic Positioning
System, Propulsion System, Central Main Computer and a small ROV for unforeseen
light jobs. The MV is fully redundant.
Each particulat operation is performed by the MV with the specific WM that has P.
T.O.
Duringt Phase 1, the engineering of the production system and the maintenance
devices has been performed.
During Phase 2 the prototype of the production system and the maintenance
vehicle wil be manufactured.
Phase 2 is split in two stages. During Stage A qualified manufacturers will be
selected and the construction engineering will be performed. During Stage B the
subsystems will be manufactured. This contract is relevant to Phase 2 - Stage A.
Integration of the systems, dry tests, installation on a live well and long term
tests are the scope of Phase 3.

STATE OF ADVANCEMENT :
****
Phase 2 - Stage A started in May 1987. Construction engineering of the
prototypes is finished.
Most constructions orders for modules have been limitted. Christmas tree
flowline frame and jumpers have been ordered.

```
******************************************************************************
* TITLE : DEVELOPMENT AND MATHEMATICAL ANALYSIS        *      PROJECT NO      *
*          OF HIGH PERFORMANCE RESERVOIR               *                      *
*          SIMULATORS BASED ON ADVANCED MODELS         *   TH./03202/86/ES/.. *
*                                                      *                      *
******************************************************************************
* CONTRACTOR :                                         *     TELEPHONE NO     *
*     REPSOL EXPLORACION S.A.                          *                      *
*                                                      *     1 274.72.00      *
*     PEZ VOLADOR, 2                                   *                      *
*     ES - 28007 MADRID                                *                      *
*                                                      *       TELEX NO       *
*                                                      *                      *
* TECHNICAL DIRECTOR :                                 *       49544          *
*     MR. L. PEREZ MANZANERA                           *                      *
*                                                      *                      *
******************************************************************************
                                                     VERSION : 01/06/89
```

AIM OF THE PROJECT :
****
The main objective of this project is to improve the present technology in oil
reservoirs numerical simulation, measured in terms of computer efficiency,
investigating the behaviour of various numerical models for multicomponent flow
equations in porous media.

PROJECT DESCRIPTION :
****
The project will develop advanced numerical simulation prototypes based on a
compositional model of the state equation (Peng-Robinson).
These prototypes will include the numerical models developed as follows :
- Investigation of time approximation methods, improving the adaptative implicit
type.
- Investigation of space approximation methods. Three methods will be
investigated :
  . Standard finite element method.
  . Mixed finite element method with method of characteristics.
  . Particle method.
Three main phases can be differenciated during the project development :
Phase 1. Numerical models development.
1.1 Analysis and development of numerical models which will be further
implemented.
1.2 Models behaviour analysis using mathematical tools.
Phase 2. Software development.
2.1 Algorithm design.
2.2 Software design.
2.3 Software implementation.
2.4 Software tests and validation.
Phase 3. Models tests.
3.1 Time approximation.
3.2 Space approximation.

STATE OF ADVANCEMENT :
****
Ongoing (phases1, 2 and 3).

RESULTS :
****
Relative to the investigation of time approximation methods, a new fully
implicit and adaptive implicit formulations for compositional models are
presented.

REFERENCES :
****
1.AZIZ, K., SETTARI, A.
PETROLEUM SI+MULATION". APPLIED SCIENCES PUBLISHERS LTD LONDON 1979.
2. BERTIHGER, SPE NR 13501 PRESENTED AT THE EIGH SPE SYMPOSIUM ON RESERVOIR
SIMULATION HELD IN DALLAS, TEXAS1985.
5. CARRILLO, J."A CONVERGENT SCHEME FOR THE EVOLUTION DAM PROBLE
. PREPRINT DPTO. MATEMATICA APLICADA UCM. 1987.
4. COLLINS, D.A., NGHIEM, L.X. AND LI, Y.K. "AN EFFICIENT APPROAACH TO ADAPTIVE-
IMPLICIT COMPOSITIONAL SIMULATION WITH AN EQUATION OF STATE". SPE NR 15133
PRESENTED AT THE 56TH CALIFORNIA REGIONAL MEETING OF THE SOCIETY OF PETROLEUM
ENGINEERS HELD IN OAKLAND C.A. 1986.
5. CHAVENT, G., JEROME, J. "MATHEMATICAL MODELS AND FINITE ELEMENTS FOR
RESERVOIR SIMULATION".

```
******************************************************************************
* TITLE : INVESTIGATION OF THE DYNAMIC              *      PROJECT NO        *
*         PERFORMANCE OF FLEXIBLE RISERS            *                        *
*                                                   *    TH./03213/86/UK/..  *
*                                                   *                        *
******************************************************************************
* CONTRACTOR :                                      *    TELEPHONE NO        *
*    BHRA THE FLUID ENGINEERING CENTRE              *                        *
*                                                   *    0234 750422         *
*    CRANFIELD                                      *                        *
*    UK - BEDFORD MK43 OAJ                          *                        *
*                                                   *    TELEX NO            *
*                                                   *                        *
* TECHNICAL DIRECTOR :                              *    825 059             *
*    MR S. LAMB/MR G. JONES                         *                        *
*                                                   *                        *
******************************************************************************
                                              VERSION : 01/08/89
```

AIM OF THE PROJECT :
****
The aim of the project is to undertake a closely controlled parametric
experimental study of the dynamic response of flexible pipes due to current and
slug flow. The information obtained will help form the basis of an improved
understanding off the behaviour of this type of pipe technology.

PROJECT DESCRIPTION :
****
The project is in two stages each taking one year to complete.
Stage 1 will investigate the vortex induced response of flexible pipes including
influence of bending stiffness and mass. Both modelled and commercially
available pipe will be tested for a range of configurations and conditions. The
influence of imposed top motions will also be investigated.
Stage 2 will investigate the effect of internal slug flow up both the modelled
and commercially available pipe. Changes in slug flow length and frequency will
be made to examine if flexible risers will/be susceptible to transient or
resonant motions.

STATE OF ADVANCEMENT :
****
Completed.

RESULTS :
****
Results showed that both the models and the commercial pipes suffer sustained
vortex induced vibrations, the amplitude of which is dependent on the amount of
structual damping. Vibrations are predominantly cross-flow. Internal slug flow
is of secondary importance and for the configuration tested caused motion in the
in-line direction only. Part way through the project the work on top-end motions
was abandoned in favour of work on tension effects and the effect of structural
mass and slenderness ratio. The experimental work conducted on these topics
showed that tension influenced the amplitude of vibration and a combination of
structural mass and slenderness ratio influenced the onset of vibration.

```
*******************************************************************************
* TITLE : SYSTEM FOR AUTOMATIC CONTROL &        *        PROJECT NO          *
*         OPTIMIZATION OF FACILITIES FOR        *                            *
*         TRANSPORT & PROCESSING OF HYDROCARBON *    TH./03216/86/DK/..      *
*         PRODUCTS                              *                            *
*******************************************************************************
* CONTRACTOR :                                  *    TELEPHONE NO            *
*     LICCONSULT A/S                            *                            *
*                                               *    45 1 132703             *
*     KOMPAGNISTRAEDE 22                        *                            *
*     DK-1208 COPENHAGEN K.                     *                            *
*                                               *    TELEX NO                *
*                                               *                            *
* TECHNICAL DIRECTOR :                          *    16505                   *
*     HENNING EGAA JENSEN                        *                            *
*                                               *                            *
*******************************************************************************
                                                        VERSION : 29/03/89
```

AIM OF THE PROJECT :
****
The operation of facilities for transportation and processing of hydrocarbon products is complicated, costly and implies potential risks to man and environment. Thus safe and cost effective operation of these facilities is of key importance for the feasibility of a hydrocarbon production scheme. The goal is to develop and test a real-time computer based system which will improve the techniques for supervision, control, operation and optimization for transportation and processing of hydrocarbon products.
The key innovative aspect in the project is the development and application of an integrated dynamic simulator of the entire process and transportation facilities.

PROJECT DESCRIPTION :
****
The central element in the system is the application of a dynamic simulator as basis for the supervision and control of the operations. The simulator can run in on-line mode as well as off-line mode. In on-line mode, the simulator will be driven by real time process data and system state data. In this mode the simulator forms basis for optimization of the operation by calculating current optimum values of set points for the local controls. This is based on real time calculations of the actual fluid dynamic and process dynamic conditions throughout the entire transportation/processing facility. Further, the simulator provides for continuous leak detection and location of leaks, as well as continuous supervision of instruments, equipment and critical points in the facility.
In off-line look-ahead mode, the simulator is used for the prediction of consequences of alternative operating strategies. Specifications of the alternative future operating strategies are entered into the simulator by the operator. The simulator then predicts the consequential conditions throughout the facility. In off-line training mode, the simulator will normally work based on a fictive set of initial conditions stored on a dedicated data base.
The total system will comprise the following main modules:
- Communication Interface
- Data Preprocessor
- Simulator
- Applications: handling the results of simulation and providing the dedicated user functions such as planning of operating strategies, optimization, monitoring of "Critical Points", leak detection etc.
The project will comprise the following main activities in five phases:
- Preparation of System Requirement Specifications.
- Development and testing of element models.
- Design, development and testing of flexible program structure.
- Development and testing of communication interface module, data preprocessing module and application modules.
- Set-up and demonstration of prototype systems for typical transportation/processing facilities.

STATE OF ADVANCEMENT :
****
Ongoing : preparation of System Description, System Requirements Specification and Software Design and Test Description. Definition of plant configuration, process flow and process control for sample applications. Implementation of basis program modules.
Completed : prototype pictures with simulator driven dynamic updating.

RESULTS :
****
The project is at present in the Design and Development phase and so far the following results have been obtained: Implementation in electronic data processing system of physically based mathematical models of the following types: Proportional, Integral and Derivative Regulator elements, controlled valve elements, pipe and pipe node elements, operating optimizer element for heat exchange.
Major parts of the basic structure of a flexible integrated simulator have been

178

designed and is subject to prototype testing.
Basic Design of a graphical based man/machine interface for parameter oriented
configuration of simulator and application modules has been performed.

REFERENCES :
****
"COMPUTER SYSTEM FOR SIMULATION AND ON-LINE OPTIMIZATION OF THE OPERATION OF A
LARGE DISTRICT HEATING TRANSMISSION SYSTEM", PROPER PRESENTED ON THE 23RD
UNICHAL CONGRESS, BERLIN, JUNE 1987, AND DESCRIBING A NUMBER OF BASIC ELEMENTS
INVOLVED IN THE PRESENT SYSTEM.
LECTURES:APPLICATION OF INTELLIGENT SUPERVISORY SYSTEM IN MOBILE DRILLING RIGS,
TECHNICAL CONFERENCE SPONSORED BY THE DANISH MINISTRY OF ENERGY, DEPARTMENT OF
TECHNOLOGY, 1987 COPENHAGEN.
NOVEMBER 1987 - PRESENTATION OF THE PROJECT ON ANNUAL NATIONAL TECHNICAL
CONFERENCE ON ENVIRONMENTAL DEVELOPMENT, HERNING, DENMARK.

```
**********************************************************************
* TITLE : GA-SP PROJECT                          *    PROJECT NO     *
*                                                *                   *
*                                                *  TH./03217/86/UK/..*
*                                                *                   *
**********************************************************************
* CONTRACTOR :                                   *   TELEPHONE NO    *
*    GOODFELLOW ASSOCIATES LTD                   *                   *
*                                                *   01 821 1377     *
*    71 ECCLESTON SQUARE                         *                   *
*    UK - LONDON SW1V 1PJ                        *                   *
*                                                *   TELEX NO        *
*                                                *                   *
* TECHNICAL DIRECTOR :                           *   929272          *
*    RICHARD LAWRENCE                            *                   *
*                                                *                   *
**********************************************************************
                                          VERSION : 06/02/90
```

AIM OF THE PROJECT :
****
The aim of the GA-SP Project is to develop and test full scale, a subsea
processing system that will facilitate economic development of small hydrocarbon
reservoirs. The system will include well stream commingling, multiphase
separation and pumping, and will be engineered to be suitable for a wide range
of reservoir fluid properties, shallow or deep water application, with control
from the surface from either a low cost in-field floating platform or from an
existing remote fixed installation.
Innovative aspects include :
a) Subsea closed loop modulating control of process parameters
b) Hybrid diverless maintenance philosophy with vertical retrieval of insert
components or complete modules.
c) Single datum alignment of multi-port interconnections.

PROJECT DESCRIPTION :
****
The GA-SP Project, which is also supported by a number of oil companies, is
being executed in stages. This first stage, Front End Engineering and
Preliminary Design, is concerned with outline definition of field scenario
concepts, based on a study of market requirements. From the 6 field scenario
concepts, a total capability configuration is adopted as a base case, including
the following functions :
1) Chocking and commingling a total flow of 30.000 b/d from 4 wells
2) Multiphase wellstream metering (as an alternative to well test)
3) 2 stage 3 phase separation of oil gas and produced water in two parallel
trains.
4) Oil export pumping to 140 bar for direct export including fiscal metering and
line pigging capability
5) Produced water pumping to surface facility for clean up and disposal
6) Gas export to surface facility including pressure boosting and line pigging
capability
7) Power distribution via high pressure water circuit
8) Modules dimensioned for running and retrieval from semisubmersible drilling
rig vessel where possible
9) Use of novel multibore connector for single datum module interconnection.

STATE OF ADVANCEMENT :
****
Completed

RESULTS :
****
System characteristics are the following :
- the subsea production and process system is housed within a compact support
structure in modularised form.
- The system consists of two identical process trains, each with a capacity to
handle 15,000 BOPD (2385 m3/D).
- The product from each pair of wells is manifolded and, via a header, enters
one of the process streams. The manifold system also includes the following :
. A bypass line and a pair of valves to divert flow from each well into the test
header and the multiphase flow meter for production testing of individual wells.
. A choke and manifold system for gas lift to enable the regulation and
distribution of gas into the four lines leading to the production wells.
. A choke and manifold system for regulation and distribution of water into the
four water injection wells.
. The chemical injection system for serving the production wells, the separators,
the gas transport line and the oil export line.
- The process system consists of two, three-phase, separators operating at
different pressures.
- The produced water is transported via a booster pump and a line to the main
platform.
- The produced low pressure and high pressure gas are commingled and are
transported via a single line to the platform.
- The oil booster pump(s) boost the pressure of the produced oil to 1500 Psia

(103 bars) for export into an existing trunkline.
- A subsea pig launching system for the gas and oil export line.
- A single phase metering system and associated sampling unit for fiscal metering.
Electric and hydraulic power pack to provide power for the subsea hydraulic power system to drive the pumps and operate subsea valves.
- The multiplex control system for monitoring and controlling the operation of the entire facilities.
Key components of the system, including all the valves are housed within the primary modules which are retrievable. The secondary module containing only the distribution piping can also be retrieved under this system but the option exists for keeping them as permanent, non-retrievable modules.

```
*****************************************************************************
* TITLE : IMPROVEMENT AND TEST OF A COMPACT    *       PROJECT NO         *
*         SEPARATOR ON AN OFFSHORE PLATFORM    *                          *
*                                              *    TH./03219/86/FR/..    *
*                                              *                          *
*****************************************************************************
* CONTRACTOR :                                 *      TELEPHONE NO         *
*    BERTIN & CIE                              *                          *
*                                              *    (33.1)34.81.85.42     *
*    BP 3                                      *                          *
*    FR - 78373 PLAISIR CEDEX                  *                          *
*                                              *      TELEX NO            *
*                                              *                          *
* TECHNICAL DIRECTOR :                         *      696231             *
*    MR MARCHAND                               *                          *
*                                              *                          *
*****************************************************************************
```

                                                    VERSION : 10/08/89

AIM OF THE PROJECT :
****
Improvement and test of a new compact vertical three phased separator for
offshore oil production reducing weight and saving floor area on top-sides.

PROJECT DESCRIPTION :
****
- Inspection of the state of the 15/25 000 bbl/d compact separator on the OBAGI
site (ELF-Nigeria)
- Definition and implementation of several modifications in order to reduce
liquid carry over in the gas and improve oil/water separation.
- Test of the modified separator on the OBAGI site
- Transport and installation of the separator on an offshore production
platform
- Long duration testing of the separator.

STATE OF ADVANCEMENT :
****
The first three phases are complted, and phase 4 is in progress.
The tests of the separator have been performed on the oil field of ELF-NIGERIA
at OBAGI.
Modifications of the equipment for long duration testing have been defined.
Testing site should be ELF-NIGERIA OLO field. Comparative studies of equivalent
conventional separator have been performed (direct costs, size, weight...).

RESULTS :
****
AS TO PERFORMANCES :
Performances were satisfactory regarding oil or gas treatment capacity (25 000
BLPD) and separation performance.
AS TO THE BEHAVIOUR OF THE SEPARATOR :
- The separator very well accepted the instantaneous flow excursions of the oil
well : the flow varied from 2 000 to 9 000 m3/hr with an average at 4 000.
- There was no incident or shut-down of the oil field due to the separator (for
instance shut-down of the gas compressors due to liquid droplets in the gas).
-The examination of the internal equipments of the separator after the test
campaign does not reveal any damages.
AS TO COMPARISON WITH CONVENTIONAL SEPARATOR :
The compact separator is about 3 times smaller than usual separator (and design
deckload 3 times smaller). Direct manufacturing cost is at most the same.
CONCLUSION :
The performances are in good agreement with what was expected from the basic
design of the separator. The separator will now be submitted to a programme of
long duration on-site testing with a view to assessing its reliability and its
ability to conform to actual oil field conditions. After revoew with ELF-
AQUITAINE, the onshore field of OLO in NIGERIAL is contempleted for testing of
the prototype under operational conditions.

```
************************************************************************
* TITLE : FLEXTECH : PROGRAM FOR ANALYSIS AND         *    PROJECT NO     *
*         DESIGN OF FLEXIBLE PIPES AND RISERS.        *                   *
*                                                     *  TH./03224/86/IR/..  *
*                                                     *                   *
************************************************************************
* CONTRACTOR :                                        *  TELEPHONE NO     *
*    MCS INTERNATIONAL                                *                   *
*                                                     *  353 91 66455     *
*    3 BUTTERMILK WALK                                *                   *
*    IR - GALWAY                                      *                   *
*                                                     *  TELEX NO         *
*                                                     *                   *
* TECHNICAL DIRECTOR :                                *  50094            *
*    MR. J.F. MCNAMARA                                *                   *
*                                                     *                   *
************************************************************************
                                              VERSION : 31/10/89
```

AIM OF THE PROJECT :
****
The overall aim of the project is the development of an integrated software
package which will allow users to model all aspects of the complex behaviour of
flexible pipeline risers under ambient offshore loading conditions. A major
innovation is that fully three-dimensional finite motions and rotations of the
flexible pipeline will be available and, also, that these global values may be
used in a local analysis for the stresses and strains in a layered pipeline
section.

PROJECT DESCRIPTION :
****
The basic mathematical technique used in FLEXTECH for modelling in flexible
riser response is the finite element method. Stiffness, mass and finite rigid
body rotation terms are assembled in a convective coordinate system that follows
the moving pipeline. All possible load terms including current, waves, ship
motions, gravity and seabed friction are incorporated and the varying seabed
touchdown point is monitored. A very efficient numerical solution method in time
is available whereby the optimum size of the time increment is automatically
selected by the program for the user; a linear frequency domain solution about a
static nonlinear position is incorporated. The mechanical behaviour of bonded
and unbonded flexible pipeline wall constructions may also be analysed and
failure states of the armouring or bonding layers predicted. Data entry and
interpretation of the generated results is facilitated by purpose-written pre-
and post-processors. The accuracy and utility of FLEXTECH has been validated by
numerous test cases covering all the major features described above.

STATE OF ADVANCEMENT :
****
All phases of the FLEXTECH project as originally specified in the proposal
document have been completed.

RESULTS :
****
A generalised suite of finite element programs with integrated pre-and post-
processing facilities for the analysis of offshore flexible riser production
systems has been developed by MCS International. Extensive testing and
validation of the FLEXTECH package has been successfully completed for an number
of realistic offshore flexible riser configurations. MCS have contributed to a
number of submissions whereby competing flexible riser packages have addressed a
set of test cases selected by interested industrial companies and these
comparisons will be published shortly. Preliminary indications are that the
FLEXTECH results are in good agreement with those produced by alternative
products. It is concluded that the development of the FLEXTECH package as a
comprehensive flexible riser analysis suite has been successful in achieving its
specified technical objectives. Commercial success wil be strongly influenced by
the ability of MCS to overcome the intense competition from other well-
positioned international engineering companies.

REFERENCES :
****
O'BRIEN, P J AND MC NAMARA J F, "ANALYSIS OF FLEXIBLE RISER SYSTEMS SUBJECT TO
THREE-DIMENSIONAL SEASTATE LOADINGS", PROCEEDINGS OF 5TH INTERNATIONAL
CONFERENCE ON BEHAVIOUR OF OFFSHORE STRUCTURES, (BOSS '88), TRONDHEIM, NORWAY,
JUNE 1988.
O'BRIEN, P J AND MC NAMARA J F, "SIGNIFICANT CHARACTERISTICS OF THREE
DIMENSIONAL FLEXIBLE RISER ANALYSIS", ENGINEERING STRUCTURES, VOL.11, OCTOBER
1989, PP 223-233.
MC NAMARA J F AND HARTE A M, "THREE DIMENSIONAL ANALYTICAL SIMULATION OF
FLEXIBLE PIPE WALL STRUCTURE". PROCEEDINGS OF THE 8TH ANNUAL OFFSHORE MECHANICS
AND ARCTIC ENGINEERING, VOL 1, MARCH 1989, PP 477-482.

```
*****************************************************************************
* TITLE : FLEXIBLE PRODUCTION BUNDLE          *        PROJECT NO         *
*                                             *                           *
*                                             *    TH./03225/87/FR/..     *
*                                             *                           *
*****************************************************************************
* CONTRACTOR :                                *       TELEPHONE NO        *
*    GERTH                                    *                           *
*                                             *    (1) 47.52.61.39        *
*    AVENUE DE BOIS PREAU 4                   *                           *
*    FR - 92502 RUEIL MALMAISON               *                           *
*                                             *      TELEX NO             *
*                                             *                           *
* TECHNICAL DIRECTOR :                        *    203050 FAX : (1) 47.*
*    MR. L. LEGALLAIS                          *    52.69.27                *
*                                             *                           *
*****************************************************************************
                                               VERSION : 31/12/89
```

AIM OF THE PROJECT :
****
The aim of the project is to group together in a single pipe both the service
line and the electrohydraulic umbilical for Formerun I.S.U. (Integrated Service
Umbilical) providing service and remote control and monitoring functions for a
wellhead. Alternatively, the aim is to group together the production line,
service line and electrohydraulic umbilical forming a SWM (Single Well Multibore)
 providing production line, service line and remote control and monitoring
functions for a wellhead.
The original features of this project lies in the placing of the various
hydraulic and electric lines around a central COFLEXIP core and protecting then
with a packing device with enables all stresses to be transferred to the central
core. They must have a high compressive, shear and bending strengh; be as light
as possible and withstand temperatures from -20 deg. C to +50 deg. C, be sea
water and sea organism resistant.
The second essential point is that simultaneous connection of all elements and

PROJECT DESCRIPTION :
****
2.1 BASIC STUDIES
Selection of base elements (flexible pipes, electric cables and packing
materials), theoretical study of the materials that can be used for packing,
stainless steel pipe fatigue when assembled around a COFLEXIP core, the
behaviour of these kind of flexible pipes (I.S.U. and S.W.M).
2.2 FEASIBILITY STUDIES
Adapting material means required to produce samples. This covers the following :
- tools required to make packing filler (extrusion heads and tooling, moulds,
etc...)
- tools required to make samples (mandrel, stiffening tube,...)
- tools required for ensheating the samples,
- adapting COFLEXIP test means to carry out preliminary tests (coefficient of
friction measurement, compression between plates and bending moment measurement).
MANUFACTURING TECHNIQUES are seven, each with a length of between 5 and 7 metres.
- Number 2. ISU type A : the peripheral elements are separated by packing in the
form of an extruded shaped bar. We have made one ISU with stainless steel
elements and another with flexible elements. With the latter we had filler heads
with 2 types of materials of different hardness (EPDM and EP048).
- Number 1. ISU type B : the stainless steel peripheral elements are laid on a
grooved band and then covered with a form strip.
- Number 2. ISU type C : one has stainless steel peripheral elements and the
other flexible elements. The elements are buried in a flexible matrix in the
form of a flat pack which is then spiral wound round a core.
- Number 1. ISU type A : the flexible peripheral elements are separated by
packing in the form of a low density, polyethylene hollow extruded shaped piece.
- Number 1. ISU type A : the flexible peripheral elements are separated by
packing in the form of low density, polyethylene, solid extruded shaped pieces.
LAYING FEASIBILITY
Traction tests with a caterpillar (determination of the coefficient of friction),
 collapse tests between plates and measurement of the bending moment enabled us
to make a selection between the various solutions as a function of different
laying and storage conditions.
2.3 ADAPTING THE MEANS TO CARRY OUT THE FEASIBILITY TESTS
Production of prototypes for dynamic and statistical tests. This basically
involves adpating the means for :
- extrusion and sheating in the production of filler and sheaths,
- adaptation of the industrial assembling machine,
- tests to measure the precise characteristics of the prototypes,
- a test bench for dynamic testing of the two prototypes selected subject to
3300000 bending cycles in a single plane.
2.4 PROTOTYPE PRODUCTION
Two prototypes were produced. The peripheral elements are in stainless steel in
one instance and hose in the other. They are both type A with solid, low density
polyethylene filler. They were fitted with end pieces grouping together all the
terminals.
2.5 STATIC AND DYNAMIC TESTS

Static tests to determine the characteristics of the prototype :
- collapse test between plates,
- caterpillar traction (coefficient of

STATE OF ADVANCEMENT :
****
Completed. All the dynamic and static tests and the studies listed in the programme were performed.

RESULTS :
****
- The behaviour of the peripheral hoses was satisfactory during the two phases of dynamic tests.
- The behaviour of the peripheral hoses made in stainless steel tube was satisfactory during the two phases of dynamic tests.
- The theoretical determination method of the stress rate on stainless steel tubes is satisfactory. This method is checked by the stress measurement in tubes with strain gauges.
- Electrical cables : In both prototypes, omly the armoured electrical cable has kept its characteristics. So, to be reliable the electric cables layed up in ISU must have a very high axial stiffness. This condition involves the use of armoured electical cables having a very low armouring angle < 15 deg.

REFERENCES :
****
A PART OF THIS PROJECT HAS BEEN PRESENTED DURING THE OFFSHORE TECHNOLOGY CONFERENCE - HOUSTON 1987 - REF. OTC 5470.

```
******************************************************************************
* TITLE : NEW TREATMENT PROCESS FOR OIL AND WATER    *      PROJECT NO        *
*         EMULSION ON OFFSHORE PLATFORMS (PHASE      *                        *
*         II)                                        *    TH./03228/87/FR/..  *
*                                                    *                        *
******************************************************************************
* CONTRACTOR :                                       *      TELEPHONE NO      *
*    GERTH (MAIN CONTRACTOR)                         *                        *
*                                                    *    (1) 47 52 61 39     *
*    4 AVENUE DE BOIS PREAU                          *                        *
*    92502 RUEIL MALMAISON - FRANCE                  *                        *
*                                                    *      TELEX NO          *
*                                                    *                        *
* TECHNICAL DIRECTOR :                               *      203050            *
*    MR. SCHRANZ                                     *                        *
*                                                    *                        *
******************************************************************************
                                                          VERSION : 30/06/89
```

AIM OF THE PROJECT :
****
A critical examination of the processes on which the treatment equipment are
based, shows that considerable savings could be made regarding weight, bulk and
performances on offshore platforms, by using a new technique for treating fine
oil/water/gas emulsions.
The results obtained within the scope of the project nr. TH 03.165/84 make the
application of a new settling process by cyclone flow a serious possibility for
treatment units such as crude dehydration, and produced water deoiling.
The project consists in:
- developing a fast-flow electrostatic coalescer, capable of handling high water
contents, coupled with a cyclone dehydrator.
- optimizing the deoiling cyclone design, in particular regarding oil recovery
and energy consumption aspects.

PROJECT DESCRIPTION :
****
a) Electrostatic dehydrator
The work carried out within the scope of the project nr. TH 03.165/84 revealed
that it was possible to reduce weights by a factor of five, by developing a
"contactor" is vertical, meaning a considerable reduction of the surface area
required. The particular design of the electrodes makes it possible to treat
water-in-oil emulsions, having water contents of as much as 50% (5 to 10% is the
maximum for traditional electrostatic dehydrators). Preliminary work showed that
it should be possible to reduce weight and bulk yet further, by separating the
"coalescence" and "settling" functions. The settling function being operated in
a specialized unit (cyclone dehyrator type) placed downstream the contactor,
the water droplets coalesced in the contactor can be driven with the oil; it is
possible to circulate faster in the apparatus.
b) Cyclone deoiler
The results obtained within the scope ot the EEC contract TH 03.165/84 led to
the design of a new type of cyclone known as a "rotary" cyclone. Improvements
both regarding separation efficiency and flow rate flexibility was proved in the
field on a 50 m3/hour prototype.
The rotation of the walls of the unit eleminates friction between the fluid and
the wall, and means that the cyclone, has hydraulic properties very similar to
those of a theoretical cyclone, which explains its level of performance
regarding deoiling.
Further work will consist in optimizing the many geometrical characteristics of
fluid inlets and outlets, which govern the hydraulic behaviour ot the deoiling
cyclone. In order to be able to operature under nearly any field conditions, the
characteristics of the cyclone should be such, that it would not be necessary to
install a pumping unit upstream the water treatment system.

STATE OF ADVANCEMENT :
****
a)Electrostatic dehydrator
An accelerated 5m3/h mockq-up coalescer has been studied and constructed. The
laboratory tests are encouraging. Due to these promising results, we were led to
carry out a theoretical and experimental study in order to make up the sizing
rules of an industrial coalescer.
At the same time, a dehydrator cyclone model, with a 2 to 5 m3:h flowrate was
constructed. Evaluation tests on loops are underway.

RESULTS :
****
a) Electrostatic dehydrator
The feasibility tests, run on the electrostatic coalescing mock-up show that
careful attention has to be given in the choice of the insulating sheath of the
central electrode in order to avoid short circuits. It is capable of converting
at appropriate voltage and frequency, an emulsion with a given size distribution
into a much coarser emulsion. The suitable residence time is very low : 1 to 10
seconds. At constant residence time, a flowrate increase could have a favourable
effect on water droplet coalescence.
Further work involved to establish the theoretical laws taking into account

186

electric or hydraulic effects on the electrocoalescence rate. To check the laws
validity, a complementary test program was completed.
Alternatively a dehydrator cyclone mock-up has been built, capable of a
throughput of 2 to 5 m3/h. At first different geometrical outlet configurations
were tested on a small loop. After this feasibility study, the dehydrator
cyclone mock-up has been installed on the crude oil test loop (in the place of
the electrostatic coalescing unit). At a 1 m3/h rate, dehydration efficiency
values of 85 to 95 % were obtained. These experimental results are in good
accordance with the theoretical value. However, at rate higher than 3 m3/h, the
dehydration efficiency becomes lower than the theoretical values. So further
works are in progress to optimize the dehydrator cyclone efficiency. Tests were
also carried out on the combination of coalescer and dehydrating cyclone mounted
in series. The first results are promising but they have to be confirmed and
completed by further experiments.
b) Cyclone deoiler
Several solutions for reducing pressure losses were found. For each industrial
application, it would now be possible to select the best efficiency - pressure
losses arrangement.
The industrial cyclones will be sealed with mechanical packing seals. Previously,
we wish to check them on the basis of a specification allowing for the special
and severe offshore operating conditions. The mechanical endurance tests were
run with success on the 50 m3/h prototype installed for this purpose at an
oilfield in the Paris region.

REFERENCES :
****
THREE PAPERS WERE PREPARED AND RESPECTIVELY PRESENTED AT :
- OFFSHORE EUROPE 87 IN ABERDEEN (GB), 11 SEPTEMBER 1987
   "NEW DEVELOPMENT IN HYDROCYCLONES"
- RECENT INTERNATIONAL TECHNOLOGICAL DVELOPMENTS CONFERENCE IN LONDON (GB), 3-4
NOVEMBER 1988
   "DYNACLEAN - A STEP FORWARD IN PRODUCED WATER TREATMENT"
- BHRA - 4TH INTERNATIONAL CONFERENCE ON MULTI PHASE FLOW IN NICE (F), 19-21
JUNE 1989 : "A NEW HIGH EFFICIENCY LIQUID:LIQUID SEPARATOR".

```
****************************************************************************
* TITLE : PLATINE II                          *      PROJECT NO        *
*                                             *                        *
*                                             *   TH./03229/87/FR/..   *
*                                             *                        *
****************************************************************************
* CONTRACTOR :                                *      TELEPHONE NO       *
*    GERTH                                    *                         *
*                                             *   (1)47.52.65.88        *
*    AVENUE DU BOIS PREAU 4-F 95502 RUEIL MALMAISON  *                  *
*                                             *                         *
*                                             *      TELEX NO           *
*                                             *                         *
* TECHNICAL DIRECTOR :                        *      203050             *
*    MR. PIERRE LEFEBVRE                       *                         *
*                                             *                         *
****************************************************************************
                                                  VERSION : 07/11/89
```

AIM OF THE PROJECT :
****
The aim of platine project is to specify a central production platform, unmanned, teleoperated from a control room located onshore, with a yearly (or half-yearly) visit for inspection and overhauling.
The intermediate aims consist in working out and testing pilot units related to sub-systems and testing new equipments.
The project has been divided into 3 phases : platine-Phase I (1986-88); Platine-Phase II (1987-89); Platine-Phase III (1988-90).
The present project deals with Phase II.

PROJECT DESCRIPTION :
****
This project aims at progressively integrating the results of studies performed in Platine Phase I project, The facilities have been divided into 7 main sub-systems : 4 production sub-systems (gas-lift, sea water injection, separation-treatment, well pumping) and 3 sub-systems non specific to oil industry (energy/rotating machinery, safety/reliability, automation systems).
This project is related to all these sub-systems except separation-treatment.
Gas lift : Two prototypes will be studied and developed, one being related to automatic start-up(in case of unique injection point gas-lift) and the other related to optimization of gas-lift injection.
Well pumping : A reliability study of the existing long stroke pumping will be carried out; then, modification of elements will be proposed and tests performed.
Water injection : Construction and tests of prototypes (or test of existing prototypes) related to specific process (chlorination, filtration, deoxygenation, sterilization) and analyzers (oxygen, chlorine, hydrogen, bacteria detection, filtration quality will be carried out). A preliminary engineering study of the whole sub-system will be performed.
Rotating machinery : Actions will be defined and undertaken in order to make existing machines more reliable. Use of new technologies (peripheral compressor, magnetic bearings, dry seals, variable high speed engine) will be assessed in connection with manufacturers.
Expert system : An off-line expert system related to diagnosis and control of crude oil desalting facilities will be developed.
Optics : Two prototype multiplexing (in time or in wave lenght) optical transmission systems will be tested. A bottom hole optical sensor will be studied : feasibility study, then if possible, prototype study.
Reliability : Reliability studies related to specific equipment, whole systems and software will be carried out.
Well safety : An electronic control of well safety devices will be studied, A prototype will be developed.

STATE OF ADVANCEMENT :
****
For water injection sub-system, new prototype processes have been developed and tested. For other sub-systems, prototypes are being studied. An off-line expert system has been developed.

RESULTS :
****
Gas-lift : The unique injection point technique has been selected. The automatic start-up has been studied; the operation sequence has been defined and equipment for prototype has been selected, including a failure tolerance automate.
The gas-lift optimization has been studied. It will be carried out in testing the wells needed, i.e. when the well behaviour has changed. This well behaviour will be assessed by a permanent bottom hole pressure sensor or, in a degraded sensor or, in a degraded mode, by permanent knowledge of well head temperature and/or well head casing pressure.
Well pumping : the long stroke pumping has been selected as being the most complementary means of gas-lift and the most suitable for platine. A reliability study of existing pumps has been carried out. Modifications of some elements have been proposed. Testing of new elements or modified existing elements will start in September 1988.
Water injection : specific sea water treatment processes (existing or new) have

been assessed and selected in platine phase I project. They are related to filtration, deoxygenation, chlorination and bacterial treatment. Prototypes have been tested in 87 in Toulon. Good results were obtained, especially with sqns filtration, catalytic deoxygenation and ultraviolet bacterial treatment. Besides, some analyzers have been tested and a robot-aided bacteria detector has been developed (and the preliminary testresults - Toulon - are satisfactory). A preliminary engineering study of the whole sub-system has been carried out.

Rotating machinery : Further to the general inquiry done in platine phase I project, detailed studies have been carried out on selected machines, considered as the most suitable to platine.

Use of new technologies (peripheral compressor, magnetic bearings, dry seals, variable high speed engine) has been assessed; Some experience in various industries have been evaluated.

Expert system : An off-line expert system related to diagnosis and control of crude oil desalting facilities has been developed. It has been assessed by several production experts and will be tested on site in the third quarter of 1988.

Optics : Two prototypes multiplexing (in time or in wavelength) optical transmission systems have been purchased and delivered. One has been installed on a production site for testing, the other will be tested in laboratory. A bottom hole optical sensor has been studied; A feasibility study and a prototype study have been carried out.

Reliability : Reliability studies have been performed on specific equipment (electrochlorinator, long stroke pump, automate, electrical equipment) and on systems (control, water injection, well safety). A software reliability inquiry has been carried out.

Well safety : Specification sheets of an electronic control of well safety devices have been worked out. Equipment has been selected (including a failure tolerance automate). A prototype test has been satisfactorily carried out.

```
********************************************************************************
* TITLE : GA-SP-STAGE 2                          *        PROJECT NO         *
*                                                *                           *
*                                                *    TH./03230/87/UK/..     *
*                                                *                           *
********************************************************************************
* CONTRACTOR :                                   *       TELEPHONE NO        *
*     GOODFELLOW ASSOCIATES LIMITED              *                           *
*                                                *      01-821-1377          *
*     71 ECCLESTON SQUARE                        *                           *
*     UK-LONDON SW1V 1PJ                         *                           *
*                                                *       TELEX NO            *
*                                                *                           *
* TECHNICAL DIRECTOR :                           *      929272               *
*     RICHARD LAWRENCE                           *                           *
*                                                *                           *
********************************************************************************
                                                        VERSION : 01/02/90
```

AIM OF THE PROJECT :
****
The aim of the GA-SP Project is to develop and test full scale, a subsea
processing system that will facilitate economic development of small hydrocarbon
reservoirs. The system will include well stram commingling, multiphase
separation and pumping, and will be engineered to be suitable for a wide range
of reservoir fluid properties, shallow or deep water application, with control
from the surface from either a low cost in-field floating platform or from an
existing remote fixed installation.
Innovative aspects include :a) Subsea closed loop modulating control of process
parameters.
b) Hybrid diverless maintenance philosophy with vertical retrieval of insert
components or complete modules.
c) Single datum alignment of multi-port interconnections.

PROJECT DESCRIPTION :
****
The GA-SP Project, which is also supported by a number of oil companies, is
being executed in stages. The first stage, Front End Engineering and Preliminary
Design, was concerned with outline definition of field scenario concepts and
identification of major areas of novelty. This second stage, Detailed Design and
Engineering, is concerned with development of a full scale prototype system
which incorporates all of the key areas of novelty required to demonstrate the
technical feasibility of the concept. The major tasks of this stage of work have
included :
1) Development of Process Flow Diagrams, General Arrangement Drawing and Piping
and Instrument Diagrams;
2) Building of Scale Models to show module details and interfaces;
3) Preparation of functional specifications for all equipment, systems,
fabrication and assembly requirements;
4) Detailed engineering including hydraulic, mechanical and structural
calculations, and preparation of drawings and schedules;
5) Specification of offshore test facility requirements including a conditioning
plant to provide a flow of simulated crude oil at 15000 b/d and 35 bar;
6) Development of a computer simulation programme, to be used as a basis for
predicting process/control response of the prototype, and subsequently, after
correlation with test results, design of commercial systems.

STATE OF ADVANCEMENT :
****
The Stage 1 Front End Engineering and Preliminary Design is complete. This has
identified 6 field scenarios for potential application of the GA-SP concept, and
defined a base case scenario for specification of equipment and system
development requirements.
Enquiries have been issued for the fabrication of PCM/SDM and Separator unit.
Third evaluation is in progress.

RESULTS :
****
During the course of this work some significant design modifications were made
to the Stage 1 concept, including :
a) Basic module size increased from 3,5 m sq x 5,5 m high to 4 m sq x 6 m high.
b) First stage separator to incorporate oil/water separations, and comprise
   3 individual vessels, fabricated from line pipe.
c) Second stage separator vessel to be included in the same module as the first
   stage separator.
The main components of the prototype system include :
a) Dummy Production Choke Module (PCM), with a single bay Secondary D
   Distribution Module (SDM) incorporating the prototype valved Multibore
   Connector (VCM)
b) Separator Module, for 2 stage 3 phase separation, including 4 types of level
   detector from 3 manufacturers, and 4 sets of modulating control valves with
   hydraulic actuators from 2 manufacturing groups.
c) Ejector skid unit, with 2 alternate ejectors for gas commingling, flow,
   pressure and temperature sensors, and (manual) recycle capabilities to the

second stage separator and/or booster pump suction.

d) Power system, comprising 3.3 kV transformer, switchgear and cable with a sea water charge pump driven by 500 H.P. 3.3 kV synchronous motor. The high pressure water drives a turbine, coupled to the oil export booster pump. A modulating control valve bypasses water to control speed of the turbine/ booster pump in response to separator level control.

e) Control and data monitoring system, comprising a topsides process controll computer, a topsides data logging computer and printer a topsides hydraulic power unit, a subsea pod for hydraulic fluid distribution and data transmission together with a variety of transducers, cables, connectors and junction boxes.

```
****************************************************************************
* TITLE : DIVERLESS SUBSEA PRODUCTION SYSTEMS          *       PROJECT NO      *
*         (DISPS) TEMPLATE SYSTEM TESTING              *                       *
*         PROGRAMME STAGES 1 AND 2.                    *   TH./03233/87/UK/..   *
*                                                      *                       *
****************************************************************************
* CONTRACTOR :                                         *    TELEPHONE NO       *
*    BP PETROLEUM DEVELOPMENT LTD                      *                       *
*                                                      *    041 204 2525        *
*    301 ST VINCENT STREET                             *                       *
*    UK-GLASGOW G2 5DD                                 *                       *
*                                                      *    TELEX NO            *
*                                                      *                       *
* TECHNICAL DIRECTOR :                                 *    777633             *
*    N. RODDA                                          *                       *
*                                                      *                       *
****************************************************************************
                                                      VERSION : 31/12/88
```

AIM OF THE PROJECT :
****
The objective is to develop the principles, concepts, designs, techniques and
equipment for a DISPS subsea seabed production template, which will permit
exploitation of hydrocarbon reserves in 350m-750m water depths utilising subsea
diverless techniques. The Testing Programme's objective is to prove the novel
principles of the DISPS template design, and verify the feasibility.

PROJECT DESCRIPTION :
****
BP's DISPS concept for a subsea diverless production template is a modular one
configured for 8 slots, in which the interconnected equipment modules are
retrieved to the surface for repair/replacement.
Under Stage 1 it is proposed to design and build a Test Rig representing a
section of the DISPS template consisting of a mock up of one bay of the template,
 including the equipment modules. The Test Rig structure size is 10m x 8m x 10m
and it weights 130 t.
Under Stage 2, land testing using the Test Rig will be carried out to verify the
DISPS template design principles which do not require an aqueous environment.
Module installation and retrieval tests, ROV intervention tests using a mock up,
hard landing tests, and module connector integrity tests will be undertaken.

STATE OF ADVANCEMENT :
****
Stage 1 complete, a test rig and associated equipment has been designed and
fabricated.
Stage2, in the land testing, is virtually successfully complete.

RESULTS :
****
The testing has provided confidence in the principles of the of the DISPS
modular designs proposed. Specifically :
- the template and modules can be fabricated to normal tolerances;
- the DISPS guidance system successfully captured, aligned and guided the
modules to their location without damage;
- measured stresses less than calculated;
- soft landing system demonstrated;
- ROV tooling operated as an integrated system;
- Components successfully operated from a cabin;

```
*****************************************************************************
* TITLE : THE STRUCTURAL INTEGRITY MONITORING OF    *      PROJECT NO       *
*          GROUT/STEEL ANNULI ON OFFSHORE           *                       *
*          INSTALLATIONS SUBSEA                     *   TH./03249/87/UK/..   *
*                                                   *                       *
*****************************************************************************
* CONTRACTOR :                                      *    TELEPHONE NO        *
*    AV TECHNOLOGY LIMITED                          *                       *
*                                                   *    061 491 2222        *
*    AVTECH HOUSE-BIRDHALL LANE-CHEADLE HEATH-STOCKPO *                     *
*                                                   *                       *
*                                                   *    TELEX NO            *
*                                                   *                       *
* TECHNICAL DIRECTOR :                              *    669028              *
*    DR. I. S. SOUTHERN                             *                       *
*                                                   *                       *
*****************************************************************************
                                            VERSION : 08/08/89
```

AIM OF THE PROJECT :
****
To develop a non-destructive measurement and analysis technique and associated
instrumentation to assess the integrity of the grouted annular connections
between the piles and legs of offshore jackets.
The proposed technique will use impulsive excitation and dynamic response
measurements to detect the presence of voids in the grout.
The technique will be developed into a product suitable for application by
divers, with associated instrumentation to provide on-site assessments of grout
integrity.

PROJECT DESCRIPTION :
****
The project will be carried out in 5 phases as follows:
Phase 1: Development of a test facility
A purpose designed test facility will be manufactured and tubular test specimens
representing a range of skirt-piled structures will be procured. A data
acquisition system and data analysis solftwqre will be developed.
A pre-prototype instrumented hammer will be manufactured to be used in Phase 2
to determine the required parameters of the subsea hammer and instrumentation
system..
Phase 2: Determination of Dynamic Behaviour of Preliminary Test Specimens
Tests in Phase 2 will be carried out using un-grouted and grouted test specimens.
 The dynamic response of the test specimens will be investigated both in and out
of water.
These tests will enable a specification for the prototype subsea hammer-
accelerometer assembly to be formed and the device can then be manufactured.
These results will influence the range of test specimens for Phase 3.
Phase 3: Development of Grout-Void Monitoring Techniaues
 Impulsive tests using the prototype subsea assembly will be conducted on a new
range of test specimens as defined from Phase 2. The specimens will model a
range of void sizes to enable the limiting resolution of the techniaue to be
quantified.
The computer software will be reviewed and optimised with a view to a final
product.
Phase 4: Offshore trials
Subject to the co-operation of the UK Oil and Gas Industries, a prototype system
will be tested on an offshore jacket under realistic operating conditions.
Phase 5: Product development
Based upon the results of the laboratory and offshore trials, the technique will
be further developed into a product to provide on-site assessments of grout
integrity to be made offshore using inspection divers.

STATE OF ADVANCEMENT :
****
Ongoing. The project is in the monitoring stage in Phase 2 with tests being
carried out on un-grouted and grouted test specimens. However the prototype
hammer is still in the design stage.

RESULTS :
****
No results have been obtained for Phase 1 as this involved the design and build
of the test facility.
Some problems have been encountered with respect to the hammer system. A
prototype impulse hammer/accelerometer unit cannot= be made until certain
parameters have been set ie. magnitude of input force, frequency range, 'shape'
of forcing function etc.
Consequently, a pre-prototype impulse hammer has been made and various tests in
Phase 2 are in progress to set the above parameters and other required
characteristics of the instrumentation and computer software system.

REFERENCES :
****
1. DEPARTMENT OF ENERGY "OFFSHORE INSTALLATIONS : GUIDANCE ON DESIGN AND
CONSTRUCTION": HMSO (1984).

2. BLEVINS, R.D. "FORMULAE FOR NATURAL FREQUENCY AND MODE SHAPE" VAN NOSTRAND, 295-318 (1979).
3. WARBURTON, G.B. "VIBRATION OF THIN CYLINDRICAL SHELLS" J. MECH. ENG. SCI, 399-407 (1965).

```
*********************************************************************************
* TITLE : WAVE FEED-FORWARD DYNAMIC POSITIONING      *     PROJECT NO        *
*         OF LARGE SHIPS                             *                       *
*                                                    *   TH./03250/87/NL/..  *
*                                                    *                       *
*********************************************************************************
* CONTRACTOR :                                       *   TELEPHONE NO        *
*    MARIN                                           *                       *
*                                                    *                       *
*    HAAGSTEEG 2                                     *                       *
*    PO BOX 28                                       *                       *
*    NL-6700 AA WAGENINGEN                           *   TELEX NO            *
*                                                    *                       *
* TECHNICAL DIRECTOR :                               *                       *
*                                                    *                       *
*                                                    *                       *
*********************************************************************************
                                                        VERSION : 04/10/89
```

AIM OF THE PROJECT :
****
MARIN HAS RECENTLY COMPLETED AN INNOVATIVE DESIGN RESEARCH PROJECT TO
INVESTIGATE THE POTENTIAL IMPROVEMENT WHICH A WAVE-FEED-FORWARD DYNAMIC
POSITIONING METHOD MAY HAVE ON THE DYNAMIC POSITIONING OF A TANKER TYPE SHIP.
IN PRESENTLY AVAILABLE DYNAMIC POSITIONING SYSTEMS THE ABOVE MENTIONED WAVE-FEED-
FORWARD CAPABILITIES ARE NOT INCORPORATED.
THE RESEARCH PROGRAM PROVIDES ESSENTIAL INFORMATION FOR THE DESIGN OF A WAVE-
FEED-FORWARD DP CONTROL SYSTEM.

PROJECT DESCRIPTION :
****
Dynamic positioning of tankers at sea is becoming a viable alternative to
mooring of floating production units or tankers by conventional means.
However, in the harsh environment of, for instance, the northern North Sea the
capability of the DP system will be taxed to the limits due to the relatively
large influence of second order wave drift forces on the vessel.
In such cases the capability of a DP system to sustain higher sea conditions can
be increased using real time information which incorporates knowledge of :
- the direction of the incoming waves with respect to the ship's heading;
- the magnitude of the wave drift forces.
This information may be obtained from real time measurement of the relative wave
elevation at the waterline of the vessel. The utilization in the DP control
system of the information on the wave direction and the drift force magnitude as
derived from the relative motion measurement is termed "wave-feed-forward".
The report of the research program for wave-feed-forward provides information on
the possibilities to obtain instantaneous information on the drift forces acting
on a tanker and on the direction of the waves relative to that tanker.
Furthermore, it provides a description of how this real time information can be
incorporated in the DP control system and the results of dynamic positioning
experiments using wave-feed-forward carried out on model scale, showing the
improvements in positioning accuracy and the effect on power consumption.
Information about the availability of the report can obtained at the Marine
Research Institute Netherlands. (Adress : P.O. Box 28, 6700 AA Wageningen, The
Netherlands).

STATE OF ADVANCEMENT :
****
After the start in April 1988, calculations to provide a data base for the
selected ship have been completed and reported in August, 1988. The results of
model measurements of the relative motions along the ship sides have been
reported in November 1988. The final model tests with working DP system have
been reported in April 1989.

RESULTS :
****
Calculations data report No. 47929-1-OE.
 Relative motion and drift force measurement report No. 47929-2-GT Summary Phase
1 report No. 47929-3-OE
Final "wave-feed-forward" DP report No. 47929-4-GT
(Distribution restricted to participants).

REFERENCES :
****
A.B. AALBERS AND U.NIENHUIS : "WAVE DIRECTION FEED-FORWARD ON BASIS OF RELATIVE
MOTION MEASUREMENTS TO IMPROVE DYNAMIC POSITIONING PERFORMANCE", OFFSHORE
TECHNOLOGY CONFERENCE, HOUSTON 1987, O.T.C. PAPER NR. 5445 (PAPER DESCRIBING A
PRE-STUDY).
A.B. AALBERS AND U. NIENHUIS : "DYNAMIC POSITIONING OF TANKERS : DESIGN ASPECTS
AND RECENT DEVELOPMENTS", 5-TH INTERNATIONAL CONFERENCE ON FLOATING PRODUCTION
SYSTEMS, LONDON, DECEMBER 1989. (PAPER REVIEWING DEVELOPMENTS FOR D.P.
APPLICATIONS).

```
****************************************************************************
* TITLE : DOWNTIME ANALYSIS FOR MARGINAL FIELD        *    PROJECT NO      *
*         PRODUCTION SYSTEMS (DAMPS)                   *                    *
*                                                      *   TH./03258/87/IR/.. *
*                                                      *                    *
****************************************************************************
* CONTRACTOR :                                         *   TELEPHONE NO     *
*    EOLAS, THE IRISH SCIENCE AND TECHNOLOGY AGENCY    *                    *
*                                                      *   01-370101        *
*    GLASNEVIN                                         *                    *
*    IR-DUBLIN 9                                       *                    *
*                                                      *   TELEX NO         *
*                                                      *                    *
* TECHNICAL DIRECTOR :                                 *   32501            *
*    MR GERARD KEANE                                   *                    *
*                                                      *                    *
****************************************************************************
                                                    VERSION : 10/08/89
```

AIM OF THE PROJECT :
****

'DAMPS' is a system which will analyse and predict the downtime of offshore marginal fields operated via various types of floating production systems. The emphasis is on a practical approach. Results are validated by means of hindcasting of the real time physical environment and comparing this to the actual production experience of marginal fields. This study combines environmental studies, marginal field systems analysis (component breakdown), identification of environment/structure interaction, optimisation of production systems and validation of the results of the project using actual production histories of marginal field systems and environments.

PROJECT DESCRIPTION :
****

The system incorporates a number of Knowledge Data Bases (KDB's) which may be defined as :
KDB1: Structural Data;
KDB2: On-Site Condition Data;
KDB3: Environmental Data;
KDB4: Motion Analysis Package;
KDB5: Simulation Package;
KDB6: Optimised Results File.
The package is innovative in its aim to encompass all relevant operational and environmental data for the various system in one file.
Those systems which have already been used to exploit hydrocarbon deposits and which could have widespread application to marginal field development are being examined. To carry out this task in a logical and systematic manner, considering global and sub-system response to the environment, a solftware package was initiated and is being augmented which will accept all the relevant data for each system, stored in individual Knowledge Data Bases (KDB's) as follows :
KDB1: Structural Data
- These data define the physical details of the various sub-systems involved (production unit, riser, loading buoy/column, export tanker), their location, inter-connections, dimensions, mooring systems.
KDB2: On-Site Condition Data
- This data set defines the water depth, operating criteria, throughput and the global downtime level of each system.
KDB3: Environmental Data at each Location
- This data base contains the data necessary to define the georaphic location for each marginal field. All available information on the wind, wave and current regime at location will be fed into this unit.
KDB4: Analysis Package
- This package uses data from KDB1, KDB2 and KDB3 to determine the response modes of the elements of each marginal field system under consideration.
KDB5: Simulation Package
- The response data are fed into a simulation package, where an estimate is made of downtime. The simulation model is tuned using real data obtained from designers and operators. The results of each analysis from KDB4 and KDB5 are incorporated into KDB6, the Results File.
KDB6: Results File
- This file contains limit values with respect to the environment for each sub-system and interconnected sub-systems. On each iteration, using modified structural and operational data from KDB1 and KDB2, this file will be up-dated until an optimum result is achieved, and thus the most efficient marginal field production system, in terms of minimal downtime, field characteristics and operator requirements.
The project managers believe that this package, 'DAMPS', will provide an innovative self-contained information and analysis tool for downtime analysis of marginal field production systems in Community waters. The package is also envisaged as a potential export product for use in remote developments worldwide.

STATE OF ADVANCEMENT :
****

Ongoing.
The "DAMPS" project was commenced officially on 28th November 1986. The first

Interim Report, covering the period 28.11.86 to 30.04.88, was submitted to the Commission during 1988. The second Interin Report, covering the period 01.05.88 to 31.12.88, was issued January 1989.

RESULTS :
****
Verification of the simulation package has indicated that the principles employed are accurate. Further system data will be gathered as considered necessary.
Environmental downtime is only one parameter in the overall downtime which can be experienced offshore. Mechanical failure downtime, repair downtime, maintenance downtime and operator error must also be considered. EOLAS is at present investigating the work done to date on this subject (e.g., Det norske Veritas, Lloyd's Register of Shipping, designers, oil companies, etc.).
A routine to simulate this downtime is being added to the existing package defining the complete list of controlling factors. Probability levels will be assigned to these factors from data sourced from such studies as the OREDA Survey.
Project results will be available on an ongoing basis. To date, the project has been successful in determining downtime levels for numerous offshore exploration activities in Irish waters. Also the project has enabled EOLAS to advise the Department of Energy on predicted downtime levels for various potential production systems under various weather conditions offshore Ireland.

```
*******************************************************************
* TITLE : COLUMN STABILIZED PRODUCTION PLATFORMS      *    PROJECT NO       *
*                                                     *                     *
*                                                     * TH./03261/87/FR/..  *
*                                                     *                     *
*******************************************************************
* CONTRACTOR :                                        *   TELEPHONE NO      *
*    SEAMET INTERNATIONAL                             *                     *
*                                                     *  33(1)45.34.85.23   *
*    61 RUE DE LA GARENNE                             *                     *
*    FR-92310 SEVRES                                  *                     *
*                                                     *   TELEX NO          *
*                                                     *                     *
* TECHNICAL DIRECTOR :                                *   206111            *
*    A. REY-GRANGE & J.M. BOSGIRAUD                   *                     *
*                                                     *                     *
*******************************************************************
                                                       VERSION : 18/04/88
```

AIM OF THE PROJECT :
****
Unlike drilling platforms which involve highly standardized procedures and
equipment, floating production systems must fit an array of special requirements
and equipments in a combination which is specific to each oil or gas field.
Standards are far from being set whilst operators have so far considered a wide
variety of concepts.
The aim of the project was to concentrate on typical areas or systems of a
column stabilized production platform and to provide for each area or system a
set of guidance drawings, specifications and operation procedures which could be
combined to meet any given circumstances.
Areas or systems of interest were, but not limited to : subsea and riser systems,
riser suspensions and handling systems, drilling/workover/servicing rig,
production processing system, water/gas injection systems, flare, export system
and temporary storage, auxiliaries, living quarters, structural patterns,
mooring systems, etc.

PROJECT DESCRIPTION :
****
A program extending over a 27 month period was carried out. It included a market
survey : potential builders (such as naval shipyards) and oil & gas operators
received a questionnaire about their current needs, points of interest and
present technical targets. A call survey and a questionnaire analysis followed
and helped in finalizing the design criteria and workscope of the project. Then,
the research and design development phase started up. Innovation was focused on
platform arrangement in close relation with subsea equipment (rig layout,
servicing & diving systems layout, production & auxiliaries layout).Special
attention was paid to buffer storage and export systems as well as to guidelines
for stability and ballast systems. Improvement of primary structure, living
quarters & permanent mooring systems was sought. Writing of operation manuals
completed this.
The next step of the study consisted of designing three new typical platform
configurations and layouts (respectively : 4000 tons pay load/4 piles, 7500 tons
pay load/6piles, 10000 tons pay load/8 piles) taking into account the material
produced in the previous phases. A related cost assessment was then established.
Finally, a set of documents (specification, drawings, procedures was assembled
in a report book.

STATE OF ADVANCEMENT :
****
Completed

RESULTS :
****
PHASE 1 - MARKET SURVEY :
24 Mailed questionnaires were sent to Major Oil and Gas European Companies. 11
replied & showed zone interest. A bibliographic survey of existing F.P.S.
facilities complemented the analysis of Operators' expressed trends, 20 Europena
Shipyards received a specific questionnaire; 12 answered. To suit most shipyards
building capacities, the following platforms iddimensions were enforced : max.
width = 63 m - max. Lightship draft = 7 m.
PHASE 2 - RESEARCH & DEVELOPMENT WORK :
Regulations & Research criteria were set up.., Riser configurations, Rig layouts, process equipments lau-outs as well as related nomenclatures are completed.
 Primary structure sampling, living quarters study, crude oil buffer storage and
mooring system design were carried out, too.
PHASE3 - SUGGESTED PLATFORM CONFIGURATIONS & LAY-OUTS :
3 suggested configurzation sketches were drawn up (ghost or typical) for 3
prodution cases:
- 60 000 BPSO }
- 30 000 BPSO with or without workover rig, gas lift or gas injection/gas
export, water injection
- 100 000 BPSO }
The payload capacity of each platform was slightly beefed-up with regards to the
original project aim.

The 60 000 BPSO is at detailed lay-out stage with workover, gas injection and water injection options, while the 30 000 PSD is drawn with gas lift and handling option.
The relevant computer check of structural analysis, stability and behaviour at sea was carried out, cost assessment of subsystems & systems was finally done. The research and Development work is available as a set of drawings, computations, specifications and nomenclatures. It will be assembled, for commercial purposes and presented in an A-3 format binder to potential users.

```
********************************************************************************
* TITLE : THE DEVELOPMENT OF NEW TYPES OF MULTI-      *       PROJECT NO       *
*         BORE UMBILICALS FOR CONTROL AND             *                        *
*         TRANSPORT IN OFFSHORE PRODUCTION            *   TH./03262/87/DE/..    *
*         SYSTEMS                                     *                        *
********************************************************************************
* CONTRACTOR :                                        *   TELEPHONE NO         *
*    NORDDEUTSCHE SEEKABELWERKE AG                    *                        *
*                                                     *   0211-658030          *
*    MEERESTECHNIK 1                                  *                        *
*    IN DEN DIKEN 16                                  *                        *
*    D-4000 DUSSELDORF 30                             *   TELEX NO             *
*                                                     *                        *
* TECHNICAL DIRECTOR :                                *   (17)211-4450 POF     *
*    DR. H. DRESENKAMP                                *                        *
*                                                     *                        *
********************************************************************************
                                                        VERSION : 06/02/90
```

AIM OF THE PROJECT :
****
Within the project a new type of riser-umbilical will be developed combining
transport-, operating- and control-function. A flexible pipe-design with
incorporated cables (energy- and data-transfer) will be developed and
investigated by static as well as dynamic testing of the test samples and
theoretical modelling.

PROJECT DESCRIPTION :
****
The project is a joint programme with Norddeutsche Seekabelwerke AG and TU
Hamburg-Harburg and can be subdivided in three phases: the first phase consists
of the processing of the specification and possibilities of incorporating the
cables into flexible pipes and the development of cable connection.
In the frame of the second phase- the development phase- a new manufacturing
procedure will be developed and the manufacturing equipment has to be modified.
Also a first test pipe will be built and static tests will be performed.
This phase will be followed by the test phase. Within this phase static as well
as dynamic test will be performed with the new riser-umbilicals. Furthermore a
theoretical model for computation of riser-umbilicals will be developed and the
test results will be compared with the computational results.

STATE OF ADVANCEMENT :
****
Project is still within time schedule and will probably end in August 1990. At
present the literature study and formulation of pipe and cable specification is
finished. Presently component tests are being performed concerning heat,
chemical and creep resistance of the cable coating as well as its bonding to
rubber. Furthermore the first umbilical design has been developed. At present
the manufacturing of a first test pipe. has been manifactured and tested.

```
****************************************************************************
* TITLE : REMOTE MULTIPHASE PUMPING STATION          *      PROJECT NO    *
*                                                     *                    *
*                                                     *  TH./03265/87/UK/.. *
*                                                     *                    *
****************************************************************************
* CONTRACTOR :                                        *   TELEPHONE NO     *
*    HAVRON LTD                                       *                    *
*                                                     *   01 242 6644      *
*    33 JOHNS MEWS                                    *                    *
*    UK-LONDON WC1N 2NS                               *                    *
*                                                     *   TELEX NO         *
****************************************************************************
* TECHNICAL DIRECTOR :                                *   8812287          *
*    R.K. SAINSBURY                                   *                    *
*                                                     *                    *
****************************************************************************
                                                  VERSION : 01/02/90
```

AIM OF THE PROJECT :
****

The purpose of the project was to investigate and demonstrate the viability of a
remote-multiphase pumping station based on a high stability taut-moored buoyant
structure. The station is intended to control and monitor a remote seabed
wellhead or wellhead complex and to pump multiphase fluids from the wellheads to
a production facility possibly 30 kilometers away.
The project aims were to prepare sufficient engineering detail to identify major
problem areas or areas of new technology, to address such problems and either
provide a solution or a path to a solution and to provide a realistic estimate
of cast. A further requirement was to identify areas of further work.
The concept represented an innovative approach to this topical problem providing
a benign and accessible environment for the operation of the new generation of
multiphase pumps and providing a realistic vehicle for this type of development.

PROJECT DESCRIPTION :
****

The project comprised the engineering design of a remote multiphase pumping
station based on a taut-moored high stability buoyant structure housing pumps,
metering, power systems, supervisory control and monitoring equipment, house-
keeping systems and all other equipment necessary to pump a multiphase fluid
from a seabed wellhead complex (single or multiple) to a production platform
remote from the wellhead complex itself.
The structure is normally unmanned and designed for total remote control.
The riser assembly (part rigid part flexible, is designed for reliability and
longevity and provides for conducting fluids to and from the buoyant structure
and for the support of monitoring and control umbilicals.
The gravity base is site-dependent but in the exhibited design has been based on
a self-floating gravity structure which assumes a reasonably firm seabed. No
attemps has been made to accommodate all soil conditions as it is felt that this
is a matter of specific design tailored to a particular site, using proven
techniques.
The project was implemented in three distinct phases as follows :
Phase 1 : This comprised a definition of all systems and layouts following
comprehensive discussions with major oil companies and potential equipment
suppliers. Throughout these discussions the design was substantially redefined
to refloect latest research or client requirements. An outline design was
implemented to form the basis of the more detailed Phase 2 design.
Phase 2 : This comprised the detailed design of all major structural elements,
systems and system components. This involved visits to vendors and to various
constructions with similar applications to assess state-of-the-art development
and experience. A detailed study of the reliability of the entire system was
also carried out to establish the overall operability and limits of downtime
likely to apply.
Phase 3 : This comprised the completion of the detailed report.

STATE OF ADVANCEMENT :
****

The feasibility study in wholly complete together with model testing implemented
at Heriot-Watt University.

RESULTS :
****

The main conclusions of the study are that the concept is very feasible and can
be constructed for a cost of approximately UKL 7 million. Deployment costs would
be approximately UKL 1 million and annual maintenance costs would be of the
order of UKL 8800,000 including unschedules visits.
Conventional construction technology could be used throughout and good access
provided to ensure a high level of maintenance access for the multiphase pumps
which represent the only relatively unproven equipment item.
It was concluded that a high level of reliability could be achieved with an
operational availability of 95.7%. This equates to a total downtime for the
system of approximately 15.5 days per annum. Reliability overall could be
improved byt it was not considered necessary given these relatively high figures.
The study has demonstrated the availability of a new tool for the enhancement of

throughput to existing production facilities or the viable exploitation of existing but hitherto difficult fields.

```
****************************************************************************
* TITLE : SEA TESTS ON RISERS FOR FLOATING          *    PROJECT NO       *
*         PLATFORMS - PHASE 2                        *                     *
*                                                    *  TH./03266/88/IT/.. *
*                                                    *                     *
****************************************************************************
* CONTRACTOR :                                       *  TELEPHONE NO       *
*    AGIP SPA                                         *                     *
*                                                    *  025205969          *
*    TEIN DEPT - I - 20120 MILANO                    *                     *
*                                                    *                     *
*                                                    *  TELEX NO           *
****************************************************************************
* TECHNICAL DIRECTOR :                               *  310246             *
*    MR PIERANTONIO TASSINI                          *                     *
*                                                    *                     *
****************************************************************************
```
                                                  VERSION : 19/06/89

AIM OF THE PROJECT :
****
The scope of this research project is to organize and carry out sea tests on
instrumented risers, in order to acquire experimental data on the
dynamicbehaviour of operating risers suitable for verifying the capability of
the available theoretical calculation procedures of compliant structures.
Moreover, aim of the project is to carry out a direct experimental verification
of the conditions which cause hydroelastic vibrations induced by vortex shedding,
 both in the presence and absence of a second riser and the consequences which
such phenomena have on the stresses and the displacements of the riser.

PROJECT DESCRIPTION :
****
This project constitute the phase two of a two phase project. The phase one was
dedicated to design of the experimental set up; the phase two will be dedicated
to the building of the test rig, to the sea tests and to data analysis.
The basic objective is to install two instrumented risers on an already existing
structure (a fixed platform) in water depth of about 75 metres. Such a structure
will act as a support to the equipment required to carry out the tests.
The test rig will consist of : . support structure
                                . motion simulation trolley supporting a rotating
table
                                . tensioner.
The instrumentation and data acquisition system will consist of three parts : .
instruments to measure risers'dynamic behaviour
                . instruments to measure environmental conditions
                . equipment for data acquisition and storage.
 The measurement of the risers' dynamic behaviour will be achieved by means of a
sufficient number of instrumented pipe sections placed on each riser.
The signals from all the sensors will be collected by a data acquisition system
based on a computer.
The data acquired during the tests will be suitably processed and compared with
theoretical simulations.
On the basis of this comparison, verification of the theoretical calculation
procedures will follow, in order to establish whether or not the available
procedures are adequate and to produce modifications of the existing calculation
procedures or to develop new ones, where appropriate.

STATE OF ADVANCEMENT :
****
Ongoing. The phase two of the research is at the beginning. The detailed design
of the system is completed, the request for bid to the vendors have been issued.
After the construction of the system, the sea tests will start.

RESULTS :
****
The project will develop and validate the calculation procedures needed to
design production risers for floating systems suitable for exploitation of deep
water hydrocarbon fields located in the European continental shelf.
Development will refer mainly to :
. Instrumentation systems
. Data acquisition and analysis procedures
. Calculation procedures
During phase one of the project the system was designed in detail. Next relevant
results will be obtained after the sea test campaign.

REFERENCES :
****
AGIP MOVES ON DEEP WATER FACILITY DESIGNS.
OCEAN INDUSTRY, APRIL 1986 pp 156-158.
SEA TESTS ON RISERS FOR FLOATING PLATFORMS
P.CAMPELLI, M. BERTA, A.K. BASU
DEEP OFFSHORE TECHNOLOGY CONFERENCE, MONTE-CARLO.
SEA TESTS OF RISERS FOR FLOATING PLATFORMS
P.CAMPELLIm M.BERTA, A.K. BASU
3rd HYDROCARBON SYMPOSIUM, LUXEMBOURG.

```
*******************************************************************************
* TITLE : PLATINE (PHASE III)                         *      PROJECT NO       *
*                                                     *                       *
*                                                     *   TH./03269/88/FR/..  *
*                                                     *                       *
*******************************************************************************
* CONTRACTOR :                                        *     TELEPHONE NO      *
*    AGELFI C/O GERTH                                 *                       *
*                                                     *    1 47.52.61.39      *
*    4, AVENUE DE BOIS PREAU                          *                       *
*    FR - 92502 RUEIL-MALMAISON                       *                       *
*                                                     *      TELEX NO         *
*                                                     *                       *
* TECHNICAL DIRECTOR :                                *     203 050           *
*                                                     *                       *
*                                                     *                       *
*******************************************************************************
                                                        VERSION : 31/12/89
```

AIM OF THE PROJECT :

The aim of PLATINE project is to specify a central production platform, unmanned, teleoperated from a control room located onshore, with a yearly (or half-yearly) visit for inspection and overhauling. The intermediate aims consist in working out and testing pilot units related to sub-systems and testing new equipment.
The project has been divided into 4 phases : PLATINE - Phase (1986-88); PLATINE - Phase II (1987-88); PLATINE - Phase III (1988-89) and PLATINE - Phase IV (1989-90).

PROJECT DESCRIPTION :

The project aims at progressively integrating the results of studies performed in PLATINE Phase I and Phase II projects. The facilities have been divided into 7 main sub-systems : 4 production sub-systems (gas-lift, sea water injection, separation-treatment, well pumping) and 3 sub-systems non specific to oil industry (energy/rotating machinery, safety/reliability, automation systems).
2.1 Pilots
Eight pilots are related to these sub-systems, or part of them.
a) Gas-lift
Two pilots will be carried out, installed on site and tested, one being related to automatic start-up (in case of unique injection point gas-lift) and the other related to optimization of gas lift injection.
b) Water injection
Further to tests of prototypes related to specific processes (chlorination, filtration, deoxygenation, filtration quality) carried out in PLATINE Phase II project, a sub-system pilot will be constructed, installed on site and tested.
c) Well pumping
A long stroke pumping unit pilot will be constructed, installed on site and tested.
d) Condition based maintenance
A pilot will be executed on a gas compressor set (compressor and drive); it will incorporate an on-line event generator and an expert-system.
e) Desalination monitoring by on-line expert system
A pilot, including instrumentation, on-line expert system and connection interface to ad digital control-monitoring system, will be constructed, installed on site and tested.
f) Bottomhole optical sensor
A permanent bottomhole optical pressure and temperature sensor will be built, installed in a well and tested. The pilot will incorporate the transmission and connection system, as well as the signal processing on the surface
g) New-technology rotating machine
A pilot will be constructed, installed on site and tested.
h) Well safety electronic control
A pilot including hardware and software will be constructed, installed on site and tested.
2.2 Prototypes
They will be related to oil separation-treatment, discharge water deoiling and gas-treatment.

RESULTS :

a) GAS-LIFT
One pilot related to automatic start-up has been developed and installed on a platform. An other one related to optimization of gas-lift injection is being developed.
b) WATER INJECTION
Further to tests of prototypes related to specifc processes (chlorination, filtration, deoxygenation, sterilisation...) carried out in PLATINE Phase II project, a pilot sub-system has been designed.
C) WELL PUMPING
A long stroke pumping unit is being designed.
d) CONDITION BASED MAINTENANCE
A pilot will be executed on a gas compressor set (compressor and electric engine drive); the instrumentation is installed, the on-line event generator and expert

H\

system are being developed.

e) DESALINATION MONITORING BY ON-LINE EXPERT SYSTEM

A pilot, including instrumentation, on-line expert system and connection interface to a digital control monitoring system, has been constructed.

f) BOTTOM-HOLE OPTICAL SENSOR

Sensor prototypes have been constructed and are being tested in laboratory (bottom-hole conditions).

g) NEW TECHNOLOGY ROTATING MACHINE

A self lubricated oil pump is being constructed.

h) WELL SAFETY ELECTRONIC CONTROL

A pilot including hardware and software has been constructed and installed on a platform. It is being tested.

The "gas-lift automatic start-up" and "well safety electronic control" pilots have been installed on platform end of 1989; they are being tested (no problem occured during the first two months of testing).

```
********************************************************************
* TITLE : OPTIMISATION OF DELINEATION AND CONTROL    *     PROJECT NO      *
*         OF WATER INJECTION BY FINE MEASUREMENT     *                     *
*         OF SURFACE MEASUREMENTS (PHASE II)         *   TH./03270/88/FR/.. *
*                                                    *                     *
********************************************************************
* CONTRACTOR :                                       *    TELEPHONE NO     *
*    GERTH                                            *                     *
*                                                    *    33-147526139     *
*    AVENUE DE BOIS PREAU 4                           *                     *
*    F-92502 RUEIL MALMAISON                          *                     *
*                                                    *    TELEX NO         *
*                                                    *                     *
* TECHNICAL DIRECTOR :                               *    615700           *
*    MR D. DESPAX                                     *                     *
*                                                    *                     *
********************************************************************
                                              VERSION : 19/06/89
```

AIM OF THE PROJECT :
****
This project consists in measuring subsidence movements induced by pressure
changes associated to reservoir depletion in view of evaluating that depletion.
These subsurface movements are then converted into reservoir pressure changes
versus space and time using a F.E. model of all relevant layers i>e> overburden,
reservoir and underburden. Finally this pore pressure map is converted into a
permeability map in both reservoir and aquifer.

PROJECT DESCRIPTION :
****
The general purpose of this study consists in detecting the closure type (fault,
water drive, etc...) of the Villeperdue geological structure (Parisian Basin,
France) on the east and north sides. An other connected target is to monitor the
injected water front.
Phase 1 : Technological studies
1.1 Optical fiber tiltmeter development
1.2 Feasibility study of a horizontal tilt log
1.3 Development of a levelling method using laser interferometry
1.4 Development of numerical inverse methods in order to derive :
     - pore pressure variations from subsidence measurements
     - permeability map pore pressure variation map
Phase 2 : Subsidence measurement survey around Villeperdue area
2.1 Positioning of level beacons for IGN (Institut Geographique National)
levelling along a 60 km profile - IGN surveys
2.2 Pilot subsidence measurements using new developments described in Phase
I
Phase 3 : Measurement interpretation
3.1 Selection of geometrical parameters and mechanical properties
3.2 3D numerical simulation of subsidence process
3.3 Physical output interpretation

STATE OF ADVANCEMENT :
****
Phase 1 (Feasibility study) has been completely accomplished. Four topographic
surveys have been so far performed by IGN (Institut Geographique National) along
a 16 km long profile across the Villeperdue field :
- July 1987 (refeerence survey)
- November 1987 (1st survey)
- March 1988 (2nd survey)
- January 1989 (3rd survey)
Three new surveys are schedule for 1990.
The measurement interpretation of the first two surveys is developed in the
final report within the scope of the previous TH/5076/86

RESULTS :
****
Measures recorded by the optical fiber tiltmeters were not usable mainly because
of weather conditions. This method is now abandonned. On the other hand, the
levellings by IGN were very satisfactory. The accuracy is around 0.2 mm/Km and
the maximum amplitude of subsidence (7 mm) is in good agreement with the
predictions. An other profile would be necessary to obtain indications about the
closure type. However, it can be asserted that the movements measured along the
existing profile reflect the variations of pressure and that it is possible to
derive them from the subsidence measurements.

```
****************************************************************************
* TITLE : OFFSHORE TEST OF COMPOSITE PRODUCTION      *     PROJECT NO      *
*          RISER ON TENSION LEG PLATFORM             *                     *
*                                                    *   TH./03271/88/FR/.. *
*                                                    *                     *
****************************************************************************
* CONTRACTOR :                                       *    TELEPHONE NO     *
*    GERTH                                           *                     *
*                                                    *   (1)47.52.61.39    *
*    4, AV. DE BOIS PREAU                            *                     *
*    FR - 92502 RUEIL MALMAISON                      *                     *
*                                                    *    TELEX NO         *
*                                                    *                     *
* TECHNICAL DIRECTOR :                               *   20350 FAX: (1)47.52.*
*    MR. SPARKS                                      *   69.27            *
*                                                    *                     *
****************************************************************************
```

VERSION : 31/12/89

AIM OF THE PROJECT :
****
Offshore test of a production riser made of glass and carbon fibers, integrated
within a steel production TLP riser. Objectives of the offshore test are the
following :
- run a resistance test over a period of about 1 year on composite pipes for TLP
risers.
- determine the degree of solicitations to which pipes are submitted, with the
help of instrumentation existing on the platform.
The main purpose of this test is to check the fiability of these pipes for
tension leg platforms, thus representing a decisive step in the use of new high
performance components for the bottom-surface transfer of oil effluents.

PROJECT DESCRIPTION :
****
The project is divided into four phases :
PHASE 1 - Engineering
- Accurate calculation of pipe dimensions, particularly of the thickness of
carbon and glass fiber layers and of their optimal layout
- Study of piping ends and of their connection to extremity connectors.
- Study of equipment required for pipe layout and, eventually, of procedures
specific to the implementation of composite pipes.
PHASE 2 - Manufacture
- Manufacture of connecting pipe ends
- Manufacture of a 8 1/2" mandrel to wind pipes
- Manufacture of pipe layout equipment
- Manufacture of pipes.
PHASE 3 - Layout
- Transportation of pipes and transfer on TLP
- Assembling of the steel/composite pipes riser; connection at the template and
tensioning.
PHASE 4 - Offshore test
- Periodic surveillance and pipe internal pressurization operations
- Analysis of solicitations to which riser is submitted with the aid of
instrumentation existing on TLP
- Riser retrieval operations in case of failure and valuation.

STATE OF ADVANCEMENT :
****
Ongoing.
Phase 1 : completed
Phase 2 : Manufacture of pipe ends ongoing

RESULTS :
****
Report on calculated behaviour of composite riser on HUTTON.
During a promotional campaign, oil companies advice a reorientation of the
programme.

```
*****************************************************************************
* TITLE : LINEAR-MOTOR-POWERED PUMPING UNIT FOR     *     PROJECT NO       *
*          PETROLEUM WELL OPERATION                 *                      *
*                                                   *   TH./03272/88/FR/..  *
*                                                   *                      *
*****************************************************************************
* CONTRACTOR :                                      *     TELEPHONE NO     *
*    M.A.P.E                                        *                      *
*                                                   *    40 43 34 11       *
*    BD DU MARECHAL JUIN 14                         *                      *
*    F-44100 NANTES                                 *                      *
*                                                   *     TELEX NO         *
*                                                   *                      *
* TECHNICAL DIRECTOR :                              *    711436            *
*    MR CHARDONNEAU                                 *                      *
*                                                   *                      *
*****************************************************************************
                                                   VERSION : 31/12/89
```

AIM OF THE PROJECT :
****
Development of a linear-electric-motor powered pumping unit directly installed
on the wellhead, mechanically driving the pumping rods in the alternate pumping
with traditional rods.
This pumping unit is particularly intended for future automated production
installations, both onshore and offshore.
Its "all electric" design offers the following major advantages, specially when
compared to similar existing hydraulic equipment :
- Maintenance reduced to minimum
- Maximum reliability
- Easy remote control of all its parameters.
These advantages, plus its increased compactness at the wellhead, make it
particularly suitable for offshore pumping operations.

PROJECT DESCRIPTION :
****
1) The linear electric motor
1.1 Elaboration of specifications, comparatively with existing equipment
1.2 General theoretical study and development of a reduced scale test model
1.3 Fabrication and test of reduced scale test model, indispensable for the
optimization of real size motor characteristics
1.4 Detailed studies of motor conception
a) Technology
b) Electric supply
c) Cooling system
d) Conformity with standards required in explosive atmosphere
e) Monitoring and power electronics
1.5 Execution
a) of the motor
b) of its power and monitoring electronics
1.6 Qualification tests
2) Adaptation of linear motor to well
Design and construction
2.1 Mechanical supports for the motor in relation to wellhead
2.2 Mechanical connections of its mobile part with the pumping rod string
2.3 Adaptation of sensors required for the in-situ measurement of pumping
parameters (pumping load, travel distance, law of movement, cyclic instantaneous
power)
2.4 Balancing of pumping load (weight of pumping string and of the half column
of fluid) thanks to remote-controlable device.
3) Satisfactory pumping operation test on a test well
with measurement of all significant parameters
4) On-site endurance test on full-scale model
If possible, previously equipped with a standard pumping beam
4.1 Measurement and recording of all pumping parameters of the standard
equipment before its retrieval
4.2 Measurement and recording of same pumping parameters of the linear motor
pumping equipment
4.3   4.3 Comparison between the two systems
4.4 Results of endurance test after six-months service
5) Elaboration of calculation method for pumping parameters and development of a
computer adapted calculation program.

STATE OF ADVANCEMENT :
****
Specifications under revision in order to reduce the installed electric power,
judged much higher compared to that initially given in the specification.
General theoretical study completed. Execution and tests of modeling completed.

```
******************************************************************************
* TITLE : ON SITE TESTING OF A CENTRIFUGAL OIL-    *      PROJECT NO        *
*          WATER SEPARATION                        *                        *
*                                                  *    TH./03275/88/FR/..  *
*                                                  *                        *
******************************************************************************
* CONTRACTOR :                                     *    TELEPHONE NO        *
*    BERTIN & CIE                                  *                        *
*                                                  *    134818500           *
*    B.P. 3                                         *                        *
*    78373 PLAISIR CEDEX FRANCE                    *                        *
*                                                  *    TELEX NO            *
*                                                  *                        *
* TECHNICAL DIRECTOR :                             *    696231              *
*    RENE BOURASSIN                                *                        *
*                                                  *                        *
******************************************************************************
                                                 VERSION : 14/11/89
```

AIM OF THE PROJECT :
****
Since 1978, a water-oil separation high performance disk centrifuge has been
developped by Bertin & Co.
The proposed program mainly consists in on-site testing.
The purpose of these tests, which are in relation to previous prototypes is to
test the reliability of the equipment, as requested by the oil companies.
After the first separation tests, the centrifuge will be reviewed for the
correction of any defects which may have been observed during testing, and
adapted to the constraints of the worksite chosen.
Together with a preliminary appraisal of the performances and the reliability of
the centrifuge, these tests will supply the information necessary for defining
operations for maintenance during operation.
It will be possible to define an industrial centrifuge once these tests have
been completed.

PROJECT DESCRIPTION :
****
The program includes :
- modification of the machine for the sludge separation according to the
previous tests conclusions
- laboratory tests
- modification of the machine for on site testing
- on-site performance tests
- reliability tests on site.
The ultimate goal is to verify centrifuge compatibility and reliability in oil
production plant conditions.
To reduce the time required for fields tests, which would be costly and
detrimental to production, a maximum number of tests will be previously run in
laboratory with emulsions representative of oil field conditions. These
emulsions will be made in cooperation with SNEA(P) who will verify their
characteristics at centrifuge feeding and the grade of effluent at discharge.
Besides the Ashart crude used for previous tests, emulsions will be prepared
from gasoline as well as corrosion inhibiting chemicals like CK 337, NORUST 720
and SOLAMINE 129 and suspended solids : bentonite, calcium carbonate,
Fontainebleau limestone.
Such tests will help investigate the machine operating limits by varying
rotating speed, flowrates and contents in oil and solids.
After completion of the tests the machine will be inspected, overhauled and
modified and reengineered for oil field conditions; the oil field has to be
designated by SNEA(P) who will bring their assistance to BERTIN & Cie for
performing the tests.
The main anticipated modifications are :
- adaptation of the oil and water discharge nozzles to make them suited to filed
flowrates
- conditioning to gas-proof safety measures, which implies replacing the motor
and producing a frequency converter for starting and speed control (in
replacement of the DC motor)
- engineering for safety
- fitting of control and measuring devices according to operating constraints.
After reassembly the machine will be installed on the oil dield with the
associated equipment for control and measurement, and tested again in simulated
operating conditions.

STATE OF ADVANCEMENT :
****
A high flowrate centrifuge prototype had been assembled, mechanically finalized,
and run in process on a laboratory facility. These first tests had shown re-
entry of the finest particles by the water flow. Therefore it had been decided
that the centrifuge should be modified in order to overcome these difficulties.

RESULTS :
****
The work performed during the period from August 88 to July 89 consists of in
the main technical tasks of the first phase of the project (as explained in

pages 4 and 5 of Annex I to the contract TH./03275/88/FR) :
- Modification chafts of the existing centrifuge have been achieved.
- The centrifuge has been modified and reassembled.
- The existing test bench has been adapted to the new machine configuration.
- The mechanical operation checkout has been performed.
- The laboratory separation tests have been done, with the cooperation of SNEA-P.
Some work still remain to be done in order to complete Phase I :
- Results analysis (to be done by SNEA-P)
- Centrifuge expertise
- Results synthesis and possible modifications to be done with respect to
further on-site testing.
The modified version of the centrifuge has been tested with three-phase mixtures
(water, oil, solid). The mechanical behaviour is satisfactory.
The emulsions made for the tests were judhed good enough to simulate real
conditions of secondary treatment. Some defects have been identified and they
will be corrected for on-site testing.
Concerning the possible creation of mass unbalance due to heterogeneous deposits,
 the existing cemented artificial slope will be strengthen by jackets made of
welded plate.
Once all the modifications are done, the solid deposits located in the
separation stage will be reduced. Moreover, the solid separation efficiency will
be improved, due to the increased size of the solid separation stage so that the
two separation stages will have comparable sizes.
The running full expertise of the machine will make possible to define more
precisely exhaustively the operation checkout.

```
************************************************************************
* TITLE : GA-SP PROJECT STAGE 3               *      PROJECT NO       *
*                                             *                       *
*                                             *  TH./03281/88/UK/..   *
*                                             *                       *
************************************************************************
* CONTRACTOR :                                *      TELEPHONE NO     *
*    GOODFELLOW ASSOCIATES LTD                *                       *
*                                             *    018211377          *
*    71 ECCLESTON SQUARE                      *                       *
*    UK LONDON SW1V 1PJ                       *                       *
*                                             *      TELEX NO         *
*                                             *                       *
* TECHNICAL DIRECTOR :                        *    929272             *
*    RICHARD LAWRENCE                         *                       *
*                                             *                       *
************************************************************************
                                              VERSION : 14/12/89
```

AIM OF THE PROJECT :
****
The aim of the GA-SP Project is to develop and test full scale, a subsea
processing system that will facilitate economic development of small hydrocarbon
reservoirs. The system will include well stream commingling, multiphase
separation and pumping, and will be engineered to be suitable for a wide range
of reservoir fluid properties, shallow or deep water application, with control
from the surface from either a low cost in-field floating platform or an
existing remote fixed installation.

PROJECT DESCRIPTION :
****
The GA-SP Project, which is also supported by a number of oil companies, is
being executed in stages, The First Stage, Front End Engineering and Preliminary
Design, was concerned with outline definition of field scenario concepts and
identification of major areas of novelty. The second stage, Detailed Design and
Engineering, was concerned with development of a full scale prototype system.
This third stage comprises realisation of the prototype separation and boosting
facility, and provision of a floodable dry dock test facility, complete with
simulated produced fluid conditioning plant. It also includes testing of the
prototype in a dry environment.
The major tasks of this stage of work have included :
1) Procurement of major and minor equipment and placement of contracts for
fabrication, installation, test facility etc.
2) Liaison with vendors and contractors and incorporation of vendor data in the
detailed design.
3) Development of detailed programme of tests.
4) Procurement of additional instrumentation etc for recording and analysis of
test results.
5) Simulation of test runs for comparison with prototype results.
6) Execution of performance trials in a dry dock (not flooded).

STATE OF ADVANCEMENT :
****
Stage 3 was running approximately 3 months behind schedule. This was due to
delay in manufacture the turbomachinery, and the control system. Those problems
have involved additional engineering, expediting and site activities by GA.
All equipment is on site at the test facilities in Middlesborough, and
precommissioning activities are in progress, primarily relating to the control
and instrumentation system.
Although the start of testing has been delayed, the test schedule has been
reviewed

RESULTS :
****
Dry testing is now scheduled to commence in January 1990 with completion of the
wet testing by end of May 1990.

REFERENCES :
****
TESTING FOR THE GA-SP PROJECT, K.BOND - GOODFELLOW ASSOCIATES, G.CHEW - NORTHERN
OCEAN SERVICES, CONFERENCE FOR THE TESTING OF UNDERWATER EQUIPMENT.

```
************************************************************************
* TITLE : DIVERLESS MODULAR PULL-IN & CONNECTION    *      PROJECT NO       *
*         SYSTEM                                    *                       *
*                                                   *   TH./03282/88/UK/..   *
*                                                   *                       *
************************************************************************
* CONTRACTOR :                                      *    TELEPHONE NO        *
*    ALPHA OFFSHORE ENGINEERING & MANAGEMENT SERVICES *                      *
*                                                   *    01 821 1288         *
*    71 ECCLESTON SQUARE                            *                        *
*    GB - LONDON SW1V 1PJ                           *                        *
*                                                   *    TELEX NO            *
*                                                   *                        *
* TECHNICAL DIRECTOR :                              *    929272              *
*    RICHARD LAWRENCE                               *                        *
*                                                   *                        *
************************************************************************
                                                        VERSION : 11/10/89
```

AIM OF THE PROJECT :
****
The objective of the project is to undertake engineering design and detail of a
simplified diverless modular pull-in and connection system which can be
universally applied to a variety of subsea oil and or gas production scenarios.
The aim is to use a fresh approach by generating new concepts, which will allow
for a modularised system to be developed around the basinc concepts conceived.
By this method it will be possible to design a system having a universal
application which will suit various combinations of subsea installations,
including single, multiple and bundled flowline/pipeline applications.

PROJECT DESCRIPTION :
****
The project involves conceptual design using modern idea generating techniques.
These concepts will then be developed into a preliminary embodiment design of
all the main functions of the system.
The proprietary equipment needed to package the system will be sourced and the
interfaces designed to ensure overall system integrity.
During design stress analysis of the various structural elements will be
undertaken and three dimensional models will be produced in paralell to the main
design activities.
Once the overall design schemes and layouts have been completed the detail
design will commence which will produce detail drawings ready for manufacture.
The final assembly drawings will be used to assist in dimensional checking of
the system and once the parts lists/bill of materials have been completed
appropriate manufacturers programme prepared.
This stage of the work will produce a prototype design package.

STATE OF ADVANCEMENT :
****
The project is in an early stae of design development and commenced on December
8th 1988 by defining the objectives and generating conceptual designs. Ongoing
work for this stage will result in a Prototype design package and final report.

RESULTS :
****
Results will be reported during the project as they become available and within
the contract requirements. They will include :
- Conceptual design layouts
- Minutes of meetings
- Preliminary design schemes
- List of OEM suppliers
- Calculation sheets
- Design scheme & layout drawing
- Design report including photographs of model
- Detailed Manufacturing Drawings
- Assembly and Sub Assembly drawings and Parts Lists
- Detailed Manufacturing Programme for Stage 2
- Prototype design package and final report.

REFERENCES :
****
REPORTS WILL BE PUBLISHED AND PRESENTATIONS GIVEN AT TECHNICAL MEETINGS AT
STRATEGIC POINTS OF THE PROJECT.

```
****************************************************************************
* TITLE : DEVELOPMENT OF A LARGE DIAMETER PILE        *    PROJECT NO      *
*         CONNECTOR                                   *                    *
*                                                     *  TH./03283/88/UK/.. *
*                                                     *                    *
****************************************************************************
* CONTRACTOR :                                        *   TELEPHONE NO     *
*    HUNTING OILFIELD SERVICES LTD                    *                    *
*                                                     *   0224 877487      *
*    BLACKNESS ROAD                                   *                    *
*    ALTENS INDUSTRIAL ESTATE                         *                    *
*    UK - AB9 8SY ABERDEEN SCOTLAND                   *   TELEX NO         *
*                                                     *                    *
* TECHNICAL DIRECTOR :                                *   73178            *
*    W.A. TURNER                                      *                    *
*                                                     *                    *
****************************************************************************
```

VERSION : 14/12/89

AIM OF THE PROJECT :
****
Due to increasing water depth of oil reserves, platforms pile diameter is
becoming larger and jointing times for each section of pile, by welding,
increases correspondingly. To replace offshore welding, reduce the cost of
installation and lessen weather dependency mechanical connectors have been
proved viable.
This project is to demonstrate connector compatibility with the tubular pile
material by carrying out fatigue and static load tests. Scale models will be
designed and produced for verification of manufacturing feasibility, testing and
assembly trials. The connector is not threaded and requires no rotation during
make-up. Axial loads are transferred through a series of rings and grooves with
radial interference, assembly being effected by the injection of an interface
fluid between Pin and Box. The connector incorporates the varying tooth pitch
principle, resulting in a more even distribution of pre-load and axially applied
external forces between connector teeth.

PROJECT DESCRIPTION :
****
The project involves the design and production of scale model connectors for
static and fatigue testing. Computer package analysis will continue throughout
the testing phase, both complimenting and determining the trials requirements
for the component testing. A specification package will be developed, including
a Quality Plan, Material and Manufacturing Specifications and Welding and
Inspection Procedures. Tooling will be developed to enable connectors to be
machined with minimal rejection rates. In conjunction with potential connector
machinists and the Quality Department a gauging philosophy will be developed,
with the emphasis on in-process and in-machine gauging systems. The connector
design of Phase I will be re-evaluated and refined as necessary and the
optimised connector produced in scale models of 30" and 15" diameters. The
impact of scaling upon test loadings will be examined and computer analysis will
indicate locations for strain gauge attachment. Raw material forgings will be
machined to the scaled dimensions, welded to test pipe, prepared for testing,
and trials test requirements will be produced to cover each test. Assembly
tests will be carried out on both the 30" and 15" models. On the 30" specimen
static tension, compression and bending tests will be conducted, the tension
test to failure being done on the same specimen as the compression test. Axial
fatigue tests on five 15" connectors to failure will be carried out to examine
the performance under (a) varying mean stress, constant stress range, and (b)
constant mean stress, varying stress range. An axial fatigue test on a 30"
connector to failure at positive mean stress will provide results to cross-
correlate with those from the 15" tests and will be used to examine the effects
of connector scaling. Bending fatigue tests on two 15" specimens to failure,
conducted about zero mean stress will complete the scale model testing phase.
The loading and number of cycles will be selected to demonstrate a fatigue life
performance better than the Department of Energy "F2" mean curve for the
equivalent pipe to pie weld.

STATE OF ADVANCEMENT :
****
The majority of the test 15" fatigue actuator tooling, interface adapters and
associated components are on order. Raw material for the 15" diameter fast track
is ready for machining. Fatigue trial procedure and likely durations were agreed
in principle.

RESULTS :
****
Detailed designs for the self contained fatigue actuator tooling and adaptors
have been produced.

```
**********************************************************************
* TITLE : PLAIM (PLATFORM LIFETIME ASSESSMENT      *      PROJECT NO      *
*         THROUGH ANALYSIS,INSPECTION &            *                      *
*         MAINTENANCE)                             *   TH./03285/88/UK/..  *
*                                                  *                      *
**********************************************************************
* CONTRACTOR :                                     *    TELEPHONE NO      *
*     ADVANCED MECHANICS & ENGINEERING LTD         *                      *
*                                                  *    0483301219        *
*     4 FREDERICK SANGER ROAD                      *                      *
*     SURREY RESEARCH PARK                         *                      *
*     GUILDFORD, SURREY GU2 5YJ                    *    TELEX NO          *
*                                                  *                      *
* TECHNICAL DIRECTOR :                             *    8950511           *
*     DR C.P. ELLINAS                              *                      *
*                                                  *                      *
**********************************************************************
                                               VERSION : 15/09/89
```

AIM OF THE PROJECT :
****
The aim of the project is to generate databases, and algorithms necessary to execute a validated fatigue crack growth/fracture analysis of a welded tubular joint which forms part of a fixed offshore platform. T This will enable the consequences of the detection of a surface defect to be determined by tracing crack growth, the consequential reduction in joint stiffness, severance, and overall collapse. The loading applied will be representative of service loading including extreme conditions. This will be encapsulated within a structural reliability analysis procedure so that both intact and damaged safety levels can be determined. In turn the procedure will be set within an expert system to provide sophisticated data and algorithm file manipulation, and guidance, help and advice as to the options available regarding maintenance and/or repair strategies through the Knowledge Base.

PROJECT DESCRIPTION :
****
The project will be executed in 5 phases. Phase I involves the development of databases covering all facets of the project such as stress concentration factors, stress distributions, stress intensity factors, and fatigue crack growth in welded tubular joints, material properties for fracture considerations, joint stiffnesses, reliability levels, inspection results on crack length and depth, and waves, wind, and current including their joint distributions. Considerable effort will be directed towards database screening to ensure only sound data are incorporated and statistical characterisation in preparation for reliability analysis.
Phase II involves the development of algorithms to execute substantiated fatigue crack growth life and fracture calculations, elastic and collapse 3-D frame analysis which includes joint stiffness and strength modelling, and member modelling. Reliability analysis will be tackled within this activity..
This project will concentrate on a four column three-bay high jacket description whensubjected to typical North Sea conditions. Sub-contractor input to algorithm development and verification will be significant here.
Phase III involves the development of Knowledge Bases each of which relates to a specific facet of the PLAIM system such as environmental description, load assessment, fatigue life analysis, fracture calculation, safety factors, reliability levels, maintenance strategies, repair options, and design in the general and detailed algorithm sense.
Phase IV seeks to develop rational inspection, repair, and maintenance strategies. This will be on the basis of results generated by sophisticated software which enable accurate assessments of jacket ultimate strength of either intact or damaged structures to be conducted. By also determining reliability levels, judgements concerning short and long needs can be made.
Phase V involves development of the expert system in its detailed and general form. Shell selection is part of this as is the generation of expert algorithms to exploit engineering judgement on data reliability, algorithm appropriateness and accuracy, interpretation of environmental data, etc. This phase will also involve implementation of the PLAIM software.

STATE OF ADVANCEMENT :
****
Ongoing. Project has been underway for nine months. Some 60 % of the work is estimated as completee. Work is progressing simultaneously on all Phases.

RESULTS :
****
Databases have been established on turbulent wind spectra, steady state wind forces, member strengths both local and overall, tubular joint strengths, material properties, fatigue strengths, fatigue crack growth, and underwater detection success. Information on wave load effects has been collated paving the way for the development of an appropriate database.
A reliability analysis algorithm including multi-path analysis and Bayesian updating have been established : some preliminary results relating to inspection scheduling following the inspection but non-detection of a crack have been produced. A fracture mechanics algorithm has been developed using a geometry

factor derived to give a close fit to factors inferred from tubular joint fatigue test results. An additional benefit of this particular piece of work was the opportunity to derive a simplified procedure for remaining life estimation based on crack length measurements. Both this and the fracture mechanics solutions can be expressed as means or lower bounds depending on user preferences.

The feasibility of developing the PLAIM System has been established through the creation of a prototype demonstration expert system/knowledge base package. An industrial strength shell with advanced interfacing, editing and encoding capabilities has been identified for further development of the PLAIM System. A prototype System will be produced over the next few months.

REFERENCES :
****

FRIEZE, P A WICKHAM, A H S, NIU, X, LANGDON, A J AND AHMED K "EXPLOITING A KNOWLEDGE-BASED SYSTEM OF OFFSHORE INSPECTION AND MAINTENANCE", TO BE PRESENTED AT THE ROYAL INSTITUTION OF NAVAL ARCHITECTS ANNUAL SPRING MEETINGS, APRIL 1990.

```
****************************************************************************
* TITLE : ITLPS#INTEGRATED TENSION LEG PLATFORM    *     PROJECT NO       *
*         SIMULATOR                                *                       *
*                                                  *  TH./03288/88/IR/..  *
*                                                  *                       *
****************************************************************************
* CONTRACTOR :                                     *  TELEPHONE NO         *
*    MCS INTERNATIONAL                             *                       *
*                                                  *  3539166455           *
*    3 BUTTERMILK WALK                             *                       *
*    IR-GALWAY                                     *                       *
*                                                  *  TELEX NO             *
*                                                  *                       *
* TECHNICAL DIRECTOR :                             *  50094 FAX: 353916645*
*    MR. J. MCNAMARA                               *  7                    *
*                                                  *                       *
****************************************************************************
                                                      VERSION : 08/02/90
```

AIM OF THE PROJECT :
****
MCS propose the development of a general purpose simulation package, named ITLPS
for the integrated analysis and design of the motions and forces on the complete
TLP structure including platform, risers and tethers. The structural system will
be discretised by the finite element technique in three dimensions with platform
and tether/riser elements as one integrated spaceframe. An additional innovation
will be the capability of modelling the risers as flexible beam elements
attached to the platform. All wind, wave and fluid current loading functions,
including higher order terms such as slow drift and damping forces, will be
included in a computationally efficient form based on a consistent set of
hydrodynamic assumptions. Pre-and post-processors including a customised
database will be attached such that the system may be readily accessed by design
engineers.

PROJECT DESCRIPTION :
****
The various phases of the project may be listed as follows :
Phase 1   Background : Review of TLP technology, collection of applicable codes
of practice and selection and sizing of TLP model dimensions.
Phase 2   Finite Element Spaceframe : Finite element structural model of
integrated platform and riser/tethers.
Phase 3   Modification of Morison's Equation : Wave length dependent added mass
and Froude-Kryloff coefficients, Keulegan-Carpenter dependent drag components
and vortex shedding riser/tether model.
Phase 4   Drift Forces and Potential Damping : Mean wave drift and slowly
varying drift forces and potential damping coefficients based on closed-form and
approximate diffraction analysis solution techniques.
Phase 5   Wind Loads : Expressions for steady wind force and dynamic wind input
spectrum including slowly varying wind drift force.
Phase 6   Stochastice Linearisation : Stochastic linearisation of velocity
squared viscous forces in three dimensions due to a random directional sea with
arbitrary current, and also for velocity squared viscous wind forces.
Phase 7   Equilibruim Equations : Synthesis of frequency domain equilibrium
equations. Integration of MCS riser/tether equations with platform equations.
Phase 8   Solution Procedures : Development of static linear and nonlinear
solution and frequency domain dynamic solution schemes.
Phase 9   Pre-and Postprocessor : User-friendly keyword input, customised
database for output, spectral postprocessor for mean, standard deviations and
extreme values, fatigue and transfer functions postprocessor. A special
postprocessor for estimating the probability of motion interference in tether
and riser bundles under random directional seas. Associated graphics and printed
formats.
Phase 10   Testing : Execution of test cases and comparisons with available
published results and model basin tests. Liaison with independent interested
organizations and potential client companies.
Phase 11   Site - Specific Demonstration : Collection and processing of
environmental data for Porcupine site in accordance with code requirements.
Execution of full demonstration example and application of available code cheks
to processed results.
Phase 12   Validation : Validation of ITLPS system by Offshore Certification
Authority and the solicitation of feedback from other interested independent
parties.

STATE OF ADVANCEMENT :
****
Ongoing

RESULTS :
****
An integrated spaceframe finite element model of a TLP including superstructure,
deck, columns, pontoons, tethers and risers may be assembled using three-
dimensional beam-column finite elements. Semi- analytical solutions have been
obtained for the wave potentials due to diffraction and radiation about vertical
open-bottom columns and rectangular submerged pontoons. Static and dynamic

spectral wind loading from an arbitrary direction may be specified for an element set or group; this set models the superstructure as a spaceframe with special element properties relating to the wind area and local parameters. Similarly a formulation for lift forces due to vortex shedding on tethers and risers is available. The overall equilibrium equations in the frequency domain for the TLP may be solved about the original undeformed equilibrium position and also about a static nonlinear configuration due to large drift motions of the TLP. A procedure for the stochastic linearisation of noncollinear random waves and current is developed for the hydrodynamic forces on the tethers and riser sections. Data is presently being collected for the testing and verification phases of the project.

The major result to date is the derivation of the semi-analytical functions for the hydrodynamic loads on the TLP columns and pontoons. These are not available elsewhere for the configurations in question and are expected to overcome the excessive computational problems associated with competing computer-based numerical methods. The other results are associated with the implementation and testing of the load modules for wind loading, vortex shedding, spaceframe motions and various solution schemes. The major remaining step is to integrate all the load modules into the finite element package and execute the ITLPS suite using realistic field data.

```
********************************************************************
*  TITLE : EXTERNAL PRESSURE VESSEL FRAMING          *   PROJECT NO    *
*          CONCEPT FOR TLP'S AND SEMI-SUBMERSIBLES   *                 *
*                                                    *  TH./03290/88/UK/..  *
*                                                    *                 *
********************************************************************
*  CONTRACTOR :                                      *   TELEPHONE NO  *
*      BILLINGTON OSBORNE-MOSS ENGINEERING LTD       *                 *
*                                                    *   0990 872323   *
*      TECHNOLOGY TRANSFER CENTRE                    *                 *
*      SILWOOD PARK - UK ASCOT BERKSHIRE             *                 *
*                                                    *   TELEX NO      *
*                                                    *                 *
*  TECHNICAL DIRECTOR :                              *   9312 100278   *
*      D M OSBORNE-MOSS                              *                 *
*                                                    *                 *
********************************************************************
                                                    VERSION : 31/05/89
```

AIM OF THE PROJECT :
****
It is generally recognised and accepted that by the end of this century, oil
production from offshore North West Europe (and, indeed, the Mediterranean) will
be in water depths of up to 1000m. It is also a recognised fact that many of
these fields will be 'marginal' in nature, and development decisions will depend
heavily on the economics of various schemes. Increasingly, the use of tension
leg platforms, semi-submersibles and other compliant structures is being
specified as part of the development of deepwater fields. Cost reductions in
capital expenditure for such systems are of paramount importance.
This proposal is concerned with the development of a new and innovative
structural framing concept for external pressure vessels which will eliminate
the need for ring stiffeners for components such as TLP and semi-submersible
columns and pontoons. The concept is simple and relies on membranes in tension
supported by tubes spaced around the perimeter of the section.

PROJECT DESCRIPTION :
****
The proposed structural framing concept has the potential to bring an entirely
new technology to the oil industry. External pressure vessels are used as
buoyancy tanks for fixed jackets, as columns and pontoons for floating semi-
submersibles and tension leg platforms (TLP's), as buoys for offshore tanker
loading and for the support of risers or mooring lines, as extra buoyancy tanks
near the top of articulated columns and towers and as subsea habitats and diving
vessels. External pressure vessel technology is also extremely important in
structures for the Arctic which much resist the pressure of ice. It is clear,
therefore, that if the new technology proposed for framing external pressure
vessels can be proved to offer cost and/or weight reduction benefits over
existing methods employing stiffened compression shells, it could benefit the
design of virtually every type of structure now being used or proposed in the
offshore industry.
The new technology proposed eliminates all of the internal ring stiffeners
normally found in external pressure vessels and is, therefore, purposely
designed to lend itself more readily to cheaper fabrication technology making
extensive use of automatic welding methods. Furthermore with the major
structural behaviour being tensile the plating thickness is reduced therefore
resulting is a significant reduction in material weight and cost as well as
fabrication costs.
The two key technical questions with this concept are whether it has an adequate
reserve of strength at collapse and whether it can be detailed to eliminate
fatigue problems. These two subjects are the main activities of this research
programme and they will be thoroughly investigated both theoretically and
experimentally. It is the opinion of the proposer, based on their extensive
offshore design and construction experience, that both of these subjects will be
shown to be within the normally accepted limits for design.

STATE OF ADVANCEMENT :
****
Ongoing. Project engineering commenced in October 1989, following a one year
period of discussions with North Sea operators on their participation in the
project. Now there is some limited industrial support for the first phase of the
project which is to prove the theoretical and financial viability of the concept.

RESULTS :
****
The design study has commenced with an external pressure vessel structure for a
submerged buoy to be utilised as part of a deepwater mooring system. The
structure is aprx 7 metres in diameter and 10 metres in length. It is similar in
size and complexity to the inner columns of the latest design of semi-
submersible drilling vessels. Preliminary cost studies indicate that the design
is more cost effective than conventional sing-stiffened compression cylindres.

REFERENCES :
****
OFFSHORE ENGINEER JANUARY 1988 'LATERAL THINKING UNSTIFFENS STEEL STRUCTURES'.
PAGES 24-25.

```
**********************************************************************
* TITLE : COMPOSITE CONSTRUCTION TECHNOLOGY FOR    *    PROJECT NO      *
*          FLOATING AND SUBSEA STRUCTURES          *                    *
*                                                  *                    *
*                                                  *    TH./03291/88/UK/..*
*                                                  *                    *
**********************************************************************
* CONTRACTOR :                                     *    TELEPHONE NO     *
*    BILLINGTON OSBORNE-MOSS ENGINEERING LTD       *                    *
*                                                  *    0990 - 872323    *
*    TECHNOLOGY TRANSFER CENTRE, SILWOOD PARK      *                    *
*    BUCKHURST ROAD                                *                    *
*    UK - ASCOT SL5 7PM                            *    TELEX NO         *
*                                                  *                    *
* TECHNICAL DIRECTOR :                             *    9312100278 FAX: 0990*
*    MR. LALANI                                    *    -872433         *
*                                                  *                    *
**********************************************************************
                                               VERSION : 11/05/90
```

AIM OF THE PROJECT :
****
This project concerns the development of composite construction technology for
application to those components of TLP, compliant and sub-sea systems which have
hitherto been designed using the highly expensive, time-consuming, problematic
and awkward form of stiffened plate and steel construction. The development of
composite technology relies on concrete acting integrally with steel in the form
of a sandwich type section. In this way the construction materials are used
efficiently and in application best suited to their mechanical properties.
The objective of the project is to prepare validated design guidance and
methodology for the application of composite structural components to floating
production and sub-sea systems. The guidance will cover these components
subjected to static hydrostatic loads and wave induced static/fatigue loads as
experienced in deep water and 'near-surface' conditions respectively.

PROJECT DESCRIPTION :
****
(a) General
The project objectives will be achieved by performing the following technical
activities :
Activity description
State-of-the art review of stiffened plate technology and design methods as
applied to floating and subsea systems (activity number 1)
State-of-the-art review of the current applications of composite sections and
associated technology/design practice (activity number 2)
Numerical computer analyses of the composite geometries proposed as part of the
physical test programme, using enhanced existing non-linear software package
(number activity 3)
Static and fatigue testing of a representative set of scale composite sections
(activity number 4)
Calibration of numerical analyses against physical test programme results
(activity number 5)
Parametric study performed on test results and additional data generated by
computer analyses (activity number 6)
Preparation of design logic, methodology and equations using information
generated from previous activities and from an assessment of safety requirements
(activity number 7)
Definition of case study scenario (activity number 8)
Design of case study scenario using conventional stiffened plate solution
(activity number 9)
Design of case study scenario using developed composite section design method
(activity number 10)
Design of case study scenario using developed composite section design
'software' tool (activity number 11)
Comparison and assessment of case study scenario solutions (activity number 12)
Reporting and design dossier (activity number 13).
(b) Laboratory Work
The objective of the physical model tests is to provide experimental data to be
used for calibrating the numerical computer model and to from part of the
database for the parametric study and design equation development activities.
The overall diameter of the test specimen has been initially selected as 2000 mm
which corresponds to an approximate model diameter scale of up 25%.
The test programme will investigate the following parameters :
* ratio of overall composite section diameter to thickness
* total steel thickness to composite section thickness ratio
* length of composite section between support positions
* the influence of shear connectors (either in the form of discreet studs or
continuous shear connectors)
* loading mode
    * static radial hydrostatic pressure
    * cyclic radial hydrostatic pressure
A structural lightweight concrete will be used in the composite section at a
density approximatly 1800 kg/m3.
Static specimens will be loaded by incremental increases in pressure until
collapse of the specimen occurs. Extensive instrumentation will be deployed and

deformations of the composite sections will be measured throughout with strains being monitored in the steel tubulars and at the concrete/steel interface.
The detailed work will include the
* design of the test equipment
* preparation of specifications for the supply of the test pieces and tests equipment
* placing of strain gauges on the test pieces

STATE OF ADVANCEMENT :
****
Ongoing. Start date scheduled for May 1990.

```
*******************************************************************************
* TITLE : THE TORTOISE PROJECT                       *        PROJECT NO      *
*                                                     *                        *
*                                                     *    TH./03293/88/UK/..  *
*                                                     *                        *
*******************************************************************************
* CONTRACTOR :                                        *    TELEPHONE NO        *
*    AKER ENGINEERING                                 *                        *
*                                                     *    01 630 7811         *
*    EGGINTON HOUSE                                   *                        *
*    25/28 BUCKINGHAM GATE                            *                        *
*    UK-LONDON SW1E 6LD                               *    TELEX NO            *
*                                                     *                        *
* TECHNICAL DIRECTOR :                                *    915779              *
*    K.P. WOOD                                        *                        *
*                                                     *                        *
*******************************************************************************
                                                     VERSION : 01/10/89
```

AIM OF THE PROJECT :
****
The objective of the TORTOISE project is to investigate the requirements for
subsea equipment protection, and to develop an innovative, low-cost energy
absorbing system in line with these requirements. The design of such a system
must also allow for ease of installation (including retrofit) and convenience of
maintenance and workover of the protected equipment.

PROJECT DESCRIPTION :
****
The initial phase of the TORTOISE project is directed towards obtaining
sufficient information to determine logical design parameters for the protection
system. This data-gathering phase will evaluate the risks and sources of damage,
the types of installation to be protected, the behaviour of dropped objects
underwater (experimental and theoretical analyses), impact mechanics underwater,
etc., to provide a full understanding of design requirements. This work,
combined with an initial evaluation of potential materials and structures and
various conceptual design proposals, will enable targets and preferred routes
for the subsequent phases of the project. Phase 2 will be directed towards the
development of the preferred concept(s) and materials testing as required, as
phase 3 is envisaged as the detailed, design construction and testing of the
prototype protection system.

STATE OF ADVANCEMENT :
****
Project started in March 1989.

RESULTS :
****
There is a need for subsea protection. A number of new concepts for protection
systems have been evolved. Preference at present is on systems to absorb initial
impact and then deflect. A theoretical analysis of objects free-falling through
water has been developed and refined. Results compare well with other models but
correlation with experiment is awaited. Initial analysis of underwater impact
has proved difficult.

```
*******************************************************************
* TITLE : SAF -#DEEPWATER SUBSEA            *      PROJECT NO      *
*         PRODUCTION#SYSTEM AND MAINTENANCE *                      *
*         SYSTEM#PHASE2 - STAGE B           *   TH./03294/88/IT/.. *
*                                           *                      *
*******************************************************************
* CONTRACTOR :                              *   TELEPHONE NO       *
*      AGIP S.P.A.                          *                      *
*                                           *   02 250 24059       *
*      P.O. BOX 12069                       *                      *
*      IT - 20120 MILAN                     *                      *
*                                           *   TELEX NO           *
*                                           *                      *
* TECHNICAL DIRECTOR :                      *   310246             *
*      MR. C. CHIMISSO - TEIN/TEIS DEPT     *                      *
*                                           *                      *
*******************************************************************
                                            VERSION : 22/06/89
```

AIM OF THE PROJECT :
****
Aim of this project is the development of a new generation of subsea system
specifically conceived for hydrocarbon production in deepwater (200-600 m w.d.)
and very deep water (beyond 600 m w,d,) including a dedicated remote controlled
installation and maintenance system.
During Phase 2 - Stage B, the prototypes of both the production system and the
maintenance system will be nanufactured.

PROJECT DESCRIPTION :
****
During Phase 1, the engineering of the production system and the maintenance
device has been performed.
During Phase 2 the prototype of the production system and the maintenance
vehicle will be manufactured.
Phase 2 is splitted in two stages. During Stage A qualified manufacturers will
be selected and the construction engineering will be performed.
During Stage B the subsystems will be manufactured. The captioned project (TH/03.
294/88) is relevant to Phase 2 - Stage B. Integration of the systems, dry tests,
installation on a live well and long term tests are the scope of Phase 3.

STATE OF ADVANCEMENT :
****
Phase 2 - Stage B will start on 01/01/89

RESULTS :
****
A prototype will be installed and tested on a live well. This constitutes the
Phase 3 of the project.

```
**************************************************************************
*  TITLE : DIVERLESS SUBSEA PRDUCTION SYSTEM      *       PROJECT NO     *
*          (DISPS). TEMPLATE SYSTEM TESTING       *                      *
*          PROGRAMME STAGE 3 (INSHORE TESTING).   *   TH./03296/88/UK/.. *
*                                                 *                      *
**************************************************************************
*  CONTRACTOR :                                   *   TELEPHONE NO       *
*     B.P. PETROLEUM DEVELOPMENT LTD              *                      *
*                                                 *   041 2252769        *
*     301 ST VINCENT STREET                       *                      *
*     UK - GLASGOW G2 5DD                         *                      *
*                                                 *   TELEX NO           *
*                                                 *                      *
*  TECHNICAL DIRECTOR :                           *   777633             *
*     N. RODDA                                    *                      *
*                                                 *                      *
**************************************************************************
```

VERSION : 01/03/90

AIM OF THE PROJECT :
****
Development of any offshore hydrocarbon reserves discovered in water depths in
excess of 350m on the European Continental Shelf is considered to be of crucial
importance to Europe. In order to sustain exploration in such areas with the
objective of increasing European recoverable reserves it is essential that cost
effective diverless subsea production technology is developed.
BP Petroleum Development Ltd's Diverless Subsea Production Systems (DISPS)
Project has the objective of providing diverless technology for natural drive
oil or gas production, oil production with gas lift or water injection duty by
the mid 100's when development of deeper water reservoirs will be necessary to
maintain European reserves of hydrocarbons.

PROJECT DESCRIPTION :
****
The project is a staged one following on from Stages 1 and 2 being undertaken
under Contract TH/03233/87. Stage 3 consists of a series of tests at an inshore
sheltered location using the Test Rig manufactured under Stage 1. Stage 4 will
consist of testing at an offshore location. These tests are aiming to
demonstrate in progressively mogre onerous operating conditions that the
components and systems, envisaged by the BP DIOSPS development programme for a
multiwell modular template, will work.
The BP DISPS development programme is developing the principles, concepts,
designs, techniques and equipment to exploit hydrocarbon reserves in 350m-750m
water depths utilising subsea diverless techniques. The principal system under
development is a multiwell (2-18 slot) modular template able to be installed and
maintained without the use of divers. The template is capable of catering for a
manifolded or non manifolded arrangement with satellite wells tied in, and
produced through either a remote or local, fixed or floating production facility.
 The project's objective is to verify the inshore and offshore testing to prove
the design basis of the interconnected equipment modules, which are retrieved to
the surface for repair/replacement of components.
The two sequential stages of more realistic testing than the land testing stage
are described below :
Stage 3 - Inshore Testing Stage (Included in Contract TH/03296/88)
The inshore testing Stage is required to verify those aspects of the design
which cannot be simulated on land, e.g. ROV manoeuvrability. Thus the tests are
confined to those which are sensitive to a change from air to a water
environment. It is proposed that tests be carried out in a sheltered water
inshore location after the test rig used in the land tests has been set on the
seabed. A crane (on a barge) will be used to deploy the RGV and modules into the
water. Good underwater visibility and a water depth of approximately 150m are
required. Loch Linnhe in the West of Scotland is proposed as the site to
undertake the tests.
Stage 4 - Offshore Testing Stage (Not included in Contract TH/03296/88).
The requirements are presently preliminary, but the objective of this testing
would be to give confidence of the system operations in the real harsh
environment.

STATE OF ADVANCEMENT :
****
The Test Rig used on the Land Testing has been installed in Loch Linnhe (Fort
William) in 150m water.
The RGV (Remote Guidance Vehicle) was used to deploy modules into the Test Rig,
and ROV intervention activities completed.
The Test Rig was retrieved to the surface, thus completing the testing which has
successfully proved the techyniques and principles of the DISPS template.

RESULTS :
****
Evaluation of results is in progress.

```
*****************************************************************************
* TITLE : SUCTION PILE PLATFORM CONCEPT OF MONO        *      PROJECT NO     *
*         TOWER TYPE                                   *                     *
*                                                      *  TH./03297/88/DK/.. *
*                                                      *                     *
*****************************************************************************
* CONTRACTOR :                                         *    TELEPHONE NO     *
*    RAMBOLL & HANNEMANN                                *                     *
*                                                      *    4542856500       *
*    TEKNIKERBYEN 38-DK 2830 VIRUM                      *                     *
*                                                      *                     *
*                                                      *    TELEX NO         *
*                                                      *                     *
* TECHNICAL DIRECTOR :                                 *    37108            *
*    KAI B. OLSEN                                       *                     *
*                                                      *                     *
*****************************************************************************
                                                       VERSION : 19/06/89
```

AIM OF THE PROJECT :
****
General. The purpose of the project is to demonstrate the feasibility of a three-
legged platform with suction pile foundation. The piles will be an integrated
part of the structure. In particular, the possible utilization of the suction
effect arising when a closed-top pile is exposed to tension, will be examined.
The platform concept is of the monotower type. The monotower has three legs,
each of which supported on a suction pile.The platform will be designed for
dynamic loads from waves and vortex shedding.
The topside facilities will be minimal. The platform will be unmanned and most
of the traditional process and safety equipment will be left out. Access to the
platform will be from boat. The topside facilities will not be examined in this
project.

PROJECT DESCRIPTION :
****
In this R&D project the intention is to investigate the feasibility of a
platform consisting of a monotower supported on a 3-legged subbase, where the
conventionally driven piles will be replaced by suction piles.
The new development in this project is to use the suction principle when
installing the three foundation piles of the monotower platform and also to take
into account the vertical suction effect being developed when the foundation
pile(s) are subjected to short time vertical tension (pull out) forces due to
dynamic loads from waves and vortex shedding on the superstructure.
This project is divided into five main phases of which :
Phase 1 :
Technical Analysis covers collection of state of the art data in order to
establish an initial platform concept and a geotechnical calculation model as a
basis for the detailed structural analyses.
Phase 2 :
Econmical Analyses covers a comparison between a traditional platform concept
and a suction pile concept with respect to fabrication, installation and removal
costs.
Phase 3 :
Reporting reports on Phase 1 and 2.
Phase 4 :
Laboratory test supports the geotechnical calculation model with model results
and makes reevaluation of the theoretical calculation model possible.
Phase 5 : Re-calculation and reporting is the summarizing phase in which the
platform is re-designed, platform costs are re-evaluated and the final report is
written.

STATE OF ADVANCEMENT :
****
Ongoing
Phase 1 : conceptual stage of design.
Phase 4 : execution of primary tests initiated.

RESULTS :
****
Preliminary results show that it is possible to formulate a geotechnical
calculation model for the breakout capacity of suction piles embedded in pure
sand or clay profiles.
This is despite the fact that an extensive literature study showed tat the
subject only scarcely was covered and none of the sources tried to quantify the
mechanism of a load induced suction and its favourable effect on the pile pull-
out resistance.
It is believed that within the next project period, a simplified model for the
breakout resistance of a suction pile will be formulated and incorporated in the
ongoing design calculations and thus enable more detailed calculations.

```
************************************************************************
* TITLE : SUBSEA VALVES                           *      PROJECT NO    *
*                                                 *                    *
*                                                 *   TH./03299/89/IT/..*
*                                                 *                    *
************************************************************************
* CONTRACTOR :                                    *    TELEPHONE NO     *
*    GROVE ITALIA SPA                             *                    *
*                                                 *    (0383) 6911     *
*    STRADA CAMPOFERRO 15                         *                    *
*    IT - 27058 VOGHERA (PV)                      *                    *
*                                                 *    TELEX NO        *
*                                                 *                    *
* TECHNICAL DIRECTOR :                            *    320567          *
*    MR. G. BIANCHI                               *                    *
*                                                 *                    *
************************************************************************
                                                   VERSION : 30/04/90
```

AIM OF THE PROJECT :
****
To obtain an emergency shut down valve to isolate the platform in case of
accident. The product is also easy maintenable in a deep sea installation
diverless.

PROJECT DESCRIPTION :
****
To complete two valves (20" 900 and 36" 900) with different seat design. To
complete the relevant repair tool and maintenance frame. To test valve with
external load applied (piping bending moment) and to demonstrate valve
maintainability by shallow water test.

STATE OF ADVANCEMENT :
****
Ongoing. Design and construction.

RESULTS :
****
The contractor has already manufactured a 20" 900 prototype, on which the
relevant repair tool, maintenance frame and tests will be applied.

REFERENCES :
****
TECHNICAL PAPER RELEVANT TO THE SUBSEA VALVE HAS BEEN PRESENTED TO THE SUBSEA
CONFERENCE HELD IN LONDON ON APRIL 1989.

```
*********************************************************************
* TITLE : SAF - DEEPWATER SUBSEA PRODUCTION        *     PROJECT NO    *
*          SYSTEM AND MAINTENANCE SYSTEM PHASE 3    *                   *
*                                                   *  TH./03300/89/IT/..  *
*                                                   *                   *
*********************************************************************
* CONTRACTOR :                                      *   TELEPHONE NO    *
*    AGIP SPA                                        *                   *
*                                                   *   520 24059       *
*    P.O.BOX 12069                                  *                   *
*    I - 20120 MILANO                               *                   *
*                                                   *   TELEX NO        *
*                                                   *                   *
* TECHNICAL DIRECTOR :                              *   310246          *
*    C. CHIMISSO                                     *                   *
*                                                   *                   *
*********************************************************************
                                              VERSION : 15/01/90
```

AIM OF THE PROJECT :
****
Integration of the various subsystems and experimental testing in real operative
conditions of the diverless and guidelineless subsea production system and its
maintenance system, in order to ascertain the operative behaviour, reliability
and safety.
The project will use and develop advanced technologies in several fields, such
as, for example, metallurgy, robotics, electronics, applied software, subsea
telemetry and electric-hydraulic connections.

PROJECT DESCRIPTION :
****
During the previous phase, the project has een developed in two main items, that
are independent but strictly correlated : the production system and the
maintenance system. Following the good experience in the previous phase, the
project will be developed by a task force in order to assure the best
integration between the two above systems.
After the integration work, the systems will be extensively tested in yard and
shallow water in order to ascertain their operability and to perform possible
improvements before going offshore. After the installation on Laura gas well,
Ionian sea, long term tests in real operative conditions will be performed and
possible improvements for a second generation will be pointed out.

STATE OF ADVANCEMENT :
****
The project will start on 02.04/90

RESULTS :
****
A prototype will be installed and tested on a live well in real operative
conditions.

```
*******************************************************************************
* TITLE : DEEPWATER AUTONOMOUS MULTIWELL          *        PROJECT NO        *
*         PRODUCTION SYSTEM D.A.M.P.S.            *                          *
*                                                 *    TH./03304/89/IT/..    *
*                                                 *                          *
*******************************************************************************
* CONTRACTOR :                                    *      TELEPHONE NO        *
*    TECNOMARE SPA                                *                          *
*                                                 *      041 796711          *
*    S.MARCO 3584                                 *                          *
*    I - 30124 VENEZIA                            *                          *
*                                                 *      TELEX NO            *
*                                                 *                          *
* TECHNICAL DIRECTOR :                            *      410484              *
*    L. BECCEGATO                                 *                          *
*                                                 *                          *
*******************************************************************************
                                                     VERSION : 15/01/90
```

AIM OF THE PROJECT :
****
The aim of the project is :
a) the development of an autonomous multiwell production system for hydrocarbons
fields including the fabrication and testing of an underwater power generator
prototype
b) development of elastic waves signal transmission technique for system
controlling
c) development of a diverless guidelineless technique for integration and
maintenance operations
The system is autonomous i.e. the use of an electrohydraulic umbilical cable is
not foreseen being the system controlled by signals transmitted on the flowlines
(by means of elastic waves) and being powered by an autonomous in situ generator
(which exploits the difference of temperature between the field production and
the environment as conversion principle).

PROJECT DESCRIPTION :
****
This research project involves the development of the sub-systems of the
submarine production plants, which can be installed and maintained without
guidelines, and suitable for the exploitation of oil and gas fields in water
depths up to 1000 m, but also valid from an economic point of view for medium
deep waters (200-500 m). It is a modular system which, utilizing the same basic
modular components, can be set up in two configurations : the production unit
and the maniforld unit exhibiting for both configurations an autonomous
behaviour capability. The research project foresees the carrying out of :
. The design of the entire submarine production system, installation/maintenance
procedures and modules (with the integration of technologies and equipment
derived from S.A.F. project)
. The design and development of an autonomous control system.
It will also include the realization and testing of an underwater power
generator prototype.
The project proposes to utilize as far as is possible and compatible with the
particular requirements of a multiwell production system, those components
developed as part of the S.A.F. project (master vehicle, modules, and equipment)
financed by the E.E.C. for a single production tree allowing an extension of the
S.A.F. concept to a more complex and multifunctional systems.

STATE OF ADVANCEMENT :
****
Ongoing

RESULTS :
****
- Development of the topology of a production or manifolding template in order
to rationalize and simplify inspection or light maintenance operations for
presently existing vehicles.
-The development of a heavy maintenance system consisting of a vehicle and
suitable working rools capable of satisfying the requirements of a guidelineless,
multiwell, subsea system and suitable to operate even within a "closed"
template architecture
- Identification and development of a control technology with autonomous power
generation and signal and command transmission via flowlines
- Definition and design of the autonomous control system, and testing of
critical components : namely the in-situ electric power generator (the
development of such generator exploiting for energy conversion the thermal
energy of field production fluid allows to have technological spin-out in
various field of applications also outside the offshore one).

REFERENCES :
****
"THE ROLE OF AUTONOMOUS CONTROL SYSTEM IN SUBSEA PRODUCTION" OFFSHORE TECHNOLOGY
CONFERENCE, HOUSTON (USA),1989
"TSPS - A DRY PRODUCTION SYSTEM FOR VERY DEEP WATERS", COPPE-UFRJ OFFSHORE
SYMPOSIUM, RIO DE JANEIRO (BRAZIL), 1987

"SUBSEA WET PRODUCTION SYSTEM... - A SYSTEMATIC APPROACH", COPPE-UFRJ OFFSHORE SYMPOSIUM, RIO DE JANEIRO (BRAZIL), 1987

"ACOUSTIC WELLHEAD CONTROL SYSTEM DEVELOPMENT AND DRY TEST PHASE", CONGRESSO AIOM, VENICE (ITALY), 1986

"SISTEMA DI PRODUZIONE SOTTOMARINO PER GIACIMENTI DI IDROCARBURI", EEC SYMPOSIUM, LOUXEMBOURG, 1985.

```
********************************************************************************
* TITLE : PLATINE IV                          *        PROJECT NO          *
*                                             *                            *
*                                             *    TH./03305/89/FR/..      *
*                                             *                            *
********************************************************************************
* CONTRACTOR :                                *       TELEPHONE NO         *
*    GERTH                                    *                            *
*                                             *       47 52 61 39          *
*    AVENUE DU BOIS PREAU 4                   *                            *
*    FR - 95002 RUEIL-MALMAISON.              *                            *
*                                             *       TELEX NO             *
*                                             *                            *
********************************************************************************
* TECHNICAL DIRECTOR :                        *                            *
*    MR P. LEFEVRE                            *                            *
*                                             *                            *
********************************************************************************
                                                   VERSION : 15/01/90
```

AIM OF THE PROJECT :
****
The aim of the PLATINE is to specify a central production platform, unmanned,
teleoperated from a control room located onshore, with a yearly (or half-yearly)
visit for inspection and overhauling.

PROJECT DESCRIPTION :
****
The purpose of the project mis to work out, construct, install and test pilots
related to main sub-systems, either specific to oil industry or not.
Those specific to oil industry will be related to oil and gas treatment :
deoiling, separation and treatment on oil fields, gas treatment, automatic pig
launcher/receiver.
Those non specific to oil industry will be related to telecommunication and
control, rotating machinery, safety and surveillance.
The telecommunication and control pilots will allow the platform to operate
unmanned, they will include real-time on-line expert systems, for a gas field or
an oil field.
The rotating machinery pilots will be related to high speed electric rotot, a
gas lift compressor an auxiliary equipment.
The safety and surveillance pilots will be related to safety equipment,
including techniques such as robot, automaton, acoustic device, video camera...,
and a platform shock detection system.
Moreover, the final goal will consist in writing all general or particular
PLATINE specifications related to the various branches of engineering :
electricity, instrumentation, mechanics...
Besides, this will requires that local authorities modify the existing
regulation on specific matters.

STATE OF ADVANCEMENT :
****
Starting on 01/08/89.

```
*******************************************************************
* TITLE : METAL/METAL SEALING SYSTEMS          *    PROJECT NO      *
*                                              *                    *
*                                              *  TH./03306/89/FR/.. *
*                                              *                    *
*******************************************************************
* CONTRACTOR :                                 *    TELEPHONE NO     *
*    MOTI                                       *                    *
*                                              *  59 33 90 60        *
*    Z.A. DE MONTARDON                          *                    *
*    FR - 64121 SERRES CASTET                   *                    *
*                                              *    TELEX NO         *
*                                              *                    *
* TECHNICAL DIRECTOR :                          *                    *
*    MR ALAIN BASSE                             *                    *
*                                              *                    *
*******************************************************************
                                          VERSION : 15/01/90
```

AIM OF THE PROJECT :
****
Increasing the lifetime of bottomhole equipment used for drilling or completion
of wells constitutes a very important objective within the oil industry, as it
may imply a significant cost reduction, regarding both the drilling phase and
the completion phase during the life of the wells.
The lifetime of this equipment is presently limited by their sealing systems,
based on elastomers whose failure may not only lead to very costly work-over
operations, but also have an impact on the safety of wells.

PROJECT DESCRIPTION :
****
The object of the project is to develop metal/metal sealing systems, without
elastomers in order to solve the so-called "dynamic" sealing problems on
bottomhole equipment, such as :
- subsurface safety valve, sliding circulating valve, expansion joint, for the
completion of wells;
- downhole motors and jars for the drilling of wells.
The proposed technological development is based on the utilization of ceramic
materials, offering the following characteristics :
- very low dry friction coefficient in order to eliminate the seizing and wear
risks;
- indifference to well bottomhole conditions : high temperature, corrosion,
erosion, pressure...;
- high accuracy machining and grinding in order to optimize the clearance and
the contact pressures.
Tribological tests on models of different metal/metal sealing systems will allow
to select the best ceramic couples (boron carbide/boron carbide for instance),
with dry friction coefficients (0.1) four times superior to those of steel.
Metal/metal sealing systems will be adapted to two types of bottomhole equipment
:
1. volumetric downhole motor, comprising a metal stator, without elastomers;
2. hydraulically controlled safety valve (VSF), whose "dynamic" sealing between
the valve body and the sliding tube inside the valve must be perfect, whether in
opened or closed position.
After the tests under simulated conditions, prototypes of these equipments will
be tested under real operating conditions in an experimental well (in France in
ELF Aquitaine Test Center at le Fourc and in Scotland at Montrose).

STATE OF ADVANCEMENT :
****
Ongoing

```
*****************************************************************************
* TITLE : CONTROL OF WATER PRODUCTION IN          *     PROJECT NO        *
*         PETROLEUM RESERVOIRS UNDER EXTREME       *                       *
*         CONDITIONS                               *   TH./03307/89/FR/..  *
*                                                  *                       *
*****************************************************************************
* CONTRACTOR :                                     *   TELEPHONE NO        *
*    GERTH                                         *                       *
*                                                  *   47 52 61 39         *
*    AVENUE DU BOIS PREAU 1 ET 4                   *                       *
*    FR - 92500 RUEIL-MALMAISON                    *                       *
*                                                  *   TELEX NO            *
*                                                  *                       *
*****************************************************************************
* TECHNICAL DIRECTOR :                             *                       *
*    MR N. KOHLER                                  *                       *
*                                                  *                       *
*****************************************************************************
                                                    VERSION : 01/01/90
```

AIM OF THE PROJECT :
****
The aim of the project is to reduce the exploitation costs of reservoirs
affected by water production in production wells and under difficult
environmental conditions : offshore, heterogeneities, high temperature; low or
very high permeability of reservoirs.

PROJECT DESCRIPTION :
****
The purpose of the project is to develop processes for the control of water
production in petroleum reservoirs under extreme conditions.
Phase 1 : The foreseen solutions to be developed in the lab are based on :
- low hydrodynamic volume polymers, for low permeability reservoirs (K<100 mD)
- formulations containing polymers and thermal stabilizers, for high temperature
reservoirs (T>95 deg.C)
- low crosslinked polymer systems or alternative gel systems for high
permeability reservoirs (K>1000 mD) and low clay content.
Phase 2 : A methodology for in-well tests will be developed according to the
following steps :
- diagnosis of water production from exploitation data
- study of sensitivity towards reservoir parameters
- study of the mechanisms of action
- risks of well plugging
- definition of the well test programme.
Phase 3 : A technical and economic evaluation will allow to define the
application domains for processes developed within the project.

STATE OF ADVANCEMENT :
****
Ongoing work on phase 1 since 01/08/89

```
***********************************************************************
* TITLE : FLAW VISUALISATION USING A.C. FIELD      *     PROJECT NO     *
*         MEASUREMENT.                             *                     *
*                                                  *   TH./03309/89/UK/.. *
*                                                  *                     *
***********************************************************************
* CONTRACTOR :                                     *    TELEPHONE NO     *
*     TECHNICAL SOFTWARE CONSULTANTS LTD           *                     *
*                                                  *   0908 669411        *
*     34 LINFORD FORUM                             *                     *
*     ROCKINGHAM DRIVE                             *                     *
*     LINFORD WOOD                                 *    TELEX NO          *
*                                                  *                     *
* TECHNICAL DIRECTOR :                             *    82304             *
*     M.C. LUGG                                    *                     *
*                                                  *                     *
***********************************************************************
                                                   VERSION : 15/01/90
```

AIM OF THE PROJECT :
****
The aim of the project is to develop the non-contacting a.c. field measurement
technique for use with an array of probes to enable rapid scanning for defects
around tubular welded intersections. The results from a complete scan around a
weld well be used to produce a graphic display of the positions of any cracks
found together with a depth estimate. The project will result in the production
of a probe for either hand-held or robotic arm manipulation. This innovative
technique will then provide significant benefits over MPI in terms of speed of
use for crack detection, and over eddy currents due to its crack sizing
capability.

PROJECT DESCRIPTION :
****
The project involves several interdependent areas of work. Electronic
development will be undertaken to produce an ultra-low noise fast multiplexer
and input amplifying circuit to accurately sample a.c. signals at the sub-
millivolt level. An array probe incoproating 100 separate sense coils will be
designed containing internal springing to allow the weld toe profile to be
foolowed. The final piece of hardware to be produced is an electro-magnetic
position encoder to be wrapped around a weld line and read by the array probe.
The complete system will result in the accumulation of large amounts of data to
be analysed and reduced. To do this, software routines will be written to detect
the signals due to cracks using theoretically-derived algorithms. Further
routines will estimate the depths and lengths of any cracks found and will
produce a graphical display of a complete weld scan. A series of experimental
trials will validate each stage of the project, culminating in tests on a range
of fatigue cracked tubular welded intersections.

STATE OF ADVANCEMENT :
****
The project is currently in the planning stage in preparation for the start date
of 1st January 1990.

REFERENCES :
****
M.C.LUGG, A.M.LEWIS, D.H. MICHAEL AND R. COLLINS, IN "ELECTROMAGNETIC
INSPECTION", I.O.P. SHORT MEETINGS, VOL.12, I.O.P. PUBLISHING, BRISTOL (1988) PP
41-48.
A.M. LEWIS, D.H. MICHAEL, M.C. LUGG AND R. COLLINS, J. APL.PHYS., VOL. 64, PP
3777-3784, (1988).4.

```
*********************************************************************************
* TITLE : GEOMETRY OF HYDRAULIC FRACTURES        *      PROJECT NO         *
*                                                *                         *
*                                                *    TH./03311/89/NL/..   *
*                                                *                         *
*********************************************************************************
* CONTRACTOR :                                   *    TELEPHONE NO         *
*    COLLEGE VAN BESTUUR                          *                         *
*                                                *    (015)785104/1328      *
*    P.O.BOX 5                                    *                         *
*    NL - 2600 AA DELFT                           *                         *
*                                                *    TELEX NO             *
*                                                *                         *
* TECHNICAL DIRECTOR :                           *.*    38151               *
*    MR. C.J. DE PATER                            *                         *
*                                                *                         *
*********************************************************************************
                                                   VERSION : 10/11/89
```

AIM OF THE PROJECT :
****
The aim of this project is to improve the efficiency of hydraulic fracture
treatments, by determining the geometry of hydraulic fractures in the vicinity
of the borehole, when the final propagation plane differs from the initial
propagation direction. This will be accomplished by physical and numerical model
studies of the fracture process.
The research will aid in innovation of the design of hydraulic fracture
treatments, with regard to fracture reorientation.

PROJECT DESCRIPTION :
****
A) Acoustics
The first steps will be to prepare the acoustic technique for application to a
model test with a simple fracture. We will develop sensors, assess resolution,
model the acoustic diffraction and process the data obtained in the tests.
After the first phase the acoustic technology will be applied to both
transmission and reflection measurements in the model tests.
B) Model tests
Model tests will be performed on blocks of rock and cement with a cased borehole
from which fractures are propagated by applying fluid pressure in the borehole.
The experimental set-up will be made and the design of the model tests will be
developed :
- construction and testing of injection system, casings
- preparation of blocks
- dimension analysis and design of models
  . relation model to field application
  . scale effect on fracture propagation
The following model tests will be conducted :
- tests with two different borehole diameters (scale effect)
- reorientation tests with variation of : inclination, stress condition,
perforation position and rock type.
C) Material property tests
Advanced material property tests will be used for the interpretation of the
model tests. The following experiments/activities will be performed :
- direction tension tests
- mixed mode testing
- study of fracture reorientation
- construction of load frame for Iosipescu-beam shear tests (specialized four
point bending tests)
D) Numerical modelling
We will perform computations to :
- study initiation of cracks
  . 30 finite element analysis of model tests (DIANA finite element program)
  . smeared crack approach
- modelling with DIANA to obtain material parameters for fracture reorientation
- simulation using a simple hydraulic fracture model.

STATE OF ADVANCEMENT :
****
Ongoing. The project is in the design phase.

```
**********************************************************************
* TITLE : DEVELOPMENT OF COMBINED                 *    PROJECT NO     *
*         ELECTRIC/HYDRAULIC CONTROL AND          *                   *
*         TRANSPORT UMBILICALS IN LONG LENGTHS    *  TH./03329/89/DE/..*
*         EMPLOYING AN STRANDING METHOD           *                   *
**********************************************************************
* CONTRACTOR :                                    *   TELEPHONE NO     *
*    NORDDEUTSCHE SEEKABELWERKE AG                 *                   *
*                                                 *   04731/82340      *
*    POSTFACH 1464                                *                   *
*    DE - 2890 NORDENHAM                          *                   *
*                                                 *   TELEX NO         *
*                                                 *                   *
* TECHNICAL DIRECTOR :                            *   238315           *
*    MR. W. HELLE                                 *                   *
*                                                 *                   *
**********************************************************************
                                              VERSION : 30/04/90
```

AIM OF THE PROJECT :
****
For the connection of satellite wells with a control production platform a combined electrical/hydraulic umbilical without fabric joints or junction boxes is to be developed. For safety reasons it is necessary to manufacture the umbilicals in unbroken length up to 10 km or longer. Whereas the electrical, optical and hydraulic elements can be manufactured cost effectively using the production facilities generally available today, a new method of stranding must be used to produce such combined electric/hydraulic umbilicals.
In this project, a version of a stranding method with alternating stranding direction (also known as SZ-stranding) which has been known for some considerable time is to be developed further. The demand for this stranding method is to strand the sensitive and expensive elements in such a way that the quality of these elements will not decrease, neither during the stranding process nor in operation.

PROJECT DESCRIPTION :
****
In this project, a version of a stranding method with alternating stranding direction (also known as SZ-stranding) which has been known for some considerable time is to be developed further.
To date, this method has been used primarily only for stranding of smaller gauge telecommunications cables, switchboard cables and control cables. More recently, this method has also been used in the production of optical fibres telecommunications cables.
For the production of large-volume combined umbilicals, the method using the tubular stranding process would appear to be the most promising under the technical and economic aspects.
The objectives of this project will be to further develop this stranding method and to employ the method for the production of combined electric/hydraulic umbilicals. It is important to obtain further information on the mechanical behaviour of the stranding elements at the points at which the stranding changes direction.
The project will be performed in three parts :
1. Planning and design of an electric/hydraulic umbilical.
2. Production of an electric/hydraulic umbilical.
3. Theoretical and experimental examination of the electric/hydraulic umbilical.
1. PLANNING AND DESIGN OF AN ELECTRIC/HYDRAULIC UMBILICAL
The demands made on an electric/hydraulic control umbilical with regard to the transport, energy and data transmission properties were discussed with the operators of production platforms and satellite wells.
The umbilical will be designed in line with these demands. The requirements are the intended application, the laying and removal are to be taken into consideration.
The characteristics of the chosen stranding method, in particular here the behaviour of the stranding elements with their differing elasticities (high-voltage conductors, hydraulic hoses and optical fibres) at the reversal points, are to be determined in a preliminary experiment. The number of lays in each direction, the lengths and positions of the reversal points, the type of lacing, etc. are to be optimized.
2. PRODUCTION OF AN ELECTRIC/HYDRAULIC UMBILICAL
After evaluation of the results of the preliminary experiment and the detailed design documentation, a test umbilical will be produced.
3. THEORETICAL AND EXPERIMENTAL EXAMINATION OF THE ELECTRIC/HYDRAULIC UMBILICAL
During the course of the project, existing theoretical models of the umbilical used to calculate the mechanical properties must be adapted to incorporate the particular characteristics of this method of stranding. The aim is to be able to determine the mechanical properties of the umbilical as early as the draft stage. The theoretical models must be examined on existing and/or modified testing facilities. Proof must be obtained that this umbilical is able to withstand the stresses to which it is exposed during laying, trenching and pulling into the production facilities as well as during its subsequent operation.

STATE OF ADVANCEMENT : Ongoing

SECONDARY AND ENHANCED RECOVERY

```
******************************************************************************
* TITLE : EMERAUDE - EVALUATION OF RESULTS.          *      PROJECT NO       *
*                                                    *                       *
*                                                    *    TH./05050/83/FR/..  *
*                                                    *                       *
******************************************************************************
* CONTRACTOR :                                       *    TELEPHONE NO       *
*    GERTH                                           *                       *
*                                                    *    1 47 52 61 39      *
*    AV. DE BOIS PREAU 4                             *                       *
*    FR - 92502 RUEIL-MALMAISON                      *                       *
*                                                    *    TELEX NO           *
*                                                    *                       *
* TECHNICAL DIRECTOR :                               *    203 050 F          *
*    MR. B. COUDERC                                  *                       *
*                                                    *                       *
******************************************************************************
                                                    VERSION : 01/01/88
```

AIM OF THE PROJECT :
****
This project concerns a steam injection pilot for the EMERAUDE heavy oil field,
offshore Congo (ELF-CONGO/AGIP), in a water depth of 65 m.
The important accumulation (575 million tons) and the low recovery rate (3%)
obtained by primary production explain the efforts that were undertaken to make
a pilot and prove that an industrial development is possible by injecting steam
in spite of existing problems (shallow beds, reservoir, nature, oil quality...).

PROJECT DESCRIPTION :
****
Engineering works, started end 1980, within the scope of contract TH 5033/81,
have made it possible to define the platforms and equipment necessary for this
pilot. The equipment was set in-place between December 1981 and June 1983.
Drilling operations started in July 1983, without the help of the Community,
continued in 1984. The first water injection tests began in October 1983, within
the scope of contract TH 5042/82. Steam injection tests began in 1985.
The present contract is divided into three phases :
- preparation of a numerical model
- analysis of measures taken during tests
- final results of the pilot.

STATE OF ADVANCEMENT :
****
Completed

RESULTS :
****
A new thermal model to suit the specific characteristics of the EMERAUDE field
was modified. Calibration of the state of reservoirs observed during the EMV
drilling operations, primary depletion period and of different tests ran in 1983,
 has made it possible to test these modifications.
Interpretation of the results after a 30 months continuous steam injection in
five spots R1 and R2 and the two cycles ran in the huff and puff well R3, led to
the following conclusions :
- At level R1 :
  After perforating most layers at this level, a sharp increase in oil
production has been noticed (from 30 m3/d to 200 m3/d) and thermal reactions
have been observed on two different wells. A radial circular model has been used
which showed that some tertiary oil could be produced.
- At level R2 :
  Very encouraging results have been obtained at this level. Three wells reacted
to steam injection. Two are still on an increasing phase as the last one is on a
decreasing phase after having produced at a rate four times bigger than primary
production rate. Computed results and measurements are in good agreement, steam
drive is a technical success at tis level.
- At level R3 :
  Abandon cyclic steam injection this process is not suitable for this
geological pattern.
By end 1987 it was impossible to draw all the conclusions. A technico-economic
report seems premature, particulary in this degrading economical environment
ever since realisation of the pilot has been decided. However, extension of the
pilot to all or part of the Emeraude field seems totally excluded.

REFERENCES :
****
EMERAUDE VAPEUR : A STEAM PILOT IN AN OFFSHORE ENVIRONMENT-PAPER SPE 16723
PRESENTED AT THE 62ND ANNUAL SPE MEETING AT DALLAS - 27TH-30TH SEPTEMBER 1987.
EMERAUDE VAPEUR : AN OFFSHORE STEAM PILOT. PAPER PRESENTED AT THE 4TH SYMPOSIUM
ON ENHANCED OIL RECOVERY AT HAMBURG - 27TH - 29TH OCTOBER 1987.

```
*********************************************************************
* TITLE : NITROGEN GAS INJECTION                 *     PROJECT NO    *
*                                                *                   *
*                                                *  TH./05053/84/NL/.. *
*                                                *                   *
*********************************************************************
* CONTRACTOR :                                   *    TELEPHONE NO   *
*    TECHNISCHE HOGESCHOOL DELFT, AFD. MIJNBOUWKUNDE *               *
*                                                *   015 781617       *
*    POSTBUS 5028                                *                   *
*    NL - 2600 GA DELFT                          *                   *
*                                                *    TELEX NO       *
*                                                *                   *
* TECHNICAL DIRECTOR :                           *    38151          *
*    PROF.DR.IR. J. HAGOORT                       *                   *
*                                                *                   *
*********************************************************************
                                          VERSION : 31/12/88
```

AIM OF THE PROJECT :

The aim of the project is to study the possibility of miscible displacement in North Sea hydrocarbon reservoirs by nitrogen, and to study the recovery efficiency during nitrogen injection.
The advantages of nitrogen with respect to carbon dioxide are :
- Low production costs of the injection gas
- Availability
- Nitrogen is inert with respect to the wells and production facilities.
The advantage of nitrogen injection with respect to methane reinjection is that the methane is available for commercial exploitation.

PROJECT DESCRIPTION :

The project comprises the following items :
a. A theoretical study on gas injection in oil reservoirs. A literature study has been performed on Miscible gasinjection.
b. A theoretical study on N2-oil phase behaviour. A literature study on hydrocarbon systems with nitrogen has been performed.
c. Development of a computer model. A fully 1-D compositional reservoir simulator has been modified and calibrated for the simulation of slim tube experiments.
d. Design and construction of experimental rigs. A high pressure (1000 bar) and temperature (120 C) slim tube apparatus, including a two phase sample facility has been designed and constructed. A high pressure (1000 bar) and temperature (120 C) vapor liquid equilibrium cell, including an in situ sample facility has been designed and constructed. A high pressure (1000 bar) and temperature (250 C) pendent drop cell, for interfacial tension measurements, has been designed and constructed.
e. Investigation of phase behaviour of oil/nitrogen. Numerous calculations have been carried out. Interaction coefficients have been experimentally determined.
f. Conducting displacement experiment. Numerous displacement experiments have been carried out:
- atmosferic: miscible, immiscible and partly miscible fluid-fluid displacements, gravity drainage experiments in the presence of three phases (oil, water and air);
- 50-100 bar, 50 deg.C: Slim tube experiments using carbon dioxide as injection gas and pentane, decane, butylbenzene or a mixture as the reservoir fluid;
- 320-420 bar, 100 deg.C: Slim tube experiments using nitrogen as injection gas and a simple synthetic oil, representing a Statfjord fluid, as reservoir fluid.
g. Field test proposaland economic evaluation. An introductionary economic evaluation has been performed on nitrogen injection vis-a- vis a depletion scenario for a North Sea field.

RESULTS :

Fluid phase equilibria of nitrogen and aromatic compounds have been measured and interpretated.
An introductionary economic evaluation has been performed.
A high pressure slim tube apparatus has been developped.
A two phase in situ sampling device has been designed and constructed. With this device the compositional path can be measured under reservoir conditions.
A fully compositional 1-D reservoir simulator has been modified and calibrated to describe slim tube experiments. With the help of this simulator the number of slim tube experiments, necessary to determine the minimum miscibility pressure can reduced.
The results of the simulation agree with the experiments.
Gravity drainage experiments pointed out that capillary forces are dependent on the interfacial tensions between all the phases present.
A high pressure pendent drop cell has successfully been designed and constructed.
The measurements on a carbodioxide system indicate that the parachoor value can be seen as a correlation coefficient. Once determined, the interfacial tension can be estimated.

REFERENCES :

238

****
- HAGOORT J. AND D.M. BOERSMA: FINAL REPORT OF EEC-PROJECT NR TH/05053/84
"NITROGEN GAS INJECTION", REPORT DELFT UNIVERSITY OF TECHNOLOGY, FACULTY OF
MINING AND PETROLEUM ENGINEERING, JUNE 1989.
- HAGOORT J., BRINKHORST J.W AND VAN DER KLEIN, P-H.:"DEVELOPMENT OF AN OFFSHORE
GAS-CONDENSATE RESERVOIR BY NITROGEN INJECTION VIS-A-VIS PRESSURE DPLETION":
PROCEEDINGS EUROPEAN PEROLEUM CONFERENCE FACING THE 90's NEW CHALLANGES - NEW
SOLUTIONS. LONDON 20-22 OCTOBER 1986. SOCIETY OF PETROLEUM ENGINEERS, RICHARDSON,
 TEXAS, 1986, BLZ 233-242, SPE PAPER 15873.

```
****************************************************************************
* TITLE : TERTIARY RECOVERY BY CYCLIC STEAM        *      PROJECT NO      *
*         INJECTION IN THE DEPLETED HEAVY OIL       *                      *
*         FIELD OF TOCCO CASAURIA                   *    TH./05057/84/IT/.. *
*                                                   *                      *
****************************************************************************
* CONTRACTOR :                                      *      TELEPHONE NO    *
*     AGIP SPA                                      *                      *
*                                                   *     2.520.5884       *
*     CP 12069                                      *                      *
*     IT - 20120 MILANO                             *                      *
*                                                   *      TELEX NO        *
*                                                   *                      *
* TECHNICAL DIRECTOR :                              *     310 246          *
*     DR.G.SLOCCHI                                  *                      *
*                                                   *                      *
****************************************************************************
                                              VERSION : 30/08/89
```

AIM OF THE PROJECT :
****
To evaluate an old, abandoned oil reservoir in Central Italy through geophysical,
geological and drilling investigations. To study rock and oil samples obtained
from drilling in order to evaluate the potential of alternate (cyclic) steam
injection and production as an EOR process in this fractured carbonate reservoir
rock. To carry some cycles of huff-n-puff in order to validate the study.

PROJECT DESCRIPTION :
****
1. Shooting of shallow seismic and interpretation in order to evaluate the shape
and dipping of the structure and to locate the best position for the wells to be
drilled.
2. Drilling of two wells, coring and sampling of the fluids. Production testing
of the wells.
3. Study of reservoir rock and fluids and lab evaluation of the potential of
alternate (cyclic) steam injection.
4. Carrying out of some cycles of steam injection in at least one well and
interpretation of the results obtained.
5. Preparation of the final report.

STATE OF ADVANCEMENT :
****
Abandoned after the completion of Phase 1.

RESULTS :
****
The description of the structure from seismic interpretation (Phase 1) was of an
highly faulted and very dipping (gt. 60 deg.) reservoir.
The main consequence of such result was the availability of a very limited
number of drilling locations for the two wells forecasted to run the steam
injection project. All possible locations were duly inspected on-site, taking
into account the space needed for a safe drilling. Many of them were eliminated,
because of the orography of the zone, morphologically highly irregular. The few
locations remaining were marginal with respect to the reservoir. The high risk
to drill out of the structure and the impossibility to deviate the wells because
on the shallow depth (max 200 m) added to the risk of technical failure of the
experimental programme. Therefore the project was abandoned.

```
**********************************************************************
* TITLE : PREPARATION OF GAS INJECTION IN GRAND    *      PROJECT NO   *
*         ALWYN DEPOSIT.                            *                   *
*                                                   *   TH./05058/84/FR/..  *
*                                                   *                   *
**********************************************************************
* CONTRACTOR :                                      *   TELEPHONE NO    *
*    GERTH                                           *                   *
*                                                   *   1 47 52 61 39   *
*    AVENUE DE BOIS PREAU 4                          *                   *
*    FR - 92502 RUEIL-MALMAISON                     *                   *
*                                                   *   TELEX NO        *
*                                                   *                   *
* TECHNICAL DIRECTOR :                              *   203050          *
*    MR. G. AUXIETTE                                *                   *
*                                                   *                   *
**********************************************************************
```

VERSION : 31/12/88

AIM OF THE PROJECT :
****
The project consists in developing a set of thermodynamic models to provide a
really reliable restitution of the PVT laboratory experiments, and hence give a
better simulation of injections in a reservoir model, particularly as regards
miscible gas. This is essential to optimize the implementation of a pilot phase
in a project, which can be extremely expensive, especially in the North Sea
(e.g. ALWYN).

PROJECT DESCRIPTION :
****
This project consists of two phase :
PHASE 1 - MEASUREMENTS AND ANALYSES
The thermodynamic model at present available, requires data on the critical
properties of the components and the interaction coefficients between the
heaviest specific cuts and the lightest components. An apparatus has been
developed in order to determine the interaction coefficients between the heavy
components and the CO_2, by experimental determination of the phase envelopes
under given pressures and temperatures.
The heavy cut is represented by a pseudocomponent to which are attributed
critical properties, based on experimental data, such as density, molecular
weight, boiling point. The NMR determination of a typical average molecule,
makes it possible to know the molecular weight of the mixture with greater
accuracy, and the use of group contributions in the models, allowing a better
representation of this type of components.
PHASE 2 - MODELLING STUDY
The most commonly used models are derived from Van der Waals' equation (the best
known is that of Peng-Robinson). The adjustment of the parameters (Tc, Pc, W,
kij) of the pseudocomponent(s), which was almost manual, has been systematized,
determining from all the data contained in the PVT laboratory reports, the
properties of the heaviest pseudocomponent. The method uses the principle of
maximum probability, and integrates the experimental errors through variance and
covariance matrices.
Other representations of heavy components, such as that by group contributions,
or continuous distributions, are being studied. With group contributions, the
properties of the cuts will be determined, and in particular, the interaction
coefficients can be calculated. In a first phase, the continuous distributions
will be used to make the best choice of the number of pseudocomponents
representing the heaviest cut, and to determine their characterisitics.

STATE OF ADVANCEMENT :
****
COMPLETED

RESULTS :
****
Phase 1 : the methodology apparatus for determining the measurement of pressure,
volume, temperature, has been built and is widely used. It was proven on CO_2-
isobutane mixtures. The CO_2 n-octane studies have been completed, and the CO_2-
eicosane studies are underway, plus research on the dew point, for strict
methods, is underway (dew point detection by image analysis, luminous intensity).
Phase 2 : during the development of the parameter adjustement method, the
algorithms for calculating the liquid-vapor equilibrium have been made
thouroughly reliable; an algorithm for calculating three-phase equilibria (L1-L2-
G) has been developed to allow the simulation of CO_2 injection into shallow
reservoirs. An industrial version will be available in October 1987.

```
*****************************************************************************
* TITLE : ADDITIVES TO IMPROVE SWEEPING DURING      *     PROJECT NO       *
*         GAS INJECTION (PHASE I)                    *                      *
*                                                    *   TH./05059/84/FR/.. *
*                                                    *                      *
*****************************************************************************
* CONTRACTOR :                                       *     TELEPHONE NO     *
*    GERTH                                            *                      *
*                                                    *   1 47 52 61 39      *
*    AVENUE DE BOIS PREAU 4                           *                      *
*    FR - 92502 RUEIL-MALMAISON                       *                      *
*                                                    *     TELEX NO         *
*                                                    *                      *
* TECHNICAL DIRECTOR :                               *   203 050            *
*    MR. ROBIN                                        *                      *
*                                                    *                      *
*****************************************************************************
                                                    VERSION : 01/01/87
```

AIM OF THE PROJECT :
****
When using an enhanced recovery method requiring gas injection, a low oil
residual saturation can be obtained in the swept areas. Unfortunately, due to
its density, this gas tends to flow in the upper part of the formation. Also, in
the case of heterogeneous or cracked reservoirs, the important mobility of the
gaseous phase favours an instability of the displacement and the passage of the
gas in higher permeability drains, thus provoking a premature breakthrough. Once
the injected fluid has reached the production well, the volume of the reservoir
subjected to sweeping only increases slightly.
   To improve the sweeping coefficient, it is envisaged to block the areas
preferentially flooded. This can be obtained by a combined injection of gas and
additives likely to form a foam within the porous medium. The gas would thus be
diverted towards non-flooded areas, which would allow to increase the recovery.

PROJECT DESCRIPTION :
****
This project comprises 3 phases:
Phase 1 - Methodology
The object of this phase is to evaluate the foaming capacity of commercial
products, or of original products provided by the chemical industry.
A physical property of surfactant solutions, characteristic of their aptitude of
forming foam is being sought. This property is being observed under pressure
and temperature up to values representing those existing in the reservoirs.
 This phase is valid both in porous medium and out of porous medium.
Phase 2 - Combined injection of steam and foaming agents
A specific study has been carried out concerning the evolution of the foaming
properties with the temperature and the pressure. Operating conditions retained
are of 20 to 100 bars for the pressures and of 200 to 300 deg.C for the
temperatures (that is to say temperatures corresponding to the vaporization of
the water at the above mentioned pressures).
The experimental programme includes foam injection experiments in porous medium
run in cylindrical laboratory cells,containing homogeneous or non-homogeneous
porous media, formed by non-consolidated sand. The experiments will be run
either under isothermal conditions or under adiabatic conditions.
Phase3 - Combined injection of gas and foaming agents
A specific study has been carried out concerning the compatibility of surfactant
solutions with reservoir fluids. We are more particularly interested in the
water salinity and in the influence of the presence of hydrocarbonate phase in
the porous medium.
The experimental programme includes expriments in porous medium ran in
cylindrical laboratory cells, containing consolidated porous media, whether
homogeneous or not. These experiments are run under isothermal conditions.

STATE OF ADVANCEMENT :
****
Completed

RESULTS :
****
A methodology for the evaluation of foaming properties of surfactant solutions,
and to folllow-up of the evolution of this property under pressure and
temperature has been finalized.
The effect of the gravity or of the contrast of the permeability on the gas
sweep under ambient conditions has also been evidenced.Various types of
surfactants have been used, some being of commercial origin. The influence of
the foam sensitivity in the presence of a hydrocarbonate phase has been more
particularly evidenced. Devices and technique allow to evaluate, under pressure
and temperature conditions, the efficiency of these different surfactants on oil
recovery.
Experience shows that foam flows in porous medium are very long to stabilize
(several ten of volumes of pores to inject before stabilization). Hence, it was
not possible to exploit some of the tests owing to a lack of permanent flowing
pattern.
Studies have been pursued on products likely to foam under pressure and

temperature condition, but the important fact was not sensitive to the presence of a hydrocarbonate phase.

REFERENCES :
****
1 M.ROBIN
UTILISATION OF FOAMING AGENTS TO IMPROVE THE EFFICIENCY OF THE VAPOUR INJECTION.
3RD EUROPEAN COLLOQUE ON THE IMPROVEMENT OF ENHANCED OIL RECOVERY. ROME 16-18
APRIL 1985. TECHNIP PUBLICATIONS.
2. J.BURGER, P.SOURIEAU AND M.COMBARNOUS
ENHANCED OIL RECOVERY BY THERMAL METHODS.
TECHNIP PUBLICATIONS,PARIS 1985.
3. J.H.DUERKSEN
LABORATORY STUDY OF FOAMING SURFACTANTS AS STEAM DIVERTING ADDITIVES.
CALIF.REGION.MEETING OF SOC.PETROLEUM ENGRS.SPE PAPER 12785.APRIL 1984.
4.J.P.HELLER
RESERVOIR APPLICATION OF MOBILITY CONTROL FOAMS IN CO2 FLOODS.
SPE/DOE 4TH SYMP.ON ENHANCED OIL RECOVERY.SPE PAPER 12644;TULSA,APRIL 1984.

```
*************************************************************************
* TITLE : ENHANCED OIL RECOVERY BY MISCIBLE GAS      *    PROJECT NO     *
*         INJECTION IN TERTIARY CONDITIONS           *                   *
*                                                    *  TH./05061/85/FR/..*
*                                                    *                   *
*************************************************************************
* CONTRACTOR :                                       *   TELEPHONE NO    *
*    GERTH                                           *                   *
*                                                    *  1 47.52.61.39    *
*    AVENUE DE BOIS PREAU 4                          *                   *
*    FR - 92502 RUEIL-MALMAISON                      *                   *
*                                                    *   TELEX NO        *
*                                                    *                   *
* TECHNICAL DIRECTOR :                               *   203 050         *
*    MR K. MADAOUI                                   *                   *
*                                                    *                   *
*************************************************************************
                                             VERSION : 06/02/90
```

AIM OF THE PROJECT :
****
Many reservoirs have long been subjected to water drive. Hydrocarbon gas
injection in conditions of multiple contact miscibility is one of the most
promising processes for recovering "tertiary oil" (i.e. residual oil after
waterflooding or water drive).
The aim of this research project is to specify the conditions of optimization of
injecting dynamically miscible gas injection into a reservoir already
waterflooded, through experimental, numerical and economical approaches.
A detailed knowledge of the mechanisms (involved in the creation of growth of
the oil bank) is first needed to obtain reliable predictions from numerical
simulation of this process. Once the numerical model is validated, it will be
used to study the feasibility of a pilot injection project (including both
reservoir engineering and economical aspects).
This new tertiary enhanced oil recovery process should enable 25 to 30 % of the
residual oil to be recovered.

PROJECT DESCRIPTION :
****
The efficiency of miscible gas injection under tertiary conditions is related to
the formation of an oil bank by remobilization followed by coalescence of the
drops of oil trapped in dispersed fashion.
The following questions arise :
- how do the mechanisms of diffusion, dispersion, mass transfers implement the
creation and the growth of the oil bank?
- what is the influence of the petrophysical properties of the reservoir?
- what procedures must be complied with so that the laboratory results can be
considered as usable for the following reservoir simualtion?
The project is to take place as follows :
- PHASE 1 : Study of the mechanisms.
The purpose is to identify for each type of dynamic miscibility (frontal, rear,
partial) the critical parameters playing the major role to implement the
creation and the growth of the oil bank :
. mechanisms of dispersion, diffusion
. polyphasic flow mechanisms
The experimental part of the work consists mainly in displacement experiments
using synthetic fluids in order to facilitate the interpretation of the results.
The numerical simulation of these experiments is to show if adaptations of the
available numerical simulators are necesary to match their history.
- PHASE 2 : Extension to actual cases.
This phase consists of verifying the statements obtained in phase 1 by studying
the sentivity of the tertiary recovery process on actual field conditions
(displacemnet experiments with fluids and rocks from different types of
reservoir).
- PHASE 3 : Feasibility of a pilot injection project.
The choice of the reservoir for this feasibility study. will be done according
to the results of the preceding steps. Much attention will be devoted in order
to select the best candidate among the different fields preferentially situated
in the European zone where the contractor, through participations or operations,
has already got a good preliminary information. This phase will consist of a
complementary laboratory study, on the selected couple reservoir engineering
study, the pre-development study and the economical aspects./injection gas, the
reservoir-engineering studies, the pre-development study and the economical
analysis.

STATE OF ADVANCEMENT :
****
Phase 1,2 and 3 are completed.

RESULTS :
****
Phase 1 :
Both frontal and rear miscibility, as tertiary recovery process, were proven to
be very promising in terms of their efficiency to displace the oil trapped
during the water drive.

The effect of the swelling of the oil by the injection gas was brought into evidence in vaporizing and condensing gas drives. From the comparison of the results of two experiments corresponding to tertiary condensing gas drive, the swelling effect being much more considerable in the second experiment, it arises that the swelling mechanism contributes largely to stabilize the advance of the injection gas. In case of vaporizing gas drive, the effect of swelling mechanism is less considerable, since the type of miscibility developed is "self-stabilizing".
Results of the numerical simulation are satisfactory.

Phase 2 :
Physical simulation of various cases of applications led to the same type of production histories as those obtained in the phenomenological study using synthetic fluids.
The tertiary multiple contact miscible process is demonstrated to be highly efficient for recovering waterflood residual oil in the various cases of application simulated physically. Water-alternating gas injection experiments (condensing and vaporizing gas drive), indicated clearly effects of slug size and slug ratio.The numerical simulation gave satisfactory match of lab experiments, confirming that SIMCO model can be considered as a reliable predictive tool for field application studies.

Phase 3 :
The Alwyn North field was selected for the pilot feasibility study.
- A first screening study performed on 2D cross-section, allowed to select the best reservoir/gas couple (associated gas to be injected in the BRENT NW reservoir as pilot project).
- Complementary laboratory experiments have been performed on the actual selected rock-fluid system. The confirmed the feasibility of the tertiary miscible gas injection, Triphasic relative permeability experiments have also been performed. The oil PVT was adjusted on several lab results, including oil swelling by different gases.
- A 3D model was then set up to define the optimum conditions of miscibles gas injection : Additional oil recovery is evaluated to be 13.2% with total recovery of the injected gas with definition of the OOIP.
- Results on the pilot project have been extrapolated to the whole ALWYN NORTH field. Reservoir performances (gas injection/production, oil production, water injection/production versus time) have been defined.
-.
- A preliminary pre-project study has been done, to describe the equipments and to evaluate technical costs of the work programme necessary to implement the project on ALWYN. Results are positive.

REFERENCES :
****
DOCTORATE THESIS FROM UNIVERSITY OF BORDEAUX BY D. HADIATNO.
C. BARDON, C. BARROUX, H. MONTMAYEUR AND J. PACSIRSZKY - "MISCIBLE HYDROCARBON GAS INJECTION OIL TERTIARY CONDITIONS AS AN E.O.R. PROCESS" - PROCEEDINGS OF 4TH EUROPEAN SYMPOSIUM ON ENHANCED OIL RECOVERY, 27TH-29TH OCTOBER, HAMBURG.
"ENHANCED OIL RECOVERY BY MISCIBLE GAS INJECTION IN TERTIARY CONDITIONS" - POSTER SESSION - 3RD SYMPOSIUM ON NEW TECHNOLOGIES FOR THE EXPLORATION AND EXPLOITATION OF OIL AND GAS RESOURCES, LUXEMBOURG, MARCH 22-24TH, 1988.

```
****************************************************************************
* TITLE : ADDITIVES FOR THE IMPROVEMENT OF        *      PROJECT NO       *
*         SWEEPING DURING STEAM DRIVE - PHASE 2    *                      *
*                                                  *   TH./05062/85/FR/..  *
*                                                  *                      *
****************************************************************************
* CONTRACTOR :                                     *    TELEPHONE NO       *
*     GERTH                                        *                      *
*                                                  *   1 47.52.61.39       *
*     4, AVENUE DE BOIS PREAU                      *                      *
*     FR - 92502 RUEIL-MALMAISON                   *                      *
*                                                  *    TELEX NO           *
*                                                  *                      *
* TECHNICAL DIRECTOR :                             *    203 050            *
*     MR. SAHUQUET                                 *                      *
*                                                  *                      *
****************************************************************************
                                                    VERSION : 30/05/88
```

AIM OF THE PROJECT :
****
Enhanced oil recovery operations through continuous steam drive are often
penalized by the irregular displacement of the injected steam owing to the
reservoir heterogeneities. Hence, this phenomenon provokes sweeping anomalies
and often a final recovery rate that is lower than expected. The solution
consists in injecting, together with the steam, a foam which blocks the
preferential passages and thus improves sweeping.
The object of this project is to determine the optimal application conditions of
this technique while setting up a real size pilot on the field.

PROJECT DESCRIPTION :
****
This project includes 3 phases :
- Preparation of the pilot
- Execution of the pilot and tests
- Interpretation of tests.
The first phase consists of choosing a test site, where sweeping anomalies have
already been observed and located during steam drive, and thus a site where it
may be easier to assess the efficiency of the process. Once this site will have
been chosen, laboratory tests will allow, under real reservoir conditions, to
test the foaming agents and their stability, and to choose the type of gas to
inject so to obtain good performances. These tests should also allow to specify
the optimal operating conditions.
The second phase will concern the realization of the pilot on the chosen site.
After a number of zone characterization tests (well tests, interference tests,
tracers...) it will be possible to begin the foam injection tests. According to
forecasts, the pilot will require a series of foam plug injections in order to
obtain a significant and lasting effect. Various formulations and operating
conditions are necessary to optimize the process.
The third phase will consist in interpreting these tests : evaluation of
sweeping and recovery rate improvement, technical and economical result of the
process.

STATE OF ADVANCEMENT :
****
Abandoned

RESULTS :
****
Only Phase 1 has been carried out. During the second term of 1986, economic
conditions have led to the interruption of the steam injection on the POSO CREEK
site in California, which had been chosen for the pilot test. A new test site
had been sought to try to valorize the work achieved during Phase 1. But not
other field among those under steam injection within the contacted European
companies presented the required conditions. Interruption of the project was
thus solicited.
Laboratory tests carried out in the POSO CREEK field conditions enabled
selection of different technical alternatives.
Some foaming agents, available on the market, presented sufficient foaming and
stability characteristics for the implementation conditions of this site.
Characteristics of solutions to be injected (product contents, foam quality ...)
have been determined. Preference was given to nitrogen rather than air as
regards the complementary gas phase of the steam; indeed, even at the POSO CREEK
injection temperature, the presence of oxygen was prejudiciable to the stability
of foaming agents.
Displacement tests in "double porositylpermeability" media allowed to quantify
the plugging capacity of foam in high permeability media.

```
***********************************************************************
* TITLE : THE BENEFICIAL EFFECTS OF DISTILLATION      *    PROJECT NO      *
*         DURING OIL RECOVERY BY STEAMFLOODING        *                    *
*                                                     *  TH./05064/85/NL/.. *
*                          .                          *                    *
***********************************************************************
* CONTRACTOR :                                        *  TELEPHONE NO       *
*    TECHNISCHE HOGESCHOOL, DELFT#AFD.MIJNBOUWKUNDE    *                    *
*                                                     *  015 781617         *
*    POSTBUS 5628                                     *                    *
*    NL - 2600 GA DELFT                               *                    *
*                                                     *  TELEX NO           *
*                                                     *                    *
* TECHNICAL DIRECTOR :                                *  38151              *
*    DR. J. BRUINING                                  *                    *
*                                                     *                    *
***********************************************************************
```

VERSION : 18/09/89

AIM OF THE PROJECT :
****
The main objective of the project is to show that steamflooding, which is
applied routinely to the recovery of heavy oil, can be applied economically to
the recovery of medium viscosity oil i.e. can compete with waterflooding.
Another objective is to assess the beneficial effect of the distillable oil bank.
 In our previous project we have elaborated at length on the competition between
film flow effects and the distillable oil bank to lower the "residual oil
saturation" in the steam zone. In our present project we focus our attention on
factors influencing the behaviour of the steam condensation front and their
effects on the vertical sweep efficiency.

PROJECT DESCRIPTION :
****
Steam displacement applied to heavy oil reservoirs, with a low primary and
secondary potential, leads to an appreciable extra oil recovery. Indeed, steam
drive is routinely applied in heavy oil reservoirs.
Application of steamdrive to medium viscosity oil reservoirs (with a lower
primary and secondary recovery potential) leads to a higher sweep efficiency
than its application in heavy oil reservoirs. Also distillation effects will
give a larger contribution to the recovery efficiency or medium viscosity oils.
These distillation effects have been attributed to the distillable oil bank,
which is formed near the steam condensation front. This leads to a lower (heavy)
oil saturation in the steam sept zone.
Furthermore, owing to an improved mobility ratio the sweep efficiency will be
enhanced by these distillation effects.
The innovating aspect of the present project is to improve, with the help of
laboratory experiments, existing mathematical/physical models with the aim to
quantify the sweep efficiency depending on oil viscosity and oil composition.
Furthermore, we expect to introduce the streamfunction approach in the
mathematical/physical modeling of the steam displacement process.
In this way we want to investigate under which circumstances steam flooding can
be applied to the recoverry of medium viscosity oils i.e. can complete with
water flooding. Where possible use will be made of field data for which we can
call upon the Nederlåndse Aardolie Maatschappij.
The project consists of the following phases :
1. Design and building of a reactor with auxiliary equipment.
2. Visual studies of the steamdrive process in a large scale model.
3. Execution of the experiments.
4. Theoretical description of the experimental results and extrapolation to
field conditions.
5. Economical evaluation of a steam drive, based on field parameters and
mathematical/physical model calculations.

STATE OF ADVANCEMENT :
****
The project is in its completion phase. A high pressure rig has been designed ;
and constructed to perform experiments at elevated pressures. A transparent
vacuum scale model has been designed and constructed. Preliminary experiments
have ;been carried out. A low-pressure experimental rig has been designed,
constructed and used. A low-pressure transparent reactor has been designed,
constructed and is in use.

RESULTS :
****
The transparent scale model, after initial technical problems particularly
related to the application of vacuum techniques, works satisfactoryly.
Experiments are carried out.
A numerical model based upon the finite element method has been developed for
this project and will be used for interpretation of experiments and
extrapolation to field conditions. The finite elements make it possible to
accurately calculate sharp front behaviour.
The model can be extended to comprise, apart from gravity and viscous forces,
capillary forces and heat exchange effects. The model will be used in the
economical evaluation for oil recovery calculations.

REFERENCES :
****
J.BRUINING ET.AL.:PROCEEDINGS OF THE 4TH EUROPEAN SYMPOSIUM ON ENHANCED OIL
RECOVERY,27-29 OCTOBER 1987,887-898.
C.T.S.PALMGREN,H.J.BRUINING,C.J.VAN DUYN:PROCEEDINGS OF THE 3RD EC SYMPOSIUM,
LUXEMBOURG,22-24 MARCH 1988,930-938.
C.T.S.PALMGREN,J.BRUINING ET.AL.:DELFT PROGRESS REPORT,(1988/1989),VOL.13,PP.99-
111.
C.T.S.PALMGREN,J.BRUINING,H.J. DE HAAN:PROCEEDINGS OF THE 5TH EUROPEAN SYMPOSIUM
ON IMPROVED OIL RECOVERY,BUDAPEST,25-27 APRIL 1989,561-573.
+ 5 REPORTS ON M.SC. THESIS WORK.

```
***********************************************************************
*  TITLE : TENSID-POLYMER PILOTPROJEKT LEIFERDE      *     PROJECT NO     *
*                                                    *                    *
*                                                    *  TH./05066/85/DE/.. *
*                                                    *                    *
***********************************************************************
*  CONTRACTOR :                                      *    TELEPHONE NO    *
*     RWE-DEA AKTIENGESELLSCHAFT FUR MINERALOEL       *                    *
*                                                    *   040.63.75.24.79  *
*     UBERSEERING 40                                 *                    *
*     D-2000 HAMBURG 60                              *                    *
*                                                    *    TELEX NO        *
*                                                    *                    *
*  TECHNICAL DIRECTOR :                              *    211513          *
*     DR ING BALRAM K                                *                    *
*                                                    *                    *
***********************************************************************
                                                       VERSION : 15/08/89
```

AIM OF THE PROJECT :
****
The objectives of the project was the development and pilot application of a
high-salinity surfactant/polymer flood system which is able to achieve a higher
ultimate recovery than zith conventional water flooding. The chemical system had
to be designed for application in a reservoir of extremely high brine salinity
of 195 g/l TDS without preconditioning of the reservoir (Loudon Pilot-Project :
104 g/l). The reservoir simulation study should establish a numerical model in
order to simulate the flood process, support and optimize the process design for
field scale-up, enable economic evaluation and study extended field
applicability of the process. Engineering of the technical facilities for
preparation of the flood solutions in a continuous in-line mixing process with
supervision and automatic control equipment had to be planned and constructed.

PROJECT DESCRIPTION :
****
The chemical flood process will be pilot-tested in the Leiferde oilfield
(Gifhorn Trough, West Germany).
During the Laboratory Phase a chemical system which can effectively mobilize
residual oil had to be tested in laboratory flood experiment. For the numerical
simulation of the chemical flood process the required input parameters had to be
established, properties of the solutions determined, suitable tracers selected,
analytical methods for control measurements developed and specifications worked
out.
In the Reservoir Simulation Study a numerical model of the reservoir had to be
developed (history match). With this model the surfactant flood process has to
be simulated using process data from laboratory measurement (surfactant flood
prediction) in order to determine flood concept (slug sizes, injection and
production rates etc..) and incremental oil recovery. A comparison of actual
production performance with the prediction will furnish information to justify
future applicability of the system.
The Project Evaluation and Assessment will be based on following observations
and measurements :
- quality of delivered chemicals and solutions prepared by the surface
facilities. Most important factors are compliance with specifications and
injectability of the solutions.
- development of displacement process
  The chemical concentrations and the oil cut in the produced fluids will be
measured.
These data will indicate the displacement efficiency of the process and have to
be compared to the predicted values of the model study. If required, the input
data will be adjusted to update the model. For final evaluation the following
investigations were planned :
- technical and economical evaluation of the pilot
- studies with larger reservoir models in order to gain information for future
applicability in other parts of the oilfield
- economic analysis for a commercial field project
- investigate applicability to other oil reservoirs.

STATE OF ADVANCEMENT :
****
Laboratory phase for the system development and the optimization of the flood
concept were completed. Technical planning of the surfaced facilities has been
completed.
Construction of the surface facilities and further tests on water clarification
were postponed after the decision to discontinue the project due to present oil
price development.

RESULTS :
****
Laboratory work
A surfactant/polymer system in reservoir brine was optimized in several flood
experiments on Berea sandstone models and Leiferde core material of varying
geometry. The microemulsion system consists of : 1.5 % surfactant mixture (alkyl
ether sulfate/alkylxylene sulfonate), 1.5 % polyethyleneglycol (sacrificial

agent), 0.8 % white oil and 750 ppm polysaccharide in a brine/fresh water mixture of 160 kg/m3 TDS.

In laboratory flood experiments the residual oil of about 40 % pore volume after water flood could be reduced to values below 10 % by chemical flooding (0.2 PV microemulsion, 0.8 PV mobility control). Laboratory activities covered optimization of chemical system and flood sequence, selection of polymer (viscosity yield, injectivity), selection and analysis of tracers (ROH, NaSCN), development of a continuous mixing process for preparation of chemical solutions, treatment of clarification brine (coalescer), specifications of chemicals, test procedures for quality control and assistance in numerical simulation (measurement of input parameters).

Reservoir Engineering and Simulation

Reinterpretation of logs of the Leiferde field wells resulted in a modified structure map of the pilot area in the south block.

The flood concept was optimized as follows : history match of the south block, simulation of laboratory flood experiments and development of a field model for the chemical flood.

The history match (water flood) used the 3-phase, 3D blackoil-simulator ECLIPSE. The calculated OOIP, cumulative oil and water production, water cut development and pressure agreed fairly well with field performance.

The laboratory chemical flood experiments were matched with the simulator UTCHEM (University of Texas, Austin), the final data set being able to match all six linear and two areal floods with acceptable accuracy.

With a selectional model of the pilot area, transferred to UTCHEM, a sensitivity study was carried out for final flood design. The result of the simulation of the pilot area led to an additional oil recovery of 14% PV over waterflooding.

Engineering work for the required surface facilities for storage, blending and control units for the preparation of the chemical solutions and all injection facilities have been completed. Bids for construction of new and modification of available equipment have been obtained. The surfactant unit allows continuous in-line preparation of the solution. A heating unit and double-walled, insulated containers guarantee required temperature for storage and operation throughout the year. The polymer unit allows continuous preparation of the polysaccharide solution in a two-step dilution and shear treatment.

```
********************************************************************************
* TITLE : CONVERSION OF HEAVY OIL, BITUMEN AND      *      PROJECT NO         *
*         REFINERY RESIDUE INTO LIGHT BOILING        *                         *
*         DISTILLATES                                *      TH./05067/85/DE/..  *
*                                                    *                         *
********************************************************************************
* CONTRACTOR :                                       *   TELEPHONE NO          *
*     VEBA OEL AG                                    *                         *
*                                                    *   0209 366-7968         *
*     DE - 4650 GELSENKIRCHEN-HASSEL                 *                         *
*                                                    *                         *
*                                                    *   TELEX NO              *
*                                                    *                         *
* TECHNICAL DIRECTOR :                               *   824881-90             *
*     DR. R. HOLIGHAUS                               *                         *
*                                                    *                         *
********************************************************************************
```
VERSION : 13/09/89

AIM OF THE PROJECT :
****
- Conversion of non-boiling crude oil fractions under hydrogen pressure into light boiling distillates.
- Upgrading of natural bitumen and refinery residues with :
- high percentage of non-boiling components
    - high asphalt -, sulfur - and nitrogen contents
    - high metal contents
    - high viscosity and density.
The hydrogenation is performed in a cascade of liquid-phase reactors and gas-phase reactors. A one-way additive in the liquid-phase reactors is normally used.

PROJECT DESCRIPTION :
****
The VCC-Process was derived from the Bergius-Pier technology and was applied in a precursor-process called "Scholven Combi-Chamber".
The VCC-Process developed by VEBA OEL in cooperation with LURGI is characterized by its extremely high conversion efficiency up to 95 wt. % based on non boiling residue.
The project will develop in the following phases :
Phase I   : Basic bench scale research
Phase II   : Design and construction of a pilotplant for a throughput of
                 1 t/h
Phase III   : Operating phase of the pilotplant
Phase IIIA   : Second operating phase of the pilotplant
Phase IV   : design and construction of an industrial-scale plant for
demonstration. The present state of development is the second operating phase of the pilotplant (Phase IIIA).
The development covers the following fields of experiments and theoretical research
- Additive optimization
- Pressure reduction
- Establishment of design data
- Establishment of scale-up factors
- Extension of feedstock basis
- Changes in process configuration
- Screening of gase phase catalysts
- Longtern tests
The central objectives of these tests is to develop design data for commercial VCC-plants and to optimize the process for each application.

STATE OF ADVANCEMENT :
****
Process is actually demonstrated in an industrial scale demonstration plant (former coal liquefaction plant at Bottrop).

RESULTS :
****
By application of the VLC/VCC-technology several vacuum residues from conventional and heavy crudes (arab heavy, bachaquero, tia juana, morichal) and from a visbreaking plant were converted to an extreme high degree (greater than 90 perc.) into light distillates, some hydrocarbon gases and a small amount of hydrogenation residue which is suitable for the production of hydrogen via partial oxidation.
Feeding a vacuum bottom from a typical venezuelan crude with the following analytical data

Carbon	84,8	WTPERC.
Hydrogen	10,4	WTPERC.
Sulphur	3,3	WTPERC.
Nitrogen	0,6	WTPERC.
Vanadium	630	PPM
Nickel	75	PPM

to the process as an example 80 WTPERC. of a VCC syncrude is produced which contains 27 perc. of naphtha, 48 perc. of middle distillates and 25 perc. of vacuum gasoil. Due to the application of the so-called gas phase hydrogenation

(a catalytic fixed bed reactor directly combined with the primary conversion step) this VVC syncrude having an relatively high hydrogen content is almost sulphur and nitrogen free (less than 200 ppm each). The VCC middle distillates can be sold directly, the VVC naphtha meets reformer feed spezification and the vacuum gasoil is an excellent feedstock for a FFC or a hydrocracker unit.
As these results can be generalized for all these residues processed up to now it can be concluded that the VLC/VCC process is a well advanced technology to convert less valuable bottoms into light distillates thus making maximum use of the "Bottom of the Barrel".

REFERENCES :
****
U. GRAESER, K. NIEMANN
OIL GAS JOURNAL 80, NR. 12, 121 (1982)
U. GRAESER, K. KRETSCHMAR, K. NIEMANN
ERDOEL UND KOHLE, ERDGAS, PETROCHIMIE 36, NR. 8, 362 (1983)
U. GRAESER, K. NIEMANN
PREPRINTS, DIV. PETR. CHEM. AM. CHEM. SOC. 28, NR. 3, 675 (1983)

```
*******************************************************************************
* TITLE : PHYSICAL AND NUMERICAL MODELLING OF IN      *     PROJECT NO       *
*         SITU-COMBUSTION                             *                      *
*                                                     *   TH./05068/85/DE/.. *
*                                                     *                      *
*******************************************************************************
* CONTRACTOR :                                        *   TELEPHONE NO       *
*    INSTITUT FUER TIEFBOHRTECHNIK                    *                      *
*                                                     *   05323/72 26 18     *
*    ABT. LAGERSTAETTENTECHNIK                        *                      *
*    AGRICOLSTR. 10                                   *                      *
*    DE - 3392 CLAUSTHAL-ZELLERFELD                   *   TELEX NO           *
*                                                     *                      *
* TECHNICAL DIRECTOR :                                *   953813             *
*    PROF. DR.MONT. GUNTER PUSCH                      *                      *
*                                                     *                      *
*******************************************************************************
                                                     VERSION : 15/07/89
```

AIM OF THE PROJECT :
****
The objective of this project is to improve the physical modelling and numerical
simulation of oil recovery supported by in situ combustion. For this purpose,
the influence exerted by the oil composition (especially the resin and
asphaltene contents) on the generation of fuel during in situ combustion is to
be investigated. The reaction kinetic data are to be determined as functions of
the oil composition under isothermal conditions. With the aid of the data thus
ascertained, a physically more exact description of the in situ combution
process is envisaged on the basis of numerical models (at IFP, Paris). The
results concerning reaction kinetics are to be employed in the investigations of
the relationship between the oil composition and propagation velocity of the
combustion front. For this purpose, combustion tests are to be conducted in a
linear combustion cell under largely adiabatic conditions.

PROJECT DESCRIPTION :
****
The research project has been subdivided into two phases.

PHASE 1
This phase comprises three stages for investigation the combustion process.
During the first stage, the relationship between the oil composition and
generation of fuel is to be investigated. During the second stage, the effect
of the oil composition on the reaction kenetics during combustion is to be
examined. During the third stage, the influence of the oil composition on the
propagation velocity of the combustion front is to be investigated with the use
of the results from stages 1 and 2.

PHASE 2
During this phase, numerical models for in situ combustion are to be elaborated
in two stages.
Stage 1 comprises the development of a model without considering the fluid
transport.
In stage 2, a model is constructed with the fluid transport taken into account.

TECHNICAL DESCRIPTION
An appropriate crude oil is enriched with its own asphaltenes and resins for
obtaining three types of model oil differing mutually in colloidal-chemical
composition. The pyrolysis of the different model oil types is performed under
inert gas, and the quantity of fuel thereby generated is determined. A
differential reactor is being constructed for the purpose.An attempt will be
made to derive a satisfactory correlation between the quantity of fuel thus
formed and the oil composition. The combustion of the coke is to be investigated
under isothermal conditions in the presence of air in a temperature range
between 623 K and 923 K. The quantitative and qualitative analyses of the
reaction gases, as well as the combustion temperature and pressure constitute
the basis for the determination of the reaction rate constant for the oxidation.
By means of linear combustion tests, the effect of the oil composition on the
propagation velocity of the combustion front is to be investigated.
The construction of the numerical models begins with a simplified radial model
without considering kinetics and transport processes. Heat exchange, reaction
kinetics, and fluid transport are then taken into account, thus resulting in a
two-dimensional model. The block size for the process description is thereby
matched to the reservoir scale. The models are to be tested by the laboratory
results of phase 1 and by means of field conduted experiments conducted in the
classical manner.

STATE OF ADVANCEMENT :
****
During phase 1, investigations on pyrolysis and oxidation have been conducted
and the kinetic data have been determined for oxidation.
During phase 2, numerical models have been developed for wet and dry combustion.

RESULTS :
****

PHASE 1

Appropriate crude oil was enriched with its own asphaltenes or its own wax-resin complex in order to prepare oil mixtures of different colloidal composition. High-pressure differential thermal reactors were developed and employed for the experiments. The relationship between the oil composition and fuel formation was investigated with different pyrolytic media (nitrogen, carbon dioxide, steam) _ under isobaric (8 MPa) and isothermal (623, 723, 823 and 923 K) conditions. Both the total quantities of colloidal components in the oil and the pyrolytic medium thereby exert a quantitative effect on the fuel formation. Steam has proved to be the most effective pyrolytic medium, since it decidedly decreases the residue of pyrolysis, as compared with nitrogen and carbon dioxide. The quantity of fuel is directly proportional to the quantity of resins, waxes and asphaltenes. Furthermore, the relationship between the oil composition and reaction kinetics of fuel oxidation has been investigated at constant pressure under isothermal and nonisothermal conditions. The kinetic parameters thus determined(K, KO, EA) indicate that the reactivity of the fuel toward oxygen at 623 and 723 K increases with augmenting wax-resin content in the oil. At higher temperature (823 and 923 K), however, the fuel from the mixture which is richest in asphaltenes is more reactive than that from the mixture richest in wax and resin. The same trend has also been observed for the combustion of the pyrolytic residue from pure asphaltenes or waxes and resins.

A linear combustion cell has been developed for investigating the propagation velocity of the combustion front with different composition of the oil employed. With this cell, experiments can be performed under quasiadiabatic conditions at a pressure of 8 MPa. This feature was achieved by means of a transportable, external heater which moves with the combustion front and thus largely compensates for the heat losses.

The propagation velocity of the combustion front can be measured by means of thermocouples arranged axially, and the variation of the pressure gradient can be recorded with the use of differential pressure transducers arranged radially. The efficiency of the combustion and the fuel balance can be calculated from the analysis of the product gases for oxygen, carbon dioxide, and carbon monoxide. The same sock oil and mixture enriched with asphaltenes was employed for the combustion as for the kinetic investigations. Under identical conditions (air throughput : 93 sm3.m-1.h-1, pressure : 8 MPa, residual oil saturation : about 40 %, the same values of the propagation velocity were measured; the higher fuel concentration in the mixture thereby decreases the velocity, but the higher reactivity counteracts this drawback.

PHASE 2

One- and two-dimensional models have been realized for simulating forward in sity combustion in the laboratory or in the reservoir.

REFERENCES :
****
4TH EUROPEAN SYMPOSIUM ON ENHANCED OIL RECOVERY, 27TH - 29TH OCTOBER, 1987, HAMBURG, POSTER SESSION
PAPER : MEETING FOR STATUS REPORTS 1988, "GEOTECHNIQUE AND RESERVOIRS" 8TH OF MARCH IN CELLE, ORGANIZED BY THE FEDERAL MINISTERY OF RESEARCH AND TECHNOLOGY (BMFT) AND NUCLEAR RESEARCH CENTER JULICH (KFA, PBE)
"3TH HYDROCARBON SYMPOSIUM LUXEMBOURG", LUXEMBOURG, 22ND - 24TH MARCH 1988, POSTER DISPLAY

```
*****************************************************************************
* TITLE : A FLOATING BARGE-MOUNTED PLANT,              *     PROJECT NO      *
*          PRODUCING HIGH-PRESSURE N2/CO2 GASES        *                     *
*          FOR EOR PROJECTS IN THE NORTH SEA           *   TH./05069/85/DE/.. *
*                                                      *                     *
*****************************************************************************
* CONTRACTOR :                                         *   TELEPHONE NO      *
*    SALZGITTER AG                                     *                     *
*                                                      *   030 88 42 97 26   *
*    ABTEILUNG FORSCHUNG UND ENTWICKLUNG               *                     *
*    POSTFACH 15 06 27                                 *                     *
*    D - 1000 BERLIN 15                                *   TELEX NO          *
*                                                      *                     *
* TECHNICAL DIRECTOR :                                 *   185 655           *
*    DR-ING H.J. WESSEL                                *                     *
*                                                      *                     *
*****************************************************************************
```

VERSION : 20/09/89

AIM OF THE PROJECT :
****

As the world's reserves of crude oil diminish, the low yield of often less than
30% from a deposit becomes the central aspect of energy policies. The injecting
of CO_2 or N_2 represents one way of stimulating oil recovery. So far the use of
such tertiary production measures has been restricted to onshore fields. Such
measures, however, take on greater economic significance for offshore fields.
The investment necessary to develop an offshore site exceeds that required for
an onshore location by many times. This makes it even more important that the
amount of oil extracted from an offshore deposit should be as high as possible.
The proposed technical development aims at providing low-cost gas which will
then be used to recover the oil offshore by tertiary measures. In awareness of
the conditions specific to offshore sites, this means that the production
facilities will have to be installed on a mobile carrier, i.e. barges that can
be used at several locations irrespective of water depth.

PROJECT DESCRIPTION :
****

Assuming that in future large quantities of CO_2/N_2 and steam will also be
required for the tertiary production of oil offshore, a combined process is
proposed which features a particularly high degree of economy. The process is
made up of the following three main components:
a. production of an N_2/CO_2 mix by burning preferably gaseous hydrocarbons,
whereby high-pressure steam is also generated,
b. production of N_2 in an air separation plant,
c. compression of the CO_2/N_2 mix by using the high-pressure steam produced
during combustion.
The combination of combustion and air separation plant results in a substantial
decrease in the cost of injection gas when drawing comparisons with each
separate process.
The main items of the project:
- CO_2/N_2 production using a boiler and steam turbine
- CO_2/N_2 production using an air separation plant with an increased delivery
pressure for N_2
- application of gas turbines to produce CO_2/N_2 with gas turbines driving the
compressors directly
- basic engineering for the combined production of N_2/CO_2 and high-pressure
steam when using steam and gas turbines
- determination of energy requirements and comparison of the capital investment
and operating costs for each alternative
- development of a prestressed concrete barge as a floating carrier for a plant
to produce N_2/CO_2 high-pressure gas
- design of a mooring system with central articulated joint and integrated gas
conveyance for N_2/CO_2 high-pressure gas and high-pressure steam, as well as for
supplying the plant with low-pressure fuel
- summary covering capital investment and costs, and final economic analysis.

STATE OF ADVANCEMENT :
****
The project is in the final phase during which the concluding report is composed.

RESULTS :
****
The Norwegian and British oil fields in the North Sea were examined with regard
to their yearly production rate and the remaining production capacity. A
possible use of the EOR barge can be envisaged for those fields whose regular
capacity will be exhausted in the foreseeable future.
First the production capacity of the producing plant was determined. Since the
concept of an EOR gas barge aims at a multiple use of the barge in various oil
fields, the design value was selected higher than the average of the expected
cases of application. The concept of a multiple use precludes from the outset
the fixing of a determined pressure. The gas processing plant should therefore
be conceived in such a way that an adaptation to various pressures can be made
with as low an amount of alteration as possible. The plant should be conceived
in the form of a basic outfit for 200 bar injection pressure, and should be laid

out such that either and additional compressor set can be installed for compression to a higher pressure, or only one additional compressor which can be coupled onto the shaft of an existing set.

Associated gas is planned to be used as fuel which is obtained on the oil production platform.

In order to determine the most favourable shape of the barge under the aspect of its sea behaviour, two different pontoons were subjected to an investigation. The size of the deck area is governed by the processing plant which is to be housed. The plant assemblies are partly located on the pontoon's bottom and partly on the deck.

It is intended to conduct the hydrodynamic evaluation in two stages. At first the barge will be tested while freely floating during various sea states without any anchoring (calculation for two types of barge designs). Thereafter an anchoring calculation will be conducted using the optimum barge design.

The purpose of the anchoring system is, in the first instance, to anchor the concrete barge together with the gas producing plant in a permanent way, and additionally it is used to anchor the gas supply lines leading from the seabed to the barge. In recent years remarkable progress has been made as far as the production, endurance and safety of flexible underwater pipes for high pressure (up to approx. 450 bar) is concerned.

Due to the relative small dimensions of the barge with a displacement of some 15000 tons (i.e. 10% of an average shuttle tanker size) the possible solution with a tower-anchoring system was abandoned.

A turret mooring system was chosen since this can be applied as a very reliable anchoring device with 6 to 8 chains and accordingly a sufficient redundancy.

```
********************************************************************************
* TITLE : ENHANCED OIL RECOVERY FROM LOW           *      PROJECT NO        *
*         PERMEABLE CHALK RESERVOIRS (FOLKA)        *                        *
*                                                   *   TH./05070/85/DK/..   *
*                                                   *                        *
********************************************************************************
* CONTRACTOR :                                      *    TELEPHONE NO        *
*    COWICONSULT/CONSULTING ENGINEERS AS            *                        *
*                                                   *   45 45 97 21 11       *
*    TEKNIKERBYEN 45                                *                        *
*    DK - 2830 VIRUM                                *                        *
*                                                   *    TELEX NO            *
*                                                   *                        *
* TECHNICAL DIRECTOR :                              *    37280               *
*    MR. A. BJERRUM                                 *                        *
*                                                   *                        *
********************************************************************************
```

VERSION : 10/10/89

AIM OF THE PROJECT :
****
The aim of the FOLKA project is to investigate some problems related to
waterflooding and gas injections in low permeable chalk reservoirs.
The project includes 4 phases:
- Planning and preliminary studies
- Gas injection miscibility conditions
- Chalk wettability conditions
- Reservoir modelling, field studies
The FOLKA project includes a laboratory program using samples from Danish and
Norwegian chalk reservoirs. The laboratory program is carried out by IKU in
Trondheim.

PROJECT DESCRIPTION :
****
PHASE 1 : PLANNING AND PRELIMINARY STUDIES
Selection of reservoir for fluid sampling and core sampling. Preparation of
fluid PVT report.
PHASE 2 : GAS/OIL PHASE BEHAVIOUR
Phase behaviour was studied in a visual saphire cell at reservoir conditions.
Slim tube tests were carried out using lean hydrocarbon gas as the displacing
fluid.
Minimum Miscible Pressure (MMP) was determined at 45 MPa. This is about two
times reservoir pressure.
MMP was checked by correlations.
PHASE 3 : CORE DISPLACEMENT TESTS
THREE GAS TYPES WERE USED AS DISPLACING hree gas types were used as displacing
fluid
- lean hydrocarbon gas
- nitrogen
- carbon dioxide.
Lean hydrocarbon gas and nitrogen are assumed to displace the oil through
partial miscibility, whereas carbon dioxide results in a full miscible
displacement process. This assumption was confirmed by the oil recovery in the
three tests.
PHASE 4 : NUMERICAL SIMULATIONS
- The compositional PVT package, COPEC, was used to verify the oil
characterization. The reservoir simulator, CERES, which includes the COPEC fluid
model was used to simulate the slimtube displacement tests.
Good correlation between experimental and calculated MMP-values was obtained.

STATE OF ADVANCEMENT :
****
Completed

RESULTS :
****
Minimum Miscible Pressure was determined experimentally and theoretically for
Danish and Norwegian oil reservoirs.
MMP was above reservoir pressure for lean hydrocarbon gas and nitrogen. For
carbon dioxide the MMP was below reservoir pressure. This was confirmed by core
displacement tests.

REFERENCES :
****
- "ENHANCED OIL RECOVERY FROM LOW PERMEABLE CARBONATE ROCK RESERVOIRS". FINAL
REPORT. (COWICONSULT OCT.1989).
- "MISCIBLE GAS INJECTION - DETERMINATION OF OIL AND GAS MISCIBILITY
PERFORMANCE" (COWICONSULT, JUNE 1985).

```
********************************************************************
* TITLE : DEVELOPING AND TESTING A CHEMICAL        *   PROJECT NO    *
*         SYSTEM FOR POLYMER FLOODING IN OIL       *                 *
*         RESERVOIRS WITH HIGHLY SALINE FORMATION  * TH./05072/85/DE/.. *
*         WATER                                    *                 *
********************************************************************
* CONTRACTOR :                                     *   TELEPHONE NO  *
*    BASF AG, WINTERSHALL AG                        *                 *
*                                                  *  0621/60-47358   *
*    JOINT WORKING GROUP                            *                 *
*    DE - 6700 LUDWIGSHAFEN                          *                 *
*                                                  *   TELEX NO       *
*                                                  *                 *
* TECHNICAL DIRECTOR :                              *   46499-0        *
*    DR. MARTISCHIUS                                *                 *
*                                                  *                 *
********************************************************************
                                          VERSION : 09/03/87
```

AIM OF THE PROJECT :
****

The efficiency of polymer flooding in reservoirs with highly saline formation
water depends to a large extent on the polymer's stability towards salt.
Desalination or the use of a polymer that is efficient only at high
concentrations may adversely affect the economics or technical feasibility of
tertiary recovery. It is therefore necessary to develop a chemical system that
can be applied in highly saline formation waters.
Once a given field or part of a field has been selected for a pilot test, the
data relating to it must be compiled.

PROJECT DESCRIPTION :
****

A simulation study is intended to predict the optimum metering rate for the
polymer and the production pattern. For this purpose, polymer-specific data for
the field selected will be determined on samples of the reservoir rock.
POLYMERS FOR ENHANCED OIL RECOVERY. Polymers produced in the laboratory have to
be tested to determine whether they are suitable for use in reservoirs with
highly saline formation water. If they are found to be suitable, their
reproducibility must be checked on a pilot and a production scale.
PRODUCT SCREENING. The conditions for dissolving and shearing on a technical
scale are to be varied so that the optimum relationship between viscosity and
filterability is obtained under the conditions in a reservoir with a salinity of
about 150 g/l of total dissolved solids.
DETERMINATION OF THE PROPERTIES RELATING TO FLOODING ON WATER-WET MODEL CORES.
Flooding tests are performed on cores of different permeability in the range of
intended injection and production rates. Adsorption and retention effects of the
polymer are determined.
OPTIMIZATION OF THE POLYMERS' LONG-TERM STABILITY. The effects exerted by
various additives, additive concentrations and additive combinations on the
polymers' long-term stability under the conditions in the reservoir.
DETERMINATION OF THE PROPERTIES RELATING TO FLOODING ON OIL-WET MODEL CORES. The
stability of the solutions in the presence of an oil phase and the
chromatographic separation of the polymers are investigated.
SELECTION OF THE RESERVOIR. A field must be selected that might appear suitable
for polymer flooding in the light of its data, e.g. mobility ratio, average
permeability, permeability variation, formation thickness, mineralogy, stage of
depletion, and the maximum degree of depletion attainable by waterflood. The
next step is the selection of a suitable, well-delineated reservoir block, for
which purpose geological and reservoir engineering data must be reviewed. The
possibility of communication with wells in adjacent blocks must be investigated
by pulse tests and/or pressure measurements.
PETROPHYSICAL INVESTIGATIONS. Coring in unconsolidated Valendis reservoir by
special techniques and embedding the deep-frozen cores with inorganic solvent -
resistant material allow determination of permeability, porosity,
compressibility, capillary pressure, wettability, relative permeability, initial
oil saturation, and residual oil saturation after waterflooding. In addition,
mineralogical studies are performed.
SIMULATION STUDY. A history match and feasibility tests to optimize flooding
geometry and polymer slug volumes have to be carried out.
DETERMINATION OF DATA RELATING TO THE POLYMER ON ORIGINAL CORES AT THE RESERVOIR
TEMPERATURE. Mobility and permeability reduction are determined as well as
retention and adsorption effects in large cores. Based on the injectivity of the
polymer it has to be decided, whether a hydrofrac is necessary.

STATE OF ADVANCEMENT :
****

The project is still ongoing. The polymer development on a laboratory scale is
finished, the scale up is going to be realized. An area in the eastern part of
the field Dueste-Valendis is favoured, petrophysical investigations and original
core tests as well as simulation studies and tracer tests still have to be
performed.

RESULTS :
****

Anionic and nonionic synthetic polymers with different chemical characteristics show only a poor thickening behaviour in reservoir water of a high salinity. The effect exerted by polyacrylamides and acrylamide/acrylic copolymers in raising the viscosity is insufficient, even if their mol mass is extremely high. Using only nonionic monomers, e.g. vinyl pyrrolidone or acrylamide, results in a salt stable viscosity, but the absolute value of the viscosity is too low.
However, polysaccharides are very effective thickeners in highly saline formation water. In contrast to xanthanes, glucanes have no anionic components and therefore show the largest viscosities. Classification of the glucane broth is possible by the combination of a centrifugation and a filtration process.
Dueste-Valendis is selected as a field, where polymer flooding is suitable. Several geological and reservoir engineering data are obtained. Pulse tests have been performed, in order to see possible communication with wells in adjacent blocks.
Tracer tests are necessary to determine sweep geometry and length of dispersion. Basic research work resulted in the development of a tracer system, which is suitable in Dueste-Valendis.

```
************************************************************************
* TITLE : TESTING A NEW METHOD TO IMPROVE       *      PROJECT NO      *
*         SWEEPING BY INJECTION OF FOAMING AGENT *                      *
*         AND NITROGEN - PHASE 1                 *   TH./05073/86/FR/.. *
*                                                *                      *
************************************************************************
* CONTRACTOR :                                   *     TELEPHONE NO     *
*    GERTH                                        *                      *
*                                                *    1 47.52.61.39     *
*    4, AVENUE DE BOIS PREAU                      *                      *
*    FR - 92502 RUEIL-MALMAISON                   *                      *
*                                                *      TELEX NO        *
*                                                *                      *
* TECHNICAL DIRECTOR :                            *      203 050         *
*    MR. DENOYELLE                                *                      *
*                                                *                      *
************************************************************************
                                                 VERSION : 19/05/88
```

AIM OF THE PROJECT :
****
The object of this project is to prepare the test, on-field and for the first
time, of a method which will considerably increase the sweeping performance of a
reservoir while using nitrogen, a non-miscible, readily available and non
expensive gas, combined with foaming agents in small quantity.

PROJECT DESCRIPTION :
****
The project is divided into four phases.
Phase 1 : injection facilities.
Two processes are investigated :
- on-site separation of nitrogen from air using molecular sieve or membranes and
compression,
- transportation, storage and pumping of liquid nitrogen.
The study takes into account operating constraints.
Phase 2 : Tolerance study of crude oil versus oxygen.
This phase evaluates the risks of letting some oxygen to be injected with
nitrogen, especially if on-site separation is chosen.
The oxydation/combustion parameters of the crude oil are measured at two oxygen
partial pressures and two temperatures. Compositional thermal simulation of gas
injection and crude oxidation are performed.
Phase 3 : Selection of a foaming agent.
Different manufacturers have supplied samples of foaming agents. The ability of
these products to foam at 75 deg.C in the presence of 30 g/l of salt and in the
presence of oil is tested. Adsorption on rock and aging in the presence of
oxygen are studied. The products are compared with one another. The influence of
additives present in the field production water are investigated.
Phase 4 : Definition of injection mode.
This phase consists in studying the injection mode of the foaming agent, wether
simultaneously with the gas or in several slugs. Surface facilities are reviewed.
Recommendations on a measurement and a tracer program are made.

STATE OF ADVANCEMENT :
****
Completed. The feasibility of foam injection has been demonstrated and the basic
design on the surface facilities has been performed.

RESULTS :
****
After consulting foaming agent manufacturers and reviewing literature, 8
products were selected for the study. These products are either anionic
surfactants (mostly sulfonates or sulfates) or amphoteric surfactants used for
foam drilling or for manufacturing cleansing agents.
PHASE 1 :
On-site nitrogen generation is feasible at a cost varying from 0.35 FF/Nm3 for
PSA to 0.47 FF/Nm3 for membranes both for a 5 year project. Buying liquid
nitrogen would yield a final cost of 0.82 FF/Nm3 but access roads were not build
to withstand the subsequent heavy truck traffic.
PHASE 2 :
The oil from Coulommes field presents a low oxidation rate. The risks of auto-
ignition or of explosion in the production wells are negligible with less than
10 % of oxygen in the injected fluid.
PHASE 3 :
Preselected foaming agents include sulfates (Neopon LOA/F and Fenopon CD 128),
a sulfonate (Sulframin AOS) and amphoteric surfactants (Siponfor S and
Foramousse 110). The sulfates and the amphoteric surfactants show a good
resistance to aging at reservoir conditions. Adsorption tests indicate that the
Fenopon has the lowest adsorption among all products. Foam displacements in
porous medium at reservoir conditions and in the presence of oil stress the good
performance of the sulfonate and amphoteric surfactants and the poor performance
of sulfates. Compatibility tests with formol and anti-emulsion product present
in the water used to make foam, indicate that amphoteric surfactants are not
affected, while the sulfonate performs slightly worse and sulfates are very
affected.

PHASE 4 :
Surface facilities have been designed, measurements to be performed during the test have been listed, a tracer program has been set up. The injection mode will consist in a 60 m3 slug of foaming agent solution followed by a simultaneous injection of gas and foaming solution.

REFERENCES :
****
NO PUBLICATION NOR CONFERENCES EXCEPT FOR DOCUMENTS AND WORK SESSIONS BETWEEN ASSOCIATES.

```
*************************************************************************
* TITLE : HYDROCONVERSION OF HEAVY OILS ON        *      PROJECT NO      *
*         PRODUCTION FIELDS                        *                      *
*                                                  *   TH./05074/86/FR/.. *
*                                                  *                      *
*************************************************************************
* CONTRACTOR :                                     *     TELEPHONE NO     *
*    GERTH                                          *                      *
*                                                  *   1 47.52.61.39      *
*    4, AVENUE DE BOIS PREAU                        *                      *
*    FR - 92502 RUEIL-MALMAISON                     *                      *
*                                                  *     TELEX NO         *
*                                                  *                      *
* TECHNICAL DIRECTOR :                              *   203 050            *
*    MR. J.F. LE PAGE                               *                      *
*                                                  *                      *
*************************************************************************
                                                   VERSION : 31/12/88
```

AIM OF THE PROJECT :
****
The first object of this project is to render heavy oils transportable by
reducing their viscosity and their density through hydropyrolysis with the
appropriate catalyst. The second object is to carry out technico-economic study
to estimate the costs of heavy oil refining and to compare the various
processing schemes aiming at a syncrude production.

PROJECT DESCRIPTION :
****
This project involves two separate phases :
- the first one aims at demonstrating the technical feasibility of a new
catalytic hydropyrolysis using an original and very active homogeneous catalyst
as hydrogenating agent.
- the second one covers a technico-economic synthesis involving all the
processes and processing schemes implemented on the ASVAHL pilot in Solaize to
make the heavy oils transportable from the production field and them processable
in the existing refineries.
Concerning the first phase, tests have been carried out in the ASVAHL
minirefinery from the beginning of March to the end of May and then in September
(1987). Heavy oils which have been desalted and topped for these tests are
Boscan, Athabasca, Pilon, etc. Catalyst precursor was prepared in IFP.
Economic studies involved the four following items:
- achievement of detailed process books for every process (primary or secondary)
operated on the ASVAHL platform.
- estimation for each process of the investment and operating costs for
representative yearly capacities.
- comparison of various processing schemes with the aim to make them pipe-
lineable or able to be refined in existing refineries.
- sensitivity studies specially to determine at which production price heavy
oils have to be produced to obtain in the various processing schemes
investigated a syncrude the price of which would be competitive with that of
conventional oils.

STATE OF ADVANCEMENT :
****
The work programme has been achieved by the end of 1988.

RESULTS :
****
Experimentation
Results obtained at a demonstrative scale (1 to 6t/hour) are consistent with
those obtained in the preliminary experimentation carried out in pilot plants.
The tests have been worked out without any major problem after the unit has been
transformed for suitably handling the catalytic additive used in the operation.
With minute amounts of additive the route involving the couple of units
"hydropyrolysis and pentane deasphalting" most of the heavy oils are made
transportable acording to the specifications in force in all the countries, even
Canada.
Complementary tests carried out in more severe conditions (higher pressure,
contact time and catalyst amount) have enlightened the performances of this kind
of catalytic approach to perform very deep conversions.
Experimentation has been worked out with topped crudes atmospheric resids or
vacuum resids issued from the following origins : BOSCAN, ATHABASCA, COLD LAKE,
CERRONEGRO and ARABIAN HEAVY vacuum resids.
Technico-economic feasibility studies
Process designs and estimation studies have been carried out for every ASVAHL
process : deasphalting, visbreaking, hydrovisbreaking, catalyctic
hydrovisbreaking, hydrotreatments in fixed beds or moving beds. These
estimations also involved the investments and operating costs of necessary
satellite units (distillation, desalting, Amine treatment of gas, claus unit,
hydrogen synthesis).
Various processing schemes have been checked and compared either to only make
the crude pipe-lineable.
Desalting + distillation + visbreakings + visbreaking

Desalting + distillations + hydrovisbreaking
Desalting + distillations + catalytic hydropyrolysis
Desalting + distillations + hydropyrolysis (catalytic or thermal) +
deasphalting
or to carry out a deeper conversion of the heavy oils to aim at producing a
syncrude which could be treated in a conventional refinery including a reformer
and a cat-cracker (or hydro-cracker). Desalting - Distillation - hydropyrolysis
(catatlytic) - DAO hydrotreatment
Desalting - distillation - direct hydrotreatment (fixed or moving beds).
Sensitive studies through IFP's linear refining program for a better selection
of the heavy oil refining way to produce the best syncrude in given conditions.

REFERENCES :
****

-THERMAL CRACKING UNDER HYDROGEN PRESSURE -ACS DENVER 5.4.87
-RESIDUE AND HEAVY OIL CONVERSION-SYMPOSIUM OAPEC-LYON 1986
-RESIDUE AND HEAVY OIL CONVERSION-SYMPOSIUM AOSTRA-CALGARY 1986
-ASVAHL NEW ROUTES FOR PROCESSING HEAVY OILS-UNITAR-TORONTO 1988
-ECONOMIC IMPACT OF HEAVY OIL UPGRADING-6TH EUROPEAN OIL AND GAS CONFERENCE-
AMSTERDAM 1988
-NEW IMPROVEMENT IN VISBREAKING TECHNOLOGY-ACS-TORONTO 5.1.1988
-HYVAHL PROCESS FOR HIGH CONVERSIONS-JPI-TOKYO 1988
-ASVAHL NEW ROUTES FOR PROCESSING HEAVY OILS-UNITAR-EDMONTON 4.8.1988
-HEAVY OIL PROCESSING-SYNTHESIS OF ASVAHL'S RESULTS-SYMPOSIUM CEE-MARCH 1988
-COMPARISON OF THERMAL AND CATALYTIC PYROLYSIS OF HEAVY OILS-ACFAS-OTTAWA 14.5.
1987.

```
************************************************************************
*  TITLE : PROMETHEE - PHASE 1                  *      PROJECT NO      *
*                                               *                      *
*                                               *   TH./05075/86/FR/.. *
*                                               *                      *
************************************************************************
*  CONTRACTOR :                                 *    TELEPHONE NO      *
*     GERTH                                     *                      *
*                                               *   1 47.52.61.39      *
*     4, AVENUE DE BOIS PREAU                   *                      *
*     FR - 92502 RUEIL-MALMAISON               *                      *
*                                               *    TELEX NO          *
*                                               *                      *
*  TECHNICAL DIRECTOR :                         *   203050 FAX: 47.52.69*
*     MR. AUXIETTE                              *   .27                *
*                                               *                      *
************************************************************************
                                                 VERSION : 31/12/89
```

AIM OF THE PROJECT :
****
This project consists in developing laboratory tools capable of representing the exact reservoir conditions during air injection; e.g. the gas flows, temperature conditions, pressure conditions, consolidated porous medium, etc...These tools will make it possible to obtain results easily extrapolable to a reservoir scale, or study of air injection in unexplored domains such as light oil reservoirs.

PROJECT DESCRIPTION :
****
This project consists of three phases :
PHASE 1 - STUDY OF COMBUSTION WITH OXYGEN
The work consists of comparing the conditions of propagation of combustion with oxygen and with air, and in determining the influence of the principle parameters governing combustion with oxygen so as to identify the technical domain in which the method could be applied.
PHASE 2 - STUDY ON AIR INJECTION IN SITU ON AN ELABORATED MODEL
Simulation in the laboratory of combustion in reservoir conditions and a consolidated environment implies the creation of equipment unique in the world. All the developed technologies undergo routine tests to prove them before inclusion in the equipment. When the apparatus has been debugged, comparative tests are run between the standard models and the latter, to prove apparatus, and provide indications as to its limitations and the feasibility of combustion in unusual conditions.
PHASE 3 - ECONOMIC ESTIMATION
The experimental results obtained during the tests would make it possible to make an economic assessment of combustion in non-standard applications.

STATE OF ADVANCEMENT :
****
Ongoing
Phase 1 : completed
Phase 2 : the project has been reoriented towards the study of air injection. Equipment has been reviewed to meet new objectives. Delivery, assembling and tests are underway. Test survey has been scheduled, oil, reservoir gas and rocks (Vosges and Fontainebleau sandstones) have been supplied.
Phase 3 : completed.

RESULTS :
****
In phase I, experiments have been carried out on two reference oils (specific gravity 0.96 and 0.90). The effects of total pressure, oxygen partial pressure and injected oxygen flux were studied.
For the heavy gravity oil, the oxygen requirement at the front increases with the oxygen partial pressure, slighty at constant oxygen flux, more strongly at increasing oxygen flux.
For the medium gravity oil, the oxygen requirement at the front decreases when the oxygen partial pressure increases at constant oxygen flux, and increases with the oxygen flux.
Previous sweep with a mixture of N2/CO2 during combustion is conform to results of standard combustion experiments.
As far as phase 2, is concerned, works covered the fabrication, the assembling and commissioning of some of the equipment.

REFERENCES :
****
1) SPE 16741 DALLAS 27-30 SEPTEMBER 1987
M. PETIT - "IN SITU COMBUSTION WITH O2 ENRICHED AIR"
2) 4TH EUROPEAN SYMPOSIUM ON ENHANCED OIL RECOVERY - HAMBURG OCTOBER 1987. H. PETIT - "IN SITU COMBUSTION EXPERIMENTS WITH O2 ENRICHED AIR"
3) ADA EUROPE CONFERENCE 1989 - MADRID
G. AUXIETTE, J.F. CABADI, P. REHBINDER - "PROMETHEE : DESIGNING A PROCESS CONTROL SYSTEM"

```
**********************************************************************
* TITLE : GEOPHYSICS INTERPRETATION OF SUBSIDENCE     *    PROJECT NO      *
*          MEASUREMENTS (PHASE I)                      *                    *
*                                                      *  TH./05076/86/FR/.. *
*                                                      *                    *
**********************************************************************
* CONTRACTOR :                                         *  TELEPHONE NO      *
*    GERTH                                             *                    *
*                                                      *  (1)47.52.61.39    *
*    AVENUE DE BOIS PREAU 4                            *                    *
*    FR - 92502 RUEIL MALMAISON                        *                    *
*                                                      *  TELEX NO          *
*                                                      *                    *
* TECHNICAL DIRECTOR :                                 *  (1)47.52.69.27    *
*    MR. D. DESPAX                                     *                    *
*                                                      *                    *
**********************************************************************
                                              VERSION : 30/06/88
```

AIM OF THE PROJECT :
****
This project consists in measuring subsidence movements induced by pressure
charges associated to reservoir depletion in view of evaluating that depletion.
These subsurface movements are then converted into reservoir pressure charges
versus space and time using a F.E. model of all relevant layers i.e. overburden,
reservoir and underburden. Finally this pone pressure map is converted into a
permeability map in both reservoir and aquifer.

PROJECT DESCRIPTION :
****
Before the method can be applied at larger scale it is necessary to prove that
the pressure variations in the reservoir provoke movements that can be detected
from the subsurface.
The present project thus consists :
- in a first step, in measuring the subsidence generated by the production of a
well within the VILLEPERDUE reservoir (operated by TOTAL-CFP, partner TRITON),
- in a second step, in deducing thereby the possible distribution of the
pressure variations existing within the pay zone area
PHASE 1- MEASUREMENT OF SUBSIDENCE MOVEMENTS
The VILLEPERDUE field is formed by a carbonate reservoir located in the upper
layers of the DOGGER. The production layer is located at a depth of about 1 850
m.
- a first step, performed on a "versus structure" (Hautefeuille) will allow to
optimize the display of the tiltmeters around HFE1 well,
- a second step which will allow to implement benches a sine crossing both
Villeperdue and Hautefeuille structures
This phase will include the following tasks :
- layout of inclinometres
- postioning of level beacons
PHASE 2 - ELABORATION OF A NUMERICAL MODEL
This step will include :
- Evaluation of the deformation generated by placing well HFE1 on prodution
using tiltmeters
- Measurement of a cross section of the subsidence bowl using optical devices.
PHASE 3 - INTERPRETATION OF MEASUREMENTS
Tasks related to this phase only began when a significant subsidence had been
outlined. These tasks allowed the plotting of a map of the pressure variations
around the well in relation to the time. They included :
a) Development of a method of numerical resolution of the reverse problem in
which the subsurface movements are the data enabling the pressure variations
within the reservoir to be defined.
b) Determination of the mechanical properties of the entire deformable massif
(analysis of the logs and drilling parameters - laboratory measurements).
c) Interpretation of the inclinometry measurements and the level curves, first
by means of a radial symetry model and second with the aid of a three-
dimensional model.

STATE OF ADVANCEMENT :
****
Two measuring devices have been tested :
- IPG inclinometers which should reveal terrain movements associated with a
build-up followed by production re-startup in a remote well
- high precision level measuring of the IGN along an 18 km line crossing two
neighbouring deposits.
The numerical model is completed.

RESULTS :
****
- Bad weather has induced trouble with tiltmeters.
- 3 surveys carried out by IGN have pointed out a subsidence of 1 cm/year.
Converted into pressure variations, this subsidence has allowed to derive a
pressure map in accordance with production history and geology.

REFERENCES : TECHNICAL PROGRESS REPORT NR 1 (DEC. 87)
             FINAL REPORT (JULY. 88).

```
****************************************************************************
* TITLE : SCLEROGLUCANE OF EOR QUALITY          *      PROJECT NO        *
*                                               *                        *
*                                               *    TH./05078/86/FR/..  *
*                                               *                        *
****************************************************************************
* CONTRACTOR :                                  *    TELEPHONE NO        *
*    GERTH                                       *                        *
*                                               *    1 47.52.61.39       *
*    4, AVENUE DE BOIS PREAU                     *                        *
*    FR - 92502 RUEIL-MALMAISON                  *                        *
*                                               *    TELEX NO            *
*                                               *                        *
* TECHNICAL DIRECTOR :                           *    203 050             *
*    MR. A. DONCHE                               *                        *
*                                               *                        *
****************************************************************************
                                              VERSION : 30/06/89
```

AIM OF THE PROJECT :
****
Today, the injection of polymers to sweep reservoirs is a confirmed technique in the domain of oil production. In spite of the lack of suitable products, this method is nevertheless restricted to the exploitation of reservoirs of mean salinity and a temperature below 70 deg.C. Significant progress has to be made to ensure the treatment of reservoirs in more difficult environment. Therefore, Scleroglucane, a polymer produced by fermentation, offers new prospects: its resistance to temperature and salinity makes it possible to envisage the treatment of hot and salty reservoirs, still not accessible technically. Scleroglucane is presently available at industrial scale. The purpose of the project is to provide the required quality for the injection of polymers and to define the conditions of implementation of such a product.

PROJECT DESCRIPTION :
****
The scleroglucane that is available today, already presents part of the characteristics that are required for the development of a hydrosoluble polymer under EOR. Its solutions are very viscous and highly resistant to shear. Their viscosity is hardly sensitive to salts and to the pH and little or hardly affected by the temperature.
The product only lacks a few adaptations so to meet the EOR quality in view of qualifying for a pilot test.
Therefore, a few obstacles need to be overcome:
- today, scleroglucane presents a poor filtrability which may provoke a risk of clogging the formation in which it will have been injected. Thus, it is necessary to identify the factors responsible for this phenomenon and have them neutralized
- conditioning of hydrosoluble polymers and namely polysaccharides requires a particularly well adapted technique to avoid degradation of the initial filtrability and viscosity of the product. Whether in concentrate solution or in powder form, this conditioning is essential for the development of the scleroglucane and has to be finalized.
- polymer runs many risks of being degraded during its progress in the reservoir, bacterial degradation being one of the major risks. In this case, it is necessary to evaluate the risks and define, when necessary, the means of protecting the polymer.
The project is divided into four phases:
PHASE 1: improving the filtrability of the scleroglucane
The purpose of this activity is to identify the factors that limit the filtrability of the polymer (aggregation factors) to define in a second step the modifications of the manufacturing process of scleroglucane or of other solutions (additives) to improve its filtrability. A detailed study will be carried out on the characteristics of pure polymer, on its structure and on the possible interactions with impurities to identify the endogenous and exogenous aggregation factors.
PHASE 2: adaptation of manufacture process to EOR requirements
Works carried out in Phase 1 will lead to a number of modifications in the scleroglucane manufacturing process (fermentation, finalization treatment, conditioning). These modifications will be systematically tested at the plant on a manufacturing pilot. Special care will be given to the finalization of conditioning techniques which do not alter the rheologic and filtrability characteristics of the scleroglucan obtained after process. Three types of conditioning will be studied, diluted solution directly injectable, concentrate solution and powder solution.
PHASE 3: evaluation of the scleroglucane for EOR
The purpose of this activity is to define the performances of the scleroglucane in porous media and the conditions of its implementation in EOR. Studies will first cover pure scleroglucan under model conditions to evaluate the limits of its utilisation domain then in natural porous media with improved filtrability scleroglucan to analyse its behaviour and efficiency when displacing oil.

STATE OF ADVANCEMENT :
****
The project is almost completed. Scleroglucane solutions offering good

filterability characteristics under temperature can now be prepated. Processings developed to obtain this result have been tested at pilot scale. Evaluation of the performances of this product for EOR confirms its good behaviour in hot and salty environments. Under cold temperature, besides a higher adsorption, sweepling is also satisfactory.

RESULTS :
****
PHASE 1 : IMPROVING THE FILTERABILITY OF SCLEROGLUCAN
Completed
Different treatments for the purification - desagregation of scleroglucan have been developed, allowing to produce at pilot scale a polymer presenting satisfactory characteristics under temperature - high viscosity and reduced agregation trend. Under cold temperature, there remains a residual agregation which provokes poor filterability and requires additional treatment in view of its elimination.
PHASE 2 : ADAPTATION OF FABRICATION PROCESS TO EOR CONSTRAINTS
        Completed.
The different fabrication parameters likely to agregate the scleroglucane have been systematically explored and optimized in close relation with the studies carried out in phase 1. The adjunction of purification-desagregation treatments deriving from these studies allowed us to make at pilot scale batches of scleroglucane complying with the EOR needs under hot and salty conditions.
As for the conditioning, studies are underway to prepare powder solutions and concentrate solutions which preserve the good filterability and viscosity characteristics of the initial product. Original methods are being developed.
PHASE 3 : EVALUATION OF THE SCLEROGLUCAN FOR EOR
Evaluations of the behaviour of scleroglucan under flow were continued in a satisfactory manner both in model porous media (pure scleroglucan) and in natural porous media (industrial scleroglucan of EOR quality). They confirm that this polymer is suitable for injection in hot and saline environments. Under cold temperatures encountered at the beginning of the injection, the adsorption is nevertheless reinforced but with no manifestation of a plugging feature.
PHASE 4 : PROTECTION AGAINST BACTERIAL DEGRADATION
Completed.
Studies on the bacterial degradation of the scleroglucan showed that the biodegradation risk decreases as the temperature and salinity raises. Under hot temperature, no bacterial degradation was evidenced. Protection means in risk-proven areas have been defined.

REFERENCES :
****
J.LECOURTIER,C.NOIK,P.BARBEY,G.CHAUVETEAU
SEMI-RIGID POLYSACCHARIDES FOR POLYMER FLOODI(NG IN HIGH SALINITY RESERVOIR-4TH EUROPEAN SYMPOSIUM ON EOR,HAMBURG,OCTOBER 1987
C.NOIK,J.LECOURTIER,G.CHAUVETEAU,A.DONCHE
RHEOLOGICAL BEHAVIOUR OF A NEW SCLEROGLUCAN-CONFERENCE ACS (SEPTEMBRE 1987-NEW ORLEANS)
J.LECOURTIER,C.NOIK,G.CHAUVETEAU
PROCEDE DE TRAITEMENT D'UN MOUT DE FERMENTATION RENFERMANT UN POLYSACCHARIDE DANS LE BUT D'EN ACCROITRE LA FILTRABILITE ET UTILISATION DE CE MOUT EN RECUPERATION ASSISTEE DU PETROLE.BF 87-14582 (20 OCTOBRE 1987)
J.LECOURTIER,C.NOIK,G.CHAUVETEAU
PROCEDE DE TRAITEMENT D'UN MOUT DE FERMENTATION RENFERMANT UN POLYSACCHARIDE DANS LE BUTD.N.TRUONG,J.GADIOUX
PURIFICATION D

```
*****************************************************************************
*  TITLE : RETENTION OF A MULTIPLE CHEMICAL SYSTEM      *     PROJECT NO      *
*          WHILE INCREASING THE RECOVERY FACTOR BY      *                     *
*          CHEMICAL FLOODING                            *   TH./05082/86/DE/.. *
*                                                       *                     *
*****************************************************************************
*  CONTRACTOR :                                         *   TELEPHONE NO      *
*     PREUSSAG AG                                        *                     *
*                                                       *   05176/17 247      *
*     ARNDTSTRASSE 1                                    *                     *
*     DE - 3000 HANNOVER 1                              *                     *
*                                                       *   TELEX NO          *
*                                                       *                     *
*  TECHNICAL DIRECTOR :                                 *   92655             *
*     DR. D. MENZ                                       *                     *
*                                                       *                     *
*****************************************************************************
                                             VERSION : 15/09/89
```

AIM OF THE PROJECT :
****
The goal of this development project is to determine retention on the rock
surface under simulated reservoir conditions. In the proposed investigation,
retention data for both the individual surfactants and their combined mixture as
well as data on the multichemical system containing both surfactants and
polymers should be gathered and evaluated. Based on observations on linear flood
tests it is presumed that, by post-flooding using additional surfactants and
polymers, the retention of a surfactant mixture can be considerably reduced.
This would result in a further reduction of the required concentration and slug
size of the surfactant combination.

PROJECT DESCRIPTION :
****
1. Development and optimization of the analytical process for specificaton of
both the surfactant combinations and the multiple chemical system.
The surfactants which are planned to be used in the field are technical products
which contain a broad spectrum of compounds of varying molecular weight and
reaction groups. In the past, different methods of analysis (infrared
spectrography, UV-spectrography, two phased Titration, etc were attempted in the
flood tests. The main problem is the chemical separation of the used components
like a mixture of crude oil, brine emulsions.
2. Static absorption tests
The identification of absorption isotherms on sand surfaces should supply
information on the absorption characteristics of both the various components and
the multi-chemical system.
Pressure-free-flooding
Using a model sandstone core with a porosity of 20% and a permeability of 1000
md, pressure-free flooding can be done. The testing of the flood and
displacement behaviour of the individual components, the surfactant combination,
and the multi-chemical system is done in order to determine the retention
behaviour and to show, if present, a chromatographic separation of the
components.
4. The sandpack flood-tests
The sandpack flood-tests serve mostly in showing possible chromatographic
separation.
5. High pressure flooding
The work on the high pressure flooding apparatus can be done as a supplement to
the pressure-free flood tests; here the influence of pressure on the flood-
behaviour of the chemical system can be tested under simulated reservoir
pressures.

STATE OF ADVANCEMENT :
****
The adsoprtion measurements by using a flow micro-calorimeter have been carried
out.

RESULTS :
****
The retention of the multi component system : petroleum sulphonate - non ionic
surfactant in reservoir brine showed strong interactions and a high retention
potential.
It was found that adsorption of xanthan is not automatically high on surfaces
with higher area. Here it is important that the adsorption sites are accessible
to the macromolecule, which is obviously not the case for all sites of the
reservoir rock. The reservoir rock has a specific surface of 2,6 m2/g, but only
a small part of this surface seems to be accessible to the big polymer molecules.
 The major part of the specific surface measured with the very much smaller n-
Heptane-molecules is contributed by a fine structure on the grain surfaces.
Furthermore the adsorption sites of the reservoir rock, that was originally oil
saturated and then cleaned with solvents, seems to be less active than the
adsorption sites on the clean quartz sand. Adsorption of xanthan was even higher
on the quartz sand than on the reservoir rock sample.
The results from experiments performed with glucose and pyruvate clearly show
that adsorption is influenced by the pyruvate content of the molecule, which

K

means that higher pyruvate xanthans have higher adorption, which in turn may also be an additional reason for the poor injectability that was found with these products. Unfortunately these products could not be tested in the micro-flow-calorimeter, as this requires a good injectability to avoid heat effects due to uncontrolled pressure increases.

REFERENCES :
****
1. SCRRENING OF XANTHAN-BIOPOLYMER FOR A HIGH SALINITY OIL RESERVOIR KLEINITZ,W.
LITTMANN, W;HERBST H.3FIFTH EUROPENA SYMPOSIUM ON INPROVED OIL RECOVERY,
BUDAPEST, APRIL 1989
2. RETENTION EINES GESAMTCHEMIKALIENSYSTEMS BEI DER MEHRENTOLUNG VON
LAGERSTATTEN DURCH CHEMISCHES FLUTEN
KLEINITZ, W.
STATUSBERICH, 8.-10 MARZ 1989, CELLE
3. MEASURING INTRINSIC VISCOSITIES OF XANTHAN
KULICKE, W; KLEINITZ, W; LITTMANN, W; IN PREPARATION.

```
********************************************************************************
* TITLE : DEVELOPMENT OF CELLULOSE DERIVATIVES      *       PROJECT NO       *
*         FOR THE USE AS VISCOSITY BUILDERS IN      *                        *
*         FLOODING MEDIA FOR THE ENHANCED OIL       *    TH./05083/86/DE/..  *
*         RECOVERY                                  *                        *
********************************************************************************
* CONTRACTOR :                                      *    TELEPHONE NO        *
*     WOLFF WALSRODE AG                             *                        *
*                                                   *    05161/442977        *
*     DE - 3030 WALSRODE                            *                        *
*                                                   *                        *
*                                                   *    TELEX NO            *
*                                                   *                        *
* TECHNICAL DIRECTOR :                              *    924 324             *
*     DR. REINHARD KNIEWSKE                         *                        *
*                                                   *                        *
********************************************************************************
                                                   VERSION : 01/10/89
```

AIM OF THE PROJECT :
****
The aim of the project is the development of cellulose derivatives that can be used as viscosifiers in flooding media in enhanced oil recovery. These novel cellulose polymers ought to be used in polymer flooding but also as a mobility control agent in micellar-polymer-processes.

PROJECT DESCRIPTION :
****
The project is divided into three parts.
PART A
"Modification of carboxylmethyl cellulose (CMC)" contains two phases: the improvement of the injectibility and specific viscosity of CMC insertion of a second substitutentnt into CMC to improve the electrolyte stability.
PART B
"Development of a new cellulose derivatives" is the main part of the project which is divided into four phases. Preparation of new cellulose derivatives by using unusual substitutents. Optimization of the processing and product properties of the new cellulose derivatives.
- Insertion of a second substitutent into these new derivatives. The use of bulky substitutes as a third substitutent into new cellulose derivatives.
PART C
"Screening of the products under practical aspects" is continuously done during the whole proccessing of the project.

STATE OF ADVANCEMENT :
****
The project started on 01.09.85. and terminated on 01.03.89.

RESULTS :
****
Novel cellulose derivatives with one and two substituents were prepared and characterised in accordance with the project schedule. The product properties were investigated in distilled water and synthetic formation water. The qualifications as viscosity builders in flooding media were tested at the Bayer AG using an establishment lab program. Novel cellulose derivatives with very good properties, defined by a certain combination of substituents, were comprehensively investigated and optimised. The properties of these types of cellulose products are :
- excellent salt stability/good temperature stability;
- excellent injectivity;
- deoiling efficiency of 7 to 8 % OOIP.
The deoiling efficiency of Xanthan under these conditions is between 3 and 6 % OOIP. The novel cellulose ethers can be prepared with standard processes of cellulose ether technology. One drawback, however, which could not be overcome as yet, is the insufficient viscosity yield compared with Xanthan. For a certain viscosity level, e.g. the concentration of the new developed cellulose produts has to be 2 - 3 times higher than of Xanthan. Moreover, the structure viscosity is less pronounced for the new types.

REFERENCES :
****
LECTURES WITH THE TITLE OF THE PROJECT WERE HELD
- IN CELLE, FRG, AT THE STATUS SEMINAR "GEOTECHNIK UND LAGERSTATTEN" ON MARCH 9TH 1988;
- IN LUXEMBOURG AT THE "3RD HYDROCARBON SYMPOSIUM" ON MARCH 23RD 1988.
THREE PATENT APPLICATIONS HAVE BEEN PUBLISHED : DE-OS 3742 104, DE-OS 3742 105, DE-OS 3742 106.

```
*****************************************************************************
* TITLE : IN-SITU COMBUSTION PROCESSES FOR          *      PROJECT NO      *
*          ENHANCED OIL RECOVERY                     *                      *
*                                                    *  TH./05084/86/UK/..  *
*                                                    *                      *
*****************************************************************************
* CONTRACTOR :                                       *     TELEPHONE NO     *
*    UNIVERSITY OF BATH                              *                      *
*                                                    *    0225-826826       *
*    SCHOOL OF CHEMICAL ENGINEERING                  *                      *
*    CLAVERTON DOWN                                  *                      *
*    UK - BA2 7AY BATH                               *     TELEX NO         *
*                                                    *                      *
* TECHNICAL DIRECTOR :                               *     449097           *
*    DR. M. GREAVES/DR. R.W. FIELD                   *                      *
*                                                    *                      *
*****************************************************************************
                                                VERSION : 26/09/89
```

AIM OF THE PROJECT :
****
To develop an accurate simulation model of the in-situ combustion process as
applied to heavy oil recovery.

PROJECT DESCRIPTION :
****
The development of an advanced, fully automated high pressure, oxygen-assisted
combustion tube facility represents the major part of the overall programme.
Carefully designed experiments will ensure that the data obtained has the
desired precision and selectivity so that simulation accuracy can be critically
assessed.
The stages of the works are the following:
A. EXPERIMENTAL
1. Design of combustion tube facility
2. Equipment fabrication and procurement
3. Equipment installation and commissioning
4. Combustion tube experiments at 50 bar pressur
5. Combustion tube experiments at 100 bar and 200 bar pressure
6. Detailed analysis of results
B. NUMERICAL SIMULATION
1. Crude oi properties
2. Modelling development
3. Simulation of combustion tube experiments at 50 bar pressure
4. Simulation of combustion tube experiments at 100 bar and 200 bar pressure
5. History matching
6. Detailed analysis of results.

STATE OF ADVANCEMENT :
****
Fabrication and procurement have been completed. Installation is almost finished.
Commissioning will be finished by the end of 1989

```
*******************************************************************************
* TITLE : OIL RECOVERY BY NITROGEN FLOODING          *      PROJECT NO        *
*                                                    *                        *
*                                                    *   TH./05086/88/NL/..   *
*                                                    *                        *
*******************************************************************************
* CONTRACTOR :                                       *    TELEPHONE NO        *
*     DELFT UNIVERSITY OF TECHNOLOGY                 *                        *
*                                                    *    015 781328          *
*     PO BOX 5                                       *                        *
*     2600 AA DELFT                                  *                        *
*                                                    *    TELEX NO            *
*                                                    *                        *
* TECHNICAL DIRECTOR :                               *    28151               *
*     MR. J. HAGOORT                                 *                        *
*                                                    *                        *
*******************************************************************************
                                                        VERSION : 20/02/90
```

AIM OF THE PROJECT :
****
Establishment of technical and economic design criteria and evaluation tools for oil and condensate recovery by nitrogen injection.

PROJECT DESCRIPTION :
****
A. DEPENDENCY ON OIL COMPOSITION :
Phase behaviour calculations and experiments on synthetic oil-nitrogen systems, of flooding experiments using the slim tube equipment and numerical simulations (use of simple one dimensional compositional reservoir simulators) using mathematical compositional simulators.
B. APPLICATION OF SLIM TUBES
Evaluation of pros and cons of slim tube nitrogen flooding experiments at various experimental conditions using our high-pressure slim-tube equipment.
C. MINIMUM MISCIBILITY PRESSURE
Evaluation of the concept of the MMP for its applicability to nitrogen flooding. Theoretical (based on equations of state) and experimental (based on slim tube experiments) methods for the determination of the MMP. Existing correlations will be checked, extended and/or modified if necessary.
D. SIMULATION
Analyse, test, extension and/or modification of the available one-dimensional compositional simulators. Formulation of guidelines for the simulation of nitrogen flooding by means of commercial three-dimensional simulators.
E. PHYSICAL PROPERTIES
Evaluation of the existing prediction methods by comparing their results with available experimental data and, if necessary, with specially designed experiments. Development of new correlations and/or extension of the existing correlations.
F. RELATIVE PERMEABILITY
Measurement of relative permeabilities under conditions of low interfacial tension on well defined simple sandpack columns. Based on the results, check of the heuristic models and development of new relative permeability models, if necessary.
G. LIQUID DROP OUT
Formulation of a model in which liquid drop out and the subsequent evaporation is taken into account. Tests against slim tube experiments using synthetic condensate gases.
H. NITROGEN GENERATION
Investigation by means of a literature search, the relative advantages and disadvantages of membrane separation for application in oilfields onshore as well as offshore.
I. RESERVOIR HETEROGENEITY
Definition of reservoir geological prototypes for some commonly encountered depositonal environments. Using simplified mathematical models. Simulation of nitrogen flooding in each of the prototypes.
J. GRAVITY STABLE CONDITIONS
A Conceptual reservoir engineering design of nitrogen flooding under gravity stable conditions using horizontal wells. Comparison with a conventional design based on vertical wells.
K. TECHNICAL AND ECONOMICAL EVALUATION
Technical and economic evaluation of nitrogen flooding for a realistic prototype reservoir preferably located in the North Sea. This study should address the question whether nitrogen injection is a viable recovery process and if so what are the principal factors that favour its application.

STATE OF ADVANCEMENT :
****
Ongoing. The project will start at january the 1st 1989.

RESULTS :
****
For volatile oils with a high content of methane, the (thermodynamic) MMP is equal for nitrogen and methane injection.

REFERENCES :
****
BATH, P.G.H.: "STATUS REPORT ON MISCIBLE/IMMISCIBLE GAS FLOODING", PROCEEDINGS
4TH EUROPEAN SYMPOSIUM ON ENHANCED OIL RECOVERY, HAMBURG, 27-29 OCTOBER 1987.
BOERSMA, D.M.,HAGOORT, J. : "DISPLACEMENT CHARACTERISTICS OF NITROGEN FLOODING
VERSUS METHANE FLOODING IN VOLATILE OIL RESERVOIRS"; TO BE PRESENTED AT THE
SPE/DOE 7TH SYMPOSIUM ON ENHANCED OIL RECOVERY, APRIL 22-25, 1990, TULSA,
OKLAHOMA, USA, SPE PAPER NR SPE 20187.

```
****************************************************************************
* TITLE : INJECTION OF NON MISCIBLE GAS IN     *     PROJECT NO      *
*         FRACTURED OR THICK RESERVOIRS IN THE *                     *
*         NORTH SEA                            *   TH./05088/89/FR/.. *
*                                              *                     *
****************************************************************************
* CONTRACTOR :                                 *     TELEPHONE NO     *
*    GERTH                                     *                     *
*                                              *     47 52 61 39      *
*    AVENUE DE BOIS PREAU 4                    *                     *
*    FR - 92500 RUEIL MALMAISON                *                     *
*                                              *     TELEX NO         *
*                                              *                     *
* TECHNICAL DIRECTOR :                         *                     *
*    MR C BARDON                               *                     *
*                                              *                     *
****************************************************************************
```

VERSION : 01/01/90

AIM OF THE PROJECT :
****
The purpose of the project is to evaluate the advantages to be obtained from
injecting non miscible gas in fractured or thick reservoirs, which are the types
of reservoirs the most commonly operated in the North Sea for more than some
fifteen years. Some of these deposits, whether fractured or not, are becoming
flooded with water and production rates will drop over the nexte few years if
the operating mode is not changed. As regards those being developed, the choice
of their operating mode must be decided upon within a very short delay.
Therefore, this type of evaluation is currently of prime importance.
Concerning non miscible gas injection, there is a lack of data and of physical
models in particular for describing gas injection in :
- fractured reservoirs
- thick and continuous reservoirs
However some preliminary studies showed that non miscible gas can be very
attractive for two reasons :
- hiogh recovery rates can be expected
- moderate cost of gas production is possible.

PROJECT DESCRIPTION :
****
The purpose of the project is to evaluate the injection of non miscible gas :
methane (C1), nitrogen (N2) or air, in two types of north sea reservoirs :
fractured reservoirs and thick reservoirs.
This project consists in four phases :
Phase 1 : Measurement of triphasic relative permeabilities and capillary
pressure.
The results expected from this work consist in finalizing the correlations to be
used in the numerical simulation models taking into account all the physical
laws determined at the laboratory scale.
Phase 2 : Quantitative evaluation of stripping improved by gaseous dispersion in
fractured reservoirs.
The results expected from this work are the control of process efficiency, the
evaluation of exchange kinetics, and the equivalent dispersion coefficient.
These data will help to adapt a numerical compositional model.
Phase 3 : Air injection in thick or fractured reservoir.
The main purpose of this work is to gain informations on the catalytic effect of
the rock and the oxidation kinetics by air of the oil in place. The experimental
results will be used in a numerical model to simulate hydrocarbon oxidation
reactions by air and their consequences.
Phase 4 : Simulation of non miscible gas injection in two typical types of
deposits in the North Sea.
The experimental results obtained during phases 1 to 3 will be extrapoled in
digital simulation of non miscible gas injection in two cases of North Sea
reservoirs (thick reservoir and fractured reservoir).

STATE OF ADVANCEMENT :
****
Started on 01/08/89

ENVIRONMENTAL INFLUENCE ON OFFSHORE

```
***********************************************************************
* TITLE : DEVELOPMENT OF TECHNOLOGIES RELATIVE TO    *     PROJECT NO    *
*         IMPROVEMENT OF PILE CAPACITIES BY          *                   *
*         CONTROLLED PRESSURE GROUTING               *   TH./06018/85/BE/.. *
*                                                    *                   *
***********************************************************************
* CONTRACTOR :                                       *   TELEPHONE NO    *
*    BELGIAN OFFSHORE SERVICES                       *                   *
*                                                    *   03 2318770      *
*    SCHERMERSTRAAT 46                               *                   *
*    BE - 2000 ANTWERPEN                             *     ..            *
*                                                    *   TELEX NO        *
*                                                    *                   *
* TECHNICAL DIRECTOR :                               *   34129           *
*    MR. S. DECKERS                                  *                   *
*                                                    *                   *
***********************************************************************
                                            VERSION : 20/09/89
```

AIM OF THE PROJECT :
****
The aim of the project is to extend in a reliable manner the technique of
controlled pressure grouting to the improvement of pile capacities of offshore
structures and to develop new pressure grouting technologies in order to
decrease the installation cost of offshore and possibly subsea structures.
So far grouting techniques used in the offshore industry are very conservative
and have not evolved in decades. New innovative technologies will lead to lower
installation costs and improved performances of the pile foundation.

PROJECT DESCRIPTION :
****
The analysis of the state of the art of pressure grouting and of potential
problems existing in the offshore piling industry led to the selection of
pressure grouting techniques which are developed under 3 aspects :
- theoretical analysis of pressure grouting
- detailed study of grouts for offshore applications
- development of new technologies
The theoretical research is related to the phenomenon of soil "cracking" and is
divided in three stages :
- the first stage is dedicated to a naturalistic description of the involved
mechanisms.
- the second stage consist of using an elasto-plastic consolidation finite
element computer program to study the "cracking" phenomenon.
- in the third stage "cracking" laboratory tests are to be performed on small
scale models and interpreted with the help of numerical computations in order to
produce a mathematical model of the "cracking" phenomenon.
The detailed study of grouts for offshore applications consists of :
- checking the suitability of grout used onshore for offshore application
- developing new grouts for specific offshore application
This part of the project is mostly experimental and it involves intensive
laboratory treating to determine all the grout properties which are essential
for offshore use.
The development of new technologies involves :
- reviewing all the components of existing pressure grouting systems (i.e.
pressure grouting units, non return valves, packers, grouting lines...) and
adapting them to offshore use. This phase consists in designing, building and
testing components.
- defining and testing inventive and low cost pressure grouting techniques and
technologies to solve specific offshore piling problems.

STATE OF ADVANCEMENT :
****
Completed

RESULTS :
****
It is difficult to present any brief, comprehensive conclusion to the study by
reason of the diversity of approaches and work undertaken.
It will be recalled that the study was to focus on the issue of pile grouting,
in a two-pronged approach :
1. theoretical analysis combined with triaxial testing and a computer program on
claquage in ground subjected to grout pressure,2. development, ie, design,
construction and testing, of appropriate pilegrouting technologies.
Both these approaches cover a vast area, one because of the ambitious programme
covering the subject of widespread specialist controversy for many decades, and
the other because of the multiplicity of the potential subjects of study.
The claquage tests and its computer modelling have been brought to an advanced
stage of development (although they cannot claim to be completed because of the
complexities involved), and have yielded valuable information on the conditions
to be observed for proper grouting.
Many technological developments and improvements have been generated, by means
of :
1. an exhaustive laboratory study on grout mix design, followed by validation
tests,

2. research into chck valves and packers, which are the basis for selective grouting,
3. modifications to grout mixing and pumping plant for precise and accurate proportioning on a floating platform,
4. grouting parameter monitoring and recording,
5. modifications to these techniques to suit (a) deep boreholes and (b) driven pile grouting.

REFERENCES :
****
PAPERS ON GROUTED DRIVEN PILES AND/OR MICROSOL HAVE BEEN PRESENTED AT SEVERAL CONFERENCES SUCH AS NANTES PROC. 3RD INT'L CONF. ON NUM. METHODS IN OFFSH. PILING - (86), OCT 87 (5409), OMAE 87 (319-327), OFFSHORE BRAZIL (87), LUXEMBURG 93RD INT'L SYMP. ON NEW OIL/GAS TECHN. - 87) AND PERTH 1988 (313-319).

```
****************************************************************************
* TITLE : THE HYDRA-LOK PILE STRUCTURE CONNECTING    *      PROJECT NO     *
*         SYSTEM FOR LARGE DEEP WATER OFFSHORE        *                     *
*         PRODUCTION FACILITIES                       *    TH./06019/85/UK/..*
*                                                     *                     *
****************************************************************************
* CONTRACTOR :                                        *    TELEPHONE NO     *
*    BUE HYDRA-LOK LTD                                *                     *
*                                                     *    (0229)25080      *
*    WALNEY ROAD                    .                 *                     *
*    BARROW-IN-FURNESS                                *                     *
*    UK - CUMBRIA LA14 5UG                            *    TELEX NO         *
*                                                     *                     *
* TECHNICAL DIRECTOR :                                *    65147            *
*    MR. J. MARSHALL LOWES                            *                     *
*                                                     *                     *
****************************************************************************
                                                       VERSION : 20/09/89
```

AIM OF THE PROJECT :
****
It is the aim of the current project to develop the existing Hydra-Lok
technology into a piece of offshore operational equipment suitable for pile
connection on jackets. This will be achieved by way of a review of the design of
the prototype Hydra-Lok tool and its subsequent use in a field demonstration.
The field demonstration will involve the installation of a Southern North Sea
jacket from a semi-submersible crane vessel (SSCV). This represents a totally
new approach to jacket installation which has been previously dominated by
grouting technology. The new technology offers a more economical and technically
superior connection system.

PROJECT DESCRIPTION :
****
The project is divided into a number of discrete areas of investigation. These
cover redesign of the existing prototype tools, followed by the building of an
offshore operational tool package, some further small scale testing to verify
certain design parameters and a trial of the system on an offshore installation.
The connection system is a swage pile connection system which serves as a
replacement for more conventional grouting or welding off operations. The system
offers substantial technical and economic advantages since it is performed by
equipment which is reusable. Alternative systems tend to rely on extensive
'installed equipment' and are therefore very capital cost intensive. Hydra-Lok
avoids these problems with negligible capital equipment being purchased by the
Client.
Initial work centres around the redesign of the tool and a programme of the
small scale testing to verify certain design parameters. A back-up operational
tool will also be built and suitable operating procedures developed. Subject to
the successful completion of these phases the equipment is to be used in the
installation of an offshore jacket with 72" piles in the Southern North Sea.

STATE OF ADVANCEMENT :
****
Completed

RESULTS :
****
Final trials of both 72" tools have confirmed the ability of the Hydra-Lok
equipment to satisfactorily swage 72" O.D. x 2.5" W.T. pile, into the
appropriate sleeve geometry and the built-in inspection system has demonstrated
its effectiveness in measuring the resultant swage profile.
The small scale tests are, from results to date, confirming the load/deflection,
load distribution and safe working load capacity, predicted for multiple groove
connections.

REFERENCES :
****
RISER CONVERSION ON THISTLE 'A' PLATFORM BY J.M. LOWES, BUE HYDRA-LOK LTD. AND K.
C.HUNTER AND J.M. LOVELL, BRITOIL PLC PRESENTED AT THE WORLDWIDE UNDERWATER
TECHNOLOGY 83 EXHIBITION AND SUBSEA CHALLENGE CONFERENCE, AMSTERDAM JUNE 22-24
1984.
TUBULAR CONNECTOR FINDS USES ON PIPELINES BY J.M.LOWES - 'OFFSHORE' - NOVEMBER
1985.
THE APPLICATION OF A DIVERLESS SWAGE PILE CONNECTION SYSTEM FOR THE INSTALLATION
OF SUBSEA STRUCTURES BY J.M. LOWES, BUEH, C.W. SWAIN, BP AND J.E. ASWEL, MCS
HYDRA-LOK INC - OFFSHORE OPERATIONS SYMPOSIUM 1986, NEW ORLEANS 23-27 FEBRUARY
1986.

```
****************************************************************************
* TITLE : COMMERCIAL ALTERNATIVE TO CONSTRUCT SUB-     *      PROJECT NO      *
*         SEA FOUNDATIONS AND STRUCTURES TO BE         *                      *
*         INTERFACED AND IMPLEMENTED BY ROVS           *   TH./06023/86/UK/.. *
*                                                      *                      *
****************************************************************************
* CONTRACTOR :                                         *    TELEPHONE NO      *
*    ADCO INTERNATIONAL LTD                            *                      *
*                                                      *    0324 23682        *
*    6 BOOTH PLACE                                     *                      *
*    UK - FALKIRK FR1 1BA                              *                      *
*                                                      *    TELEX NO          *
*                                                      *                      *
* TECHNICAL DIRECTOR :                                 *    777 517           *
*    MR. JOHN ANDERSON                                 *                      *
*                                                      *                      *
****************************************************************************
                                                      VERSION : 12/02/90
```

AIM OF THE PROJECT :
****
To develop the design and construct capability for subsea foundations which can
be installed and maintained by ROV, the final phase of the project involves the
full scale fabrication of a prototype foundation in deepwater using only ROV
technology.

PROJECT DESCRIPTION :
****
The project is planned in three phase :
2.1 Phase 1 involved a detailed market research study on the submarine pipeline
market in the UK and Norway.
Mini pipeline supports and structures were developed and tested in a suitable
offshore location.
2.2 Phase 2 will cover the demonstration of suitable prototype foundations
developed in phase 1 by installing them in a suitable offshore location.
From the experience gained in offshore testing the foundations will be refined
and further developed to cover a wide range of applications and also to
accommodate differing seabed conditions.
2.3 Phase 3 will involve the design and installation of a full scale foundation
to be installed at a suitable offshore location to be agreed between the
companies involved with the project.
Approval from the appropriate certifying authorities will be obtained through
directly involving them with all aspects of the design and development to ensure
final concepts comply with relevant standards.
Commercialisation of the technology is planned to commence 1990.

STATE OF ADVANCEMENT :
****
Phase 2 of the project is now complete. A final design concept for the
foundation project has been developed.
Detailed design engineering for each component is currently in progress and
should be completed by December 89.
The foundation system consists of 5 main components :
1. A central domed core fabrication using a double walled fabric form filled in-
situ with concrete. This structure can be used as a habitat/protection system
for subsea equipment.

RESULTS :
****
A report including details on sealings suitable for underwater use providing
information on techniques and results on field Trials to test the sealing
systems (inflacable, compression and grouted fabric seals, injected polymer
seals, bonding systems, rubber and foam rubber strip extrusion) have been
prepared. A concept concerning a submarine cable protection system and level
controlling supports, all in fibre composite, has been designed to incorporate
advanced composite technology for construction by diverless systems in deepwater.
 Tests for handling substantial structures (40 T in air) have been carried out
using buoyance bags and proved to be extremely effective.

```
********************************************************************************
* TITLE : STUD CONNECTIONS FOR JOINING AND REPAIR     *      PROJECT NO       *
*         STEEL OFFSHORE STRUCTURES                    *                       *
*                                                      *  TH./06026/86/UK/..   *
*                                                      *                       *
********************************************************************************
* CONTRACTOR :                                         *     TELEPHONE NO      *
*    WIMPEY OFFSHORE ENGINEERS & CONSTRUCTORS LTD      *                       *
*                                                      *     01 560 3100       *
*    FLYOVER HOUSE, GREAT WEST ROAD                    *                       *
*    BRENTFORD                                         *                       *
*    UK - MIDDLESEX TW8 9AR                            *     TELEX NO          *
*                                                      *                       *
* TECHNICAL DIRECTOR :                                 *     933 861           *
*    DR. I.E. TEBBETT                                  *                       *
*                                                      *                       *
********************************************************************************
```

VERSION : 01/08/89

AIM OF THE PROJECT :
To establish and verify a new shear connection technique for use in underwaater grouted connections; to develop a small, portable automatic system for the attachment of the shear connectors underwater using friction welding technology; and to provide a complete engineering service to secure design and construction management work utilising grouted stud connection technology.

PROJECT DESCRIPTION :
The project is split into five phase. Phase I has been completed. This included a review of stud welding methods which identified friction welding as having the most scope for further development.
The preferred subsea friction welding apparatus can be used to weld a stud "through" a pre-drilled strap, firmly attaching the stud and strap to a tubular member. With this arrangement, since the behaviour of the connection is the same as that of a traditional weld beaded connection, comprehensive large scale testing is now considered to be unnecessary and Phase II of the project has been reduced to confirmatory tests which will be carried out at a later stage.
Phase III is in progress. This includes verification of friction stud welding and development of a system suitable for rapid and accurate semi-automatic attachment of studded shear straps underwater. A number of design studies have been completed comparing the stud grouted connection with other repair schemes such as stressed grouted and mechanical friction clamps. Software has been developed to implement the design procedure. Work is being carried out to determine an optimum procedure for strap attachment and detailed design of the automatic indexing system and stud cartridge loader is in hand.
Phase IV will demonstrate the complete system by executing a large scale repair in an underwater environment, replicating a typical offshore repair.
Phase V is concerned with the marketing and commercial development of the system, and is ongoing. It was felt important to obtain some commitment from offshore operators at this stage in the project and a number of visits to potential clients have taken place.

STATE OF ADVANCEMENT :
Phase III and V of the project are in progress.

RESULTS :
The results at the end of Phase I demonstrate the viability of the stud connection concept.

REFERENCES :
SO FAR, THERE HAS BEEN ONE TECHNICAL PUBLICATION ("LOW COST JACKET REPAIRS USING NEW SHEAR CONNECTION TECHNIQUES", BY I E TEBBETT, P FORSYTH AND S CHISHOLM) AND PRESENTATION OF THE STUD CONNECTION TECHNIQUE IN A PAPER AT THE 1988 IRM (INSPECTION, REPAIR AND MAINTENANCE) CONFERENCE IN ABERDEEN IN NOVEMBER.

```
*************************************************************************
* TITLE : FEASIBILITY STUDY FOR DEEP WATER      *     PROJECT NO        *
*          ANCHORING SYSTEM                      *                       *
*                                                *  TH./06027/86/HE/..   *
*                                                *                       *
*************************************************************************
* CONTRACTOR :                                   *    TELEPHONE NO       *
*    NATIONAL TECHNICAL UNIVERSITY OF ATHENS      *                       *
*                                                *  0030-1-6521692       *
*    DEPARTMENT OF NAVAL ARCHITECTURE AND MARINE ENGI *                  *
*    42, 28IS OCTOVRIOU AVE.,                     *                       *
*    HE-106 82 ATHENS                             *    TELEX NO           *
*                                                *                       *
* TECHNICAL DIRECTOR :                            *    221682             *
*    DR. S.A. MAVRAKOS                            *                       *
*                                                *                       *
*************************************************************************
                                              VERSION : 01/06/89
```

AIM OF THE PROJECT :
****
A theoretical and experimental study of the dynamics of mooring lines will be
conducted, with particular emphasis on the effect of attached submerged buoys,
and on the effect of automatically controlled winches. This will allow imposed
designs of mooring systems for current water depths (up to 800 m) and, more
importantly, the development of mooring systems for deep water applications (of
the order of 3.000 m).

PROJECT DESCRIPTION :
****
The duration of the project was two (2) years. As described in the contract, the
tasks to be caried out within this project were the following :
(1) Study of the effect of pretension
(2) Study of wire versus chain length
(3) Study of the effect of submerged buoys
(4) Experimental study
(5) Comparison of experimental and mumerical results
(6) Study of the feasibility of automatic winches
The major findings from these tasks are described in paragraph 4.

STATE OF ADVANCEMENT :
****
All tasks have been successfully completed.

RESULTS :
****
The main results obtained in the various project tasks are summarized in the
following.
(A) REGARDING THE STUDY OF THE EFFECT OF PRETENSION :
For extreme storms, elastic stiffness is the predominant mechanism for tension
build-up and pretension must be set to a sufficiently low limit to guarantee
that the maximum tension will not exceed the breaking strength of the cable.
The dynamic tension, is directly proportional to the pretension to breaking
tension ratio. When the maximum tension (static plus dynamic) is considered,
however, it is inversely proportional to the pretension and the total tension is
influenced by a change in pretension more in shallow water than in deeper waters.
(B) REGARDING THE STUDY OF WIRE VERSUS CHAIN LENGTH :
The determination of the appropriate chain and wire lengths in a mooring line
for a specific water depth and specific storm must follow an iterative procedure
aimed at balancing the two conflicting attributes of chain :
reduction in the overall length of the line and in the amount of pretension, and
increase it in the elastic stiffness of the line.
(C) REGARDING THE STUDY OF THE EFFECT OF SUBMERGED BUOYS :
A large peak of the dynamic tension per unit length at higher frequencies and
cancellation frequency in the dynamic tension which can serve as the basis for
designing and optimizing mooring lines in deeper waters were ligtend.
(D) REGARDING THE EXPERIMENTAL STUDY :
The following general trends were deduced from the experiments :
(a) The dynamic tension amplification obtained in the case of heave excitation
is insignificant in comparison to the corresponding one for the surge excitation.
(b) Increasing the cable's pretension, results in a corresponding increase in
the dynamic tension amplification, regardless of the presence of buoys and their
location.
(c) For the case of one buoy, its location along the cable's length plays a
significant role in the amount of reduction of the dynamic tension amplification.
(E) REGARDING THE COMPARISON OF EXPERIMENTAL AND NUMERICAL RESULTS :
(a) Very satisfactory agreement has been observed between the experimental data
and the numerical.
(b) Nonlinearities in the cable's response which were observed in the computer
output were not capatured during the experiments. The main reason for this
discrepancy is thought to be the limitations imposed on proper scaling by the
finite depth of the towing tank. This requires further scaled experiment TH.06.
Dec (88).
(c) Dynamic response of the mooring line is not influenced by drag coefficient

of the cable and the buoy in low excitation frequency regions.
(F) REGARDING THE STUDY OF THE FEASIBILITY OF AUTOMATIC WINCHES :
(a) compensation poses excessive power requirements just to overcome the static tension.
(b) alternative schemes, employing passive devices are desirable, however massive equipment or pneumatic devices are required.
(c) An alternative passive control scheme is the use of submerged buoys.

```
************************************************************************
* TITLE : CLASSIFICATION OF SEA FLOOR ENGINEERING    *    PROJECT NO    *
*         PROPERTIES FOR MARINE FOUNDATIONS AND      *                  *
*         STRUCTURES                                 *  TH./06029/86/IR/.. *
*                                                    *                  *
************************************************************************
* CONTRACTOR :                                       *  TELEPHONE NO    *
*    IRISH HYDRODATA LTD                             *                  *
*                                                    *  021 311255      *
*    RATHMACULLIG WEST                               *                  *
*    BALLYGARVAN                                     *                  *
*    IR - CO.CORK                                    *  TELEX NO        *
*                                                    *                  *
* TECHNICAL DIRECTOR :                               *  75850           *
*    MR. T.N. EMERSON                                *                  *
*                                                    *                  *
************************************************************************
                                           VERSION : 10/10/89
```

AIM OF THE PROJECT :
****
The main objective of this project is to investigate the interrelationships
between acoustic reflections from the seabed measured with an Ultrasonic Signal
Processor, with in-situ values determined from a seabed instrument sled. A
successful conclusion to the project will enable compilation of a classification
table of results determined from analysis of acoustic reflections. This will
enable engineering properties and stability of the sea floor to be directly
inferred, such that more rapid and cheaper surveys of areas can be achieved
using acoustic methods.
The Sea Floor Probe itself may prove to be an innovative method of in-situ
measurement of foundation engineering properties. However, the successful
development of USP will revolutionise shallow geophysics and the results
presently obtained which are to an extent based on the individual interpretation
by the geologist/geophysicist.

PROJECT DESCRIPTION :
****
The project is designed to follow a six phase operation :
Phase 1
- Development of Sea Floor Probe
  Development of survey version of USP
  Conventionql surveys of test bad areas off Cork:North Wales
Phase 2
- Joint initial trials of probe, USP and conventional instrumentation
Phase 3
- Redevelopment of work in probe
  Redevelopment of work USP
  Further conventional surveys of test bad areas off Cork/North Sea Wales.
Phase 4
- Joint trials, marine ground truthing and surveys, using USP probe
  and conventional instrumentation including coring:grab sampling
Phase 5
- Data analysis of USP results
  Laboratory testing and comparison of probe results
  Analysis of conventional survey results for cahacterisation and control
  of sediment distribution in test-bad areas
  Mechanical testing of core samples for static date
Phase 1-5
- Mathematical analusis and numerical modelling of acoustic wave theory
Phase 6 - Reporting
The joint Irish/British project will involve three main subcontractors :
- Micro Marine (system) Limited
  CORK
- Augur Geophysical Services
  University College of Worth Wales
  ANGLESEY
- Dr R.C. Chivers
  University of Surrey
  GUILFORD

STATE OF ADVANCEMENT :
****
The present project has finished with completion of Phase 6 and a final report
presented to the Commission, October 1989. While the project is considered an
undoubted success several areas were identified as needing further development
to fully capitalise on the achievements gained thus far. (Project TH/06055/89).

RESULTS :
****
Site and platform design engineers require precise information of the seafloor
characteristics at the earliest stages of development to assess site suitability
and to input into structural designs. Similarly, resource managers of all kinds
operating in the offshore environment are employing increasingly sophisticated
models in their search for viable propositions.

To this end Irish Hydrodata Limited have combined their efforts with Auger Geophysical Services, Micro Marine (Systems) Limited and the University of Surrey to develop new techniques applicable to the offshore survey industry. By bringing together geophysical, electronic engineers, engineeringgeologists, geologists, and oceanographers and physicists, the aim was to develop methods for the rapid quantification of the seafloor sediment properties, using the opportunity of an integrated testing programme to compare individual systems and ultimately enhance data acquisition.

USP

A survey version of the original fishing USP model has been produced which can quantify the type of sea-bed underneath the vessel using the ships echosounder. USP technology currently requires extensive ground truthing to calibrate its output which is variable dependent on such factors as :
- Echosounder frequency, pulse width, transducer type, mounting on hull, etc.
- Vessel's hull shape (especially in shallow water)
- Background acoustic level (ship's engine, etc)
- Water temperature and salinity.

Ideally, USP output should be capable of being supplied in a wholly standardised format to be deployed on 'Craft of Opportunity' with minimal, if any, calibration. A parallel receiver to be developed in the project extension will achieve all expected aims.

Sledge

The sea floor sledge is the first pseudo-underway profiling system capable of measuring both seismic body wave velocities as well as formation factor. As such it represents a major advancement in geophysical measurement facilities available to the marine scientist. Its potential lies in that it is able to provide sediment data remotely and for the in-situ condition, and that it can collect this information more rapidly and economically than any other currently available technique.

Although now a working system, a number of potential improvements have been identified which would convert the instrument package into a commercially viable tool. These include some innovative developments e.g. simultaneous measurement of focused and unfocused resistivity to provide a depth variation in apparent formation factor, shear wave source design modifications to remove pre-pulse effects and whole wave form analysis of seismic data.

REFERENCES :
****

APPROXIMATELY TN PAPERS/LECTURES PRESENTED TO CONFERENCES AND SCIENTIFIC MEETINGS, INDLUDING THE SCIENCE AND TECHNOLOGY OF OCEAN MANAGEMENT 1988, CARDIFF; ULTRASONICS INTERNATIONAL 89, LOWESTOFT.

```
*********************************************************************************
* TITLE : THE CONDUCTION OF TESTS DURING THE        *      PROJECT NO       *
*         FUNCTION AND ACCEPTANCE TRIALS OF AN       *                       *
*         OFFSHORE UNDERWATER PILE HAMMER PLANT      *   TH./06030/87/DE/..  *
*                                                    *                       *
*********************************************************************************
* CONTRACTOR :                                       *      TELEPHONE NO     *
*    BOMAG - MENCK GMBH                               *                       *
*                                                    *   (0)4106-7002-0      *
*  · WERNER VON SIEMENSSTRASSE 2                      *                       *
*    D-2086 ELLERAU/POSTFACH 1165                     *                       *
*                                                    *      TELEX NO         *
*                                                    *                       *
* TECHNICAL DIRECTOR :                               *   213294 MENK D       *
*    HANS KUHN                                        *                       *
*                                                    *                       *
*********************************************************************************
                                                          VERSION : 06/02/90
```

AIM OF THE PROJECT :
****
To exploit the hydrocarbon reservoirs under the ocean beds, more and more the
new technique with underwater pile driving plants will be applied for the
installation of the necessary offshore platforms. The development of these
plants has been hectic over the last few years because of their economical
operation. Their application is spreading rapidly. Fundamental research and
establishment of a technical base have not, however, kept in step with the rapid
development of these plants. A point has thus been reached whereby missing basic
data must be obtained. Otherwise, problems in the development of such plants for
future demands will arise, By the project missing data and knowledge shall be
obtained.

PROJECT DESCRIPTION :
****
A test program is to be carried out. For This the unique and once-
onlyopportunity to use the to date largest pile hammer plant could be made
availableThis enables to come to realistic basics regarding load limit
conditions and the behaviour and control of large oil flow volumes under
pressure. Similar severe and demanding operating conditions cannot in any way
whatsoever be simulated by models.
The program comprises twelve individual test activities and several pretests.
For each the individual equipment has been prepared completely in advance in
order to shorten changing and operation times for this extraordinary expensive
plant to a minimum, but yet to obtain a maximum of data output. After finishing
the program an evaluation of measurements accompanied by several verification
tests with a small hammer plant will follow.

STATE OF ADVANCEMENT :
****
COMPLETED

RESULTS :
****
The test program has been fulfilled as to schedule. Some problems arose from
malfunction of measuring devices and a few tests components which obviously
could not withstand the very high shocks generated by the pile hammer. However,
the data obtained proved to be more than sufficient for an extensive evaluation.
The evaluation results have furnished the missing data on a large scale. They
will help to close the gaps in the knowledge and experience of this new
technique as expected resp. will give impulse to further researach work

```
*************************************************************************
* TITLE : ABANDONMENT AND REMOVAL OF STEEL        *     PROJECT NO      *
*          PLATFORMS                               *                     *
*                                                  *                     *
*                                                  *   TH./06033/87/ES/..  *
*                                                  *                     *
*************************************************************************
* CONTRACTOR :                                     *    TELEPHONE NO     *
*    REPSOL EXPLORACION SA                         *                     *
*                                                  *    (1)5747200       *
*    PEZ VOLADOR 2 - ES 28007 MADRID               *                     *
*                                                  *                     *
*                                                  *    TELEX NO         *
*                                                  *                     *
* TECHNICAL DIRECTOR :                             *    49544            *
*    MR. JOSE MARTIN BOURGON                       *                     *
*                                                  *                     *
*************************************************************************
```

VERSION : 27/10/89

AIM OF THE PROJECT :
****
An approach on abandonment and removal of steel platforms technology based on
the following topics :
- The analysis of the problems associated with decommissioning and removal of
steel platforms larger and heavier than those removed in the past.
- The study of feasible solutions from a technical and economical point of view,
taking into consideration regulatory requirements, available subsea cutting
techniques and existing high capacity crane barges.
- A comparative analysis among different methods for topsides removal : reverse
the installation procedure, remove the equipment and scrap the structure.
- A comparative analysis for the jacket removal among the following methods :
demolition, reflotation and lifting.
- Development of cost estimates for the various studied removal methods.
- Establishment of design guidelines to introduce during the project phase to
allow a simpler and cheaper platform removal.

PROJECT DESCRIPTION :
****
The main phases and activities of the project are as follows :
Phase I. Study of existing methods related to abandonment and removal of steel
platforms.
Phase II. Compile existing regulatory requirements taking into consideration
legislation and recommendations established by international organizations and
specific regulations which may be applicable in E.E.C. countries.
Phase III. Collect, analyse and log all available information from case
histories of platform installation and removal.
A platform data base using latest Data Base Management System is developed which
contains platform categorization and relevant information regarding operator
identity, field data, platform features, etc.
Phase IV. Study of underwater cutting methods.
 An analysis of underwater cutting methods is carried out. Methods under study
are cutting with explosives, waterjet, oxy-acetylene, mechanical cutting, etc.
Phase V. Study of topsides removal.
 The methods subject to study are the following :
- Reverse lifting by crane barges.
- Hanging from barges.
- Demolition.
 Phase VI. Study of jacket removal for series of typical cases.
This phase consists in a technical evaluation of different "jacket" removal
methods listed below :
- Lifting the "jacket" by crane barges.
- Floating the "jacket" with auxiliary buoyancy bags.
- Hanging the "jacket" from barges.- Toppling the "jacket" on the sea floor.
Phase VII. Develop cost estimates for the various applicable removal methods.
 In this phase a cost estimate is done for the different activities and
operations performed in every platform removal case.

STATE OF ADVANCEMENT :
****
Phase I. State of the Art is completed.
Phase II. Compilation of international legislation is completed.
Phase III. Data base programme completed. Data base is up to 95 %.
Phase IV. Underwater cutting study is up to 70 %.
Phase V. Topsides removal study progress is 65 %.
Phase VI. Jacket removal study progress is 65 % - 70 %.
Phase VII. Cost estimation progress is 30 - 40 %.

RESULTS :
****
Results obtained at this stage of the project are summarized hereinafter :
1. Phase I. Review of the state of the art on abandonment and removal of
offshore steel platforms. The study showed that there are several methods to
remove both "topsides" and "jackets" depending on which marine spread is used,
and even more using different marine support vessels for the "topsides" removal

than for the "jacket" removal.
However as a result of the study the most interesting removal methods to apply
are the following :
1.1 Topsides and decks removal : Reverse lifting by crane barges, hanging from
barges, equipment scrapping and demolition of deck structure.
1.2 Jacket removal : Lifting by crane barges, hanging from barges, refloating,
toppling.
2. Phase II. Legislation applicable to offshore platform removal enacted by
international organizations and E.E.C. countries.
The most important event regarding this subject is the issue of the Guidelines
and Standards for the Removal of Offshore Installations and structures on the
Continental Shelf and in the Exclusive Economic Zone by IMO (International
Maritime Organization). This guidelines and standards were approved by the I.M.O.
 Maritime Safety Committee at its session in April 88 and one reviewed in April
89 with comments received from different international organizations will be
submitted to the IMO Assembly for approval at its session due in October instant.
3. Phase III. Platform Data Base.
A platform data base programme has been developed, compiled data of offshore
platforms and loaded in the data base.
The programme presents different application such as listing of Platforms in
alphabetic order, or group them by jacket weight range, water depth, etc.
4. Phase IV. Study of different technologies related to platfrom members cutt
off.
A survey has been done of current and proposed methods to apply for underwater
cutting of steel members.
The first stage of this phase summarizes the "State of the Art" on this
particular subject, describing each method with its application range, its
advantages and limitations.
The study focuses the application of techniques, lists equipment and resources
requirements, and discusses efficiency and reliability of the the related
methods of cutting.
5. Phase V. Study of "Topsides" removal. "A detailed study has been done
topsides removal for three types of platforms :
Type "A". The jacket heigh is less than 100 metres and the total weight is not
exceeding 4000 tonnes.
Type "B". The jacket heigh is between 100 and 120 metres and neither the jacket
nor the topsides weights exceed 12000 tonnes.
Type "C". The jacket heigh is higher than 150 metres and both the jacket and the
topsides weight exceed 12000 tonnes.

REFERENCES :
****
See attached documents.

```
***********************************************************************
* TITLE : AN INTEGRATED APPROACH TO THE STABILITY     *     PROJECT NO     *
*         EVALUATION OF PROSPECTIVE OFFSHORE          *                    *
*         SITES                                       *   TH./06036/87/BE/.. *
*                                                     *                    *
***********************************************************************
* CONTRACTOR :                                        *   TELEPHONE NO     *
*    RIJKSUNIVERSITEIT GENT                           *                    *
*                                                     *   091/22.57.15EXT2679 *
*    GEOLOGISCH INSTITUUT                             *                    *
*    KRIJGSLAAN 281-S8                                *                    *
*    B-9000 GENT                                      *   TELEX NO         *
*                                                     *                    *
* TECHNICAL DIRECTOR :                                *   12754            *
*    Dr. J.P. HENRIET                                 *                    *
*                                                     *                    *
***********************************************************************
                                              VERSION : 14/11/89
```

AIM OF THE PROJECT :
****
Involving high-resolution, digital reflection seismic acquisition and processing,
and shear wave measurements, an alternative form of offshore site investigation
is searched for, which should be available at a fraction of the cost of
conventional offshore drilling and geotechnical programs.
The aim is the construction of a three-dimensional analytical model of the
investigated offshore site, illustrating areas of potential hazard for any
foundation and providing the basic data frame for subsequent simulation studies.
The development of an offshore shear wave refraction capability is envisaged,
making it possible to measure remotely from the surface, data relating to the
top few tens of meters of seafloor material and to gather information on
physical properties prior to drilling.

PROJECT DESCRIPTION :
****
The first phase includes the development of the data acquisition methodology. A
marine shear wave source, detectors and a signal processing system will be
designed and constructed by AUGER.
Contemporarily a specific high resolution reflection methodology will be
optimized for geotechnical applications by RCMG.
Meanwhile a preliminary literature survety and conceptual preparative works for
the subsequent 3-D acquisition, for P-wave velocity measurements and for the
processing and moddeling phases will be executed.
During the second phase initial sea trials will be carried out of the shear wave
refraction system and the reflection system. Possible redevelopment efforts will
follow these trials in the subsequent monts.
Phase 3 is a data processing and data analysis phase. The in sity measurements
will be supplemented with laboratory mechanical and geophysical tests. The
reflection data processing and analysis aims to enhance the signal-to-nopise
ratio of reflection profiles, to develop a 3-D data processing package and to
analyse the velocity information. These processing routines should be available
on the emerging generation of relatively low cost microcomputers suitable for
field use in geotechnical contracting work.
A joint sea trial of the integrated-system will be carried out during the fourth
project phase.
Phase 5 deals with mathematical analysis and modelling. The three-dimensional
reflection survey will yield the structural framework of the investigated site.
Sediment characteristics and geotechnical quantities will be translated in
geomechanical attributes on the basis of analytical or empirical relationships.
A mathematical model of the soil properties under varying stress conditions will
be produced. Ultimately a final reporting phase (phase 6) is foreseen.

STATE OF ADVANCEMENT :
****
Phases 1 (Development) and 2 (Reconnaissance surveys) and sea trials).

```
**********************************************************************
* TITLE : THE UNIFIED DESIGN OF AN ANCHOR CHAIN      *    PROJECT NO    *
*                                                    *                  *
*                                                    *  TH./06037/87/UK/.. *
*                                                    *                  *
**********************************************************************
* CONTRACTOR :                                       *   TELEPHONE NO   *
*    BILLINGTON OSBORNE-MOSS ENGINEERING LIMITED     *                  *
*                                                    *   0990 872323    *
*    SILWOOD PARK                                    *                  *
*    ASCOT, BERKSHIRE                                *                  *
*    UK-SL5 7PW                                      *    TELEX NO       *
*                                                    *                  *
* TECHNICAL DIRECTOR :                               *   9312100278     *
*    M. LALANI                                       *                  *
*                                                    *                  *
**********************************************************************
                                             VERSION : 26/10/89
```

AIM OF THE PROJECT :
****
A failure of an anchor chain mooring a floating production system could cause
the loss of the vessel, with loss of life, environmental pollution, and an
interruption in the supply of hydrocarbons. A similar mooring failure affecting
a heavy lift vessel working in close proximity to a fixed offshore structure
could cause serious damage to the platform with similar consequences to life,
pollution and to interruption in the supply. The principal objectives of this
project on anchor chains are to :
- Assess the accuracy of all existing design techniques that determine the load
in anchor chains using computerised techniques.
- Collate all existing test data on the static and fatigue strength of anchor
chains.
- Perform fatigue tests on range of anchor links in order to produced a
systematic assessment of their fatigue strength.
- Develop new and rational design guidelines which are capable of use by
practising engineers.

PROJECT DESCRIPTION :
****
Details of this workscope to meet the above objectives are given below.
a) ASSESSMENT OF EXISTING DESIGN TECHNIQUES AND COMPUTER ANALYSIS. The detailed
work includes :
- gathering of model and field test data for a variety of vessels in a range
  of environmental conditions
- identification of existing design techniques
- comparison of the model test data with the results predicted by the
  existing design techniques
- comparison of the model test data and the predicted values with the
  national and Certifying Authorities' standards
- parametric assessment of the existing design techniques
- production of an interim technical report on all the work on the existing
  design techniques
- review of the quality assurance techniques currently in use by the
  manufacturers of anchor chains
- production of an interim technical report on the existing quality control
  procedures
- preparation of new design guidelines, to be presented in the project final
  report.
b) REVIEW OF EXISTING DATA. The detailed work includes :
- identification of previous research work, worldwide
- gathering of test data from previous research projects
- gathering of test data from anchor chain manufacturers
- screening, or validification, of the existing test data
- analysis of the shortfalls of the existing test database
- preparation of a programme of laboratory tests on anchor chain
- production of an interim report on all this work on existing test data.
c) LABORATORY WORK. Although the final laboratory programme will be determined
partly on the basis of validated existing results for achieving (b) above, it is
anticipated that the laboratory work will consider the effects caused by the
following variables :
- chain size : to determine scale effects
- grade of steel : to determine variations in fatigue performance
- stud welded or left free : to determine variations in fatigue performance
- tension-bending : to determine bending effects on tension fatigue strength
A total of twenty-eight fatigue tests are scheduled, and the detailed work
includes :
- design of the test equipment
- preparation of specifications for the supply of the test pieces and test
  equipment.
- monitoring of the manufacture of the test pieces and test equipment
- placing of strain gauges on the test pieces
- static tests on each test piece to determine the stress concentrations
- fatigue tests to destruction on each test piece
- production of an interim technical report giving the full testing details and

the test results.

STATE OF ADVANCEMENT :
****
ongoing

RESULTS :
****
An assessment of available fatigue data has been undertaken. This indicates that
the mean tension effects may be significant, and the test programme is therefore
being reviewed to include investigation of these mean tension effects. The first
test results will be available in May 1990.

```
**************************************************************************
* TITLE : DEVELOPMENT OF AN EXPERT SYSTEM FOR THE    *      PROJECT NO    *
*         INSPECTION, MAINTENANCE, REASSESSMENT &    *                    *
*         DAMAGE MONITORING OF STEEL OFFSHORE        *  TH./06038/87/UK/.. *
*         STRUCTUR                                   *                    *
**************************************************************************
* CONTRACTOR :                                       *    TELEPHONE NO    *
*     THE STEEL CONSTRUCTION INSTITUTE               *                    *
*                                                    *    (0990)23345     *
*     SILWOOD PARK, ASCOT, BERKSHIRE  UK - SL5 7QN   *                    *
*                                                    *                    *
*                                                    *    TELEX NO        *
*                                                    *                    *
* TECHNICAL DIRECTOR :                               *    846843 STCON G  *
*     DR. J.K. WARD                                  *                    *
*                                                    *                    *
**************************************************************************
```

VERSION : 30/03/89

AIM OF THE PROJECT :
****
The development of an expert system for the inspection, maintenance,
reassessment and damage monitoring of steel offshore structures, is to increase
operational efficiency by integrating the activities of diving companiesm
monitoring specialists, structural engineers, and other disciplines. Integration
will be achieved by the computerisation of knowledge and data using expert
system technology, an area of computer science which is new, and is expected to
display such characteristics as understanding natural language and the ability
to learn, reason and solve problems. The proposed development is therefore
innovative and the primary economic benefit is to reduce operating costs for
existing offshore installations, whilst maintaining standards of operational
safety.

PROJECT DESCRIPTION :
****
The proposed expert system is aimed at rapid assessment of structural integrity,
damage and/or state of repair of steel offshore structures, using environmental
data, observations ;ade during inspection or by an on-site monitoring system,
the details of up-to-date platform configuration and such engineering knowledge
as necessary and consistent with the state-of-the-art technology. The system is
expected to notify the operator of any existing critical state or potential
member failure, and regularly update, refine and interpret the information held
about the platform and the environment.
The system requires the use of a database access manager to facilitate inter-
communication between the knowledge base for the primary disciplines, such as
inspection, structural analysis, defect assessment & environmental data, each
forming a domain of the expert system.
The project can be divided into the following areas :
- development of the overall expert system for damage monitoring
- identification and implementation of structural engineering systems for
elastic analysis and defect assessment
- identification and implementation of a relational database system for storing
data
- testing of system using data for a Southern North Sea gas platform.
The development within each of the above areas is as follows :
Damage Monitoring Expert System :
- Survey and identify suitable tools and languages for developing knowledge
based systems.
- Establish relationships with operators and companies involved in sub-sea
inspection and monitoring.
 Study and assess structural monitoring techniques, and specify domains of
expert system.
- Design, code and implement the expert system.
- Testing of interfaces between engineering analysis programs and the expert
system.
- Documentation of system.
Structural Engineering Systems
- Survey, select and implement elastic and non-linear frame analysis programs,
and fracture mechanics defect assessment programs.
- Implement database and programs for calculating stress distributions around
tubular joints.
Database Systems
- Survey, identification and implementation of a relational database for
engineering applications.
- Establish databases for storing all necessary data required by the damage
monitoring system.
Testing of integrated expert system
- Test the complete expert system, with its interfaces to databases and
structural analysis systems, using inspection and monitoring data recorded for a
Southern North Sea gas platform.

STATE OF ADVANCEMENT :
****
Start date scheduled for July 1988

RESULTS : Not yet available
****

```
*************************************************************************
* TITLE : IMPROVED METHODS FOR ASSESSMENT OF WAVE        *   PROJECT NO    *
*         AND CURRENT IMPACT ON FIXED AND                *                 *
*         FLOATING OFFSHORE STRUCTURES.                  *  TH./06040/87/DK/.. *
*                                                        *                 *
*************************************************************************
* CONTRACTOR :                                           *  TELEPHONE NO   *
*    DANISH HYDRAULIC INSTITUTE                           *                 *
*                                                        *  45 2 86 80 33  *
*    AGERN ALLE 5                                        *                 *
*    DK-2970 HORSHOLM                                    *                 *
*                                                        *  TELEX NO       *
*                                                        *                 *
* TECHNICAL DIRECTOR :                                   *  37402          *
*    ASGER KEJ                                           *                 *
*                                                        *                 *
*************************************************************************
                                              VERSION : 30/08/89
```

AIM OF THE PROJECT :
****
Accurate prediction of wave and current impacts is essential in any offshore
development project. Extreme wave and current scenarios must be known during the
design in order to establish survival conditions, and frequently occurring
severe situations must be known for assessing fatigue problems. The present
project aims at developing new knowledge and methods in the following fields :
- Knowledge about water particle velocities above mean water
  level in storm waves.
- Three-dimensional numerical modelling of currents impacting
  offshore structures.
- Methods for assessing joint probabilities of wave and current
  loads.
- Time-domain modelling of the full non-linear interaction
  between waves/currents and floating structures.

PROJECT DESCRIPTION :
****
PROJECT No 1 : THE WAVE PROJECT
Designing of fixed offshore structures, e.g. jackets is often performed on the
basis of loads from one extreme wave. In the North Sea this wave may typically
have a height of 20 m, out of which approx. 13 m will be above the mean water
level. It is evident that this part of the wave is of tremendous importance to
the total wave force and the overturning moment. It is therefore imperative to
have knowledge about the velocities in the wave crest.
The scope of work will comprise initial pilot test and prototype development of
an instrumentation system for laboratory measurements of wave crest kinematics,
and planning of measurements in nature on an existing platform.
PROJECT No 2 : THE CURRENT PROJECT
DHI has since 1980 carried out various feasibility studies aiming at the
development of a general three-dimensional flow model. The plan is now to
develop this model within the next 3 years. The scope of work consists of three
parallel activities :
a) general three-dimensional flow modelling. An implicit finite difference
approximation to the generalized Navier-Stokes equation will be employed
b) modelling of turbulence
c) modelling of structural response from flexible cylindrical members.
PROJECT No 3 : THE COMBINED WAVE/CURRENT PROJECT
DHI has carried out pilot tests on the effect of following a more precise design
approach based on an extrapolation of the critical key load parameter, e.g. the
overturning moment for a jacket structure. These tests have led to quite
encouraging (less conservative) results. The scope of work for the present
Project No 3 consists of a more thorough and extensive analysis of this advanced
methodology, aiming at the development of a well documented practical
calculation procedure.
PROJECT No 4 : THE STRUCTURAL RESPONSE/INTERACTION PROJECT
The calculation of forces and movements of floating offshore structures will be
solved in a two-step procedure. Step one consists of the description of the
acting wave field on basis of actual measurements or mathematical model
simulations. Step two consists of the calculation of the resulting movements in
a separate model.
The numerical model to be used in step one will be a further development of an
existing non-linear, time-domain wave model developed earlier by DHI (SYSTEM 21
MK8). A (coarse) description of the floating body will be introduced in this
model in order to resolve the wave/structure interaction in step two.
Furthermore, a boundary element time domain solution will be used in step two to
describe the wave field in the immediate viscinity of the floating structure and
the associated structural motions.This model can be used independently in case
of deep water and simple wave fields, and will be used as a "near-field"
description, coupled to the SYSTEM 21 model, in the case of complicated shallow
water wave fields.

STATE OF ADVANCEMENT :
****
Tools and concepts have been developed under all four sub-projects.

RESULTS :
****
Anticipated results are given under item 2 above. Description of the final projects will be available during 1989.

REFERENCES :
****
- MADSEN PA & LARSEN J. (1987)
  AN EFFICIENT FINITE DIFFERENCE APPROACH TO THE MILD-SLOPE
  EQUATION. TO APPEAR IN COASTAL ENGNG. 12, 1987
- BRORSEN M. & LARSEN J., (1987)
  SOURCE GENERATION OF NONLINEAR GRAVITY WAVES WITH THE BOUNDARY
  INTEGRAL EQUATION METHOD. COASTAL ENGNG. 11, 1987
- MADSEN M.N., NIELSEN J.B., KLINTING P. & KNUDSEN J. (1987)
  A DESIGN LOAD METHOD FOR OFFSHORE STRUCTURES BASED UPON THE
  JOINT PROBABILITY OF ENVIRONMENTAL PARAMETERS

```
*****************************************************************************
* TITLE : REMOTELY CONTROLLED TOWED VEHICLE              *   PROJECT NO      *
*                                                         *                   *
*                                                         *  TH./06042/87/UK/.. *
*                                                         *                   *
*****************************************************************************
* CONTRACTOR :                                            *   TELEPHONE NO    *
*   BROWN & ROOT CONSTRUCTION(UK)LTD-SURVEY DIVISION      *                   *
*                                                         *   0224-724855     *
*   WELLHEADS PLACE                                       *                   *
*   WELLHEADS INDUSTRIAL ESTATE                           *                   *
*   UK-DYCE ABERDEEN AB2 0GG                              *   TELEX NO        *
*                                                         *                   *
* TECHNICAL DIRECTOR :                                    *   73260           *
*   MARK VORENKAMP                                        *                   *
*                                                         *                   *
*****************************************************************************
                                                        VERSION : 05/10/89
```

AIM OF THE PROJECT :
****
Produce a controllable towed vehicle which will house a range of survey and
inspection sensors.
The crucial aspect of the manufacture of this vehicle will be the design of
production of the movable towing point (actuated lead screws or ball screws
under control of a microprocessor system).

PROJECT DESCRIPTION :
****
PHASE I - FEASIBILITY STUDY
This phase has already been carried out which confirmed the principal that by
moving the towing point of the vehicle, uner a dynamic conditions, it would
respond by moving both laterally and vertically.
PHASE II - SYSTEM SPECIFICATION
The object of this phase was to produce a specification for conceived operating
modes and configuration, and to attempt to anticipate future requirements for
sensors and inspection tools.
PHASE III - SYSTEM OUTLINE DESIGN
During this phase, an outline design for the control system and possible
actuator configuration was agreed. Block schematics of the system and control
flow charts were prepared.
Umbilical cable conductor requirements for the control system and power
transmission were established.
PHASE IV - CONTROL SYSTEM DETAIL DESIGN AND PROTOTYPING
The object of this phase was to produce an operational prototype control system.
The initial requirement was to establish duplex data communication along a
twisted pair cable at a suitable data rate, to permit transfer of commands and
data between the surface and underwater processors.
The towpoint monitor system operates under the control of a microprocessor which
commands the motors and monitors rotation sensors to adjust the towpoint
position. Data is also passed between the control processor and two peripheral
processors which handle other digital and analog functions.
The surface command and display system interprets control stick commands from
the operation and encodes these for transmission to the underwater unit. The
unit displays these commands together with the actual towpoint position which is
telemetered from the underwater unit so that control follow up can be observed.
The surface units also displays vehicle parameters such as pitch, roll, depth
etc and can encode commands such as camera pan and tilt and lamp intensity for
transmission to the underwater unit.
Development of the control electronics is largely complete, the next stage being
the development of power switching electronics for motot actuation.
An initial mechanical design for the towpoint support and actuation system has
been completed.
PHASE V - DETAIL DESIGN OF UMBILICAL AND UNDERWATER HOUSINGS
Some initial design work has been completed on the underwater housings for the
control electronics and actuating motors.
PHASE VI - SYSTEM INSTALLATION AND SEA TRIALS
Not started.

STATE OF ADVANCEMENT :
****
Ongoing. Phase I - FEASIBILITY STUDY - has been completed.
PHASE II - SYSTEM SPECIFICATION - completed.
PHASE III - SYSTEM OUTLINE DESIGN - completed.
PHASE IV - CONTROL SYSTEM DETAIL DESIGN - largely complete - power control
circuit design ongoing.
PHASE V - DETAIL DESIGN,UMBILICAL,WINCH,ACTUATORS,HOUSINGS - some work completed
on underwater housings.

```
****************************************************************************
* TITLE : FABRICATION AND TEST OF A PILE DRIVING    *     PROJECT NO      *
*         SYSTEM FOR 1000 M WATER DEPTH             *                     *
*                                                   *    TH./06044/88/DE/..*
*                                                   *                     *
****************************************************************************
* CONTRACTOR :                                      *    TELEPHONE NO      *
*    BOMAG - MENCK GMBH                             *                     *
*                                                   *    41 06 74812       *
*    POSTFACH 1165                                  *                     *
*    D - 2086 ELLERAU                               *                     *
*                                                   *    TELEX NO          *
*                                                   *                     *
* TECHNICAL DIRECTOR :                              *    213294            *
*    H. KUHN                                        *                     *
*                                                   *                     *
****************************************************************************
                                                        VERSION : 13/03/90
```

AIM OF THE PROJECT :
****
The water depths in which work has to be carried out and in which equipment and
structures have to be installed for the exploitation of the hydrocarbon deposits
occurring below the seabed are increasing. At present installations are being
studied for water depths between 500 and 1000 m. Exploration drilling has been
carried out in water depths of more than 1000 m whereby exploitable hydrocarbon
deposits have been located.
Thus it has to be expected that a need for a driving system in such water depths
will develop so that structures can be securely fixed to the seabed by driven
piles. The production and testing of a driving system fulfilling those needs is
the aim of this project.

PROJECT DESCRIPTION :
****
The following phases are planned :
1. Fabrication of a prototype of a deepwater driving system
2. Carrying out of function tests above water including installation of a test
pile
3. Execution of a realistic offshore test in 1000 m water depth including the
installation of a test pile at the seabed.
A prototype of a deepwater piling system should be produced including the
following components :
I.   Deepwater hammer
II.   Circular underwater powerpack
III. Electrical system for power supply to and monitoring of the underwater
hammer

IV.   Thruster or jet drive with cameras for positioning of the hammer under
water

V.   Accessories for the hammer and the related equipment
VI.   Miscellaneous
- A function test above water should be executed requiring the set-up of a test
pile. This test pile should be installed in a harbour basin near a quay in order
to permit a crane application for the test procedure either from the water side
applying a floatable or from the shore side using a telescopic crane. This
solution would permit easy access at low cost.
- A liability test in a water depth of 1000 m in the Gulf of Mexico should be
executed applying a dynamic positioned work vessel. In this case a test pile
should be installed on the sea floor. This test pile and its support and guiding
structure should be lowered to the sea floor and positioned in place. Afterwards
the hammer should follow being guided and positioned onto the test pile in order
to hammer the pile into the ocean floor.

STATE OF ADVANCEMENT :
****
Ongoing

RESULTS :
****
After assembly of all plant components, execution of the functional tests, and
transport to Morgan City, Louisiana, the equipment was brought on board the
McDermott derrick barge DB 50 where it was reassembled, connected to the board
equipment, and prepared for the test.
Two test runs took place on a SHELL oil field in the Gulf of mexico interrupted
by two short bad weather periods.
As planned a 90 m long tubular test pile of 1,50 m diameter was driven about 100
m deep into the seabed at a water depth of 1000 m. The target of this project
has thus been achieved. Several concluding works in accordance with the project
description are to be done in order to complete the project.
It has been proven successfully that pile driving operations in a water depth of
1000 m can be executed with the new developed pile driving system. Thus in the
future structures for the exploitation of hydrocarbon deposits can be founded at
the seabed in deep water by driven piles.

296

REFERENCES :
****

- ENTWICKLUNGEN IN DER UNTERWASSER-RAMMTECHNIK, 1983, OCTOBER, KUHN - BAUMASCHINE UND BAUTECHNIK.
- THE APPLICATION OF THE HYDRAULIC UNDERWATER HAMMER IN SLENDER AND FREE RIDING MODE WITH OPTIONAL UNDERWATER POWERPACK, 1987, MAY, VAN LUIPEN, OFFSHORE TECHNOLOGIE CONFERENCE, HOUSTON, PAPER 5423.
- TESTS EXTEND RANGE FOR UNDERWATER PILE DRIVING, 1988, AUGUST, KUHN OCEAN INDUSTRY.

```
*****************************************************************************
* TITLE : SEABOTTOM INSTABILITY AROUND SMALL        *      PROJECT NO      *
*         STRUCTURES IN MARINE ENVIRONMENT          *                      *
*                                                   *   TH./06045/88/IT/..  *
*                                                   *                      *
*****************************************************************************
* CONTRACTOR :                                      *    TELEPHONE NO      *
*    SNAMPROGETTI SPA                               *                      *
*                                                   *    52038929          *
*    V.LE DE GASPERI 16                             *                      *
*    I - S.DONATO MILANESE (MILANO)                 *                      *
*                                                   *    TELEX NO          *
*                                                   *                      *
* TECHNICAL DIRECTOR :                              *    310246            *
*    FAUSTO CITERNESI                               *                      *
*                                                   *                      *
*****************************************************************************
                                                        VERSION : 14/12/89
```

AIM OF THE PROJECT :
****
Aim of this project is to deepen some aspects of the stability of the sea floor
when eroded by the action of the currents in order to :
- identificate the way of behaviour of the flows near the sea bottom and thus
also of submarine systems and small structures
- study the correlation etween environment, fluid, the perturbations induced by
the structures and the sediment movement
- define the engineering parameters to adopt when designing submarine systems
envolved with these phenomena.
The results of the proposed research will allow the engineer to develop a better
design philosophy in order to reduce the effect of the scouring and consequently
to improve the reliability of any immersed structure.

PROJECT DESCRIPTION :
****
The project will be developed in 4 main stages :
- Phase 1 : Analysis of the state of the art;
In this phase existant literature and experimental data already available will
be critically analysed.
The limits of the generally adopted theories will be highlighted and the most
important parameters will be individuated in order to optimize the experimental
tests.
- Phase 2 : Design and execution of experimental scale laboratory tests;
Taking into account the results obtained in Phase 1 the number and type of
laboratory tests to be performed will be defined in technical specifications.
Thus the optimal flume and instrumentation will be selected together with the
external Contractor charges for the execution of the tests.
At last the test will be executed; the data, recorded, decoded and converted in
physical parameters will be validated and organised on a suitable magnetic tape
for the foreseen elaborations.
- Phase 3 : Design and execution of experimental full scale field tests;
Considering the results achieved in Phase 1, the field measurements to be
performed will be defined in technical specifications.
The most suitable areas will be selected, together with the necessary
instruments and the optimal season.
At last the measurements will be performed; the data, recorded, decoded and
converted into physical parameters will be validated and organised on a suitable
magnetic tape for the foreseen elaborations.
- Phase 4 : Elaboration and analysis of the experimental data, comparison and
interpretation of results, set-up and calibration of mathematical models,
assessment of a design procedure.
In this phase the collected data will be elaborated and analysed. The results
from the laboratory tests will be compared with those from the field.
The assessed mathematical models will be calibrated on selected laboratory data.
 Then the models will be checked with the results of the full scale tests.
The design philosophy will be presented in a technical report.

STATE OF ADVANCEMENT :
****
Ongoing from January 1990.

```
*********************************************************************************
* TITLE : USE OF BUOYS TO REDUCE STATIC AND        *        PROJECT NO        *
*         DYNAMIC TENSION IN DEEP WATER MOORING     *                          *
*         LINES : A PILOT STUDY                     *    TH./06046/88/HE/..    *
*                                                   *                          *
*********************************************************************************
* CONTRACTOR :                                      *    TELEPHONE NO          *
*    TECNOMARE S.P.A.#NATIONAL TECHNICAL UNIVERSITY O*                          *
*                                                   *    041/796711            *
*    S. MARCO 2091                                  *                          *
*    IT - 30124 VENEZIA                             *                          *
*                                                   *    TELEX NO              *
*                                                   *                          *
* TECHNICAL DIRECTOR :                              *    410484                *
*    MR. P. BRANDO                                  *                          *
*                                                   *                          *
*********************************************************************************
                                                         VERSION : 22/06/89
```

AIM OF THE PROJECT :
****
To develop a theory for optimizing the size and the location of buoys in mooring
lines in 1000 meter water depth, validated by scaled tests in 20-30 meter water
depth.

PROJECT DESCRIPTION :
****
The main activities are :
- Developing the optimization theory
- Testing the line in 25 meter water depth
- Interpretation of the results.

STATE OF ADVANCEMENT :
****
Completed activities
- Development of the optimization theory
- Site selection (Lake of Barcls)
- Tests specifications
- Design of the testing machine
- Purchasing of buoys and other instrumentation required for performing the
experimente.
Ongoing activities
- Assembling of the testing machine
- Preparation of the site for performing the experiments
- Preparation of software for data acquisition and data processing.

RESULTS :
****
- A computer procedure for the optimization of the size and position of buoys in
mooring lines of floating offshore systems
- The design of the machine and instrumentation for testing scaled mooring lines
with buoys.

REFERENCES :
****
PROJECT TH/06027/87 - FEASIBILITY STUDY FOR DEEP WATER ANCHORING SYSTEMS.

```
****************************************************************************
* TITLE : DECOMMISSIONING AND REMOVAL OF        *      PROJECT NO         *
*         PLATFORMS USING EXPLOSIVE CUTTING      *                        *
*                                                *    TH./06047/88/UK/..   *
*                                                *                        *
****************************************************************************
* CONTRACTOR :                                   *     TELEPHONE NO        *
*    ADVANCED MECHANICS & ENGINEERING LTD        *                        *
*                                                *    0483 301219          *
*    4 FREDERICK SANGER ROAD                     *                        *
*    SURREY RESEARCH PARK                        *                        *
*    UK - GUILDFORD, SURREY GU2 5YJ              *     TELEX NO            *
*                                                *                        *
* TECHNICAL DIRECTOR :                           *     8950511            *
*    DR C.P. ELLINAS                             *                        *
*                                                *                        *
****************************************************************************
```

VERSION : 27/09/89

AIM OF THE PROJECT :
****
The aim of the project will be to provide technical data, from the results of testing and analysis, detailing methods of decommissioning steel jacket structures using explosive charges. It will provide details of the use of suitable shaped cutting charges, the shock effects of these on the structure, attendant vessels and the environment. A design approach for explosive demolition of a typical jacket structure will be detailed with guidance on safety and reliability aspects, use of buoyancy, lifting equipment and retention of stability.

PROJECT DESCRIPTION :
****
The project will be completed in three phases each completed by the issue of a technical report. Phase 1 covers data gathering and evaluation, analysis of explosive cutting of thin walled tubulars, analysis of shock loading of jacket structures and design of a suitable testing programme.
Phase 2 will concentrate on more specific application of the information obtained in Phase 1. Tests will be conducted to confirm analytical predictions and give guidance on practical application aspects.
Phase 3 will concentrate on the specific design of cutting charges for the anticipated range of applications. Guidance on shock wave effects from the charges will be developped. Details related to the behaviour of the structure during removal and suggested methods of handling will be given. This will be extended to a complete plan for a typical North Sea jacket. Finally reports from all three phases will be edited to produce a complete project report.

STATE OF ADVANCEMENT :
****
Review work complete, analytical work on underwater blast effects substantially complete, shaped charge performance tests on flat plate in air and through water complete.

RESULTS :
****
Results of review work indicate that work in the area of underwater shock and bubble pulses, particularly related to explosions within or near structures is very necessary. Information on these subjects is quite dispersed and often dates from 1950's. This project, in collecting together as much material as possible in this area, will be of considerable value for future platform removals.
Analytical results to date indicate that the transient effects of underwater shock have less effect on the structures concerned that the gas bubble pulses. Suggestions for reduced vessel stand-off distances are being formulated especially for confined charge effects. Many benefits are expected to result from improved design of shaped charges with simplified installation techniques for which patents have been lodged.
Test results to date indicate that the shaped charges used have adequate capacity to cut cleanly and efficiently through 100mm plate. The tests through a water layer have provided useful data on scaling requirements and jet breakup behaviour.

REFERENCES :
****
THE AIMS AND CONCEPTS OF THE PROJECT WERE OUTLINED IN A PRESENTATION TO THE 1989 OFFSHORE MECHANICS AND ARCTIC ENGINEERING CONFERENCE IN THE HAGUE ENTITLED DECOMMISSIONING AND REMOVAL OF OFFSHORE STRUCTURES A-STATE-OF-THE-ART. IN APRIL 1989 A DEMONSTRATION OF STEEL CUTTING USING A SHAPED CHARGE, TOGETHER WITH A DETAILED DESCRIPTION OF THE EXPLOCUT PROJECT, WAS STAGED FOR REPRESENTATIVES OF MAJOR OIL COMPANIES, GOVERNMENT DEPARTMENTS THE CEC AND THE OFFSHORE PRESS TO ELICIT SPONSORSHIP.

L

```
************************************************************************
* TITLE : ABRASIVE JETTING APPLIED TO PLATFORM      *    PROJECT NO     *
*         DECOMMISSIONING                            *                   *
*                                                    *  TH./06048/88/FR/.. *
*                                                    *                   *
************************************************************************
* CONTRACTOR :                                       *   TELEPHONE NO    *
*    COMEX SA                                         *                   *
*                                                    *   91 23 50 00     *
*    36 BD DES OCEANS                                *                   *
*    F - 13275 MARSEILLE CEDEX 9                     *                   *
*                                                    *   TELEX NO        *
*                                                    *                   *
* TECHNICAL DIRECTOR :                               *   410985          *
*    ROBERT SURLE                                    *                   *
*                                                    *                   *
************************************************************************
```
                                                  VERSION : 01/02/90

AIM OF THE PROJECT :
****
The aims of the project are :
- The development of UHP abrasive jetting as a supplement to explosive cutting
for platform removal.
- The development of typical operational scenario for the offshore deployment of
UHP abrasive jetting and explosive cutting in order to optimise safety and
minimise costs.
- The study of methods for operationally deploying both UHP abrasive jetting and
explosive cutting as a function of different worksite characteristics.

PROJECT DESCRIPTION :
****
Initial work performed by COMEX has demonstrated the abrasive jet is a process
with major potential for use in platform removal when used as a supplement to
explosive cutting methods. However, considerable development work remains to be
performed before this technique can be used operationally at any water depth in
line the North Sea requirements.
The programme consists of three work packages :
Work Package nr 1 : Development of abrasive water jet a- An abrasive water jet
equipment development and evaluation of abrasive metering and feeding device,
(optimisation to increase duty cycle, nozzle design and/or improvement, etc...)
- Exhaustive hyperbaric tests to define cutting parameters for a range of
materials (heavy wall pipe with concrete, casing, multi wall pile, thick
reinforced concrete walls and flexible pipes) over a range of significant depths.
Work Package nr 2 : Operational Techniques related to Platform
Decommissioning
WP 2.1 : "Identification of platform to be removed"
Creation of a data bank for platform decommissioning (details of structures and
when available conditions in which they are to be decommissioned).
WP 2.2 : "Interactivity between overall removal procedures and the cutting
operations"
* Combination of cutting processes
* New development on abrasive water jet techniques
* Outline procedures for cutting operations
WP 2.3 : "Specifications of the deployment systems required for installation
remote control, monitoring of the different cutting equipment"
Definition of specifications for the system(s) to be used for deploying both
explosive cutting and abrasive water jet.
Work Package nr 3 : Basic Engineering of deployment of abrasive water jet
systems and associated equipment
- Deployment by ROV
- Deployment by a dedicated robot
Note : A revised scope of work has been submitted to the CEC for information and
approval.

STATE OF ADVANCEMENT :
****
Ongoing.

RESULTS :
****
Up to the end of 1989 the following technical tasks were performed :
A) An exhaustive inquiry about the up-to-date situation about abrasive water jet
techniques
In particular, attention was focused on :
- the actual technological trend in this field,
- the limits of the existing equipments, featuring 3 different methods of
transporting and focusing the high velocity abrasive water jet.
B) An inquiry in order to locate some specialized engineering and manufacturing
companies able to produce or adapt an abrasive metering device.
C) The adaptation of such device to our :
- water jetting systems
- evaluation and cutting test program.
D) The purchasing of such device (semi-prototype for future onshore application).

E) The mobilisation of our hyperbaric testing site
F) An inquiry concerning the activity in this field of our competitors which are operating abrasive water jet techniques different from the one that we have selected.
G) A study concerning the offshore market on potential under-water cutting operations related to offshore platform removal.
Results so far reached :
a) Abrasive material selection
* (1) abrasive material out of the 4 was eliminated
* slight difference in efficiency found between the (3) selected ones (with 2,4 mm dia nozzle)
* efficiency
b) Water depth (hyperbaric pressure)
* system as efficient at 50 m simulated and at 400 m
* no significant influence of pressure to 40 bar
c) Stand off of cutting nozzle
Not critical if standing between 5 to 15 mm approx (material thickness less than 2").
d) Number of cutting tests performed up to mid February : 153.
e) Material cut so far
1) Concrete : 125 mm thick
2) Steel plate (E34-4) : 50 mm thick
3) Flap pipe : 23 mm thick
4) Coflexip pipe : 40 mm thick
5) Miscellaneous : umbilical, chains, shackees, steel cables,...
6) Sandwich materials : under way.
f) Type of cut : straight and bevelled
g) Cutting speed : from 20 to 200 mm/mn.

REFERENCES :
****
A) DECOMMISSIONING AND REMOVAL OF OFFSHORE STRUCTURES : 19/20.04.89, CONFERENCE WHICH WAS HELD AT THE LONDON PRESS CENTER, LONDON UK.
PAPER PRESENTED: "CUTTING WITH HIGH PRESSURE ABRASSIVE WATER JETS".
B) JET TEC'89 : 14.09.89 AT CRANFIELD CONFERENCE CENTER, UK.
PRESENTED : AN UPDATED VERSION OF THE AVOVE MENTIONNED PAPER.
C) OAR'90 (OFFSHORE ABANDONMENT AND REMOVAL)
INTERNATIONAL OFFSHORE DECOMMISSIONING, PLATFORM REMOVAL AND MARINE SALVAGE EXHIBITION AND CONFERENCE : TO BE HELD 27/29.03.90, ABERDEEN.
- IN PARTICIPATION WITH AME AND TECNOMARE
- WHERE WE ARE PRESENTIN THE FOLLOWING PAPER : "REMOVAL OF OFFSHORE STRUCTURE - DEVELOPMENTS IN ENABLING TECHNOLOGY AND OPERATIONAL PROCEDURES".

```
***************************************************************************
* TITLE : WELL CONDUCTORS AND PILES UNDERWATER       *    PROJECT NO     *
*         CUTTING SYSTEM FOR THE REMOVAL OF          *                   *
*         OFFSHORE STRUCTURES                        *  TH./06049/88/IT/..*
*                                                    *                   *
***************************************************************************
* CONTRACTOR :                                       *   TELEPHONE NO    *
*    TECNOMARE SPA                                   *                   *
*                                                    *   041 796711      *
*    S. MARCO 2091                                   *                   *
*    I-30124 VENEZIA  .                              *                   *
*                                                    *   TELEX NO        *
*                                                    *                   *
* TECHNICAL DIRECTOR :                               *   410484          *
*    MR P. MINARDI                                   *                   *
*                                                    *                   *
***************************************************************************
                                               VERSION : 12/02/90
```

AIM OF THE PROJECT :
****
To identify removal procedures for offshore structures and to design a system to
be used in cutting of conductors and piles.

PROJECT DESCRIPTION :
****
The main activities of the project are:
- Identification of removal operating procedures
- Design of a diverless underwater cutting module
- Technical and market assessment of the results and specification of a
subsequent operative phase.

STATE OF ADVANCEMENT :
****
State of art, Regulations and Market analysis completed; C100 - Design Premises
completed in relation to Marine operations activity; Marine operations ongoing,
System configuration ongoing. Collaboration with AMECOMEX complete; looking for
further collaboration with AME

RESULTS :
****
Expected project results are relevant to:
-development of a low-cost removal methodology which allows safety and efficient
operations;
-operating flexibility which permits to meet the specific market needs;
- identification of robotic technics in order to perform diverless cutting of
structural members through the study of :
. Multipurpose Cutting Robot to cut skirt pile from internal side
. Use of existing ROVs to perform leg and bracing cutting.

REFERENCES :
****
NONE

```
******************************************************************
* TITLE : SELF RELEASING OFFSHORE ANCHOR          *    PROJECT NO       *
*                                                 *                     *
*                                                 *  TH./06051/89/NL/..  *
*                                                 *                     *
******************************************************************
* CONTRACTOR :                                    *  TELEPHONE NO       *
*    NEDDRILL NEDERLAND B.V.                       *                     *
*                                                 *  010-4007205        *
*    COOLSINGEL 139                               *                     *
*    NL-3012 AG ROTTERDAM                          *                     *
*                                                 *  TELEX NO           *
*                                                 *                     *
* TECHNICAL DIRECTOR :                             *  27033              *
*    MR H.F. PASMA                                 *                     *
*                                                 *                     *
******************************************************************
                                          VERSION : 14/03/90
```

AIM OF THE PROJECT :
Construction of a prototype of a remotely released offshore anchor, testing of
the prototype and determination of the feasibility of commercial exploitation of
such an anchor.
The innovation is to re-design a "Flipper" Delta anchor to incorporate an
acoustic release device to allow the shank to rotate up to an angle of 90
degrees with the flukes after receipt of a trigger signal. The anchor will come
loss then if the rig hauls the anchor line.
Construction of a "full scale" prototype of such an anchor is essential as to
test the suitability of the system in practice.

PROJECT DESCRIPTION :
Retrieval of heavy anchors for offshore units is normally done by anchor
handling tugs. Such operations are normally restricted to moderate wave
conditions due to excessive tug motions and limited winch power.
Neddrill is thinking of the possibility to use the anchor line in order to
retrieve the anchor by the offshore unit itself. Such a procedure requires that
the holding power of the anchor can be released remotely.
The innovative self-releasing offshore anchor will enable immediate release of
the anchors in case of emergencies.
Furthermore the anchor has several operational advantages when compared with
conventional offshore anchors.

STATE OF ADVANCEMENT :
A detailed technical feasibility of the system has been performed. Drawings of a
prototype of the anchor, including savings for the release mechanism, are
available. The idea of a release system with a retractable locking pin is
protected by a pending patent. Tests with scale models of the anchor have been
performed both in practice as well as in a test tank.

RESULTS :
Since the anchor is still in a design phase and as per start of the project no
prototype of the anchor is available yet details about the results and
description of the final product as well as problems encountered and solutions
adopted are not yet applicable.

AUXILIARY SHIPS AND SUBMERSIBLES

```
***************************************************************************
* TITLE : ARGYRONETE                              *      PROJECT NO       *
*                                                 *                       *
*                                                 *   TH./07050/83/FR/..  *
*                                                 *                       *
***************************************************************************
* CONTRACTOR :                                    *    TELEPHONE NO       *
*    COMEX                                         *                       *
*                                                 *    091 410170         *
*    36, BOULEVARD DES OCEANS                     *                       *
*    B.P. 143                                     *                       *
*    FR - 13275 MARSEILLE CEDEX 9                 *    TELEX NO           *
*                                                 *                       *
* TECHNICAL DIRECTOR :                            *    410 985            *
*    MR. Y. DURAND                                *                       *
*                                                 *                       *
***************************************************************************
```
                                                      VERSION : 20/09/89

AIM OF THE PROJECT :
****
Development of new technology necessary for the realisation of the first long-
range autonomous lock-out submarine, SAGA (ex ARGYRONETE).
SAGA is the prototype of a new generation of industrial submarines capable of
carrying out underwater operations using divers and robots at close proximity to
the work site without surface assistance.
This new concept of operations from the seabed overcomes surface-related
difficulties encountered by specialised surface support vessels.

PROJECT DESCRIPTION :
****
Main characteristics of SAGA :
- Length overall : 28.06
- Submerged deplacement : 545 T
- Depth capability for diver lock-out : 450 m
- Typical submerged autonomy : 3 weeks
- Crew atmospheric compartment : 6 men
- Divers in hyperbaric chamber : 4 to 6
Technological innovations of the project
The high autonomy of SAGA is mainly obtained by two air-independent closed cycle
STIRLING engines.
- Development of specific composite material, very advantageous on board
submarines with regard to their weight in water, in particular : high pressure
hooped gas cylinders, buoyant fairing, pressure tanks.
- Cryogenic storage of oxygen : the majority of efficient underwater power
sources requires storage of oxygen. The best weight/volume ratio is achieved
with cryogenic storage.
- Deep diving system : reduction of diving gas consumption increases the divers'
intervention autonomy. As existing systems are unsatisfactory at great depth a
specific closed circuit diving system has been developed.
- Diver heating system : heating of divers in chamber or locked-out is very
energy consuming for an autonomous submarine. Work carried out involved the
development of an efficient hot water suit and a reclaim system of the heat
dissipated by the STIRLING engines.

STATE OF ADVANCEMENT :
****
Completed

RESULTS :
****
- H.P. Gas cylinders (pressure : 400 bar) : qualification programme using models
and prototypes completed. Set of prototype cylinders (length :1.80, 3.0 and 6.0
m, diameter 443 mm) manufactured and installed onboard.
- Composite pressure tanks (variable buoyancy tanks of the sub, pressure both
sides : 60 bar) : qualification programme accepted. Acceptance tests passed.
Installed onboard.
- Lightweight fairing : manufactured and assembled onboard.
- Cryogenic oxygen tanks and associated evaporation system : acceptance tests
passed, installed onboard.
- Prototype diver breathing system : qualification programme in pressure chamber
: completed, prototypes manufactured.
- Low energy heating system : reclaim circuit assembled onboard and tested,
specific diver suit qualified in pressure chamber and tested omboard.
The installation of these innovations omboard was completed mid 1987. Launching
of the submarine took place in October 1987.
Verification trials and demonstration for potential users are programmed until
mid 1990.

REFERENCES :
****
- CEE SYMPOSIUM DG XVII, LUXEMBOURG 6.12.84
- SWEDOCEAN MALMOE 17.10.84
- SUBTECH ABERDEEN 25.10.84

-4e JMO MARSEILLE 24.10.86
- MAN MACHINE SYMPOSIUM GOTHENBERG 22.10.87
-CEE SYMPOSIUM DG XVII LUXEMBOURG 22.03.88
- 5e JMO MARSEILLE 24.11.88.

```
*******************************************************************************
* TITLE : DEVELOPMENT OF A REMOTELY OPERATED      *      PROJECT NO          *
*         SUBMERSIBLE                             *                          *
*                                                 *   TH./07055/85/DE/..     *
*                                                 *                          *
*******************************************************************************
* CONTRACTOR :                                    *   TELEPHONE NO           *
*    ZF-HERION-SYSTEMTECHNIK GMBH                  *                          *
*                                                 *   0711 5209 351          *
*    POSTFACH 2168 - HOEHENSTR. 21                *                          *
*    D - 7012 FELLBACH                            *                          *
*                                                 *   TELEX NO               *
*                                                 *                          *
* TECHNICAL DIRECTOR :                            *   7254 733               *
*    K. WIEMER                                     *                          *
*                                                 *                          *
*******************************************************************************
                                                        VERSION : 07/11/89
```

AIM OF THE PROJECT :
****
The project described here covers the second phase of a program referred to as
the MARS project. The final goal of the complete project is to produce a
submersible vehicle system to be used generally for tasks associated with subsea
inspection, maintenance and repair work in situations where to employ a diver
would be either too dangerous or uneconomical. The system is therefore intended
to work as far as possible under remote control from the surface. Because
certain tasks cannot however be completed using such a remotely operated system,
it is intended that the MARS vehicle should be capable of quick conversion into
a form suitable for diver assistance for use when diver intervention becomes
unavoidable.
Phase 1 was concerned with establishing the system concept and specification.
Phase 2 relates to design,development,manufacture and testing of components and
modules, with subsequent integration into a basic vehicle arrangement to form a
complete working system.

PROJECT DESCRIPTION :
****
The purpose of the MARS 2 project was to manufacture and test the MARS vehicle
as a complete system which would be designed to incorporate the knowledge and
experience gained from the MARS 1 project. The required project work waqs
carried out in various stages as follows :
Design and manufacture of individual mechanical and hydraulic components.
Pressure chamber and tank testing of components.
Integration of components in a DAVID submersible vehicle.
Study work to define system design and parameters for the vehicle control system.
Design of the control system.
Offshore trials to assess the relative performance of the various individual
components when operating together with the vehicle module.
Integration of the components into the MARS carrier vehicle.
Defining the manipulator equipment for the vehicle according to the type and
situation required for the vehicle.
Offshore trials for the complete MARS vehicle system, without the manipulator.

STATE OF ADVANCEMENT :
****
The system without the manipulator has been manufactured and tested. Test
results have indicated areas where further development is necessary, and
modifications have still to be carried out. The installation and testing of the
manipulator system and associated tools have yet to be completed.

RESULTS :
****
Components
The individual components for the MARS system were designed and manufactured
according to the specifications defined as a result of the work carried out in
Phase 1. After subsequent testing it was found necessary in certain cases to
carry out further modification work as follows :
Power supply system : Fitting of capacitors for the reduction of reactive power.
Umbilical winch : Modification to the spooling drive.
Umbilical termination : Use of alternative resins to reduce the time required to
complete a termination.
Fitting of an umbilical bending protector.
Docking claw : Improving the mechanical strength of various component parts.
Propulsion units : Redesigning the propulsion units for improved service life.
Handling system : Investigation showed that for financial and operational
reasons, the design and development of a universal handling system could not be
justified. It was found that a handling system could be constructed only to suit
particular vehicle parameters.
Manipulator : The manipulator type and arrangement was selected to suit the
requirements described by the varous offshore service companies. A 7-function
master/slave manipulator system was integrated into movable manipulator module
for attachment onto the top of the vehicle.
Tests

From the many tests carried out during the duration of the project, the most useful information was gained when the system was operated from the Seaway Condor. This trial enabled the system to be tested over a longer period of time working in actual offshore conditions. For this trial, the vehicle was arranged as a diver assistance system without the manipulator.

Trial results : The vehicle system displayed various faults which were corrected on board the vessel. Typical problems were loss of signals from the diver's control station and water jetting module, data-link faults, leaks in the hydraulic system and failure of the main electric motor. The trial duration ran through a total of 1,032 hours and within this period the vehicle was eiter operating or available for operation for a total of 818 hours to achieve 79.3 % reliability. Service work needed for fault correction required 127 hours (=12.3%) , and time waiting for spare parts amounted to 72 hours (=7%).

Faults occuring in the first half of the trial period accounted for 127 hours of non-operational time but no further significant downtime was recorded during the second half.

REFERENCES :
****

COMPANIES INVOLVED OR ASSOCIATED WITH THE PROJECT WERE :
OFFSHORE SUPPLY ASSOCIATION LTD (OSA)
AEG MARINE- UND SONDERTECHNIK
OCEANEERING A/S
COMEX HOULDER DIVING LTD
STOLT-NIELSEN SEAWAY A/S
ELF AQUITAINE NORGE A/S
SHELL UK EXPLORATION AND PRODUCTION
GKSS FORSCHUNGSZENTRUM GEESTHACHT GMBH
TECHNISCHE UNIVERSITAET HAMBURG-HARBURG
GESELLSCHAFT ZUR FOERDERUNG DER MEERESTECHNIK.

```
****************************************************************************
* TITLE : DEVELOPMENT OF A HEAVY LIFT AND          *     PROJECT NO        *
*         TRANSPORT SYSTEM FOR THE INSTALLATION    *                       *
*         OF OFFSHORE PRODUCTION PLATFORMS         *     TH./07056/85/DE/.. *
*                                                  *                       *
****************************************************************************
* CONTRACTOR :                                     *     TELEPHONE NO       *
*    BLOHM + VOSS AG                               *                       *
*                                                  *     40 3119-2328       *
*    P.O. BOX 100720                               *                       *
*    DE - 2000 HAMBURG 1                           *                       *
*                                                  *     TELEX NO           *
*                                                  *                       *
* TECHNICAL DIRECTOR :                             *     211 047-0          *
*    MR. G. O. ANDERSSON                           *                       *
*                                                  *                       *
****************************************************************************
                                                  VERSION : 29/08/89
```

AIM OF THE PROJECT :
****
Installation technique of the topsides of offshore production platforms is by
crane lifting of reasonably sized modules. considerable cost and time savings
particularly with earlier production would be achieved if the inshore proven
mating technique of platform deck and substructure could be transferred to at
sea application. The present project is a feasibility study of a specialized
transport and lifting barge design for worldwide operation combining
semisubmersible motion characteristics and simple dock type lifting capability.
Integrated decks would be picked up by the barge at the fabyard, transported to
the site, lifted by deballasting and lowered gently to the preinstalled jacket.

PROJECT DESCRIPTION :
****
In a comparative study the requirements and merits of modularized and integrated
deck installations will be summarized. Various proposals for offshore mating
devices will be analyzed especially with respect to application in North Sea
environment. The chances of a new mating device will be discussed competing in
the market with well established operators of heavy lift crane vessels. The size
of the barge and its form will be determined by an analysis of form and size of
existing jackets and the weight of the topsides they carry. Operational aspects
of load-out, transit amd mating procedure will be looked into. By weather
statistic of a relevant North Sea area the probability of encountering
sufficiently long good weather periods for transport and mating will be
evaluated. Among the studies of the technical feasibility of the barge will be a
discussion of its fabrication on a conventional shipyard. The most critical load
cases will be defined which determine the detail design of form and structure.
Ballast handling during the different loading conditions will be considered and
the compartmentation designed appropriately. Hydrostatic stability will be
checked and damage stability investigated. Motion behaviour in the decisive
loading conditions will be calculated using potential theory. Model tests will
be performed to doublecheck motion characteristics and study the limiting
conditions of the sea state for initiation of the mating procedure. Structural
design will be controlled by longitudinal strength calculations and stress
analysis using finite element method applied to a beam model of barge and
platform deck structure in the design wave. Internal forces and moments will be
also measured in model tests in regular and irregular waves. Comparison will be
performed with forces obtained from the quasi-steady loaded beam model.

STATE OF ADVANCEMENT :
****
Ongoing. Steps 1 to 4.3 of the work plan are nearly completed; final evaluation
of seekeeping tests and calculation of motion behaviour is ongoing. Step 4.3 -
Structural Analysis was carried out up to the initial design stage.

RESULTS :
****
Review of installation modes (modularized or integrated deck) shows superiority
of integrated type. A vessel for one piece installation of topsides will,
however, envisage strong competition by established heavy lift crane barges.
Therefore, the barge must be also available for platform removal. The vessel
will have U-shape in plan view with legs similar to the floaters and columns of
a semisubmersible. Open at one side, they are cross-connected by two box shaped
bars at the other side, one below and one above water to achieve good torsional
resistance. With the deck spanned across the legs, conventionally shaped jackets
will be forked in for mating. Environmental forces acting on deck and U-barge
will load the jacket after the first contact is established. Jacket has to be
designed accordingly. Dimensions of U-barge were chosen to serve jackets of a 60
m width at upper end, carrying a topsides weight of 30,000 t. Length and beam of
the U-barge are 140 m and 120 m, the depth is 28 m. Best method of fabrication
established motion characteristics of the barge - as found from calculations and
model tests - is such that mating is possible in seaways of up to 2 m
significant wave height provided the prevailing wave periods do not surpass 7 s,
maximum 9 s. The square arrangement and size of the columns warrant sufficient
intact stability. Damage stability is provided by suitably selected

compartmentation and buoyancy boxes. Though the tolerable sea state in transit, loaded with a platform deck, must be limited for safety reasons, the module support frame will be subjected to additional loads induced by the motions of the legs against each other. This will either lead to an MSF structure properly strengthened or a cross bar construction designed against large deformations at the leg ends rather than stresses.

An outline specification details general aspects, outfit, accomodation machinery and electric equipment of the barge.

REFERENCES :
****

A PAPER ENTITLED "ALTERNATIVES SCHWERLASTHEBE- UND TRANSPORTSYSTEM FUER DIE INSTALLATION VON OFFSHORE-PLATTFORMEN" WAS PRESENTED BY G.O. ANDERSSON AT THE 81ST ANNUAL MEETING OF THE SCHIFFBAUTECNISCHE GESELLSCHAFT, NOVEMBER 20 AND 21, 1986 IN BERLIN.

```
*****************************************************************************
* TITLE : INTEGRATED INTERVENTION SYSTEMS FOR    *      PROJECT NO        *
*         SUBSEA PRODUCTION EQUIPMENT            *                        *
*                                               *   TH./07059/86/UK/..    *
*                                               *                        *
*****************************************************************************
* CONTRACTOR :                                  *      TELEPHONE NO       *
*    SUBSEA INTERVENTION SYSTEMS LIMITED        *                        *
*                                               *      0224 714101        *
*    BUCKSBURN HOUSE, HOWES ROAD                *                        *
*    BUCKSBURN                                  *                        *
*    UK - ABERDEEN AB2 9RQ                      *      TELEX NO           *
*                                               *                        *
* TECHNICAL DIRECTOR :                          *      73394              *
*    MR BOWRING                                 *                        *
*                                               *                        *
*****************************************************************************
```

VERSION : 21/09/89

AIM OF THE PROJECT :
****
The objective of the IISSPE project is to develop a new generation of Subsea
Production Equipment. The design of the subsea tree and ;its associated
eauipment will be performed with regard to the demands of maintenance and
intervention by Remotely Operated Vehicle (ROV) Tooling. The project will design,
 manufacture and test the techniques and interfaces which are required to
perform effective intervention. The project brings together oil companies who
specify subsea equipment, a manufacturer of production hardware and an ROV
operator to perform maintenance intervention.
The aim of the project is to produce subsea production equipment and subsea
maintenance systems which will result in lower operating costs for subsea
production developments.

PROJECT DESCRIPTION :
****
The project will design, manufacture and test equipment as described by the
following project phases;
Task Series 1000 - SUBSEA EQUIPMENT APPRAISAL
Study present applications of subsea production systems. Compile a comprehensive
list of all functions which may be required in a subsea system and establish
intervention priorities. Compile a detailed specification of the project subsea
system and intervention tasks.
Task Series 2000 - INTERVENTION SYSTEM
Outline desing of interfaces between ROV and tree. Design of an ROV toolskid and
heavy lift facility. Interfacing and control of the intervention equipment.
Launching and operating criteria.
Task Series 3000 - COMPONENT TOOLING
Outline design of several key intervention tasks. Design of the tree component
functions and the associated tooling requirements.
Task Series 4000 - SUBSEA SYSTEM DESIGN
Arrangement of the Subsea Production Tree for intervention. General arrangement
of components for access and system considerations.]
Task Series 5000 - DETAIL DESIGN
Detail design of the intervention system, tooling and handling equipment. Detail
design of several key subsea components. Design of a dummy tree.
Task Series 6000 - MANUFACTURE
Manufacture of a fully operational subsea intervention system and a
representative dummy test tree.
Task Series 7000 - FUNCTIONAL TESTING
Onshore functional testing of the subsea intervention systems.
Task Series 8000 - UNDERWATER TESTING
Shallow water testing of the subsea intervention system. Confirmation of the
project objectives.

STATE OF ADVANCEMENT :
****
Ongoing. Manufacture of the intervention system and dummy tree has been
completed. Functional testing of the systems is now taking place in the workshop.
 Due to technical problems during the design and manufacturing phases of the
project, the expected completion date of the project is now December 1989.

RESULTS :
****
The work completed to date is fully described in the Detail Design Report issued
with the Third Interim Report.
Activities during the six month period covered by this report have been mainly
equipment manufacturing. The manufacturing completion has been documented by
Factory Acceptance Testing (FAT) procedures and acceptance certificates.
Current activity is based around Functional Testing. The results of this testing
will be fully documented in a Test Report.

REFERENCES :
****
TWO TECHNICAL PAPERS HAVE BEEN PRESENTED AND PUBLISHED ON THE PROJECT DURING THE

PREVIOUS SIX MONTHS.
1. SUBSEA INTERNATIONAL 89, SECOND GENERATION SUBSEA PRODUCTION SYSTEMS, LONDON, 25 APRIL 1989.
TITLE - LOW COS INTERVENTION., M BOWRING.
2. OFFSHORE TECHNOLOGY CONFERENCE, HOUSTON, 2 MAY 1989.
TITLE - INTERVENTION AND SUPPORT OF SUBSEA EQUIPMENT BY REMOTELY OPERATED VEHICLES., M BOWRING.

```
*****************************************************************************
* TITLE : UNDERWATER TELE-MANIPULATION SYSTEM          *      PROJECT NO     *
*         WITH SUPERVISORY CONTROL FOR INSPECTION      *                     *
*         AND MAINTENANCE OF OFFSHORE                  *    TH./07061/86/IT/..*
*         INSTALLATIONS                                *                     *
*****************************************************************************
* CONTRACTOR :                                         *    TELEPHONE NO     *
*    TECNOMARE SPA                                     *                     *
*                                                      *     041 796711      *
*    SAN MARCO 3584                                    *                     *
*    IT - 30124 VENEZIA                                *                     *
*                                                      *    TELEX NO         *
*                                                      *                     *
* TECHNICAL DIRECTOR :                                 *     410484          *
*    MR. P. MINARDI                                    *                     *
*                                                      *                     *
*****************************************************************************
```

                                                    VERSION : 03/07/89

AIM OF THE PROJECT :
****
To develop and test under simulated operating conditions a new, computer
assisted underwater telemanipulation system, aimed at the execution of tasks now
performed by divers and not feasible by means of the existing directly
controlled underwater manipulators mounted onboard ROVs (Remotely Operated
Vehicles).
Typical tasks include inspection of nodes of offshore platforms and light
manipulation on submarine wellheads.

PROJECT DESCRIPTION :
****
The telemanipulation system consists of:
- An underwater manipulator, whose kinematic and trajectory control
characteristics are optimized for operation in complex environments, like
structural nodes or wellheads.
- A supervisory control system, in which a computer system interfaces the remote
manipulator with the operator.
- A man-machine interface system, including TV monitors and real time 3-D
graphic display of the work scene.
Full scale shop test are foreseen relevant to typical underwater operations, to
be performed in accordance with industrial practices.

STATE OF ADVANCEMENT :
****
The project is in a very advanced stage. The shop trials of the system will
start september 1989.

RESULTS :
****
Expected project results are relevant to:
- development of an intelligent underwater manipulation system;
- optimization of the manipulator kinematics;
- trajectory and environment interference controls;
- systematic performance evaluation of industrial tasks in real scale and in
conditions simulating actual operational environmental;
- study of the possibility of operating with a manipulator shoulder mobile in
relation to the work scene.

REFERENCES :
****
1. SUPERVISORY CONTROLLED TELEMANIPULATION SYSTEM FOR UNDERWATER APPLICATIONS -
NOV. 89 SAN DIEGO - CALIFORNIA.
2. NEW ROBOTIC WILL IMPROVE SUBSEA WORK EFFICIENCY - OCEAN INDUSTRY, APRIL 89.
3. UNDERWATER TELEMANIPULATION SYSTEM WITH SUPERVISORY CONTROL - INTERNATIONAL
SYMPOSIUM TELEOPERATION AND CONTROL JULY 1988.

```
********************************************************************************
* TITLE : DEVELOPMENT OF KEY SYSTEM FOR AN      *      PROJECT NO          *
*         OFFSHORE SUBMERSIBLE                  *                          *
*                                               *    TH./07062/86/DE/..    *
*                                               *                          *
********************************************************************************
* CONTRACTOR :                                  *    TELEPHONE NO          *
*    THYSSEN NORDSEEWERKE GMBH                   *                          *
*                                               *    (4921) 851            *
*    AM ZUNGENKAI                               *                          *
*    P.O. BOX 23 51                             *                          *
*    DE - 2970 EMDEN                            *    TELEX NO              *
*                                               *                          *
* TECHNICAL DIRECTOR :                          *    27 802                *
*    MR. A. FREITAS                             *                          *
*                                               *                          *
********************************************************************************
                                                     VERSION : 27/11/89
```

AIM OF THE PROJECT :
****
The offshore submersible is planned to be a part of a subsea oil and gas
production system. Its operation profile is very different from that of
conventional naval submarines, being characterized by great operating depth
(over 400 m), long diving periods (up to 21 days), exact 3-dimensional
positioning and low manning levels. As current technology cannot meet these
requirements, new key systems have to be developed. This project aims at the
development and in some cases the prototype testing of 5 key systems of the
submersible and their integration in a basic submersible design.

PROJECT DESCRIPTION :
****
The project covers the development design of the following submersible systems
and their integration in the conceptual design of a prototype vessel :
ENERGY SUPPLY SYSTEM
The energy system is to operate independent of ambient air and is based on a
Cosworth closed cycle diesel engine.
This system uses sea-water for scrubbing the CO_2 from the exhaust gases and a
mixture of CO_2 and argon for the inert part of the suction air.
A special water management system provides fresh sea-water at reduced pressure
and discharges the waste water outboards by using the pressure energy of the
surrounding water.
A test rig for 120 kW prototype system will be built to test the closed cycle
engine under service conditions.
PROPULSION AND DYNAMIC POSITIONING SYSTEM
The propulsion and DP-system must have a highly efficient transmission system
with sensitive power control for economic transit and accurate 3-dimensional
positioning.
At first alternatives for the transmission system are to be investigated and a
system chosen.
Next a conceptual design of the propulsion and DP-system for the prototype
submersible will be prepared to satisfy the requirements of the planned
operation profile.
Finally a prototype thruster will be built and tested under simulated operating
conditions.
LIFE SUPPORT SYSTEM
The extended diving period requires an efficient air purifying system with
respect to size and weight. New systems have to be investigated and evaluated. A
literature study will be made to identify the possible contaminants in breathing
air. Working conditions and safety criteria will be examined. A concept for the
life support will be designed and evaluated, and a system specification prepared.
 Emergency systems will be studied.
MISSION CONTROL SYSTEM
A control and monitoring system for the submersible with a high degree of
integration and automation is to be developed in order to increase safety and
reduce the manning level.
To begin, system functions, interfaces, kind of signals and type of sensors are
to be defined. Concepts for control and monitoring of equipment, navigation and
safety will be prepared. Layout for the monitoring and control of the vessel
systems will be made. A specification of the required hard and software will be
drawn up.
CRANE WORK MODULE
A crane serviced work module with manipulators is to be designed for operation
in and from a wet cargo hold. At first the operation procedures and task
profiles are to be investigated and defined. On this basis the crane and its
components and the multi-function work unit with manipulators will be designed.
Special development is required for the machinery operating in a wet environment
and for the power supply and control systems.
CONCEPTUAL DESIGN OF THE PROTOTYPE SUBMERSIBLE will be made for a theoretical
but realistic operation profile. The work includes the integration of the key
systems, the general layout of the submersible and the specification of the main
components.

STATE OF ADVANCEMENT :

316

****
Completed

RESULTS :
****
ENERGY SUPPLY SYSTEM
120 kW prototype of the closed cycle diesel built and tested on a rig under simulated operating conditions for 240 hours. Reliability of certain components and system efficiency unsatisfactory. Further development required.
PROPULSION AND DYNAMIC POSITIONING SYSTEM
Concept design of main propulsion system and DP thruster. Layout design of the hydraulic power supply and transmission system. 20 kW prototype of the azimuth thruster built and tested under atmospheric conditions on a test rig and under simulated diving conditions in a hyperbaric chamber with good results.
LIFE SUPPORT SYSTEM
Concept for a system to monitor and control the submarine atmosphere. Layout of the ventilation system. Concept design of the regenerative CO_2 absorption plant. Concept for the emergency life support system.
CRANE WORK MODULE
Design and engineering of the hydraulic crane in the wet cargo hold. Design of the cargo hold and equipment. Concept design of the work module. Analytical investigation (computer simulation) of the motion behaviour of the work module suspended from the crane in water.
PROTOTYPE SUBMERSIBLE
Conceptual design of submarine covering :
- main dimensions
- pressure hull and outer hull
- energy supply and storage
- submarine systems
- safety and rescue
- general arrangement plan.

```
*******************************************************************************
* TITLE :  INNOVATORY TECHNIQUES FOR AN AUTOMATED     *      PROJECT NO       *
*          ROBOTIC SUBMERSIBLE SYSTEM FOR USE IN      *                       *
*          THE EXPLORATION, OF OFFSHORE               *    TH./07069/88/UK/..  *
*          INSTALLATIONS                              *                       *
*******************************************************************************
* CONTRACTOR :                                        *      TELEPHONE NO      *
*    WINCHESTER ASSOCIATES LTD                        *                       *
*                                                     *    44.224.822833       *
*    UNIT 16, DENMORE INDUSTRIAL ESTASTE              *                       *
*    DENMORE ROAD, BRIDGE OF DON                      *                       *
*    UK - ABERDEEN AB2 8JW                            *     TELEX NO           *
*                                                     *                       *
* TECHNICAL DIRECTOR :                                *   94011419 FAX: 442247*
*    MR. D. LIDDLE                                    *   02469                *
*                                                     *                       *
*******************************************************************************
                                                       VERSION : 21/05/90
```

AIM OF THE PROJECT :
The objective of the SID vehicle concept is to provide a more integrated and
dedicated package than has been previously possible. As a minimum it was decided
to take the radical approach that the vehicle had to have the ability to be
controlled in all six degrees of freedom and thus the astable or hydrobatic
vehicle concept was established. It also deliberately makes use of some
technologies which are not generally found within the subsea industry but which
are used on a regular basis in other sectors of industry. WA believe this
approach to be the first step towards the development of a truly successful
automated robotic submersible system capable of maintenance and integrity
control of offshore installations.

PROJECT DESCRIPTION :
Having studied the limitations to current systems it has been possible to
establish an overall design concept to match the requirements for maintenance
and integrity control of offshore installations. One of the main conclusion WA
have come to is that any new automated robotic submersible system will have to
be designed for the duty in an attempt to avoid being drawn into producing a
compromise based on a conventional production vehicle as has been the case to
date. Also fairly radical changes in design philosophy will have to be employed
for any new system to succeed, and the application of technologies normally
found outside the subsea industry may be necessary for the full potential to be
realized.
The ideas put forward in this application have been developed by Winchester
Associates and provides the radical approach to achieving an automated robotic
submersible system specifically dedicated to maintenance and integrity control,
the system is referred to as a Structural Inspection Device or "SID".
SID is an astable or hydrobatic vehicle designed to be able to achieve combined
pitch and roll angles of +/- 90 degrees and +/- 180 degrees respectively. This
gives the system a full six degree of freedom capability enabling movement in
any direction at any attitude of pitch, roll or yaw.
There is therefore a fundamental difference between this system and other more
conventional systems developed to date, in that traditionally ROVs have been
designed to be stable and therefore to remain more or less upright. For the
majority of applications this conventional format is ideal and there is no
reason to change it. However, for the more advanced and sophisticated tasks
where the replacement of the diver is considered economically and practically
desirable then an ability to emulate the divers freedom of movement would be an
advantage.

STATE OF ADVANCEMENT :
Completed. The next phase (Phase 3) to develop the design results and produce
the technology is ongoing.

RESULTS :
The project team has been successful in producing an astable ROV system design
that will undertake maintenance and integrity checks on offshore structural
installations.
Whilst it is impossible at this stage to completely quality the above statement,
the results obtained at this stage of the overall development programme are very
encouraging. The results so far are in the form of design data, drawings,
calculations and computer analysis. The results of the design process have not
lead to any major obstacles being highlighted that mean the programme should be
discontinued. Indeed the opposite is more justifiable in that the results
indicate every element of success. The design process has not been carried out
without any problems in its execution. Some areas of work have taken more effort
than has been anticipated, certain developing technology has meant that some of
the work scope has changed, and some areas of work has highlighted deficiencies
in the current state of the art technology.

REFERENCES :

318

****

INTERNATIONAL UNDERWATER SYSTEMS DESIGN (TECHNICAL JOURNAL), SEPT/OCT 1988 ISSUE,
 ARTICLE : STRUCTURAL INSPECTION DEVICE - SID, DICK WINCHESTER, MANAGING
DIRECTOR, WINCHESTER ASSOCIATES LTD.
ADVANCES IN UNDERWATER INSPECTION AND MAINTENANCE, SUT CONFERENCE, UK, MAY 1989,
PAPER : ASTABLE ROV DESIGN CONCEPT, R. WINCHESTER.

```
****************************************************************************
* TITLE : TOPSIDE INSTALLATION VESSEL           *     PROJECT NO       *
*                                               *                      *
*                                               *   TH./07071/88/NL/.. *
*                                               *                      *
****************************************************************************
* CONTRACTOR :                                  *     TELEPHONE NO     *
*    ALLSEAS ENGINEERING B.V.                   *                      *
*                                               *   070-612161         *
*    R.J. SCHIMMELPENNINCKLAAN                  *                      *
*    NL-2517 JN THE HAGUE                       *                      *
*                                               *     TELEX NO         *
*                                               *                      *
* TECHNICAL DIRECTOR :                          *                      *
*    MR W.P. KALDENBACH                         *                      *
*                                               *                      *
****************************************************************************
                                                 VERSION : 15/02/90
```

AIM OF THE PROJECT :
****
The objective of the project is to develop the principal innovative parts of an
installation vessel for topside units of fixed offshore platforms.
The innovative aspects are :
- A system which compensates the wave induced vessel motions.
- Mechanical structures which are used to support the topside units.
- Construction of the vessel, which is of an unconventional shape.

PROJECT DESCRIPTION :
****
The first part of the project is the evaluation of an optimal shape and
construction type of the vessel. This will involve estimates of strength and
stiffness of various vessel configurations and estimates of the amounts of steel
required in each configuration.
Once an ojptimum configuration has been chosen the main subsystems of the vessel
have to be designed. This part of the project will result in a general
arrangement of the vessel, together with detailed operating procedures.
Development of the motion compensation system forms the second part of the
project. This system comprises a mechanical construction which allows the
installation vessel to move relative to the topside unit, together with a set of
hydraulic actuators which are controlled in such a way as to hold the topside
unit stationary while the vessel oscillates due to wind and waves. A functional
model will contain all functions of the final system. The model will be used to
test the effects of failure of selected subsystems. The model will further be
used to validate a computer program which simulates the dynamic resoponse of the
system.
Further parts of the development of this compensation system are the evaluation
and selection of sensors and monitoring systems, which provide the required data
to the control system's the development of control algorithms are the selection
of electronic and hydraulic components.
As the topside unit is lifted in an unconventional way, the mechanical
componenets of this system have to be newly developed. These components comprise
:
- Large floating bodies which provide the lifting force to lift the topside unit,
  together with a system which guides the motion of these floaters.
- Ball joints for large weights.
- Lifting frames for the topside unit.

STATE OF ADVANCEMENT :
****
Structural engineering has been carried out to the extend that is required for
choosing the final vessel configurations (monohull or twinhull option).
The working scale model (1,50) has been completed. Preliminary tests with regard
to the functioning of the motion compensation system have been carried out
satisfactorily. The model has not yet been tested in a wave basin.
Pre-engineering of principal system components and layout of the hydraulic
system is completed.

RESULTS :
****
VESSEL STRUCTURE
Structural engineering has led to estimates for the amount of new steel which is
required for the construcation of the vessel.
MECHANICAL COMPONENTS
Initial designs of the main components which are part of the compensation as
well as of the supporting structures of the topside unit, have been made. This
work has established the feasibility of the construction.
FUNCTIONAL MODEL
Testing of the scale model of the motion compensation system demonstrated the
feasibility of this system. It proved possible to reduce both the vertical and
the horizontal motions of a topside unit gradually to zero, while no large
forces between the topside unit and jacket were required.
MOTION COMPENSATION SYSTEM
The initial design of the layout of the hydraulic system, which forms the heart

of this system, demonstrated that it is possible to realise the hydraulic system using existing components.

A computer simulation fo the 3-dimensional motion behaviour of the vessel and the topside unit, incorporating the motion compensation system, yielded the values for the design parameters for the hydraulic cylinders and other main components.

REFERENCES :
****
LECTURES
PROCEEDINGS OF CONFERENCE ON COST EFFECTIVE TOPSIDE DESIGN, OCTOBER 27-28, 1987, LONDON;
SEMINAR ON ENGINEERING ASPECTS OF PLATFORM REMOVAL, SOCIETY FOR UNDERWATER TECHNOLOGY, FEBRUARY 24, 1988, ABERDEEN;
FEBRUARY 15, 1990, 19.00 HRS AT THE KIVI, THE HAGUE;
SUBJECT : "TWO CONCEPTS FOR SUPER-HEAVY OFFSHORE LIFTRING"
SPEAKERS : J. MEMELINK (KIVI), H. V.D. HEIJDEN (HEEREMA), A. DE GROOT (ALLSEAS).
PATENT APPLICATIONS
FILED ON JANUARY 14, 1987, ON 8700076, THE NETHERLANDS;
"WERKWIJZE VOOR HET MANOEUVREREN VAN EEN OPBOUWELEMENT TEN OPZICHTE VAN EEN IN WATER AANGEBRACHTE VASTE CONSTRUCTIE, WERKWWIJZE VOOR HET BOUWEN VAN EEN BOUWWERKE EN BOUWWERKE GEBOUWD VOLGENS EEN DERGELIJKE WERKWIJZE";

```
********************************************************************
* TITLE : ROV DEPLOYED NODE CLEANING AND      *    PROJECT NO      *
*         INSPECTION.                          *                    *
*                                              *  TH./07073/88/FR/.. *
*                                              *                    *
********************************************************************
* CONTRACTOR :                                 *    TELEPHONE NO    *
*    COMEX SA                                   *                    *
*                                              *   91 23 50 00      *
*    BOULEVARD DES OCEANS 36                    *                    *
*    F - 13275 MARSEILLE CEDEX 9                *                    *
*                                              *    TELEX NO        *
*                                              *                    *
* TECHNICAL DIRECTOR :                          *    410985          *
*    ROBERT FOSSARD                             *                    *
*                                              *                    *
********************************************************************
                                        VERSION : 01/02/90
```

AIM OF THE PROJECT :
****
The aim of the project is to develop a node cleaning and inspection system which
can be deployed by an existing working ROV. In particular, the project will
concentrate on :
- designing attachment systems which overcome the limitations of present-day
systems,
- use of a high performance manipulator arm in either a telemanipulation of
robotic mode for deploying various NDT and cleaning end effectors,
- development of marinised MPI/Eddy Current systems for deployment by the
manipulator arm.

PROJECT DESCRIPTION :
****
The project consist of the following 6 work Packages :
Work Package nr 1 - Definition of the Overall Architecture of the System.
- Preparation of initial specifications (details of structures to be inspected,
definition of operating and environmental conditions, NDT requirements).
- Conceptual designs (inventory of possible system configurations, costs,
operating performance,...),
- Selection of architecture to be retained for detailed design phase.
Work Package nr 2 - Preliminary Development of NDT Techniques
(preliminary studies of ROV deployed MPI/Eddy Current - Problems to be solved,
alternatives to investigate).
Worrk Package nr 3 - Detailed Designs of Tooling
(detailed design of attachment system, manipulator arm system, integration of
cleaning system and integration to ROV).
Work Package nr 4 - Procurement, Manufacture and Factory Testing of
Components
(attachment system, manipulator system, teletransmission, cleaning system,...)
Work Package nr 5 - Integration to an ROV and Shallow Water Test of the cleaning
system.
Work Package nr 6 - Detailed Development of NDT Sensors
(MPI/Eddy Current...)

STATE OF ADVANCEMENT :
****
Ongoing

RESULTS :
****
The following status has been reached by the end of 1989 :
Access study and manipulator arm geometry completed on typical node
configuration (WP1)
Study on potential attachment systems completed and selection of preferred
method finalised (WP1)
Integration studies for various ROV ongoing, more than 50% completed (WP3)
Evaluation of various cleaning systems completed - System selection made (WP2)
Manipulator arm (WP3) :
- slave arm - Technical specifications completed
- Control system - Technical specifications completed
- Master arm - Definition of "commercial product" underway
High speed teletrans system specification under preparation (WP3)
Subject to results of shallow water test (WP5).

```
***********************************************************************************
* TITLE : DEVELOPMENT OF KEY SYSTEMS FOR OFFSHORE    *      PROJECT NO         *
*         SUBMARINES (PHASE II)                      *                         *
*                                                    *   TH./07076/89/DE/..    *
*                                                    *                         *
***********************************************************************************
* CONTRACTOR :                                       *     TELEPHONE NO        *
*    THYSSEN NORDSEEWERKE GMBH                        *                         *
*                                                    *   (4921) 852916         *
*    AM ZUNGENKAI                                     *                         *
*    P.O. BOX 23 51                                   *                         *
*    DE-2970 EMDEN                                    *     TELEX NO            *
*                                                    *                         *
* TECHNICAL DIRECTOR :                               *     27802               *
*    MR A. FREITAS                                   *                         *
*                                                    *                         *
***********************************************************************************
                                                          VERSION : 14/03/90
```

AIM OF THE PROJECT :
****
The overall aim of the project is to further pursue the development of the key systems of offshore submarines begun in Phase I (TH.07062) in order to attain the level of proven technology. By extensive testing of those systems which are not standard technology on test rigs and in a submarine in real operations, it is intended to obtain sufficient data to gain acceptance from potential users and to qualify the submarine for subsea work.

PROJECT DESCRIPTION :
****
The project covers the engineering, construction and testing of the systems mentioned below. The programme also includes the engineering for the installation of the test rigs in a submarine. It is planned to install the plant in a suitable submarine, e.g. the SAGA 1, and to test the systems in real subsea operations in a subsequent Phase III.
ENERGY SYSTEM
Based on the experience gained on the test rig built in Phase I, a new and improved closed cycle diesel engine of 150 kW will be engineered and built.
The improvements are expected to achieve the following :
(1) Increase in the specificiency. (2) Improvement of the system efficiency. (3) Reduction in noise level. (4) Higher endurance.
After extensive testing of the new closed cycle diesel on the test rig, it is planned to install the plant on a submarine for trials at sea. The engineering for the installation will also be done within the current project.
Corsworth Deep Sea Systems UK, is responsible for this part of the project.
LIFE SUPPORT SYSTEM
During Phase I a concept for the regenerative CO_2 absorption system was established. In the present project a full scale test plant for a full submarine crew will be engineered, built and tested on a land based rig. Later the equipment will be tested together with the energy system on a submarine in a planned Phase III.
Dornier GmbH, FRG is responsible for this part of the project.

STATE OF ADVANCEMENT :
****
Both part projects are in the engineering phase. Components for the test rigs are under manufacture.

RESULTS :
****
The test rigs are in the engineering and construction stage.

REFERENCES :
****
FINAL REPORT OF PHASE I, PROJECT NO TH./07062/86

```
****************************************************************************
* TITLE : INNOVATORY TECHNIQUES FOR AN AUTOMATED    *      PROJECT NO      *
*         ROBOTIC SUBMERSIBLE SYSTEM FO- PHASE 3    *                      *
*                                                   *   TH./07080/89/UK/..  *
*                                                   *                      *
****************************************************************************
* CONTRACTOR :                                      *      TELEPHONE NO    *
*    WINCHESTER ASSOCIATED LTD                      *                      *
*                                                   *   44 224.822833      *
*    DUNBAR HOUSE, BALGOWNIE ROAD                   *                      *
*    BRIDGE OF DON                                  *                      *
*    UK - ABERDEEN AB2 8JS                          *      TELEX NO        *
*                                                   *                      *
* TECHNICAL DIRECTOR :                              *   94011419           *
*    MR .D. LIDDLE                                  *                      *
*                                                   *                      *
****************************************************************************
                                              VERSION : 22/01/90
```

AIM OF THE PROJECT :
Is to take a completely innovative approach to automated inspection and
integrity control techniques as applied to offshore installations and to develop
a robotic submersible system which meets these parameters. The main objective is
to provide a more integrated and dedicated package than has been previously
possible. As a minimum it was decided to take the radical approach that the
vehicle should be controllable in all six degrees of freedom and to further make
use of technology transferred from other sectors of industry.

PROJECT DESCRIPTION :
The robotic submersible system proposed has been configured for structural
inspection and cleaning work and is referred to as a Structural Inspection
Device or "SID".
SID is an astable or hydrobatic vehicle designed to be able to achieve combined
roll/pitch angles of +/-180 deg and +/-90 deg. This gives the system a full six
degree of freedom capability enabling it to be driven in any attitude in any
direction. There is therefore a fundamental difference between this system and
other more conventional systems, which traditionally have been designed to
remain stable.
The project will enable a working system to be produced, work being undertaken
in design, development, manufacture, assembly and final testing. The main
scheduled phases being :
1. Conceptual Design - Complete
2. Detailed Design - Complete March 1990
3. Manufacture, Assembly and Tests - Current Phase

i.	Manufacture/assembly	(Timescale - 166 days)
ii.	Factory testing	(Timescale - 295 days)
iii.	Shallow water inshore trials	(Timescale - 60 days)
iv.	Offshore trials	(Timescale - 60 days)
v.	Final report	(Timescale - 60 days)

STATE OF ADVANCEMENT :
Ongoing. Phase 2 of the project is due for completion by the end of March 1990
and Phase 3 work follows on immediately. Phase 2 of the project is progressing
the detail design and this is approximately 85% complete as at the end of
January 1989.

RESULTS :
Phase 2 work has undertaken the detailed design required for this project. This
is due for completion by the end of March 1990. Results available at this time
include the detail design, in the form of engineering drawings, calculations and
tables of data, of the whole ROV system thus enabling Phase 3 work to continue
to develop a technology demonstrator in the form of a working prototype system
which will be complete in the second half of 1991.
One of the main problems encountered during the design phase has been one of
underwater manipulation. Whilst there exists a large range of underwater
manipulator systems none of the available systems are considered accurate enough
to deploy the necessary range of sensors and inspection equipment. This problem
has been resolved by the transfer of technology from the industrial sector where
acuracies and repeatabilities far in excess of the actual requirement are
readily obtained. Whilst an outline design has been produced for the manipulator
system a whole development programme will be executed during phase 3.

REFERENCES :
INTERNATIONAL UNDERWATER SYSTEMS DESIGN (TECHNICAL JOURNAL), SEPT/OCT 1988 ISSUE,
 ARTICLE : STRUCTURAL INSPECTION DEVICE - SID, DICK WINCHESTER, MANAGING
DIRECTOR, WINCHESTER ASSOCIATES LTD.

PIPELINES

```
*****************************************************************************
* TITLE : DEVELOPMENT OF A FLUORIMETER FOR EARLY        *     PROJECT NO    *
*          DETECTION OF HYDROCARBONS FROM SMALL         *                   *
*          LEAKAGES IN OFF-SHORE PIPELINES              *  TH./09021/85/DK/.. *
*                                                       *                   *
*****************************************************************************
* CONTRACTOR :                                          *   TELEPHONE NO    *
*    WATER QUALITY INSTITUTE                            *                   *
*                                                       *   45 6 202000     *
*    SCIENCE PARK AARHUS                                *                   *
*    10, GUSTAV WIEDSVEJ                                 *                   *
*    DK - 8000 AARHUS C                                 *   TELEX NO        *
*                                                       *                   *
* TECHNICAL DIRECTOR :                                  *   37874           *
*    MR. A. LYNGGAARD-JENSEN                            *                   *
*                                                       *                   *
*****************************************************************************
                                              VERSION : 20/09/89
```

AIM OF THE PROJECT :
****
The developing of a tool for improvement of the safety of supply of oil/gas
through off-shore pipelines implies the wish for an early detection of small
leakages. An early warning of accidentally occurred cracks in the pipeline will
improve the operation security and thus avoid sudden stop of production.
The traditional methods such as registration of the pressure headloss through
the pipeline will not be sufficient for the detection of small leakages, whereas
a determination of the content and type of hydrocarbons in the water along the
pipeline will be a valuable tool. 1st generation fluorimeters may detect oil in
water, but apart from this it is important to be able to ascertain whether the
oil detected origins from the pipeline. The sensitivity of existing fluorimeters
is also to be improved.

PROJECT DESCRIPTION :
****
Fluorescence spectra of different oil in water are not to specific for the
single type of oil, but taking into account the lifetime curves along the
spectrum, thus forming a 3-dimensional spectrum will give the specific
information of the type of oil being measured. The project is therefore divided
into two main phases :
Phase A :
Development of a 2nd generation prototype fluorimeter basd on laser-induced
single shot fluorescence. The exitation wavelength should be tuneable, due to
the use of a frequency doubler in cooperation sith a color dye laser pumped by a
nitrogen-laser. The fluorescence should be measured simultaneously in 35
channels, forming a spectrum covering the wavelengths from 250 nm to 660 nm,
with a channel width of 10 nm.
Phase B :
Development of a 3rd generation prototype fluorimeter based on the 2nd
generation fluorimeter, but including the measurement of lifetime-curves in each
channel, thus forming 3-dimensional spectra. This technique should be called
laser-induced single-shot fluorescence lifetime spectroscopy, and the developed
prototype should be tested and form the basis for industrial production of an in-
situ 3rd generation fluorimeter.
Each of the two main phases is divided into a number of stages as follows :
A1- Collection of existing knowledge concerning UVF-technology, lasers and
optical fibres. Specification of a 2nd generation system to be build in the
laboratory (wavelength, energy-requirement, transmission losses in fibres, fibre
types etc..).
A2- Construction of the first laboratory prototype model of the instrument (one
channel only). The model will be constructed in an optical bench at the
Institute of Physics (University of Aarhus).
A3- Optimization and testing of the laboratory prototype model will be performed
using samples of pure hydrocarbons. The optical parts of all 35 channels being
tested in the model channel.
A4- Construction of a final prototype of the 2nd generation fluorimeter
including all 35 channels.
B1- Development of very fast detection electronics, (samplingtime less than 10
ns) which should be tested in the model channel and then miniaturising this for
use in all 35 channels.
B2- Development of a micro-computer system for control of the lqser and
collection and storage of the fluorescence lifetime curves from the model
channel.
B3- Reconstruction of the 2nd generation prototype to form the final 3rd
generation prototype including all 35 channels.
B4- Optimization and testing of the 3rd generation prototype fluorimeter using
standard oils mixtures etc.. in seawater.

STATE OF ADVANCEMENT :
****
Stages A1,A2,A3,B2 and half of the stages A4 and B1 are finished. In other words
the measuring principle is now at function.

RESULTS :

****
The results are produced according to above mentioned project stages and described state of advancement.
 The laser system consisting of a large nitrogen laser, a dye laser (pumped by the nitrogen laser) and a frequency doubler is working according to routine. By this system the exitation wavelength may be tuned in the interval from 228 nm to 900 nm.
The 3rd generation fluorimeter principle as described above is working in a model channel, and this channel is now being multiplicated and build into the constructed 35-channel sensor head.
In the work with the 2nd generation fluorimeter the detection limit has been reduced by the use of plexiglass rods doped with a dye. (This technique is also known from scintilator rod in the atomic science). For that reason the model channel and the luorimeter is not based on classical optics, but uses only an interference filter followed by the plexiglass rod and a Photo Multiplier-tube. The channel is working very well and interference filters in the range from 250 to 600 nm with a band-width of 5 nm and a distance between filters of 10 nm has been made.
Updating the model channel from 2nd to 3rd generation gave the project heavy problems due to the demands to very fast detection electronics. The problem have now been solved, but the project has been delayed as well as the planned field testing of the resulting fluorimeter has been cancelled.

REFERENCES :
****
"PROJECT OIL SNIFFER" : PAPER PRESENTED IN INFORMATION BULLETIN FROM THE INSTITUTE OF PHYSICS UNIVERSITY OF AARHUS, NO.4 OCT.1987. (AND SENT IN COPY TO DG XVII).
POSTER PRESENTATION AT THE 3RD EEC-SYMPOSIUM IN LUXEMBOURG, 22-24 MARCH 1988, NEW TECHNOLOGIES FOR THE EXPLORATION AND EXPLOITATION OF OIL AND GAS RESOURCES.

```
************************************************************************
* TITLE : DEVELOPMENT OF AN INTEGRATED PIPELINE      * *     PROJECT NO      *
*         MANAGEMENT SYSTEM (IPMS)                    * *                     *
*                                                     * *  TH./09024/85/DK/.. *
*                                                     * *                     *
************************************************************************
* CONTRACTOR :                                        *   TELEPHONE NO        *
*    DHI/R & H PIPEDATA                                *                       *
*                                                     *   45 2 856500          *
*    TEKNIKERBYEN 38                                   *                        *
*    DK - 2830 VIRUM                                   *                        *
*                                                     *   TELEX NO             *
*                                                     *                        *
* TECHNICAL DIRECTOR :                                *   37108                *
*    JORGEN BO NIELSEN                                 *                        *
*                                                     *                        *
************************************************************************
                                              VERSION : 26/09/88
```

AIM OF THE PROJECT :
****
The objective of the present project is to develop a multiple purpose submarine
pipeline and riser data base management system containing a large variety of
pipeline and riser related data and a multitude of utility programs. The major
purpose of the system is to provide the pipeline operator with an all-inclusive
information base and a range of analysis tools for his current pipeline
integrety evaluations, and inspection/maintenance planning. Project innovative
elements include:
- integration of a large family of databases
- wide range of new analyses methods
- relational database methodology
- high degree of software portability
- posibility for offshore application on survey vessel.

PROJECT DESCRIPTION :
****
The IPMS has been divided into 12 modules to enhance flexibility, viz:
1. Network Modelling
   Description of physical configuration of the pipeline and riser system.
2. Pipeline Design
   Storage of all data generated during pipeline design.
3. Riser Design
   storage of riser design data.
4. Pipeline and Riser Construction
   Data from the construction phase.
5. Inspection Characteristics
   Storage of information related to the execution of the individuel inspection
contracts.
6. Pipeline Inspection
   Storage and analyses of data gathered during pipeline inspection.
7. Position Correction Tables
   Compensation of positioning Inaccuracies
8. Riser Inspection
   Storage and analyses of data gathered during riser inspection
9. Pipeline End Manifold
   Storage and maintenance of data related to a PLEM.
10.Seabed General
   Data related to seabed in the vicinity of the pipeline route.
11.Environment
   Storage of relevant hydrographic data.
 12. External Records
   References to external records.
the project development is divided into seven phases, viz;
a) System Analyses and Definition
b) System design
c) Programming
d) System Documentation
e) Implementation and Testing
f) Installation and Commissioning
g) Training

STATE OF ADVANCEMENT :
****
Completed

RESULTS :
****
The project has resulted in a highly flexible and portable database management
system suitable for installation in a large variety of hardware/software
environments.
The successful completion of the IPMS project has proven that it is viable to
develop an I,M&R information system based on the philosophies outlined in 3.2.
The usefulness of the system and the benefits from the integrated approach can
however of course only be established through using the system in the operating

329

environment for which it is intended. Several years of experience will be needed
before all aspects may be cleared up, and adjustments and improvements will most
certainly be needed on a number of points.
The next step will include the integration of "expert system" - modules or
"decision - support" modules. In the case of the IPMS it would be an obvious
future development to include a "decision support" module for explicitly
assisting the operator in inspection planning. This module would e.g.
- identify critical areas to be inspected according to user defined criteria
- identify other areas to be inspected according to a fixed inspection
programme
- propose inspection methodology (vessel, type of survey, sensors, etc,) on
basis of experience stored from earlier surveys
- set-up inspection schedule
- evaluate economical aspects.

REFERENCES :
****
"AN INTEGRATED PIPELINE MANAGEMENT SYSTEM" BY PETER I. HINSTRUP AND T.
SVENSSON
PRESENTED AT THE 1984 OFFSHORE COMPUTERS CONFERENCE IN ABERDEEN

```
*********************************************************************************
* TITLE : NEW TECHNOLOGY FOR INTERNAL NDT-         *      PROJECT NO        *
*         INSPECTION OF PIPELINES                  *                        *
*                                                  *    TH./09025/85/DK/..  *
*                                                  *                        *
*********************************************************************************
* CONTRACTOR :                                     *    TELEPHONE NO        *
*    RONTGEN TECHNISCHE DIENST BV                  *                        *
*                                                  *    3110-4150200 EXT 291*
*    P.O. BOX 10065                                *                        *
*    NL - 3004 AB ROTTERDAM                        *                        *
*                                                  *    TELEX NO            *
*                                                  *                        *
* TECHNICAL DIRECTOR :                             *    23366               *
*    MR. J.A. DE RAAD                              *                        *
*                                                  *                        *
*********************************************************************************
```

VERSION : 01/08/89

AIM OF THE PROJECT :
****
The present aim of the project is to use glass fibres to communicate between
equipment at an open pipe end with an ispection vehicle carrying an array of
ultrasonic probes. The intention is to measure presence of corrosion or cracks
in liquid filled pipelines to distances of 5000 metres. Communication of control
signals, measurement signals and power will be done by the cable in which the
glass fibre technology is incorporated.

PROJECT DESCRIPTION :
****
A prototype pipeline crawler will be designed for diameters of 16" and larger.
The articulated tool can pass 3 D bends and climb 20 deg. uphill. Its maximum
range will be 5000 metres. This requires the use or design of components to
build a compact but ruggedized crawler. This includes the design of an efficient
power circuitry which requires smallest possible space.
To cover long distances of 5000 metres glass fibre technology has to be used
including optical slip rings. An essential part of the development is to
incorporate optical/electrical transformers in both directions as well as all
related electronic circuitry. The project will be completed with a field test.
At first an attempt was made by the Danish project leader to apply glass fibres
in on-stream gas and oil pipelines. A preliminary study showed that this was not
feasible. The project then was directed towards inspection of open pipelines and
taken over by NDT-Netherlands. This interrupted the progress of the project by
at least 2 years.
Since early 1988 RTD made it a full time project and serious progress has taken
place.
At present all key parts of the system have been tried, designed or found on the
market and final design and construction has been started by the middle of 1989.
Assumed completion before the end of 1990.

STATE OF ADVANCEMENT :
****
Ongoing

RESULTS :
****
A considerable amount of electronic design and experiments were done in the year
that the project is in hands of RTD. A summary of achievements is :
- Cable electronics
With purchased and RTD designed electronics it was shown that a conventional
copper cable of normally up to 500 metres can be replaced by a glass fibre of
15000 metres !
- Crawler
A basic concept exists to design a crawler for 16" which can negotiate 3 D bends.
 This includes space for on board electronics and power supply of the drive
motor.
- High frequency ultrasonics
To measure wall thickness reduction to 3 mm a new pulser-amplifier circuitry was
designed which allows operation by glass fible cables. These electronics will be
used to construct a 32 channel measuring instrument "Sonolog".
- High frequency ultrasonic probes
A vacuum centrifuge has been designed and built to make the special high
frequency probes in conjunction with item 5.3.
- Control box and mixer
All control signals vice versa are digital. The ultrasonic signal is analogue.
The mixer takes care that control box and Sonolog signal are synchronized.
- Preliminary test
This showed that all essential parts work well and further design can go on.
-cable
The specification for the cable has been completed including the optical rotary
contact (= slip ring) necessary because the cable will be on a large reel.

REFERENCES :
****

- "METHODS FOR INSIDE PIPELINE INSPECTION", LECTURE GIVEN BY J.A. DE RAAD AND B. ROGGEBAND ON 17.05.89 AT THE 2ND INTERNATIONAL CONFERENCE OF THE COMMITTEE ON THE STUDY OF PIPE CORROSION AND PROTECTION".
- "VARIOUS METHODS OF ULTRASONIC PIPELINE INSPECTION" BY J.A. DE RAAD. LECTURE PRESENTED AT PIPELINE PIGGING TECHNOLOGY CONFERENCE, FEBR. 21-23, 1989, HOUSTON, TEXAS.
- "INSIDE PIPELINE INSPECTION", BROCHURE OF RTD INDICATING POSSIBILITIES OF CABLE OPERATED TOOLS WHICH ALREADY MENTION LONG GLASS FIBRE CABLES.

M

```
***************************************************************************
* TITLE : TEST OF PROTOTYPE OF A SELF-FRAGMENTING      *    PROJECT NO      *
*         INSTRUMENTED VEHICLE DESTINED FOR THE        *                    *
*         CONTROL OF PIPELINES                         *  TH./09028/85/FR/..*
*                                                      *                    *
***************************************************************************
* CONTRACTOR :                                         *  TELEPHONE NO      *
*    SYMINEX                                           *                    *
*                                                      *  91 73 90 03       *
*    2, BOULEVARD DE L'OCEAN                           *                    *
*    FR - 13275 MARSEILLE CEDEX 09                     *                    *
*                                                      *  TELEX NO          *
*                                                      *                    *
* TECHNICAL DIRECTOR :                                 *  400 563           *
*    MR B. BISSO                                       *                    *
*                                                      *                    *
***************************************************************************
                                             VERSION : 07/11/89
```

AIM OF THE PROJECT :
****
The instrumented vehicle developed by SYMINEX enables the measurements of
corrosion in hydrocarbon pipelines.
The risk of PIG permanently jamming in the pipe is eliminated by a sytem of
automatic fragmentation.
The object of the project was the construction of the vehicle with integration
of the thickness measurement system and the realization of tests on a pipeline
in operation.

PROJECT DESCRIPTION :
****
After studying the various features linked with fragmentation, the use of
syntactic foam for the body of the instrument seemed to be the best solution.
The vehicle length is 0.9 m, its weight with the electronic components is 25 Kg.
It is propelled along the pipe by the fluid flow by means of two urethane cups
at speeds of up to 0.5 m/s.
Fragmentation of the vehicle takes place with the explosion of a pyrotechnical
transmission cord embedded in the body of the vehicle.
Another solution using expansive cement is presently being tested.
Corrosion phenomena are monitored from the measurements of the quantity of metal
in a section of pipe. The operational principle is based on the deformation
undergone by an alternating magnetic field traversing a conducting surface.
Microprocessor circuits integrated into the equipment enable the measurement of
the phase shift at each period with storage of the results in RAM CMOS.
The sensor/acquisition unit assembly requires little energy and allows complete
autonomy via a distance of 20 Km.

STATE OF ADVANCEMENT :
****
An industrial prototype was built with integration of electronics.
Further improvement and further tests on site have been carried out on a 12"
diameter 10 km long (France) and on a 10" calibration test loop at Shell
Laboratorium (Hollande).

RESULTS :
****
Calibration tests were performed on a pipe test whose defects were know, a phase
shift of 5 degrees represented an attack of 3% of pipe wall for general surface
pitting.
Tests on site were performed to compare the machine with an industrial
environment.
The measurements at different speeds (20 cm/s to 40 cm/s) corresponded well.
There was a phase shift in some tubes of 30 degrees corresponding to a 20%
reduction in wall thickness.
These measurements were compared to those obtained from tests using an
ultrasonic thickness transducer.
Wall thickness was controlled at 500 points per tube, areas of corrosion
detected by the PIG corresponded to decrease in wall thickness.
The reliability of the system was established. Further improvements on
electronics have been performed offering metal loss treshold detection of 2 % in
volume with a reference log.
The dynamic measurements have been evaluated from 15 cm/s to 50 cm/s, speed
increase has been afforded by a doubled sampling rate (twice the frequency of
the alternative emitted field).

```
***************************************************************************
* TITLE : INTELLIGENT PIG                            *    PROJECT NO      *
*                                                    *                    *
*                                                    *  TH./09029/86/IT/.. *
*                                                    *                    *
***************************************************************************
* CONTRACTOR :                                       *   TELEPHONE NO     *
*    SNAMPROGETTI                                     *                    *
*                                                    *   02 520 56 90     *
*    P.O. BOX 12059                                   *                    *
*    I - 20120 MILANO                                 *                    *
*                                                    *   TELEX NO         *
*                                                    *                    *
* TECHNICAL DIRECTOR :                               *   310246           *
*    G. MANTOVANI                                     *                    *
*                                                    *                    *
***************************************************************************
                                              VERSION : 13/02/90
```

AIM OF THE PROJECT :
****
To design, construct and test an internal inspection system ("intelligent pig")
capable of carrying out and advanced monitoring of internal and external
conditions of an offshore pipeline. The main requisites of the system are the
following :
- to be used for onshore and offshore pipelines
- to record pipe data and position in a uniform and reliable way
- to be able to adapt to significant range of diameter and to pipe local
deformations.
- to be able to record and process local images in a specified position along
the pipe.
- to be able to detect, at any point of the route, if the pipe is buried or if
there is a free span. In general to be able to define the pipe/sea-bottom
configuration. To evaluate the condition of the cathodic protection system.
- to be able to carry out the inspection without obstructing fluid
transportation through the pipeline.

PROJECT DESCRIPTION :
****
The project includes the following phases :
- Basic project :
    . conceptual design of the system
    . Definition and chacterization of all the inspection functions to be carried
out
    Basic design of the pig positioning system.
- Development in critical areas : In the phase every critical aspect which
appears in the first phase will be closely examined. At this end of this phase
there is an important project check point.
The conclusion could be that it is not feasible to proceed to the other phases
because the unresolved difficulties would lead to a pig similar to existing
commercial ones.
A second conclusion, more positive than the previous one,would follow if the
main problems in critical areas are solved : in this case the output of the
phase would be the final basic cofiguration of the system.
- Detailed design : detailed design, construction drawings and diagrams will be
prepared. A detailed plan of laboratory and field tests will also be prepared.

STATE OF ADVANCEMENT :
****
The system architecture has been defined from a mechanical point of view.
Options for the hardware and the software have been chosen. The configuration of
the power generation the pig position detection system and the functions of the
outer data collection system have been defined.

RESULTS :
****
The feasibility of a pig based on the principle of neutron radiation followed by
gamma spectrographic analysis has been examined and the mechanical solution was
to be abandoned. As some prototype tests of a device wholly similar to this
nuclear pig were carried out by another company, the whole project was abandoned.

```
****************************************************************************
*  TITLE : INTERNAL PIPELINE ALIGNMENT CLAMP          *     PROJECT NO      *
*                                                     *                     *
*                                                     *  TH./09030/86/UK/..  *
*                                                     *                     *
****************************************************************************
*  CONTRACTOR :                                       *   TELEPHONE NO       *
*     HOULDER OFFSHORE LIMITED                        *                     *
*                                                     *   01 357 6001        *
*     LAFONE STREET 59                                *                     *
*     UK - LONDON SE1 2LX                             *                     *
*                                                     *   TELEX NO           *
*                                                     *                     *
*  TECHNICAL DIRECTOR :                               *   884 801            *
*     MR E.L.MEADOWS                                  *                     *
*                                                     *                     *
****************************************************************************
```

VERSION : 09/10/89

AIM OF THE PROJECT :
****

Pipeline ovality and misalignment are problems which affect the success of any
pipeline connection, whether onshore or subsea in a hyperbaric tie-in. To date,
these problems have been resolved using manually fitted external clamps which
presuppose that the pipe is readily accessible.
We believe that this problem can be overcome by an intelligent pipeline pig
equipped to undertake the task internally. The alignment pig would be inserted
in the pipe prior to the weld such that hydraulically powered radial rams can be
extended to span the weld and correct any misalignment or distorsion in the pipe,
ensuring that welds of consistent high quality are achieved. Its independent
power source allows the pig to operate inside the pipe whilst controlled by
external command signals. In addition, gamma rays of the weld can be undertaken
using an atomic isotope, ensuring improved quality of radiography as the rays
only pass through a single wall thickness of pipe.

PROJECT DESCRIPTION :
****

The proposed design of the pipe alignment pig is as follows :
The unit is made up of a central frame which will house the hydraulics,
electrics and microprocessor in a 60 bar box together with the gamma head for
gamma raying the weld. The unit is equipped with 16 radial rams, these will be
extended to span the weld and further expanded to correct the pipe distorsion
and hold the pipe in position whilst the weld is made.
For transit the pig is fitted with polyurethane wheels to ensure that there is
no metal to metal contact and the unit is designed such that it fails safe into
this position. In addition, driving discs will be fitted to facilitate the
removal of the unit on completion of the operation. The pig will be designed to
negotiate 3D bends whilst withstanding acceleration and deceleration forces of
up to 50 "G".
All internal units will be suitably protected to ensure that no damage is
sustained when recovering the pig into the pig trap.
It is intended to install a gamma ray source of approximately 20 Curies which
will enable a single panoramic scan of the weld to be taken.
This method will allow an improved gamma ray image as the rays only have to pass
through a single wall thickness.
It is intended that the alignment pig should be a totally self-contained unit
requiring no external power source. In addition it is proposed that the
hydraulic requirements will be met using a small positive displacement pump with
a pressure of approximately 700 bar.
The pig will be equipped with an "intelligence" system which will allow all
stages of the operation to be monitored. It is currently proposed that a 32
function control system should be fitted.
The project has been divided into 4 sections to ensure that each stage in the
development progresses with the minimum technical and financial risk.
STAGE 1
Research and development engineering leading to the construction drawings for a
16 inch diameter prototype alignment pig will be undertaken. In conjunction with
this work, a market research programme will be conducted to establish
requirements, trends and possible future developments.
STAGE II
A prototype 16 inch diameter alignment pig will be constructed and tested in a
service test pipeline 200 m in length with a variety of simulated pipeline bends
and conditions. Pressure testing of the unit will aslo be undertaken in the
hyperbaric chamber in Aberdeen. During this stage a marketing video and brochure
will be produced and used for a sales drive to seek customer commitment for the
use of an alignment pig for hyperbaric welds.
STAGE III
Once a client's commitment for the use of an alignment pig has been secured, the
manufacturing drawings for the required diameter of the pig to suit the clients
pipeline will be produced and fully tested as described above. The first
production unit will then be constructed and tested prior to setting to work in
the field with the client.
STAGE IV
Construction drawings for a range of pigs.

STATE OF ADVANCEMENT :
****
Stage I compteted. Stage II completed excepted for hyperbaric tests. Stages III
& IV have been suspended.

RESULTS :
****
Stage I of the Project has been totally completed. Stage II has been completed
except for tests in hyperbaric chamber and the design of the gamma source but
including manufacture of a prototype and functional testing in a test loop in
Lingen West Germany thus providing that the IPAP could successfully negotiate 5D
bends and the ability to remotely control the unit utilizing a low frequency
radio signal through the pipe wall. The design of the gamma Ray Unit has been
completed and approved by the UK Atomic Energy Authority. Tests in a hyperbaric
chamber and Stage III were not started as this necessitates the support on a
frequent user of the service. The project, following completion of Stage II, was
stopped as up to this point in time no oil company could be found that was
willing to operate this piece of equipment in their offshore pipelines due to a
changed market situation.
All high development goals have been met. Operations of this piece of equipment
will be possible following extensive hyperbaric chamber tests as soon as the
market situation changes. However, operation will be limited to pipelines 16"
and larger. Operating the equipment in a waterdepth of up to 650m could, if
necessary, be extended to a depth of 1,200 m taking the present technical
viewpoint into account.

REFERENCES :
****
PAFEC LTD - REPORT NR B/992.

```
****************************************************************************
* TITLE : DEVELOPMENT OF A DESIGN SYSTEM FOR      *      PROJECT NO       *
*          SUBMARINE PIPELINE BURIAL              *                       *
*                                                 *   TH./09034/86/DK/..  *
*                                                 *                       *
****************************************************************************
* CONTRACTOR :                                    *      TELEPHONE NO     *
*    DANISH HYDRAULIC INSTITUTE                   *                       *
*                                                 *     45 2 86.8033      *
*    AGERN ALLE 5                                 *                       *
*    DK - 2970 HOERSHOLM                          *                       *
*                                                 *      TELEX NO         *
*                                                 *                       *
* TECHNICAL DIRECTOR :                            *      37402            *
*    MR MADS B. BRYNDUM                           *                       *
*                                                 *                       *
****************************************************************************
                                              VERSION : 31/12/88
```

AIM OF THE PROJECT :
****
A design system for submarine pipeline burial will be developed to guide
submarine pipeline engineers when they are to decide on the use of pipeline
selfburial or natural backfilling of a pipeline trench as cost effective
alternatives to mechanical trenching or engineering backfilling.
Also sand wave migration and shifting of offshore sand bars along the pipeline
route are more and more considered as a problem area and the design system will
comprise methods and recommendations to optimize pipeline routing and pipeline
design.
Innovative selfburial methods and high quality mathematical models for natural
backfilling and sandware behaviour are used engineering application for the
practical design system.

PROJECT DESCRIPTION :
****
PHASE 1.
1. Review of overall requirements and design procedures.
2. Data collection and review of existing pipeline behaviour.
3. Establishment of State of the art.
4. Development of Engineering models.
   4.1 Sea bed stability (sand waves).
   4.2 Natural Backfilling.
   4.3 Pipeline Self Burial.
5. Preparation of User Manual
6. Detailed identification of work for phase 2.
PHASE 2.
1. Field monitoring programmes.
   1.1 Sea Bed Stability (sand waves)
   1.2 Natural Backfilling and Pipeline Self Burial.
2. Physical Model Tests.
   2.1 Sea Bed Stability.
   2.2 Natural Backfilling.
   2.3 Self Burial.
3. Completion of mathematical and engineering models.
5. Preparation of User Manual.
The User Manual should be looked upon as an extended users manual for the
Engineering Models.
As such the manual will guide and explain to the operator how to utilise the
models.
All theoretical documentation will be contained in this document.

STATE OF ADVANCEMENT :
****
Phase 1 of the project is close to completion. The engineering models have been
completed. The scientific documentation and final reports are almost completed.
It needs only proof reading and addition of one small section. A detailed
proposal for a Phase 2 has been sent to the Industry Sponsors for consideration.

RESULTS :
****
Preliminary reports have been prepared for :
- Stage-of-the Art review
- Engineering Models
- Data Collection
Engineering models have been developed
- Sand waves
- Selfburial and scour
- Manual to operate the models.

```
**************************************************************************
* TITLE : THOR 2: SECOND GENERATION AUTOMATIC    *    PROJECT NO       *
*         HYPERBARIC WELDING SYSTEM              *                      *
*                                               *  TH./09037/87/FR/..   *
*                                               *                      *
**************************************************************************
* CONTRACTOR :                                  *    TELEPHONE NO      *
*    COMEX S.A.                                 *                      *
*                                               *  91.23.50.00         *
*    BD DES OCEANS 36                           *                      *
*    FR - 13275 MARSEILLE CEDEX 9               *                      *
*                                               *  TELEX NO            *
*                                               *                      *
* TECHNICAL DIRECTOR :                          *  410985 FAX: (33)91 4*
*    MR. R. ROUGIER                             *  0 12 80             *
*                                               *                      *
**************************************************************************
                                          VERSION : 01/02/90
```

AIM OF THE PROJECT :
The aim of the Project is to develop, build and test a fully integrated pipe preparation and welding system, THOR 2, which can be used for both advanced diver-assisted and future diverless hyperbaric welding tie-ins.
THOR 2 will be the first automatic welding system which can be used in a diverless mode and as such it will have a major impact on both costs and personnel safety for deepwater tie-ins and repairs.

PROJECT DESCRIPTION :
The "THOR-2" system will consist of :
* A system of surface controlled pipe clamp and integrated orbital tool carrier which can be positionned on the pipe ends and can serve both to remove pipe end ovality and to carry different tools for pipe cutting/bevelling, or welding.
* A series of tools which can be placed on orbital carrier either by a manipulating arm or by the diver, and which can perform pipe cutting/bevelling, joint metrology and welding (weld heads) under remote control.
* An advanced welding system consisting of two weld heads suitable for operation without diver-assistance, two multi-process power sources (possibility TIG and MIG), and advanced weld control system which allows welding with two weld heads simultaneously (faster welding), an automatic demagnetisation system, an automatic induction pre-heating system.
* An automatic system based on a laser diod and camera which would be used for determining welding parameter as a function of the weld groove metrology, modifying weld parameters in real time as a function of weld bead shape/size and penetration, system status data logging and malfunction diagnostics.
* All equipment integrated into a containerised surface control, and a subsea module with associated umbilicals.
The "THOR-2" programme consists of :
PHASE 1
- Design and building the pipe round-up and tool support clamp, pipe machining tools, metrology systems and associated software.
- Development of the welding equipment and weld control system.
- Development of the artificial intelligence capacity of "THOR-2".
PHASE 2
- An onshore and shallow water test of the above equipment, integration into a welding spread (habitat, module, etc...)
- The full offshore test of the "THOR-2" system which will demonstrate the viability of diverless hyperbaric welding and allow the different advantages of the "THOR-2" system, to be tested in typical operational conditions.

STATE OF ADVANCEMENT :
PHASE 1
Largely completed
PHASE 2
Onshore tests underway.

RESULTS :
WORK PACKAGE NR 1 : REMOTELY CONTROLLED PIPE PREPARATION
* Development of the hardware and software for the reduction of pipe ovality : Hardware and software designed, built and tests in COMEX workshops completed.
* Development of a remotely controlled pipe cutting and bevelling tool which can be installed by a remotely controlled manipulating arm : Hardware and software built and tests in COMEX workshops completed.
WORK PACKAGE NR 2 : DEVELOPMENT OF "THOR-2" WELDING EQUIPMENT
* Development of multi-process weld heads which can be installed and serviced by either a diver or a remotely controlled manipulating arm : - Two weld heads built and undergoing workshops tests. Mechanism for remotely controlled torch replacement still under design.
* Development of multi-process welding power source which can be used efficiently down to depths of at least 500 m.s.w. : - Two power source units built, delivered and integrated into the subsea module. Workshop tests underway.

WORK PACKAGE NR 3 : DEVELOPMENT OF "THOR-2" WELD SYSTEM OF DIAGNOSTICS
INTELLIGENCE SYSTEM AND FUNCTION TESTS
* Development of an adapted weld control system (software and hardware) for
simultaneous control of two weld heads and induction preheating : - System built
and factory acceptance tests complete. Integration into the module underway.
* Development of a system status data logging and equipment fault finding
(diagnostics) facility based on the "marriage" of the weld control system amd
an Expert system (artificial intelligence) : - Work concentrating on diagnostics
function. Tested on "THOR-1", will be transferred on "THOR-2" during test phases.
* Function tests of the complete "THOR-2" weld system (weld control, diagnostics,
 clamps/orbital carrier and weld heads, welding power sources). Determine the
procedures for two weld heads operations will start upon completion of subsea
module integration.
WORK PACKAGE NR 4 : AUTOMATIC REAL TIME GENERATION OF WELDING PARAMETERS
* Development of existing laser sensors for use in hyberbaric conditions : -
Marinised prototype built for 300 m.s.w. and commissioning tests completed.
* Consolidation using algorithms of a welding parameters databank as a function
of joint metrology gained through welding tests, interfaced with the weld
control system software under completion.
* Major advances in hyperbaric welding to both inter-run inspection of weld
passes and possible real time interactive control under final development of
software. Tests will start during onshore integration.
WORK PACKAGE NR 5 : ONSHORE AND OFFSHORE TESTS OF COMPLETE SYSTEM
* Umbilicals linking the subsea module to the surface control cabin and the
habitat, are under manufacturing.
* Integration of all the surface control equipment into a container is underway.
* Onshore test of the complete system, work-up of operational procedures will
start upon completion of subsea module and control cabin.

REFERENCES :
****

'OPERATIONAL AUTOMATIC HYPERBARIC WELDING', D.O.T. CONFERENCE (MONACO, OCTOBE
1987) - S.U.T. CONFERENCE (ABERDEEN, NOVEMBER 1987).
"THE "THOR-2" DIVERLESS HYBERBARIC WELDING SYSTEM", T.W.I. CONFERENCE (LONDON,
NOVEMBER 1987).
"WELDING FOR THE OFFSHORE AND NUCLEAR INDUSTRY", I.I.W. CONFERENCE (HELSINKI,
SEPTEMBER 1989).

```
*******************************************************************
* TITLE : METHODOLOGIES FOR ADAPTING SUBMARINE   *    PROJECT NO    *
*          PIPELINES TO VERY UNEVEN SEA BOTTOM   *                  *
*                                                *  TH./09040/87/IT/.. *
*                                                *                  *
*******************************************************************
* CONTRACTOR :                                   *  TELEPHONE NO    *
*    SNAMPROGETTI SPA                            *                  *
*                                                *  0721-8811       *
*    VIA TONIOLO 1 - FANO (PS) - ITALY           *                  *
*                                                *                  *
*                                                *  TELEX NO        *
*                                                *                  *
* TECHNICAL DIRECTOR :                           *  560279          *
*    MR. ROBERTO BRUSCHI & MR L. VITALI          *                  *
*                                                *                  *
*******************************************************************
                                        VERSION : 06/03/90
```

AIM OF THE PROJECT :
****
To overcome the difficulties deriving from crossing uneven sea beds, by means of
new technologies based on the following principle : to adapt the pipeline to the
shape of the sea bottom, rather than modify the sea bed to adapt it to the
pipelines as proposed by current technologies. This aim will be achieved with a
theoretical/experimental research which will utilize the results of a preceding
research on hydroelastic behaviour of pipeline spans under vortex-shedding
action (EEC Contract TH 1038/83).

PROJECT DESCRIPTION :
****
Phase 1000 : Critical review of current technologies
            1010 - State of the art
            1020 - Critical review of current design criteria
            1030 - Critical review of current criteria on elastoplastic coll
                   apse
            1040 - Critical review of current criteria for acceptance of fre
                   e spans
            1050 - Critical review of current methods for the evaluation of
                   stresses on pipelines during operation
            1060 - Analysis ofr geographical areas with uneven sea beds
            1070 - Development of the proposed new concepts
    Phase 2000 : Feasibility analysis of proposed solutions
            2010 - First method : Plastic behaviour of pipe
            2020 - Second method : Flexible spools inserted in pipe
            2030 - Third method : Suspended spans
            2040 - Fourth method : Crossing with submarine bridges
            2050 - Other methods, based on new materials applications
            2060 - Economic review of the above methods and selection
    Phase 3000 : Basic design of experimental tests
            3010 - Basic design of laboratory (basin) tests
            3020 - Basic design of field tests
            3030 - Basic design of tests on mechanical devices
            3040 - Economic analysis of the individual projects
            3050 - Selection of subcontractor for execution of tests
    Phase 4000 : Laboratory tests
            4010 - Design and fabrication of models
            4020 - Design and arrangement of the instrumentation
            4030 - Setting up of basin
            4040 - First set of laboratory tests
            4050 - Definition of parameters to be studied in more details
            4060 - Second set of laboratory tests
            4070 - Analysis of results and theoretical formulation of all
phenomena related with the problem studied all phenomena related with the
problems studied

            Phase 5000 : Resume and issue of final report
            5010 - Summary and connection of results
            5020 - Issue of final report.

STATE OF ADVANCEMENT :
****
Ongoing

RESULTS :
****
Phase 1000 has been concluded
A detailed scope of work is being prepared regarding the following main lines :
- advanced strength strain - based creteria
- very long free span - multi span behaviour
- alternative pipeline configurations
- monitoring systems
Phase 2000 has begun
Main lines 1 : advanced strength strain-based criteria

A numerical model has been developed to simulate the elaso-plastic behaviour of pipe..
Main lines 2 : very long free span – multi span behaviour.
A numerical model has been developed to simulate the multi span behaviour.
The preliminary results have pointed out the advantages of an advanced strain-based criteria, where alternative pipeline configurations and monitoring systems are utilized.

```
********************************************************************
* TITLE : ONE-SHOT PIPELINE WELDING            *    PROJECT NO     *
*                                              *                   *
*                                              * TH./09044/87/UK/..*
*                                              *                   *
********************************************************************
* CONTRACTOR :                                 *   TELEPHONE NO    *
*    EXPLOWELD LIMITED                         *                   *
*                                              *    224 592476     *
*    DEVANHA HOUSE, RIVERSIDE DRIVE            *                   *
*    ABERDEEN                                  *                   *
*    UK-AB1 2SL, SCOTLAND                      *    TELEX NO       *
*                                              *                   *
* TECHNICAL DIRECTOR :                         *                   *
*    MICHAEL C.V. HEY                          *                   *
*                                              *                   *
********************************************************************
                                             VERSION : 07/02/90
```

AIM OF THE PROJECT :
****
To develop and commercialise the Exploweld/Hotforge explosive one shot welding
technique for pipelines by perfecting the welding technology to the point of
good repeatability, qualifying the welding procedures to Oil Industry and
regulatory body standards, developing operational equipment and preparing
operating procedures for application of the welding procedure to variously; 'J'
laying of flowlines and tie-backs for subsea production systems with particular
attention to special alloys pipe, overland remote location pipe construction,
subsea installation, tie-ins and repair of pipe.
Innovation in the use of explosive welding technology to the specific demands of
hydrocarbons transportation pipelines construction subsea, offshore, overland.
Innovation will be realised in the commercial development of new one shot
pipeline welding procedures, high energy absorbent explosion proof clamps and
sophisticated explosion sound proofing equipment.

PROJECT DESCRIPTION :
****
The project essence is contained in three major phases :
- Qualification to industry standards of the explosive welding procedure
- Development of field equipment
- Establishment of operating procedures.
Peripheral to this is :
- Essential welding technique development
- Marketing combined with field trials and demonstrations
and controlling and progressing the project will be :
- Project Management Team.
Thus the major phases of the project may be defined :
QUALIFICATION : The exercise to qualify to Oil Industry, Det Norske Veritas and
Lloyds standards the explosive one shot welding procedure for hydrocarbon
pipelines in three size ranges, 2"-8", 8"-20", 20"-40" for pipe material
including Duplex (SEC 2205) X52, X52 Q & T, X70, X65, X60 for application subsea,
 offshore and overland.
DEVELOPMENT OF FIELD EQUIPMENT : Design, development and manufacture of high
energy absorbing clamps, sound proofing equipment, inspection tools, offshore
'J' lay frames and onshore handling frames.
OPERATING PROCEDURES : Drafting of procedures for the application of the
explosive one shot pipe welding procedure to pipelines overland, offshore 'J'
lay, subsea tie-in and repair.

STATE OF ADVANCEMENT :
****
A series of 6" and 1" duplex flowline and hydraulic line welds for high pressure
transmission have been completed and witnessed by Det Norske Veritas for pre-
qualification. A failure rate of about 3% disallowed full qualification but D.N.
V. do not view this as a serious problem and strongly favour a re-run of the
exercise.

RESULTS :
****
Although ongoing the development work is approaching completion. A satisfactory
clamping system with a good life is designed and tested. A vacuum box has been
shown to be an effective sound reducer. A set of procedures for 'J' laying from
the back of a supply boat or d.s.v. has been prepared. A successful
demonstration for major oil company engineers is completed. A first order for 40
km overland pipe is expected for 1990.

REFERENCES :
****
A PRESENTATION WAS MADE TO AN EXCLUSIVE SEMINAT OF SENIOR CORROSION ENGINEERS
FROM THE MAJOR OIL OPERATORS AND ENGINEERING COMPANIES ORGANISED BY THE NICKEL
DEVELOPMENT INSTITUTE>.

```
***************************************************************************
* TITLE : WET WELDING FOR STRUCTURAL REPORT          *    PROJECT NO      *
*                                                    *                     *
*                                                    *  TH./09045/88/DE/FR *
*                                                    *                     *
***************************************************************************
* CONTRACTOR :                                       *    TELEPHONE NO     *
*    GKSS FORSCHUNGS                                  *                     *
*                                                    *   41 52 87 1921     *
*    ZENTRUM                                          *                     *
*    GEESTHACHT GMBH                                  *                     *
*    MAX PLANCKSTRASSE                                *    TELEX NO         *
*                                                    *                     *
* TECHNICAL DIRECTOR :                               *    218712           *
*    PROF. SCHAFSTALL                                *                     *
*                                                    *                     *
***************************************************************************
                                                   VERSION : 01/10/89
```

AIM OF THE PROJECT :
****
The aim of this project is to develop the necessary technology (equipment, consumables and techniques) for successfully using Wet Welding as a repair technique for both mild steel and high yield structures down to depths of at least 100-125 msw.
Present technology limits the use of Wet Welding for structural applications on high yield steel to depths of 5-10 msw and mild steel to depths of 20-25 msw. Although Wet Welding has been performed down to a depth of 170 msw, the quality obtained is currently not acceptable for structural repair work.

PROJECT DESCRIPTION :
****
The programme originally consisted of 2 phases :
Phase 1
Development of consumables, equipment and procedures in unmanned and limited manned "onshore" tests. Determination of the maximum depth to which good quality welds can be made using the consumables etc... which will be developed during the programme.
Phase 2
Verify and confirm the developed technology during a full size simulation of major offshore structural repairs in operational conditions. Demonstrate the validity of Wet Welding for repairing both mild steel and high yield steel structures down to the maximum depth identified during Phase 1.
The EEC has confirmed approval for co-financing Phase 1.
The detailed programme for Phase 1 consists of :
- bead on plate tests at ever increasing depths using an unmanned simulator at GKSS. Tests will be performed using equipment and consumables developed by both COMEX and GKSS. The results of the tests will be used to improve both equipment and consumables;
- single pass and multipass groove and fillet weld tests in the flat position using the unmanned simulator. The results will be used to optimise the chemical analysis of the electrodes and characteristics of the equipment.
- manned dive at a simulated dpth corresponding to the maximum depth at which successfull results were obtained during the unmanned trials. This dive will be used to produce test welds in non-flat positions (note that the unmanned simulator can only be used in the flat position).
At the end of this Phase, COMEX/GKSS will have developed and optimised consumables and equipment for use at the maximum depth and would have proven the valisdity of this technique on small size coupons in simulated conditions.
The following Phase 2 would allow the above results to be fully tested on full size structural compoments in operational conditions.

STATE OF ADVANCEMENT :
****
Ongoing.

RESULTS :
****
A first series of unmanned tests were performed using
  - the unmanned simulator
  GKSS pressure chamber complex
  GKSS and COMEX developed consumables and equipments
A second series of tests using COMEX and GKSS welder divers was performed at simulated depths of 55 msw and 100 msw in November 1988.
Initial results obtained both on the unmanned and manned tests are highly satisfactorily. They have confirmed that good quality welds can be made in the flat position on mild steel at 100 msw using the MCOMEX/GKSS developed technology. More work is required for out-of-flat welding high yield steels.

```
*************************************************************************
* TITLE : LEAK DETECTION OF OIL PIPELINES BY        *    PROJECT NO    *
*         MEANS OF AN ACOUSTIC METHOD.              *                  *
*                                                   *  TH./09047/88/IT/.. *
*                                                   *                  *
*************************************************************************
* CONTRACTOR :                                      *    TELEPHONE NO  *
*    POLYCONSULT SERVIZI SRL                         *                  *
*                                                   *    0721804314    *
*    VIA DEL TEATRO 8-I 61032 FANO                  *                  *
*                                                   *                  *
*                                                   *    TELEX NO      *
*                                                   *                  *
* TECHNICAL DIRECTOR :                              *    560315        *
*    ING.FRANCESCO FERRINI                          *                  *
*                                                   *                  *
*************************************************************************
                                                    VERSION : 19/06/89
```

AIM OF THE PROJECT :
****
The purpose of the project is to develop a measuring process and a measuring
instrument for the continuous leak detection of pipelines.
At the present state of art many accurate and sophisticated instruments are
available for measuring flow-rates, densities, pressures and temperatures.
However no device exists for the measure and the continuous mnonitoring of the
pressure wave celerity in the pipeline.

PROJECT DESCRIPTION :
****
The idea of the present work is to study, to test and to develop an
instrumentation suitable to be installed on pipelines and which shall be capable
of detecting on continuous having a length of some kilometers.
This hardware could be installed at several locations on the pipeline (for
instance where pressure and temperature ;easures are taken) and its outputs
could be transmitted by SCADA to the supervisory computer.
In this way a leak detection program would enable to compute the inventory
changes with much more accuracy than made so far.
Actually, if we consider a theoretical steady condition the continuity equation
is : Q1 - Q2 = q2 and the problem of detecting a leak is confined to the one of
measuring with suitable accurancy the flow-rates. Unfortunately transients occur
quite often in most pipelines and even a non-severe transient may result in a
sensible change of inventory. Therefore the capability of computing these
changes becomes an important feature for a leak detection system in order to
fulfill the following objectives :
a) maintaining an acceptable level of accuracy in presence of transients
b) adjusting the alarm threshold according to the amount of the inventory change
being detected, i.e. to the consequent lack of accuracy.
The above is valid for both liquid and gqs pipelines.] This work is however
addressed to the development of a technique intended for liquid pipelines. In
this case it is useful to refer to the liquid volumes at reference conditions
for accounting the flow-rates and the inventory.
The reference conditions usually are 1 Atmosphere pressure and 15 deg. C.
temperature.
The causes for change of the pipeline inventory are substantially the two
following :
1) change of the pressure profile, which results in an inventory change
depending on the combined elasticities of the liquid and of the pipe;
2) change of the temperature profile, which results in an inventory change
depending on the combined thermal expansions of the liquid and of the pipe.
The determination of the inventory changes due to pressure transients is usually
achieved using a certain number of pressure measurements taken along the
pipeline and running a computer model to build up the complete pressure profile
basing on those measures.
Various computer models are available, but all ot them operate integrating the
waves differential equations for liquid flow in a pipe.

STATE OF ADVANCEMENT :
****
Phase 1 which deals specifically to the feasibility study of the method, was
been completed.

RESULTS :
****
4~e studies carried out in the phase 1 confirm the theoretical feasibility of
the method. It is expected that the detection of sound is easily achievable by
means of commercially available transducers. Phase1 suggest to include some
tests in al flowing liquid : definite conclusions could not be drawn from simple
static tests.

```
*****************************************************************************
* TITLE : PIPELINE TRANSPORT OF NATURAL GAS IN      *     PROJECT NO       *
*         THE PRESENCE OF CONDENSATE,RESERVOIR      *                      *
*         WATER AND SUSPENDED HYDRATES              *  TH./09048/88/FR/..  *
*                                                   *                      *
*****************************************************************************
* CONTRACTOR :                                      *   TELEPHONE NO       *
*    GERTH                                          *                      *
*                                                   *   1.47.52.61.39      *
*    AVENUE DE BOIS PREAU 4                         *                      *
*    F-92502 RUEIL MALMAISON                        *                      *
*                                                   *   TELEX NO           *
*                                                   *                      *
* TECHNICAL DIRECTOR :                              *                      *
*    DR EMMANUEL BEHAR                              *                      *
*                                                   *                      *
*****************************************************************************
```

VERSION : 30/06/89

AIM OF THE PROJECT :
****
The purpose of this project is to test chemical additives selected in a new
class of inhibitors, in order to present the clogging of natural gas lines in
the presence of hydrates, by limiting the growth and/or the coalescence of
hydrate crystals.
The benefit of these new additives is their ability to be used in very small
amounts; indeed, their specific property is that they insert themselves at the
water-hydrocarbons interface whereas the drawback of presently used solvents,
such as methanol, is that lowering the hydrates formation temperature is
directly linked with their concentration in the aqueous phase, which requires
injection of solvent amounts comparable to the amount of water produced together
with the gas.
The project consists in studying the formation conditions of the hydrates and
the effect of this formation on the evolution of flow parameters.
These studies will be conducted in test loops with dry and condensate synthetic
natural gas, under typical conditions.

PROJECT DESCRIPTION :
****
This project comprises two phases :
Phase 1 : laboratory testing
Phase 2 : Testing with a pilot scale loop and validation with a semi-industrial
scale loop.
PHASE 1 will consist mainly in a "study of hydrate formation kinetics" in a
first step, followed by the "selection of new classes of inhibitors".
The hydrate formation kinetics will be studied in a laboratory installation.
This study will enable to analyse the various operating parameters :
supercooling degree, water-oil ratio (WOR), composition of the gaseous
hydrocarbon mixtures, gas-liquid ratio (GLR) and pressure influence.
Further, various inhibitors preventing the growth and/or coalescence of hydrates
will be tested in a laboratory installation in order to select those able to
prevent clogging of the flowlines and of the pipelines, even under conditions a
priori favourable to the formation of crystals.
PHASE 2 will consist in "tests with a pilot scale loop" and further "validating
these tests with a semi-industrial scale loop".
The first tests will help to screen on a limited size loop the additives and
materials previously selected and to adapt the semi-industrial installation to
conditions typical of gas production wells.
The validation tests will be conducted on this last semi-industrial loop which
is 40 meters long and with a minimum diameter of 50 mm.
The inhibitors screening tests will be performed on the pilot scale loop with
various conventional materials for internal coating, in order to determine the
possible adhesive effects on the one hand and the possible corrosive effects due
to the additives on the other hand. This limited-size installation will also
help to test the sensitivity of the equipment functionning to the operating
parameters, as well as to define the data necessary to be measured in view of
interpretation and modelling work (positioning of the measuring instruments).
Finally, this pilot scale loop will help to define the hardware and software
specifications for raw data acquisition and processing.
The validation of the above-described tests under semi-industrial conditions
will be performed in pressure and temperature ranges, with gas and liquid phases
velocities and water content close to those prevailing in actual gas field
flowlines and pipelines. These validation tests will be used to determine the
influence of the nature of the additives selected (control of crystal growth, of
crystal shape, of hydrate coalescence, blocking of the gas-water interface) and
to look for the optimal concentration of each inhibitor in order to determine
the cost effectiveness of the process.

STATE OF ADVANCEMENT :
****
The work performed since the beginning of this project (1.10.88) until the
present period (30.06.89) is related to Phase 1 "Laboratory testing".
Over this starting period only item 1.1 of the work programme could be
considered : "Study of hydrate formation kinetics".

RESULTS :
****

In order to determine the relative importance of various operating parameters on hydrates crystallization rate, the first results describe the influence of turbulence, pressure and sub-cooling.
Turbulence at the hydrocarbons (liquid or gaseous)-water interface is controlled by the rotation rate of a mechanical stirrer between 200 and 1000 rpm.
For (dry) water-gas systems, there is a strong influence of turbulence on the speed of hydrates formation when transition from a planar interface to foaming occurs. If condensate is present, emulsions are created at moderate stirring rates the results of which are poorly reproducible data.
Concerning pressure influence, it has been observed that for (dry) gas-water systems, it depends strongly on the sub-cooling level : at 2 kelvin it is almost negligible while the initial rate of hydrates formation increases with pressure at 4 kelvin.
For condensate gas-water mixtures, there seems to be an influence of pressure on hydrate formation kinetics when the sub-cooling is at 2 kelvin; however, the reproducibility of the measured data is very poor.The most important parameter for the rate of hydrates formation is sub-cooling. The obtained results indicate a strong increase of this rate above 3 kelvin. It has been observed visually that this is correlated to crystals coalescence which occurs under these conditions.

REFERENCES :
****

1. E.BEHAR, A.SUGIER,A.ROJEN
HYDRATES FORMATION AND INHIBITION IN MULTIPHASE FLOW. OPERATIONAL CONSEQUENCES OF HYDRATES FORMATION AND INHIBITION OFFSHORE (CRANFIELD,NOVEMBER 88).
2. E.BEHAR
NEW WAYS FOR PREVENTING PRODUCTION FACILITIES PLUGGING BY HYDRATES. MULTIPHASE FLOW TECHNOLOGY AND CONSEQUENCES FOR FIELD DEVELOPMENT (STAVANGER,MAY 89).
3. P.BOURMAYER,A.SUGIER, E.BEHAR
A NEW PROPOSAL FOR SOLVING HYDRATES PROBLEM IN MULTI-PHASE FLOW. 4TH INTERNATIONAL CONFERENCE ON MULTIPHASE FLOW (NICE,JUNE 89).

```
************************************************************************
* TITLE : EXPLOSIVE FUSION LINING OF TUBULARS FOR     *    PROJECT NO    *
*         OILFIELD DEVELOPMENT                         *                 *
*                                                      *  TH./09051/88/UK/.. *
*                                                      *                 *
************************************************************************
* CONTRACTOR :                                         *  TELEPHONE NO   *
*    HOTFORGE LIMITED                                  *                 *
*                                                      *  0224 592476    *
*    DEVANAH HOUSE                                     *                 *
*    RIVERSIDE DRIVE                                   *                 *
*    UK-ABERDEEN                                       *  TELEX NO       *
*                                                      *                 *
* TECHNICAL DIRECTOR :                                 *  738817         *
*    MICHAEL C.V. HEY                                  *                 *
*                                                      *                 *
************************************************************************
                                               VERSION : 01/10/89
```

AIM OF THE PROJECT :
****
To develop an explosive fusion bonding process to line full length of oilfield
tubulars or oilfield pipe with corrosion resistant alloys such as stainless
steels, nickel or chrome alloys.
To provide the global oil industry a lower cost alternative to solid alloys and
offer them greater choice in the type of alloy they can specify. This is because
the steel body of the tubular or pipe will provide the necessary strength and
thus a highly corrosion resistant alloy unsuitable for use as a solid due to
poor strength may now be considered seriously.
Technical innovation lies in the development of a special low gas explosive, the
development and application of a high energy absorbent clamp, full understanding
of explosively generated shock waves, the development of a vacuum chamber at the
central part of the manufacturing process and the special application of
explosive welding techniques.

PROJECT DESCRIPTION :
****
The project phases may be summarised as follows :
1) Development and Testing of Explosives
    Develop an explosive that is non-toxic, has a good shelf life,
    is safe and stable with a detonation velocity that can be
    varied between 2,400 and 3,200 m/s to order, has a low gas/mass
    ratio.
2) Testing and Development of Clamp Design
    An energy absorbing clamp is required to support the pipe or tubular during
firing.
    This part of the development exercise will include;
    - Design, testing and final specification of body
    - Strain gauge monitoring of pressure distribution, shock
       loading and strain rates
    - Empirically leavened computer simulation of process
    - Construction of a full size clamp for linig an lim long
       tubular of 7" diameter.
3) Design and Development of Vacuum Chamber
    A vacuum chamber will absorb the noise of an explosion and the gas pressures
generated if the ratio of vacuum level to evacuated volum is correctly set for
given explosive quantities and types.
    However there are dangers posed to executing process in a vacuum chamber.
    The activities within this section therefore include :
    - Conceptual design of vacuum chamber production facility.
    - Research into local planning and environmental barriers.
    - Research and design of fail-safe enclosure building.
    - Testing of various vacuum level/volum ratios.
    - Development and design of fail safe vacuum explosive
       initiation interlinked systems.
4) Qualification and Testing
    A wide range of tubular diameters, and material variants must be tested to oil
industry, D.N.V. and Lloyds standard to ensure product acceptance.

STATE OF ADVANCEMENT :
****
The project is 60 % complete. Small orders for short lengths of oilfield pipe
and tubes havebeen placed by several major oil companies. Two of these orders
have been completed and a third is being processed.

RESULTS :
****
Lengths of tubing 3 metres long have been lined to oilfields standards with 1/2
mm to 3 mm thick of corrosion resistant alloys.
The explosive mixture develpment and the clamping system for sustaining the
process is complete and proven on a production basis. A trial of pilgering a 3
metre length of 6" clad pipe has successfully produced a 5 1/2 metre length of
4" clad pipe. An arrangement is made for 1990 with British Steel to run a 2 1/2
m billet, internally lined with stainless steel, through the hot rolling mill

and to see if this will roll out to a full length 13 metres long production tubing. Outline QA/QC and production procedures are in place. A conceptual layout for a factory has been completed.

```
*****************************************************************************
* TITLE : MULTIPHASE PIPE FLOW MODELLING IN          *     PROJECT NO       *
*         NATURAL GAS AND OIL.SYSTEMS                 *                      *
*                                                     *   TH./09052/89/HE/.. *
*                                                     *                      *
*****************************************************************************
* CONTRACTOR :                                        *    TELEPHONE NO      *
*    ALFAPI SA                                        *                      *
*                                                     *    301 6532579       *
*    MESSOGHION AVENUE 304                            *                      *
*    HE - 155 62 ATHENS                               *                      *
*                                                     *    TELEX NO          *
*                                                     *                      *
* TECHNICAL DIRECTOR :                                *    223296            *
*    MR. D.G. PAPANIKAS                               *                      *
*                                                     *                      *
*****************************************************************************
```
                                                          VERSION : 01/05/90

AIM OF THE PROJECT :
****
The objective of the project is the computer aided prediction of main transport
properties of two-phase, multicomponent fluid media produced and transported in
Natural Gas and Oil Systems gas-liquid, fluid-solid, liquid-liquid and vapor-
liquid equilibria analysis will be performed. Methods and formulae which will be
derived, will be modelled by algorithmic formulation.
Validation and reliability of the modelling will be proved by technical and
physical results in comparison with data from real design and operating
environments. The objective modelling along with Transport Properties Estimation
System (TPES, Contract TH./10070/89) will integrate a significant tool needed
for the relative fluid analysis of hydrocarbons piping systems.

PROJECT DESCRIPTION :
****
The project has been divided in the four following stages :
1. Compilation and Classification of Methods in this stage modelling methods,
analysing two-phase flow in Hydrocarbons (Natural Gas + Oil) piping systems,
will be compiled and classified.
2. Adaptation and Programming and Methods and Formulae Deliverables of the stage
are computer assisted selected methods studied in previous stage predicting main
transport properties of two-phase, multicomponent hydrocarbon fluids.
3. Testing and integration two tubes of testing will be considered :
 (1) bottom testing unit and module test and
 (2) thread testing
The testing procedures must fulfil quality requirements in certainty of the
technical and physical results through test referred in literature and real
application cases.
5. Documentation and Project Management Deliverables of this stage are internal
reports documenting technical and scientific progress of the project and EEC
Technical and Financial Reporting.

STATE OF ADVANCEMENT :
****
Ongoing. The project works begin in the first days of January 1990.

REFERENCES :
****
THE PROGRESS OF THE WHOLE PROJECT IS DOCUMENTED BY INTERNAL REPORTS AND IS
PRESENTED BY INTERIM REPORTS TO THE COMMISSION PRESENTATION OF RESULTS AND
DISSEMINATION OF KNOWLEDGE IS PERFORMED BY SPECIAL LECTURES AND PUBLICATION OF
PAPERS IN SCIENTIFIC PERIODICALS.

```
*********************************************************************
* TITLE : REMOTE UNDERWATER EXCAVATOR - ENHANCED    *    PROJECT NO    *
*          DEVELOPMENT SYSTEMS                       *                  *
*                                                    *   TH./09057/89/UK/..  *
*                                                    *                  *
*********************************************************************
* CONTRACTOR :                                       *   TELEPHONE NO   *
*    CONSORTIUM RESOURCE MANAGEMENT LIMITED          *                  *
*                                                    *   0224 703094    *
*    19 DENMORE                                      *                  *
*    INDUSTRIAL ESTATE, BRIDGE OF DON                *                  *
*    UK - ABERDEEN AB2 8 JW                          *   TELEX NO       *
*                                                    *                  *
* TECHNICAL DIRECTOR :                               *   739008 FAX: 0224-824*
*    MR. N.V. SILLS                                  *   277/824739     *
*                                                    *                  *
*********************************************************************
                                                     VERSION : 02/05/90
```

AIM OF THE PROJECT :
Develop seabed and midwater transportation systems for the "RUE" underwater excavation system together with ancillary improvement of the energy supply system by the introduction of seawater hydraulics. The innovations encompass a completely new type of seabed crawler to manoeuvre the buoyant excavator around the seabed; an ROV deployable excavator for use where seabed access is impossible and will introduce the new technology of seawater hydraulics to the offshore industry.

PROJECT DESCRIPTION :
A three stage development programme comprising :
1) Underwater Seabed Transportation System
The unique manner in which the RUE underwater unit is used, that is buoyantly from a weight deployed on the seabed, requires an innovatory approach to its transportation system. A caterpillar type tracked system is favoured for pipeline excavation of novel two in line design rather than the conventional side by side arrangement as in bulldozers and tanks etc.
2) Neutrally Buoyant Reactionless System
To allow the underwater unit to be carried to areas inaccessible from the seabed, such as inside steel platform structures and onto cell tops of concrete structures, a "flying" system is required. Presently available Remote Operated Vehicles would be used to fly the excavator unit underwater, an operation which is only feasible if the unit is neutrally buoyant and reactionless. The innovative development here is to design out reactive forces when operating the excavator rather than changes to the buoyancy system which are within known technology. A combination of the present thrust compensation system (where suction from the reversed water intake opposes thrust) and a limited use of retroactive thrust could achieve this objective. This will require a detailed study and redesign of the RUE underwater unit.
3) Conversion from Oil to Seawater Hydraulics
To double the power of the present oil hydraulic powered units to cope with very large excavations, and to increase the depth at which present RUE's can operate, a conversion to seawater hydraulics is required.
The use of natural seawater as a hydraulic fluid would allow at least a twofold increase in power at the excavator without increasing the present "handleable" size of the umbilial. The present return line could be replaced with a second supply line (no return line is necessary). The exhaust seawater is simply vented back into the sea. Additionally, conversion to seawater hydraulics would eliminate all risk of oil pollution from burst hoses or leaking seals. The use of seawater hydraulics to power a large excavator is innovative because it will be the first application of this technology. However it will require redesign of power transmission system, new umbilical, power pack and control system. An immediate commercial advantage resulting from this innovation is a 50% reduction in the cost of umbilicals. The use and experience gained with seawater hydraulics will benefit the subsea engineering industry generally as there are many other uses for such applied technology.

STATE OF ADVANCEMENT :
Ongoing.
Stage 1 - Fabrication - Drawings have been approved and tenders are being sought for manufacture.
Stage 2 - Jet Prop concept (see Note on page 5 Annex 1 Work Programme now ready for testing.
Stage 3 - In abeyance pending results of Stage 2 testing.

TRANSPORT

```
***********************************************************************************
* TITLE : HYDROELASTIC PHENOMENA ON SUBMARINE      *      PROJECT NO        *
*         PIPELINES.                               *                        *
*                                                  *                        *
*                                                  *    TH./10038/83/IT/..  *
*                                                  *                        *
***********************************************************************************
* CONTRACTOR :                                     *    TELEPHONE NO        *
*     SNAMPROGETTI                                 *                        *
*                                                  *    0721 - 8811         *
*     C.P. 12059                                   *                        *
*     IT - 20100 MILANO                            *                        *
*                                                  *                        *
*                                                  *    TELEX NO            *
*                                                  *                        *
* TECHNICAL DIRECTOR :                             *    560 279             *
*     DOTT. MATTIELLO                              *                        *
***********************************************************************************
```

VERSION : 19/09/89

AIM OF THE PROJECT :
Aim of the project was to improve knowledge of vortex induced oscillations which occur on submarine pipelines when seabottom unevenness or sediment transport cause large free spans. Theoretical analysis and experimental works were experimental out in order to achieve new methodologies and experimental data which might be included in a certification procedure for submarine pipelines.

PROJECT DESCRIPTION :
A peculiarity of SVS project was to take into account the overall behaviour (environment, structural behaviour, hydrodynamics) and to take into account the scenario of the submarine pipeline resting on the sea bottom.
The project included four phases :
- theoretical (Phase 1000)
- basin tests (Phase 2000)
- in field tests (Phase 3000)
- processing of experimental data (Phase 4000)
As for the theoretical phase, the physical phenomenon have been analyzed and a critical review of known theoretical and experimental literature has been carried out.
Some analytical modelling techniques of phenomenon have been analyzed, particularly concerned of structural behaviour of pipelines freespans relating to specific equilibrium conditions in presence of unevenness or sedimentological instability of seabottom. As for the basin test phase experimental situations,. previously not sufficiently investigated, have been performed, specifically it concerned supercritical regime, wave induced vortex shedding, seabottom proximity free-stream turbulence and flow tridimensionality, cabletype behaviour of pipes for large freespans and so on; as for the field test phase a pipeline free-span, installed in vortex shedding induced vibrations favourable location (as concerned of mean currents velocity and/or wave regimes), has been instrumented, and hydrodynamic response has been monitored together with the meteomarine environment beside meteo-oceanographic situation has been carefully investigated.

STATE OF ADVANCEMENT :
The project activities have been completed the final summary report are being preparing.

RESULTS :
The project was conducted with the philosophy that :
- structural behaviour
- incoming flow conditions
- hydrodynamic loads
were strictly correlated when the hydroelastic response of a system, like a free span under impinging currents, is to be defined.
It was demonstrated that :
- The structural behaviour influences the compliance of the span to hydrodynamic loads and, as a result of the strict correlation with the excitation patterns from the incoming flow conditions it may modify the hydroelastic instability mechanism depending on whether the free span nature.
- The hydrodynamic loads influence the dynamic response of free span, as a result of the strict correlation with the near seabed scenario
- The environmental conditions influence the excitation patterns as a result of the strict correlation of the hydrodynamic loads with the incoming flow conditions.
As a summary, the results are twofold :
- A new analysis methodology tailored to the spanning typology, whether due to erosion or caused by seabed uneveness, and water depth i.e. whether oscillatory or steady current conditions are predominant and related hydrodynamic force models;
- A new data base resulting from the elaboration of experimental data obtained from more than 3000 tests under laboratory conditions and relevant validation

through field monitoring.
As far as hydrodynamic loads are concerned, the project demonstrated the dramatic influence of interaction of steady currents with oscillatory induced by surface waves on the forcing mechanism, as the impinging hydrodynamic field and the structural response of the suspended lengths of sealines in the near seafloor scenario are strictly linked. These results move to an extensive development of new approaches to characterize the hydrodynamic environment which take into account the coexistence of oscillatory and steady flows from a probabilistic point of view at various level of predominance.

REFERENCES :
****
1. 3RD INTERNATIONAL SYMPOSIUM ON NEW OIL AND GAS TECHNOLOGIES - MARCH 1988
2. BRUSHI, R. ET AL. : "SUBMARINE PIPELINE DESIGN AGAINST HYDROELASTIC OSCILLATIONS : THE SVS PROJECT" : OFFSHORE MECHANICS AND ARCTIC ENGINEERING SYMPOSIUM, ASME, HOUSTON 1988.
3. BRUSCHI,R. AND VITALI, L. : "LARGE AMPLITUDE OSCILLATIONS OF GEOMETRICALLY NON LINEAR BEAM SUBJECT TO HYDRODYNAMIC EXCITATION" : PROCEEDING OF A INTERNATIONAL CONFERENCE ON COMPUTER MODELLING IN OCEAN ENGINEERING, VENISE, 19-23 SEPT. 1988.
4. TASSINI, P.A. ET A. : "THE SUBMARINE PIPELINE VORTEX SHEDDING PROJECT : BACKGROUND, OVERVIEW, FUTURE FALL-OUT ON PIPELINE DESIGN"; OFFSHORE TECHNOLOGY CONFERENCE, OTC PAPER No. 6157, HOUSTON 1989.

```
****************************************************************************
* TITLE : OFFSHORE FLEXIBLE HOSES                 *      PROJECT NO       *
*                                                 *                       *
*                                                 *   TH./10039/83/IT/..  *
*                                                 *                       *
****************************************************************************
* CONTRACTOR :                                    *     TELEPHONE NO      *
*    INDUSTRIE PIRELLI S.P.A.                      *                       *
*                                                 *   2-64421             *
*    VIALE SARCA 222                               *                       *
*    IT - 20126 MILANO                             *                       *
*                                                 *     TELEX NO          *
*                                                 *                       *
* TECHNICAL DIRECTOR :                             *   310135 PIRELLI      *
*    MR. MANCOSU                                   *                       *
*                                                 *                       *
****************************************************************************
                                                      VERSION : 25/05/87
```

AIM OF THE PROJECT :
****
Planning of offshore hose through the Finite Element Analysis (FEA), reducing
the experimentation time and optimising the structure for the single application.

PROJECT DESCRIPTION :
****
1. Design and construction of a test tank for experimental models.
2. Preparation of some structures in 1/3 scale in respect to full size scale.
3. Static and dynamic tets on models with collection of data on behaviour.
4. Realization of a mathematical model for the various structures.
5. Simulation of tests mentioned under point 3.
6. Comparison between experimental data and calculation results.
7. Improvement of the mathematical model through iteration of points 3-4-5-6.
8. Extrapolation of 1/3 model to full size through realisation of a prototpype.
9. Practical test of the prototype.
10. Final adjustment of the mathematical model to be used for new projects.

STATE OF ADVANCEMENT :
****
Completed

RESULTS :
****
The three dimensional FEA method for model examination was successfully worked
out. The tank for dynamic tests was used with good results. At the end of this
period of research we unfortunately have to conclude that the success of the
work undertaken has only been pastial and the goals set have not been reached in
the period of time established.
Difficulties were encountered during calibration of measuremtntent system and in
the interpretation of strain valves.
Results obtained up to now are, however, of great interest and we see
possibilities of applying some of the structural modifications suggested by our
tests to the produced material. The studies will be continued outside this
research contract.

```
*********************************************************************
* TITLE : UNATTENDED UNDERWATER BOOSTING STATION   *    PROJECT NO     *
*         FOR TWO PHASE FLOW.                       *                   *
*                                                   *  TH./10045/84/IT/.. *
*                                                   *                   *
*********************************************************************
* CONTRACTOR :                                      *   TELEPHONE NO    *
*    SNAMPROGETTI SPA                               *                   *
*                                                   *     02 5201       *
*    C.P. 12059                                     *                   *
*    IT - 20100 MILANO                              *                   *
*                                                   *    TELEX NO       *
*                                                   *                   *
* TECHNICAL DIRECTOR :                              *     310246        *
*    ING. G. ORLANDO                                *                   *
*                                                   *                   *
*********************************************************************
                                               VERSION : 07/02/90
```

AIM OF THE PROJECT :
****
Aim of the project is to define and demonstrate the feasibility of a Subsea
system for the collection, pumping and dispatching of the production of an oil
field to a gathering station either onshore or on a platform in shallow waters.
The System, which is unmanned, must be suitable for installations up to depth of
1000 m and at a distance of about 100 Km from the coast.

PROJECT DESCRIPTION :
****
The project includes the following phases:
PHASE 1000 - Definition of operating scenarios and performance of the system.

PHASE 2000 - Design of the system and analysis of components subsystem.
Based on define the results of phase 1000, architecture of the system, select
the subsystems and start the reliability analysis.
PHASE 3000 - Two phase pump subsystem
Investigations, experimentation and setting-up of two phase pump which can cover
the operating scenario identified in phase 1000.
PHASE 4000 - Power generation subsystem.
To identify and develop the power generation and transmission subsystem.
PHASE 5000 - Instrumentation and control subsystem
To identify and develop the control subsystem and the data
acquisition/transmission.
PHASE 6000 - Prototype construction and onshore tests
To evaluate the performance of the developed system, the construction of a
complete prototype and its tests in oil-land pilot plant are foreseen in this
phase.
PHASE 7000 - Tests on system components in offshore simulated environment.
The project started on October 1st, 1985 and it is expected to end by September
30th, 1989.

STATE OF ADVANCEMENT :
****
Ongoing - The project is in the "prototypes construction and testing" phase.
Phases 1000, 2000, 3000, 4000 and 5000 have been completed. Phases 6000 and 7000
are 90 % and 60 % completed respectively. The total progress of the project is
87 %.

RESULTS :
****
The main performaces of the S.B.S. system have been confirmed according to the
selected scenario :
- capacity of handling the two phase mixture with gas content variable in time.
Therefore two reference operating areas have been found :
- low GOR : up to 50 % of vacuum degree;
- high GOR : up to 90 % of vacuum degree.
- Capacity of pumping a flowrate of liquids variable in time relevant to a max
production of 30,000 BOPD.
- Capacity of giving to the pumped mixture a pressure rise in the order of 90
kg/cm2 (on the assumption of a depth of 1,000 m, of a distance of 100 Km far
from the coast, of a pipeline with diameter 16").
The use of a S.B.S. system in the exploitation of marginal offshore oil fields
led to conceive the plant in a modular way.
The production capacity of the basic unit has been fixed in 10,000 BOPD.
The 4-stage centrifugal pump prototype is sized (impeller diameter of 350 mm)
for a production capacity of 10,000 barrels/day, with a suction max void
fraction of 60 %.
The construction of the two-phase hydrocarbon testing circuit has been completed.
The detailed engineering and prototype construction of system components,
"critical" from the standpoint of their suitability to operate in the actual
environment (1000 meters of water depth) such as :
- high power, subsea mateable electric connector,
- 30 KW power cable penetration,
- High delta P electric penetration,

- fiber optic signal transmission system,
- Motor/pump sealing system,
- Auxiliary system for the pump bearings,
has been completed.
The testing campaign is in progress.

```
*********************************************************************
* TITLE : REPAIR SEA TRIALS ON DEEP WATER    *        PROJECT NO      *
*         PIPELINES (600 M WATER DEPTH)       *                        *
*                                             *    TH./10047/85/IT/..  *
*                                             *                        *
*********************************************************************
* CONTRACTOR :                                *      TELEPHONE NO      *
*    SNAM SPA                                  *                        *
*                                             *     39 2 5205716       *
*    PIAZZA VANONI1                            *                        *
*    IT - 20097 SAN DONATO MILANESE            *                        *
*                                             *      TELEX NO          *
*                                             *                        *
* TECHNICAL DIRECTOR :                         *      310 246          *
*    MR. A. LOLLI                              *                        *
*                                             *                        *
*********************************************************************
                                              VERSION : 16/08/89
```

AIM OF THE PROJECT :
****
Maintenance and repair of submarine pipelines gets harder and harder when depth
increases.
For this reason SNAM, joint owner of the world's deepest pipeline (600 m), has
developed (with other Companies of ENI Group) new techniques for automatic
maintenance and repair of such pipelines.
The new system is called S.A.S. (Stazione Autonoma Sottomarina) and, at the
present time, its construction is completed.
Although a series of deep water test was already scheduled, it is now considered
that a full demonstration of the capability of the system is necessary,
including the repair of a 20" sealine in deepwater.
The innovations and know-how expected from these trials consist in proving for
the first time the feasibility of such operation at great dephts in complete
automatic mode (all steps are computerized), and safety as the human presence is
not required on the sea bottom.

PROJECT DESCRIPTION :
****
2.A Main project phases
   The project is developed in five main phases:
      1) construction of connection systems
      2) construction of a pipeline section (140 m)
      3) transport of the pipeline offshore and laying at 600 m awater depth
      4) repair of the pipeline on the sea bottom at 600 m
      5) analysis of results and final report.
2.B Project phases
   For each project phase the work schedule can be subdivided into the following
groups of operations:
      B1) construction of connection systems
         . modifications and improvements to the connection system developed by
Nuovo Pignone
         . preliminary tests on reduced and full size scale
         . construction of connection systems
      B2) Preparation of the testing pipeline section
         . setting up of onshore yard for transporting the pipeline offshore
         . purchase of 20" pipe lenghts, welding and installation of floats
      B3) Transport of the pipeline offshore and laying 600 m depth
         . mobilization and demobilization of two ocean-going tugs
         . towing of the pipeline to the site of installation and laying on the
seabed
      2.C) Repair of the pipeline
      . Mobilization of a dinamically positioned ship and of a support vessel
      . Modifications in the dinamically positioned ship for the installation on
board of a part of the repair system
      . Leasing of a complete R.O.V. system (SCORPIO class)
      . Repair of the pipeline consisting in:
         - removing any artificial overlays from the pipeline (to be defined in
details later on)
         - raising the pipeline from the seabed
         - cutting and cleaning the two ends to be connected
         - recovery the damaged pipe section by lifting it to the surface
         - transporting the base frame into the seabed and measuring the spool to
be connected
         - installation of connectors on the pipe ends, sealing tests
         - installation of the spool-piece, locking and sealing test of components
         - recovery of modules from the seabed.

STATE OF ADVANCEMENT :
****
Deep water tests have been carried out in Gulf of Taranto (Southern Italy) in
350 m waterdepth in Summer 1988.

RESULTS :
****

358

The unexpected very poor bearing capacity (0,1 kg/cm2) and visibility of the
seabed turned out to make harder testing the system. Furthermore, the
malfunction of sensors compromised sommetimes the use of total automatic
procedures (for example the acoustic navigation system). At any rate, concrete,
polyethylene coating removal and pipe cutting were automatically carried out by
the pipe preparation module, which, in addition recovered to the surface the cut
pipe section - 12,25 m long simulating the damaged section. The base frame was
also installed and clamped between the two prepared pipe ends; their distance
and orientation were automatically measured and stored int the process computer
8on the surface). The first end connector was forged on the pipe end, problems
encountered during extraction of the press caused damage to the lifting trolley.
The seatrials were interrupted.
At present time a serie of modifications, ameliorations and studies are under
way and a new deep sea trials campaign is foreseen for 1990.

REFERENCES :
****
STATUS OF SNAM DEEPWATER AUTOMATICAL PIPELINE REPAIR SYSTEM - D.O.T. 2 VALLETTA
MALTA - OCT. 83
DATEX-PACK "REPAIR SEA TRIALS ON DEEP WATER PIPELINES" - 3 HYDROCARBON SYMPOSIUM
LUXEMBURG - MARCH 1988.

```
*****************************************************************************
* TITLE : DEVELOPMENT OF A RANGE TWO-STAGE          *      PROJECT NO      *
*          CENTRIFUGAL PUMPS FOR THE TRANSPORT OF    *                      *
*          CRUDE AND ITS DERIVATES                   *   TH./10049/85/NL/.. *
*                                                    *                      *
*****************************************************************************
* CONTRACTOR :                                       *      TELEPHONE NO    *
*     STORK POMPEN B.V.                              *                      *
*                                                    *      074-404000      *
*     LANSINKESWEG 30                                *                      *
*     P.O. BOX 55                                    *                      *
*     NL - 7550 AB HENGELO (O)                       *      TELEX NO        *
*                                                    *                      *
* TECHNICAL DIRECTOR :                               *      44324           *
*     IR. E.J. BUSSEMAKER                            *                      *
*                                                    *                      *
*****************************************************************************
                                                           VERSION : 04/09/89
```

AIM OF THE PROJECT :
****
To develop a new range of two-stage centrifugal pumps for the transport of crude
oil and derivates thereof. The range will be equipped with a rotor which is
supported by a bearing at each end of the pump. Compared with the traditional
design equipped with an overhung pumprotor, the new pump design incorporates a
number of innovative aspects resulting in the following characteristics: low
rotorvibration, low sealleakage, long life, low power consumption and low weight.

PROJECT DESCRIPTION :
****
The project is divided into the following six phases:(1) orientation phase, (2)
innovation phase, (3) development phase, (4) testphase, (5) design and
production of initial series casting patterns and (6) production and sales.

Description of the phases:
(1) choice of manufacturing techniques, choice of materials,
    evaluation of hydraulic features, planning.
(2) further development of hydraulic principles and manufacturing
    methods, alternative designs.
(3) implementation of results from phases (1) and (2), detailed
    design work, choice and specification of materials.
(4) production and test of prototype pump, analysis of test results and
    and implementation of resulting design changes in the design
    of phase (3).
(5) detailed design drawings for the complete range, production of
initial casting patterns and pumps, initial test of every pump-
type and implementqtion of resulting design changes.
(6) normal production and sales.

The main technical aspects of this development project are the following:
optimal hydraulic design using CAE/CAD, introduction of very smooth hydraulic
surfaces, very high efficiency, modular set-up, capability to cope with every
pumped fluid, good mechanical pump behaviour at off-design conditions.

STATE OF ADVANCEMENT :
****
The planned pumprange was split up in a range A with pumping heads up to 300 m
and a range B up to 400 m (range B). Range A is in the fase of detaildrawings
and making of initial casting patterns, the further development of range B has
been proponed indefinitly.

RESULTS :
****
Thanks to optimized diffusordesign power consumption savings up to 20 % were
achieved. Motorweight savings are in accordance with power decrease figures.

REFERENCES :
****
--

```
*****************************************************************************
* TITLE : SUBSEA OIL LOADING SYSTEM (SOLS) FOR       *    PROJECT NO        *
*         TANKERS                                    *                      *
*                                                    *  TH./10050/85/DE/..  *
*                                                    *                      *
*****************************************************************************
* CONTRACTOR :                                       *    TELEPHONE NO      *
*    TELEFUNKEN SYSTEM TECHNIK                        *                      *
*                                                    *    040 3616-1        *
*    STEINHOEFT 9                                     *                      *
*    DE - 2000 HAMBURG 11                             *                      *
*                                                    *    TELEX NO          *
*                                                    *                      *
* TECHNICAL DIRECTOR :                               *    211 868           *
*    DR.-ING K. WILKE                                 *                      *
*****************************************************************************
```

VERSION : 01/09/89

AIM OF THE PROJECT :
****
After a feasibility study has shown the advantages of the SOLS-concept in a
second phase the main component - the re-entry unit (REU) with its water jet
thruster control - was fabricated as prototype and successfully tested in
shallow water.
During the phase III the system is to be completed with loading hose and whinch
and to be tested in deep water to prove operability under realistic conditions.

PROJECT DESCRIPTION :
****
SOLS is to be operated from a dynamically positioned tanker, which carries a
hose storage winch on deck over a moonpool.
The re-entry unit is lowered through the moonpool hanging on the hose to the
Pipeline And Manifold (PLEM) on the seabed. The position of the REU can be
controlled by waterjets and a hydroacoustic measuring system. After having
reached the final position inside the cone of the PLEM the valves are to be
switched into loading position so the oil can be pumped from the production
platform through REU and hose into the tanker.
The phases of the project are:
- preparation of re-entry unit and pressure test
- safety analysis
- design and manufacture of PLEM adapter
- design and manufacture of winch
- factory tests and fabrication of hose
- simulations and preparation of tests programs
- deep water trials (120 m)
- evaluation of tests and dismounting of equipement
- documentation and project management
After a milestone at this point it is provided to install the system for sea
trials and repeat the test program under different seastate conditions.

STATE OF ADVANCEMENT :
****
All working steps of the project (a.m. under 2.) have been carried out
successfully. The equipment has been dismounted. Phase III has been completed
with the final report.

RESULTS :
****
The evaluation of deep water trials has been completed with good results :
The shallow water trials of phase II have shown the manoevrability of the re-
entry unit. Simulations and pre-tests show that the system will be operable
under the seastate conditions designed to. A pressure test with the re-entry
unit was successfully carried out in Dec. 1986.
In March 1987 extended testing in shallow and deep water in Norway confirmed the
functioning of the system under operating conditions. Seastate was successfully
tested with simulations.

REFERENCES :
****
"SUBSEA OIL LOADING SYSTEM FOR TANKER" 1984, EUROPEAN PETROLEUM CONFERENCE (SPE
12980)
- UNTERWASSER-OLUBERNAHMESYSTEM FUR TANKER - SOLS
SCHIFF & HAFEN, HEFTE 4/5, 1986
- "SOLS - SUBSEA OIL LOADING SYSTEM FOR TANKERS" 1986, INTERVENTION, WEMT
SYMPOSIUM : ADVANCES IN OFFSHORE TECHNOLOGY
- "SOLS - SUBSEA OIL LOADING SYSTEM" 1988, INTERVENTION "88/UTC (NO. 1605).

```
*******************************************************************************
* TITLE : QUALIFICATION TRIALS OF A DEEPWATER        *     PROJECT NO        *
*         DEVICE FOR THE CONNECTION OF FLEXIBLE       *                       *
*         LINES (PHASE 1)                             *    TH./10052/86/FR/.. *
*                                                     *                       *
*******************************************************************************
* CONTRACTOR :                                        *    TELEPHONE NO       *
*     GERTH                                           *                       *
*                                                     *    1 47.52.61.39      *
*     4, AVENUE DE BOIS PREAU                         *                       *
*     FR - 92502 RUEIL-MALMAISON                      *                       *
*                                                     *    TELEX NO           *
*                                                     *                       *
* TECHNICAL DIRECTOR :                                *    203 050            *
*     MR. H. LECOMTE                                  *                       *
*                                                     *                       *
*******************************************************************************
                                                         VERSION : 30/06/89
```

AIM OF THE PROJECT :
****
The aim of the project is to qualify a specific diverless method for connecting
flexible lines to a subsea structure or another rigid or flexible line : a heavy
duty manipulator arm, fitted on either a crawler vehicle or a running tool
lowered on the subsea structure takes the extremity of the flexible line on the
seabed and barings it to the locking position on the structure. The arm is
fitted with a specific acoustic detection system which permits to gain the
geometric configuration of the two parts to be connected. Thanks to this system,
the final connection system is performed in automatic mode, thus granting the
necessary accuracy which could not be gained with a manual mode.

PROJECT DESCRIPTION :
****
The manipulator arm has been manufactured, outside the scope of this project.
Phase 1 consists in developing the other equipments necessary for the
performance of a connection (mainly the control-command of the arm) and to
qualify the method by running extensive shallow water trials. Later on, phase 2
will consist in running deepwater trials.
The project is to be developed in two main tasks :
TASK 1 : Engineering and trial preparation :
The works on the acoustic operational detection system are now started, together
with the development of the command of the manipulator arm. This command has to
take into account the various phases of a connection with their specific
characteristics :
- first preliminary approach toward the structure i.. manual mode, and then,
- final approach and simulation of an entry of the connector in robot mode,
- manual approach of the arm toward the end of the flexible line,
- gripping the end of the line in robot mode,
- installation of gasket in the connector,
- approach toward the structure at the previous position, final approach and
entry of the connector in robot mode,
- closure of the connector and hydrostatic test,
- unlocking of the arm,
- contingencies and other phases for disconnection.
The acoustic detection system as been redesigned for operational use. It
includes three different acoustic nets, one for the gripping phase, another for
the final approach and the last one for the entry. This system, fitted at the
end of the arm, includes also video cameras and other ancillaries.
Other works pertaining to this phase include the preparation of all other
equipments and procedures necessary for the performance of the trials, namely
the adaptation of the crawler vehicle, the design and manufacture of the subsea
structure and of auxiliairy equipment, the pre-design of the running tool.
TASK 2 : The procedures qualification trials related to deepwater use are
delayed : trials will be run in a pool to qualify the telemetry system which
remains the first priority.
A large of connections and disconnections will be performed. These trials will
first demonstrate the feasibility and interest of this method, as well as
confirm the parameters to be taken into account for the preparation and
realization of a connection job.

STATE OF ADVANCEMENT :
****
Trials are now performed in a pool near Toulouse. The realization of the
telemetry system of the arm as well as the new acoustic system, are completed
and used to control the 1700 daN arm.
The grabbing tool has been designed, built in IFP workshops qualified in Logabex
Laboratory and fitted on the strenght-sensor at the end of the arm.

RESULTS :
****
FEASIBILITY STUDIES OF THE RUNNING TOOL
The MDC can be as well fitted on a running tool which ca be lowered and clamped
on guide posts on the subsea structure. Its design is finalized : particular
attention has been paid in order that most structures can accomodate it without

requiring major modifications. Meanwhile, a lowering procedure has been studied, to avoid the use of guidelines and thus permit the operation with a light dynamically positioioned surface vessel.

ACOUSTIC DETECTION SYSTEM

It comprises three nets of acoustic captors : the first is used for grabbing the end of the pipe, the second for localization of the hub on the structure and approach of the pipe towards it , the last one for the insertion.

CONTROL-COMMAND OF THE ARM

An analysis of the different tasks of a connection involving the manipulator arm has been performed (taking the end of the flexible, removing the blind flange, installing the gasket, pulling toward the structure, penetration then unlocking of the arm and retrieval).

For each of above tasks, the elementary steps have been defined, and the various parameters to be taken into account have been identified. From these, the specifications have been written up, which includes :

- The study of the system safeties and contingencies
- The global architecture of the system
- The realization of the command, hardware and software
- The controls (internal and surface controls).

The command of the arm must include both a feedback on the control of position and one on the control of effort : the efforts on the arm must be accurately known on a real time basis, and the power of the arm minimized to diminish the impact should a collision occur. A sensor, located between the arm and the grabbing device, permits to continually know the efforts and to modify the arm trajectory, and to control the sliding friction during the stroking phase. This sensor has been manufactured and tested by Logabex and fitted at the end of the arm.

REFERENCES :
****

THE PROJECT HAS BEEN PRESENTED DURING THE ANNUAL CEP&M CONFERENCE HELD IN PARIS. ON THIS OCCASION, IT HAS BEEN GRANTED THE "LES CREATEURS" AWARD.

```
****************************************************************************
* TITLE : THERMAL INSULATION OF SUBSEA PIPELINES    *      PROJECT NO      *
*          FOR HOT FLUIDS (110 DEG.C) IN DEEP        *                      *
*          WATERS (400 M)                            *  TH./10053/86/UK/..  *
*                                                    *                      *
****************************************************************************
* CONTRACTOR :                                       *    TELEPHONE NO      *
*    REGAL TECHNOLOGY (UK) LTD                        *                      *
*                                                    *   31-5540391         *
*    IMPERIAL DOCK                                   *                      *
*    LEITH                                           *                      *
*    UK-EDINBURGH EH6 7DR                            *    TELEX NO          *
*                                                    *                      *
* TECHNICAL DIRECTOR :                               *    727109            *
*    IAIN SMITH                                       *                      *
*                                                    *                      *
****************************************************************************
                                              VERSION : 25/10/89
```

AIM OF THE PROJECT :
To develop and produce at an economically acceptable price, a subsea insulation
and anti-corrosion coating for application onto oil and gas flowlines. The
coating will be suitable for installation by the normal pipe lay methods. The
coating will be capable of operating at depths in excess of 400 m and oil
temperatures of greater than 110 deg. C.

PROJECT DESCRIPTION :
The project will be split into seven phases :
PHASE 1 :   DESIGN OF SYSTEM
PHASE 2 :   DEFINING THE INSULATION MATERIAL
PHASE 3 :   DEFINING THE OUTER COATING
PHASE 4 :   FIELD JOINT SYSTEM
PHASE 5 :   NON-DESTRUCTIVE TESTING
PHASE 6 :   PROTOTYPE MANUFACTURING
PHASE 7 :   QUALIFICATION TESTING
Phase's 1,2 and 3 cover the design of the system, the screening of materials and
the final formulations.
Phase 4 covers the on site joining and insulation of the weld area.
Phase 5 covers the factory QC testing.
Pase's 6 and 7 cover the building and testing of the system to the oil
industries specification.

STATE OF ADVANCEMENT :
The project is now in the construction phase.
The materials have been defined and formulated. The initial processing of the
materials has been completed successfully. The bonding system between metal and
elastomer has been defined and shown to work successfully.

RESULTS :
At this stage in the development programme all the indications are that the
product will meet the project aims. Pipelines carrying hot sour oil and gas
require to be made from semi exotic metals, which are susceptible to corrosion
in salt water. Our solution is based on a syntactic elastomer, the syntactic
nature being derived from incorporation of glass micro beads into the elastomer.
The modified elastomer is fully bonded to the pipe with a bonding system which
resists the high temperatures without degrading. The glass micro beads have very
high compressive strengths sufficient to withstand the hydrostactic forces of up
to 600 M. To prevent seperation and coagulation of the beads they are treated
with a surface active agent, this also increases the compatibility of the beads
and elastomer.
Because of the high bead level (40-60%) required to generate the insulation
properties of the material (k=0.12-0.15 W/m/k) a special extrusion technique had
to be developed to process the compound without damaging the glass beads. This
method gives a high ouhput without excessive shearing in the extruder.
The integrity, strength and flexibility of the system has been designed to
withstand the forces and loads encountered when the pipe sections are jointed
together and laid. The system can be laid by either conventional lay barge or by
reel barge, in both cases compressive loads of upto 60 tonnes can be applied to
the coating to prevent the pipe from buckling.
The field jointing system (site or offshore application of insulation to the
weld area) has been designed in principle but the full test programme is still
to be completed.

N

```
********************************************************************************
* TITLE : TURBO PIG                                        *      PROJECT NO     *
*                                                          *                      *
*                                                          *   TH./10055/86/FR/.. *
*                                                          *                      *
********************************************************************************
* CONTRACTOR :                                             *     TELEPHONE NO     *
*    CHALLENGER S.O.S.                                     *                      *
*                                                          *    1 43.59.12.11     *
*    49 BIS, AVENUE FRANKLIN ROOSEVELT                     *                      *
*    FR - 75008 PARIS                                      *                      *
*                                                          *     TELEX NO         *
*                                                          *                      *
* TECHNICAL DIRECTOR :                                     *     642 477          *
*    MR. P. SCEMAMA                                        *                      *
********************************************************************************
```

VERSION : 01/10/89

AIM OF THE PROJECT :
****
Design, build and test a pipe cleaning device (pig) based on new dynamic concepts; this device will perform specially difficult internal pipe cleaning, its components being specifically designed according to requirements, and will incorporate a new and unique by-pass system, avoiding the traditionnal risk of choking.

PROJECT DESCRIPTION :
****
PHASE 1
This phase is a general design phase, including:
- technical definition
- environmental characterisation
- system general architecture
- specifications
- laboratory simulation and tests
- drawings
PHASE 2
- construction of the prototypes
- field testing

STATE OF ADVANCEMENT :
****
Phase 1 and Phase 2 have been completed

RESULTS :
****
General concept of the tool has been defined with the design criteria of the cleaning and propulsion sub-assemblies.
General Procedure of a cleaning operation based on the Turbo Pig concept has been drawn and tried.
Tests in full scale operations with various parameters of the procedures to optimise the design of the components have been performed.
Conclusion is that the Turbo Pig cleaning principle is successful although each cas will require an adequate configuration of the pigs components.

```
****************************************************************
* TITLE : VOYAGEUR                        *     PROJECT NO      *
*                                         *                     *
*                                         *  TH./10056/86/FR/.. *
*                                         *                     *
****************************************************************
* CONTRACTOR :                            *     TELEPHONE NO    *
*    CHALLENGER S.O.S.                     *                     *
*                                         *   1 43.59.12.11     *
*    49 BIS, AVENUE FRANKLIN ROOSEVELT     *                     *
*    FR - 75008 PARIS                      *                     *
*                                         *     TELEX NO        *
*                                         *                     *
* TECHNICAL DIRECTOR :                     *    642 477          *
*    MR. P. SCEMAMA                        *                     *
*                                         *                     *
****************************************************************
                                          VERSION : 01/10/89
```

AIM OF THE PROJECT :
****
Design, build and test a pipeline inspection device (pig) based on new
instrumental concepts; this "pig" will perform traditionnal internal inspection
(gauging/calipering) and also gather all necessary "navigation" data to compute
and draw:
- the pipe path or trajectory
- the critical mechanical constraints (free spans)
providing a full geometrical survey of the pipelines.

PROJECT DESCRIPTION :
****
PHASE 1
This phase is a general design phase, including:
- technical definition
- environmental characterisation
- system general architecture
- specifications
- drawings
PHASE 2
- construction of the prototype
- writing of all necessary computer programs
- laboratory tests
- environmental qualification
PHASE 3
- field testing

STATE OF ADVANCEMENT :
****
Phase 1 , 2 and 3 have been completed as forecast..

RESULTS :
****
Detail engineering has defined both the mechanical construction (scraper, body,
chainage sensor arm and monitoring bracket, mechanical link between units) and
the sensor instrumentation pack. Various tests on components have proven a gross
precision better than 1 % of the distance travelled which can lead after
computer procession to less than 5.10-4.
Tests on site have shown a very high degree of super imposition of recorded and
as-laid drawings.

```
**********************************************************************
* TITLE : PIPELINE INTEGRITY MONITORING EXPERT    *     PROJECT NO    *
*         SYSTEM                                   *                   *
*                                                  *  TH./10057/86/UK/..*
*                                                  *                   *
**********************************************************************
* CONTRACTOR :                                     *   TELEPHONE NO    *
*    J.P. KENNY AND PARTNERS LTD                   *                   *
*                                                  *   -01 831 6644    *
*    BURNE HOUSE                                   *                   *
*    88-89 HIGH HOLBORN                            *                   *
*    UK - LONDON WC1V 6LS                          *   TELEX NO        *
*                                                  *                   *
* TECHNICAL DIRECTOR :                             *   21823           *
*    MR. T.S LUNN                                  *                   *
*                                                  *                   *
**********************************************************************
```

VERSION : 04/09/89

AIM OF THE PROJECT :
****
The aim of the project is to develop a computer system to aid subsea pipeline
operators in the assessment of the pipeline condition, and to estimate its
remaining useful life. Such a computer system will assist in reducing
maintenance costs, improve safety and extend the useful operating life of
submarine pipelines.
The project relies upon the application of the latest technologies to analyse
pipeline inspection data and flow condition measurements, produce grahical
outputs, model operator expertise in diagnosing the pipeline condition, and
advise on further actions. New approaches need to be adopted to combine these
requirements into a useful working system.

PROJECT DESCRIPTION :
****
The project is made up of three stages. At the end of each stage, an operational
version of the expert system will have been developed, ready for demonstration
and customisation for pipeline operators. Each version will be a building block
for the next version and so define the scope of development for the next stage.
Stage I was concerned with defining the data requirements; capturing engineering
knowledge in the forme of technical notes; evaluating programming languages and
expert system shells for their suitability to engineering expert system
applications; implementing pipe stability, pipe support condition and corrosion
modules for demonstrations and to assess the suitability of the language or
shell used. The expert system modules developed in Stage I are concerned with
pipe stability, pipe support condition and corrosion.
In Stage II, the design for engineering knowledge bases were produced based on
the Stage I technical notes. An integrated expert system with extended
functionality was developed which addresses pipe stability, cathodic protection
and external damage.
The system interfaces with conceptual operator data files and recommends
possible further remedial action. The final design included identification of
potential problems, analysis of these areas and recommendation for remedial
action.
Stage III is specifically concerned with internal corrosion which was found to
be an area of particular interest to pipeline operators. The expert system
module developed in this stage calculates corrosion rates and corresponding
material losses. The functions of this system are similar to those developed in
the previous stages but also include graphical outputs and hypertext
explanations. The Stage III system will have a demonstratable version which can
be modified and extended to suit individual pipeline operator requirements.

STATE OF ADVANCEMENT :
****
Engineering knowledge has been documented for the following aspects of pipeline
integrity monitoring : pipeline stability, internal and external corrosion, free-
spanning, cathodic protection, concrete coating loss, anode potential and
pipeline mechanical damage. Expert system programs have been developed for
subjects using various expert system shells and programming languages.
Stage III of the project, which specifically addresses pipeline internal
corrosion, is nearing completion.

RESULTS :
****
Engineering expertise has been acquired for various aspects of subsea pipeline
integrity monitoring. Technical engineering notes have been produced on the
following topics :
. Pipeline support evaluation
.External corrosion evaluation.
. Pipeline stability evaluation
. External damage evaluation
. Pressure testing and leakage evaluation
. Pipeline condition assessment
. Engineering and environmental data requirements
A survey of pipeline operator's and inspection contractors' pipeline integrity

monitoring practices was carried out which identifies the most applicable areas
for computerisation.
The coded program modules are listed below with the language or shell used.
. Pipeline stability assessment module coded in PC XiPlus shell
. Pipeline external corrosion assessment module coded in PC XiPlus
. Pipe free-span assessment module coded in PC PROLOG programming language
. Integrated engineering analysis of pipeline stability, mechanical damage,
concrete coating loss and cathodic protection coded in Leonardo PC expert system
shell. Inspection, maintenance and repair advice partially developed for
mechanical damage. Modules are not readily demonstratable
. Demonstrator for pipeline internal corrosion condition assessment coded in
Smalltalk PC object oriented programming environment.
The APOLLO workstation was originally chosen for program development over the PC
to make use of the state of the art sorftware engineering environment DSEE, to
utilize the high resolution graphics and superior user interface capabilities,
and to use the UNIX operating system. However, no suitable expert system shells
were available on the APOLLO so a port of ESSAI, and later Leonardo, expert
system shells was commissioned. This work was never completed due to hardware
and software problems, and closure of Alcatel's UK branch. Consequently, all
development has had to be done on the PC which has shown severe limitations of
memoray capacity, speed, poor user interfacing and weak data management
facilities and database interfaces, for an engineering project of this size.
Programming languages have shown the drawbacks of longer development times and
future maintenance complications. Recently, the adoption of object oriented
programming techniques have shown faster prototype development times with
inherent data protection and re-usable code generation.

REFERENCES :
****
"DATABASE APPLICATIONS FOR INTEGRITY MONITORING", PIPELINE INTEGRITY MONITORING
CONFERENCE, PIPES AND PIPELINES INT., ABERDEEN, OCTOBER 1986.
"THE FUTURE OF INTEGRITY MONITORING", 2ND ANNUAL SEMINAR ON PIPELINE PROTECTION
AND MAINTENANCE, BERGEN, OCTOBER 1987.
"PIPELINE INSPECTION DATA MANAGEMENT AND DIAGNOSIS", PIPELINE PIGGING AND
INTEGRITY MONITORING CONFERENCE, ABERDEEN, FEBRUARY 1988.

```
*****************************************************************************
* TITLE : BUNDLE REPAIR HABITAT                         *      PROJECT NO        *
*                                                       *                        *
*                                                       *  TH./10058/86/UK/..    *
*                                                       *                        *
*****************************************************************************
* CONTRACTOR :                                          *      TELEPHONE NO      *
*    APT LTD                                            *                        *
*                                                       *  01   748 4600         *
*    3RD FLOOR, TRAFALGAR HOUSE                         *                        *
*    HAMMERMSITH                                        *                        *
*    UK - LONDON W6 8DW                                 *  TELEX NO              *
*                                                       *                        *
* TECHNICAL DIRECTOR :                                  *  262227               *
*    MR. C. BAXTER                                      *                        *
*                                                       *                        *
*****************************************************************************
                                                        VERSION : 06/02/90
```

AIM OF THE PROJECT :
****
The aim is to define and design a one atmosphere work habitat to carry out
repairs on pipelines, bundles and umbilicals where hyperbaric repair by diver is
impossible.

PROJECT DESCRIPTION :
****
The purpose of the habitat is to provide an environment in which technicians can
work on the sea bed to carry out repairs to lines containing hydraulic,
electrical or complex flowlines.
The project will define and design the habitat. It will define its operating
technique and working environment. The build and operating costs will be
produced.

STATE OF ADVANCEMENT :
****
Works have been concentrated on the first phase only.

RESULTS :
****
Design has been based on a solutions already adopted for the communication and
power cable. Two main technical problems have been individualised : sealing
technique and handling system. A preliminary economic analysis has shown the
advantage of subsea repair of bundles and embilicals against the alternative of
relaying them.

```
*+******************************************************************************
*  TITLE : AN EXPERT SYSTEM FOR THE ENGINEERING        *      PROJECT NO        *
*          DESIGN OF HYDROCARBONS TRANSMISSION AND      *                        *
*          DISTRIBUTION PIPE NETWORKS                   *   TH./10059/86/HE/..   *
*                                                       *                        *
*+******************************************************************************
*  CONTRACTOR :                                         *     TELEPHONE NO       *
*      INSTITUTE OF COMPUTATIONAL ENGINEERING SA        *                        *
*                                                       *     01 6532579         *
*      304 MESSOGHION AVE.                              *                        *
*      GR-15562 ATHENS                                  *                        *
*                                                       *     TELEX NO           *
*                                                       *                        *
*  TECHNICAL DIRECTOR :                                 *     223296             *
*      PROF DR.-ING PAPANIKAS/ANGELOPOULOS              *                        *
*                                                       *                        *
*+******************************************************************************
                                                            VERSION : 25/10/89
```

AIM OF THE PROJECT :
****
To develop an expert interactive CAE system for use in the analysis and design
of any kind of pipe network required for hydrocarbons transmission and
distribution Covering any possible offshore, onshore industrial applications,
this CAE system is characterized by its expert ability incorporating linear and
non-linear analysis for possible loading cases along whith the use of knowledge
accumulated in various industrial data banks.

PROJECT DESCRIPTION :
****
Following the consideration of the present state of the art of such CAE systems
the project was divided into the following main stages :
I.   Interactive User Monitoring System
II.   Sea Loading Processor
III.   Pipe Laying Processor
IV.   Data Bank's and Processing
V.   Input/Output Processors
VI.   Stress Analysis Processors
VII.   Fluid-Thermo Analysis Processors
VIII. System Testing, Interpretation and the necessary documentation.

STATE OF ADVANCEMENT :
****
Stages I,III are in their initial phases where basic principles and structure
have been established.
Stages II,IV,V,VI,VII are ongoing close to be tolattly completed. Stress and
fluid-thermo analysis processors are in present working indipendently each other.
Stage VIII and management are going along with the other stages.

RESULTS :
****
The final product of the project will be a user friendly CAE system which will
be a significant tool in the design of complex pibing systems used for
transporting and distributing of hydrocarbons.
The system will guide the designer during the solution of complex problems as
well as to cover such pipe network analysis computational needs of a designing
team, as Fluid-Thermo analysis, Stress analysis, Design critera, Operational
analysis, etc.
Fluid-Thermo and Stress analysis processors compose the core of the system
cooperating with I/O processors and Data bases. Modern mathematical tools such
as Finite Elements, Finite Differences, fast iteractive methods, mesh generators,
etc. are used to make algorithms and system modules accurate and valid through
special tests which are made for the system results precision and worthiness.
At last, as a large scale software package, the system resulting includes : ·
automatic network topological description, free format input, solution
procedures controlled by the user or working automatically for standard cases,
start-stop and restart capabilities, extensive error diagnostics, installation
in the most wellknown computer sustems, central data management and overall
modularity, multi-type output capabilities and adaptation capabilities for
different output periipherals.
The system will be backed by the technical and operational manuals which consist
of an inseparable part of the whole project and it will also be intended for
this interactive system to demonstrate the application of specific knowledge
against problem complexity.

REFERENCES :
****
PRESENTATION OF RESULTS AND DISSEMINATION OF KNOWLEDGE WILL BE PERFORMED BY
SPECIAL LECTURES AND THE PUBLICATION OF PAPERS IN SCIENTIFIC PERIODICALS.

370

```
*****************************************************************************
* TITLE : LAYING TEST OF AN 8" FLEXIBLE PIPELINE      *      PROJECT NO     *
*         MADE OF LIGHT ARMOURS, IN 1 000 M OF        *                     *
*         WATER (ENGINEERING AND MANUFACTURING)       *   TH./10063/87/FR/.. *
*                                                     *                     *
*****************************************************************************
* CONTRACTOR :                                        *    TELEPHONE NO     *
*    GERTH                                             *                     *
*                                                     *   (1) 47.52.61.39   *
*    4, AVENUE DE BOIS PREAU                           *                     *
*    FR - 92502 RUEIL-MALMAISON                        *                     *
*                                                     *    TELEX NO         *
*                                                     *                     *
* TECHNICAL DIRECTOR :                                *                     *
*    MR.B. CHOISNE                                    *                     *
*****************************************************************************
```

                                                    VERSION : 16/11/89

AIM OF THE PROJECT :
****
The aim of the project was to qualify a new concept of flexible pipeline in
which classic steel armours are replaced by lighter armours made of composite
materials, thus to reduce the weight of the pipeline when keeping or improving
its performances regarding service or installation obligations. Typical
application for such flexible pipelines concerns the developments of offshore
oil fields characterized by deepwater beyond 600 meters.
This project was based on preliminary results obtained through a research
programme carried out by COFLEXIP assisted by the INSTITUT FRANCAIS DU PETROLE
which had proved the flexible pipeline lightening feasibility.

PROJECT DESCRIPTION :
****
The project consists in engineering and manufacturing of the 1000 m long 8"
flexible pipeline made of light armours. It was splitted in two main phases :
PHASE 1 concerns project engineering and implementation. It comprises the
following :
MATERIAL ENGINEERING
- Detailed specification, selection of materials, detailed manufacturing process
of the light armours,
- Study of composite material mechanical behaviour,
- Study of facilities and methods required for industrial production of the
composite armours.
PIPELINE STRUCTURE DESIGN
- Specification of flexible pipe (working pressure, fluid carried waterdepth,
laying and operating obligations, dynamic constraints life),
- Static and then dynamic analysis of behaviour under wind, swell and current
when being laid and during operation,
- Study of new deepwater effects on the pipeline structure,
- Static and dynamic tests performed on light flexible pipe samples
- Design of pipeline end-fittings.
LAYING ENGINEERING
- Detailed specifications required for laying an 8" flexible pipe in 1000 meters
of water; study of any necessary adaptations of existing equipment
- Procedures applicable for laying, operating and cases of emergency.
PHASE 2 concerns the manufacture of the flexible pipe. It comprises the
following :
- Design and assembly of an automated pilot production line to manufacture
composite sections,
- Manufacture of composite armours and quality control,
- Adaptation of existing armouring mahines for correct placement of light
armours into flexible pipe structure,
- Manufacture of 1000 meters of 8" flexible pipe, control of the process,
- Assembly of end-fittings,
- Hydrostatic test for 24 hours at 1.5 times working pressure,
- Winding of the pipe on drum,
- Completion of all documents relating to production and quality control of the
flexible pipe.

STATE OF ADVANCEMENT :
****
Completion of the two flexible pipe lengths of 500 m. Both lengths were equipped
with end fittings, pressure tested and packed, ready for laying tests. The
second length was equiped with the new generation of end fittings.
Engineering for laying these two lengths by 1000 m water depth and upgrading of
the laying equipment has been developed.

RESULTS :
****
FLEXIBLE PIPE MANUFACTURING
The two lengths of 500 m equipped with new type of end fittings has been subject
to hydrostatic tests. The characterization of the armours had been performed
allowing use in very severe environment. A 8" light armour sample has been
tested, pull tested over 100 tons, bursting tested at 44,2 MPa, allows the use

of such flexible pipe for deep sea applications.

ENGINEERING FOR THE LAYING TEST

The engineering for the modification of the actual laying equipment has been performed, manufacturing of modification is under process, and a specific engineering for laying of the 2 lengths of 500 m flexible by 1000 m water depth has been developed.

REFERENCES :
****
THE PROJECT HAS BEEN PRESENTED DURING THE ANNUAL CEP & M CONFERENCE HELD IN PARIS. AND DURING THE DOT DEEP OFFSHORE TECHNOLOGY CONFERENCE IN MARBELLA 5SPAIN)

```
*******************************************************************************
* TITLE : INSTRUMENTED VEHICLE FOR THE CONTROL OF     *     PROJECT NO        *
*         PIPELINES-MULTICOIL THICKENESS              *                       *
*         TRANDUCER                                   *   TH./10065/87/FR/..  *
*                                                     *                       *
*******************************************************************************
* CONTRACTOR :                                        *   TELEPHONE NO        *
*    SYMINEX                                          *                       *
*                                                     *   91 7390 3           *
*    BOULEVARD DE L'OCEAN, 2                          *                       *
*    FR-13275 MARSEILLE CEDEX                         *                       *
*                                                     *   TELEX NO            *
*                                                     *                       *
* TECHNICAL DIRECTOR :                                *   400 563             *
*    MR CRIADO                                        *                       *
*                                                     *                       *
*******************************************************************************
                                                    VERSION : 05/04/90
```

AIM OF THE PROJECT :
****
The aim of this project is to study and design an instrumented vehicle for the
control of pipelines with a multi coil thickness transducer, which will detect
corrosion accurately and distinguish severe local corrosion from general
corrosion.
The research carried out to date through the contract nr 09028/85 " "Test of
prototype of a self-fragmenting vehicle destined for the control of pipelines"
has enabled us to define and design an instrumented vehicle with a single coil
thickness transducer.
This PIG allowed detection of general corrosion in a pipe section, but did not
localize corrosion on a tube circumference. A multi coil PIG would enable the
measurement of local corrosion such as corrosion pits.

PROJECT DESCRIPTION :
****
The main characteristics of the multi-coil pig would be :
- detection of severe and local corrosion in the inner and outer sides of the
pipe wall,k
- construction in syntactic foam,
- self fragmenting vehicle
- microprocessor electronics, storage of data in CMOS memory.
The specific study concerns the multi-coil thickness transducer and the
realization and integration of the electronics in the vehicle.
The transducer will be composed of one transmitting coil and 4 or 6 receiving
coils situated at the rear of the vehicle.
The transmitting coil produces a sinusoidal magnetic field. The voltage induced
in each receiving coil presents a phase shift on the induced field,
proportionnal to the thickness of the surface, in each generator.
The project is divided in 3 phases :
1. study and construction of a multicoil thickness transducer
2. electronic study and integration
3. tests.

STATE OF ADVANCEMENT :
****
Completed

RESULTS :
****
Phase 1, 2, 3 have been completed
The instrumented multicoil corrosion conveyor is composed by :
- One emitting coil
- Six receiving coils
- Acquisition and measures data storage electronics in between the emitter and
receiver modules
- One odometer for distance measurements of the conveyor.
Phase 1 :
- study and modelling of the sensor has been achieved
- Several realizations and trials of diverses sensors have been made (different
impedances and different ferromagnetic materials)
Phase 2 :
Study and integration of the electronics have been achieved :
- Multichannels conditioning electronics
- Increase of memory to 4 Mbytes
- Integration of low consumption CMOS components
Phase 3 : Trials :
- Development of a test station at Syminex
- Development of hardware and software for data transfer and results
interpretation
- Complementary external station tests
- Optimization of the multicoil electronics and odometer
- Trials of signal deconvolution and interpretation
Results: the whole development has lead to a compact multichannel corrosion tool
with :

- an equivalent sensitivity compares to the monocoil tool (1 to 2 of phase shift
for 1 % of section loss)
- one punctual defect influences several channels (1 to 3).
Though diverse problems raises for detection without reference log. One may
observe variations of 30 on a log induced by the variation of the pipe magnetic
permittivity compares to 10 phase shift for severe defects limiting drastically
the use of such tools for corrosion control without any previous reference.

374

```
**********************************************************************
* TITLE : INSTRUMENT-FITTED PIPELINE SCRAPER FOR      *    PROJECT NO      *
*          INSPECTION OF CATHODIC PROTECTION           *                    *
*          (FEASIBILITY STUDY)                         *   TH./10067/88/FR/.. *
*                                                      *                    *
**********************************************************************
* CONTRACTOR :                                         *   TELEPHONE NO     *
*    CHALLENGER SPECIAL OIL SERVICES                   *                    *
*                                                      *   33 1 43591211    *
*    49BIS AV.FRANKLIN D.ROOSEVELT - F - 75008 PARIS   *                    *
*                                                      *                    *
*                                                      *   TELEX NO         *
*                                                      *                    *
* TECHNICAL DIRECTOR :                                 *   642477           *
*    MR PATRICK SCEMAMA                                *                    *
*                                                      *                    *
**********************************************************************
                                               VERSION : 01/06/89
```

AIM OF THE PROJECT :
****
Considering the costs and difficulties involved in the external submarine
cathodic protection inspection techniques, the surveys are executed only in the
vicity of the offshore platforms or close to the shores, and they are completed
by spot readings along the pipeline route where the pipe can be reached easily.
The objective of the project is to circumvent the constraints and draw-backs
associated with the use of the external methods and marine facilities for this
type of inspection, by using an instrumented pig (pipeline internal scraper)
running through the pipeline, enabling the pipe potential to be reconstituted
and protection defaults to be located and measured.

PROJECT DESCRIPTION :
****
This instrument - fitted scraper for cathodic protection survey will bring about
undisputed advances with respect to existing methods as to speed and ease of
execution of this type of inspections, moreover providing continuous measurement
of potential all along the pipeline.
The potential gradient and current density between 2 points of the line, as well
as the electromagnetic field changes will be continuously recorded during the
pig run.
An external work station will analyse and shape the recorded data. Following
processing, potential and current curves along the line will be plotted, with
indication of special points marked by a protection flaw.
Gradients of potential and electromagnetic fields to be recorded and analyzed
are very little, except where coating is defective, and the measuring systems
shall have very good performances.
Potential or current measurements will be made by contact on the pipewall. It
will be therefore mandatory to know at any time the values of contact
resistances.
Also, it will have to be coped with electromagnetic interferences (though these
phenomena (thouare limited in the caseof submarine pipelines), and with the
measuring electronics drift, especially with temperature.
These complicated parameters will form the subject of comprehensive theoretical
and experimental studies to secure the project viaability.
The feasibility study, for which EEC funds have been granted will include the
following phases:
. Detailed theoretical study of the electrical and electrochemical phenomena
associated with cathodic protection of underwater pipelines.
. Definition and pre-project of the measurement electronics (electronics not
fitted on board the scraper in view of preliminary bench-testing).
. Design and construction of the shop test-bench, comprising tube and scraper
with pulling device for travel through the tube.
. Construction of the measurement electronics and instrumentation for the test-
bench.
. Bench testing and experimental determination of the measurement parameters and
conditions.
This feasibility study is planned for 15 to 18 months.

STATE OF ADVANCEMENT :
****
Theoretical studies have been initiated in August 1988.

```
***********************************************************************
* TITLE : REPAIR SEA TRIALS ON DEEPWATER        *    PROJECT NO       *
*         PIPELINES - PHASE III                 *                     *
*                                               *   TH./10069/89/IT/.. *
*                                               *                     *
***********************************************************************
* CONTRACTOR :                                  *    TELEPHONE NO      *
*    SNAM SPA                                    *                     *
*                                               *    5205359           *
*    PIAZZA VANONI 1                            *                     *
*    IT - 20097 S.DONATO MIL.SE (MI)            *                     *
*                                               *    TELEX NO          *
*                                               *                     *
* TECHNICAL DIRECTOR :                          *    510246            *
*    ING.STEFANO VENZI                          *                     *
*                                               *                     *
***********************************************************************
                                              VERSION : 01/01/90
```

AIM OF THE PROJECT :
****
To incorporate to the S.A.S. (Stazione Autonoma Sottomarina - Automatic Repair
System) modifications and improvements derived from experiences gained durin
deepwater test campaign in 350 m waterdepth carried out in spring-summer 88. The
final goal is to have available a rugged, efficient, reliable system to on
bottom repair sealines laid in 650 m waterdepth. The system is designed for 20" -
series pipelines, concrete and polyethylene coated.

PROJECT DESCRIPTION :
****
The S.A.S. system is composed of different "service Modules" but the ones
requiring important modifications (EEC financially supported) are :
1) surface handling system (heave compensator, cables lashing device, additional
automations)
2) pipe preparation and cutting module (additional tools to clear worksite from
fallen concrete shells, more powerful jacks for concrete removal, additional
emergencies)
3) Thruster Module (Improvement of the Acoustic Navig. System, Telemetry System,
and some ROV Modification).
4) Base Frame and Hydraulic press interface (Software modifications, alignment
system modifications, electronic system rearrangement)
5) Study of a new mechanical connector to be forged on polyethylene - coated
pipe.
The complete system will be succesively tested in dry conditions and some
modules tested in shallow water (10 m). The possibility to repeat a deep zater
test compaign in spring 1991 is still under evaluation.

STATE OF ADVANCEMENT :
****
Project started on Aug. 1st 1989. At present some activities have started
(longer or critical phases), in particular the new connector design and the
Acoustic Navigation System investigation to eliminate troubles arisen during
tests.

REFERENCES :
****
REPAIR SEA TRIALS ON DEEPWATER PIPELINES - M. BRAIT, A. COLOMBO - 3RD
HYDROCARBON SYMPOSIUM - LUXEMBURG 22-24 MARCH 1988.

```
*******************************************************************************
* TITLE : TRANSPORT PROPERTIES ESTIMATION SYSTEM      *      PROJECT NO       *
*          (TPES) FOR NATURAL GAS AND OIL             *                       *
*    -     PRODUCTION AND TRANSPORT                   *   TH./10070/89/HE/..   *
*                                                     *                       *
*******************************************************************************
* CONTRACTOR :                                        *   TELEPHONE NO        *
*    ALFAPI SA                                        *                       *
*                                                     *   301 6532579         *
*    MESSOGHION AVENUE 304                            *                       *
*    HE - 155 62 ATHENS                               *                       *
*                                                     *   TELEX NO            *
*                                                     *                       *
* TECHNICAL DIRECTOR :                                *   223296              *
*    MR. D.G. PAPANIKAS                               *                       *
*                                                     *                       *
*******************************************************************************
                                                         VERSION : 01/05/90
```

AIM OF THE PROJECT :
****
The development of a multipurpose Transport Properties Estimation System TPES
including a Data Bank and Prediction Subroutines Library for the determination
of Hydrocarbons properties data ie the objective of the project. The area of
application extends to liquid and gaseous hydrocarbons and some inorganic
elements and compounds (pure or one phases mixtures) used in Natural Gas and Oil
industry.
The engineering tool produced will be designed to be used in high technology
applications (as the most recent Computer Aided Hydrocarbons Applications are)
and works complementary to the existing engineering methods of analysis and
design.

PROJECT DESCRIPTION :
****
The best description of the project can be given by its stages structure.
The project will be performed in six main stages :
1. Compilation of Properties data will be collected, evaluated, classified and
prepared to be introduced into the data bank
2. Properties Prediction Methods and Processors for Data Estimation. The product
of this stage is to build the subroutines library which will be able to produce
transport and physical properties data for pure hydrocarbons and some inorganic
compounds or one-phase mixtures of them in Oil and Natural Gas systems.
3. Properties Data Base Deliverable of this stage is a Data Base and DB data
field containing all necessary properties, coefficients and parameters involved
in the previous stages
4. Testing, Integration, Validation and Applications. The testing and validation
procedures have to fulfil quality requirements in certainty of the technical and
physical results through verification tests based on computational examples from
literature and real application cases
5. Documentation. This stage includes the prepair and production of internal
Reports, project interim Reports and publications documenting all the technical
and scientific work done.
6. Project Management.

STATE OF ADVANCEMENT :
****
Ongoing. The project works begin in the first days of January 1990.

REFERENCES :
****
THE PROGRESS OF THE HOLE PROJECT WILL BE DOCUMENTED BY INTERNAL REPORTS AND WILL
BE PRESENTED BY INTERIN REPORTS TO THE COMMISSION. PRESENTATION OF RESULTS AND
DISSEMINATION OF KNOWLEDGE WILL BE PERFORMED BY SPECIAL LECTURES AND PUBLICATION
OF PAPERS IN SCIENTIFIC PERIODICALS.

NATURAL GAS TECHNOLOGY

```
****************************************************************************
* TITLE : NEW PROCESSING TECHNIQUES FOR OFFSHORE        *   PROJECT NO      *
*         GAS LIQUEFACTION                              *                   *
*                                                       *                   *
*                                                       *  TH./12009/85/FR/..*
****************************************************************************
* CONTRACTOR :                                          *  TELEPHONE NO     *
*    GERTH                                              *                   *
*                                                       *  1 47 52 61 39    *
*    4 AVENUE DE BOIS PREAU                             *                   *
*    FR - 92502 RUEIL-MALMAISON                         *                   *
*                                                       *  TELEX NO         *
*                                                       *                   *
* TECHNICAL DIRECTOR :                                  *  203 050          *
*    MR. LARUE                                          *                   *
*                                                       *                   *
****************************************************************************
```
                                                          VERSION : 31/12/88

AIM OF THE PROJECT :
****
The object of this project is to study and develop a new integrate gas treatment
and liquefaction process adapted to offshore production particularly on a mobile
support. The integration of these two functions in a single process will allow
to achieve simultaneously and in a sole phase, the dehydration, deacidyzation
and extraction of liquids from a natural gas and constitutes the first step
towards the transformation in LNG. Compared to the more conventional processes
of deacidyzation, dehydration and extraction of liquids from a natural gas,
which are generally implemented in a successive and distinct manner, this new
method should provide a gain in investment, in weight and in surface utilisation.
 Furthermore the innovations implemented in this new process make it
particularly suitable for offshore production.

PROJECT DESCRIPTION :
****
In theory, the process studied in this project, consists in refrigerating the
gas to be treated with the appropriate solvent, thus making it possible to run
in a single phase, both the treatment (dehydration, deacidyzation) and the
refrigeration (extraction of liquids from natural gas) of the gas. Technological
developments envisaged thus concern the treatment and the cooling of gas and
their integration in a sole process.
The treatment function allows to reduce the water and acid gas contents (CO_2,
H_2S). Therefore these constituents need to be separated owing to problems of
hydrates, crystallization, corrosion and specifications to which they can lead.
Classical dehydration and deacidyzation techniques resort to the use of solvents
in absorption columns (glycol, amines) followed by conditioners, thus leading to
heavy units poorly adapted to offshore production. The proposed technique
resides on a new and optimized implementation of solvent allowing a greater
compactness and a reduction of the number of columns.
The cooling function allows the refrigeration of gas, which leads to the
condensation of natural gas liquids. Traditional cooling techniques imply heavy
rotary machines (compressors, turbines or motor engines) and are hence poorly
adapted to offshore production. Refrigeration techniques studied within the
scope of this project and associated with the treatment function are of the
static type such as absorption cooling. They allow the production of cold
directly from heat which can be supplied by recovering thermal discharges or by
combustion. These techniques also increase the operating reliability and
flexibility and offer more sophisticated possibilities of integration with the
treatment section.
The project is divided into two phases:
PHASE 1: Study of the integrated treatment-liquefaction process. Both, the
treatment function (search of appropriate solvents and corresponding data,
evaluation of different patterns) and the refrigeration function will be first
studied separately. Their integration in a single process will be examined in a
second step.
PHASE 2: Experimental tests of the various functions of the process.
Experimental units for the treatment and refrigeration functions will be
designed and tested over a wide range of operating conditions. A document
relative to the process pilot will then be drafted.

STATE OF ADVANCEMENT :
****
The process study (Phase 1) included study of the processing functions (phase 1.
1) and of the refrigeration functions (phase 1.2) which are now completed; the
integration and layout study have been achieved.
The experimental test phase (phase 2) comprised the design and the fabrication
of experimental units (phase 2.1), the test and experimentation of these units
(phase 2.2) as well as the pilot preliminary design (phase 2.3); all these
phases have been achieved.

RESULTS :
****
Within the scope of the process study (phase 1), the processing and
refrigeration functions have been carried out and the integration and layout of

the process has been applied onto concrete examples.
The process function (phase 1.1) comprises two parts : a first step on
dehydration + extraction of natural gas liquids and a second step on
deacidyzation. New patterns concerning these two steps (patents) have been
developed, tesed on a laboratory loop and appropriate simulation models have
been set up.
The refrigeration function (phase 1.2) concerns the cycles of static
refrigeration such as absorption and thermocompression cycles. Different cycle
configurations have been studied and corresponding simulation models have been
set up. A comparative cost evaluation has shown the advantage of static
refrigeration cycles on traditional cycles with mechanical compression.
An integration and layout study (phase 1.3) has been applied to three concrete
applications of gas treatment : results show a gain of cost and compactness
(weight and surface area) for the integrated process compared to traditional
processing methods.
In order to experiment the different process functions (phase 2), two layouts
have been set up in view of testing at a predevelopment scale two main
constituents of the process : a gas-liquid contactor of semi-industrial size
operating under representative conditions and an ejector test bench.
These two facilities have been tested (phase 2.2). Experimentation of the gas-
liquid contactor confirmed the good working of this type of apparatus in the
particular operating conditions (very low liauid-gas ratio), different types of
contactors have been tested and specific correlations have been set-up. For the
gas-gasejector, the size of the test bench was not sufficient, and because of
the size effect the measured performances of the ejector were under calculated
values.
The preliminary study (phase 2.3) of the process pilot (later development phase)
has been completed.

REFERENCES :
****
J. LARUE - "REFRIGERATED PROCESSING FOR OFFSHORE NATURAL GAS" OCEAN INDUSTRY -
APRIL 1987 - P. 96

```
****************************************************************************
* TITLE : DEVELOPMENT OF A SCRUBBING PROCESS FOR        *    PROJECT NO     *
*         THE SEPARATION OF (ACID) COMPONENTS           *                   *
*         FROM (NATURAL) GAS                            *   TH./12010/85/NL/.. *
*                                                       *                   *
****************************************************************************
* CONTRACTOR :                                          *   TELEPHONE NO    *
*    NEDERLANDSE GASUNIE N.V.                           *                   *
*                                                       *    050 219111     *
*    P.O. BOX 19                                        *                   *
*    LAAN CORPUS DEN HOORN 102                          *                   *
*    NL - 9700 MA GRONINGEN                             *    TELEX NO       *
*                                                       *                   *
* TECHNICAL DIRECTOR :                                  *    53448          *
*    DR. G.E.H. JOOSTEN                                 *                   *
*                                                       *                   *
****************************************************************************
                                                VERSION : 15/09/89
```

AIM OF THE PROJECT :
****
The aim of the project is to develop an efficient and compact scrubbing process
for the removal of acidic components (like CO_2 and H_2S) from natural gas. The
innovative aspect of the project is, that the absorption fluid is directly
injected into the gasstream in co-current flow. In contrast to counter-current
absorbers this allows very high gasflows, in a compact installation.

PROJECT DESCRIPTION :
****
The research in this project was especially focussed on the removal of acidic
components from natural gas, in a co-current absorber. The following phases were
distinguished within this project :
PHASE 1.
Investigation of the hydrodynamic behaviour in the equipment. The co-current
absorption unit was installed and experiments were performed to examine the
dispersion of the liquid in the gasstream under various conditions.
Experiments were done to determine the efficiency of the liquid separation from
the gasstream using the Gasunie separator.
PHASE 2.
Determination of the absortion/desorption behaviour of various absorption fluids.
 In a stirred cell reactor the kinetic and equilibrium data of the various
absorption fluids were studied.
PHASE 3.
Investigation of mass transfer with selected absorption liquids in the co-
currently operated contacting equipment.
Regeneration of the solvent after gas/liquid separation.
PHASE 4.
Economic evaluation.
An economic evaluation was based upon the data obtained in phase 3. This
evaluation had to result in a "go or no go" decision with regard to phase 5.
PHASE 5.
Pilot plant tests under real operating conditions.

STATE OF ADVANCEMENT :
****
The project is completed. Phase 1 to 4 have been realised. The technical
feasibility was demonstrated, however, because of the only marginal economic
advantages (esp. on on-shore application) compared to traditional treatment
systems, the pilot plant phase was cancelled.

RESULTS :
****
1. HYDRODYNAMIC BEHAVIOUR OF THE EQUIPMENT
- several techniques have been used to determine the gas/liquid interfacial area
under variation of gas and liquid velocities. The influence of the use of liquid
injectors on interfacial area and the mass transfer was studied. Experiments
have shown that the liquid can be separated from the gasstream with an
efficiency of > 99.95 % in the operating region (gas velocities < 50 m/s).
2. ABSORPTION/DESORPTION EXPERIMENTS WITH VARIOUS LIQUIDS
- Absorption/desorption equilibria and kinetics were determined for several
(commercially available) absorption liquids. The results of these experiments
determined which liquids were tested in the model experiments in phase 3. MEA
(Mono Ethanol Amine) and a-MDEA (activated Methyl Di Ethanol Amine) were the
most promising liquids in this test.
3. MODEL EXPERIMENTS IN THE TEST UNIT
- In the co-current test unit with "groningen gas" artificially enriched to 1 %
CO_2 as a test gas, experiments were performed at various temperatures (20-40 deg)
 and pressures (20-40 bar). The technical feasibility of the process was shown
with MEA and a-MDEA as absorption fluids. Also was shown that sufficient CO_2
absorption could be reached within an acceptable lenght (50 m) of the absorber.
Because of the high regeneration duty needed for MEA, a-MDEA was choosen for an
extensive economic evaluation.
4. ECONOMIC EVALUATION
- The economic evaluation demonstrated that for the on-shore situation the

advantage in investment costs of the compact installation is not enough to
compensate the increase in energy costs in case of the co-current process.
Off-shore economics completely depend on specific circumstances. Because of this
the pilot plant phase was cancelled.

5. PILOT PLANT
- because of the results of the economic evaluation the pilot plant phase was
cancelled.

REFERENCES :
****
PERRY, R.H.; CHILTON, C.H.; CHEMICAL ENGINEERS HANDBOOK, FIFTH ED.
MC GRAW-HILL BOOK COMPANY, NEW YORK
KOHL, A.; RIESENFELD F.; GAS PURIFICATION, SEC. ED. 1974, GPC, HOUSTON, TEXAS, U.
S.A.
MADDOX, R.N. GAS AND LIQUID SWEETENING; CAMPELL PETROLEUM SERIES.
BLAUWHOFF, P.P.M.; DISSERTATION 1982. T.H. TWENTE, THE NETHERLANDS; SELECTIVE
ABSORPTION OF H2S FROM SOUR GAS BY ALKANOLAMINE SOLUTIONS.
VERSTEEG, G.F; DISSERTATION 1987; T.H. TWENTE, THE NETHERLANDS, MASS TRANSFER
AND CHEMICAL REACTION KINETICS IN ACID GAS TREATING PROCESSES.
WESTERTERP, K.R..; SWAAIJ, W.P.M. VAN; BEENACKERS A.A.C.M. VAN; CHEMICAL REACTOR
DESIGN AND OPERATION, 2ND ED., 1984, JOHN WILEY AND SONS, NEW YORK.

```
****************************************************************************
* TITLE : DEVELOPMENT OF A FLOATING LPG TERMINAL       *    PROJECT NO      *
*         BASED ON A CONCRETE BARGE WITH WET WALL       *                    *
*         INSULATION OF THE STORAGE COMPARTMENTS        *  TH./12012/86/DE/.. *
*                                                       *                    *
****************************************************************************
* CONTRACTOR :                                          *  TELEPHONE NO      *
*    LGA GASTECHNIK GMBH                                 *                    *
*                                                       *  02228-15204       *
*    BONNER STRASSE 10                                   *                    *
*    POSTFACH 604                                        *                    *
*    DE-5480 REMAGEN 6                                   *  TELEX NO          *
*                                                       *                    *
* TECHNICAL DIRECTOR :                                   *  8869              *
*    MR. H.BACKHAUS                                      *                    *
*                                                       *                    *
****************************************************************************
```

VERSION : 01/10/89

AIM OF THE PROJECT :
****
Since the erection of land-based storage tanks for LPG in coastal terminals will
face increasing difficulties (space requirements, safety consideration) it is
the aim of the project to develop a storage unit which can be operated either
floating in fore-shore areas or grounded at-shore.
A mobile floating unit can be towed to the final site of operation, which, in
many cases, may be an economical alternative to a fixed founded structure
especially in regions with insufficient infrastructure.
Another advantage is the fact that the whole equipment, including the
reliquefaction plant and the storage tanks, can be installed at the construction
site of the barge where all facilities are available in order to guarantee a
high standard of fabrication. Once completed, the LPG terminal is towed to the
location and anchored.

PROJECT DESCRIPTION :
****
The construction elements of the storage facility are concrete for the floating
body and polyurethane foam for the insulation of the load bearing concrete walls
of the containments. The special aspect of the project is the so-called wet wall
insultion. Whilst the conventional insulation of low-temperature storage tanks
is applied on the outside of the storage tank (and therefore not in contact with
the liquid), a wet wall insulation shall be fitted inside the storage tank. This
means that the insulation material is wetted by the stored liquid. Consequently,
the effects of liquid/gas absorption, desorption, possible cracks and other
defects have to be taken into account, when it comes to the selection of
suitable PU foams.
The target of the proposed development is to achieve a wet wall insulation
consisting of :
- a vapour barrier (liner) directly applied on the concrete walls
- sprayed polyurethane foam built up in several layers
- possibly reinforcement layers and/or LPG barriers incorporated in the PUF or
applied into the PUF surface.
The wet wall insulation will have the following main functions and
characteristics :
- thermal insulation to limit boil-off
- a liquid barrier to prevent contact between concrete and cold LPG and to
prevent LPG leakage
- the vapour barrier prevents water vapour penetrating into the insulation
- above functions must be fulfilled in a floating barge, in which the concrete
barge as well as the LPG are to be considered as dynamic (liquid impact
resistance)
- insulation has to be load bearing
The development is aimed to achieve a technically acceptable and economically
attractive floating LPG terminal. The wet wall insulation applied against the
inner face of the barge's concrete compartments and in direct contact with the
LPG is the essential part, because
- it saves space and thus cost
- it eliminates a double containment (e.g. insulated steel tanks placed inside
or on a concrete barge).

STATE OF ADVANCEMENT :
****
Concrete barge : General design, type and sizes of storage compartments, outer
double wall construction, stress analyses. Insulation : Selection of suitable PU-
Foams and liner material, testing of various materials regarding gas and water
vapour density and adhesion of liner on concrete and insulation on liner.

RESULTS :
****
CONCRETE BARGE :
The design features of 2 types of storage barges, i.e. for the beached and
floating version are available.
For the beached storage barge, the results of the analysis show that double-
walled floors undergo deformations in the floor and deck in the "floating state

without ballast" which are too large. Such a design is, therefore, not suitable.
The version with a double-walled floor and with the arrangement of corner
strengthening in the deck at the barge walls reduces deformations during
transports to an acceptable amount, i.e. the applied insulation material will
not be affected.
The floating barge, i.e. the investigated version with 3 tanks and a
longitudinal bulkhead as permeable sloshing wall, also shows that expected
deformations are within the permissible range.

INSULATION MATERIAL :
A vast number of tests with various insulation and liner materials have been
carried out and are still going on, in particular :
PU-foamtest versus pentane
Tests were aimed to define chemical resistance of the PUF, absorption and
desorption.
Results :$
- It was demonstrated that the sprayed foams tested (type MDIIS 50 and HDIIS 90)
show a very low absorption for pentane under the conditions of this tests.
- The desorption is very rapid.
- From these results and by changing the surface/value ratio of the samples it
is becoming evident that the pentane penetrates only the cut cells on the
surface but does not penetrate the core of the PUF.

PU-FOAMTEST VERSUS LPG
The absorption and desorption are tested and the possibly different behaviour of
different PUF-samples are investigated. Foams of densities 35-60-90-160 kh/m3
were tested comparitively. Tests were started end of 1988 and are still
continuing.
The first tests have shown already a very significant difference in behaviour
between absorption under pressure - without pressure and low, medium and high
density foams.

```
*************************************************************************
* TITLE : PROTOTYPE STEP BY STEP SWIVEL JOINT FOR    *    PROJECT NO    *
*         HIGH PRESSURE GAS                          *                  *
*                                                    *                  *
*                                                    *   TH./12013/87/FR/..  *
*************************************************************************
* CONTRACTOR :                                       *   TELEPHONE NO   *
*    TECHNIP GEOPRODUCTION (TPG)                     *                  *
*                                                    *   (1) 47 78 21 21 *
*    TOUR TECHNIP                                    *                  *
*    FR-92090 PARIS LA DEFENSE CEDEX 23              *                  *
*    FRANCE                                          *   TELEX NO        *
*                                                    *                  *
* TECHNICAL DIRECTOR :                               *   613235          *
*    G. DELAMARE                                     *                  *
*                                                    *                  *
*************************************************************************
                                           VERSION : 01/09/89
```

AIM OF THE PROJECT :
****

The purpose is to carry out the detailed design, construction and testing of a prototype 4" diameter swivel joint for high pressure gas (5,000 psi) adapted to the offshore conditions and enabling a floating oil production support to weathervane about its mooring point.
The concept adapted by TPG resoluted discards the use of sliding lip joints for confining high pressure fluids. The sliding lip joint technique is not perfectly reliable and rapidly reaches its limits when the fluid is gas and the pressure very high (over 350 bars) as is increasingly the case in oil production.
The new concept proposed consists in confining the fluid by means of an inflatable joint which is static but allows limited rotation (+/- 15 deg) through a torsionally deformable sleeve made of a rubber-metal laminated assembly.

PROJECT DESCRIPTION :
****

The TPG step by step swivel joint will necessarily include two identical devices, each consisting of an inflatable joint associated with a torsionally deformable sleeve so as to obtain a rotation of more than 15 deg., even several full rotations, by actuating the devices step-by-step one after the other. This tightness transfer from one joint to the other will be achieved by a hydraulic distributor actuated through a cam by the rotation of the floating support about its mooring point.
The use of a swivel joint of this type is particularly suited to the conditions of a tanker weathervaning about a mooring buoy where 99 % of the oscillations is smaller than +/- 15 deg. Most of the time, the fluid confinement will thus be obtained by a fully static method, without friction, the hydraulic actuator being operated only in the case of a sharp in the wind direction.
The project includes 4 phases :
1. Finalisation of design work associated with the preliminary project for the prototype, compilation of construction drawings and engineering documents, and design work in connection with the test facilities.
2. Procurement, fabrication and assembly of the prototype and test equipment.
3. Tests (partial testing of components, operating tests on the whole assembly and minimum endurance test).
4. Studies for application of the concept to concrete cases.

STATE OF ADVANCEMENT :
****

Ongoing. The project is in the test phase.

RESULTS :
****

- Preliminary tests have been done to select the critical components.
- Basic design work has been performed.
- Several patents have been obtained worldwide.
- Construction drawings are completed.

REFERENCES :
****

LECTURE : 1986 HOUSTON. PRESENTATION OF THE CONCEPT DURING A TECHNICAL SEMINAR.

```
*****************************************************************************
* TITLE : NEW PROCESSING TECHNIQUES WITH A VIEW     *     PROJECT NO       *
*         TO THE OFFSHORE LIQUEFACTION OF GAS       *                      *
*         (PHASE II:PILOT PROCESSING)               *    TH./12014/88/FR/.. *
*                                                   *                      *
*****************************************************************************
* CONTRACTOR :                                      *    TELEPHONE NO      *
*    GERTH                                          *                      *
*                                                   *    1 47.52.61.39     *
*    4, AVENUE DE BOIS PREAU                        *                      *
*    FR - 92502 RUEIL-MALMAISON                     *                      *
*                                                   *    TELEX NO          *
*                                                   *                      *
* TECHNICAL DIRECTOR :                              *    203 050 - FAX 1 47 5*
*    MR JOSEPH LARUE                                *    2 69 27           *
*                                                   *                      *
*****************************************************************************
                                                    VERSION : 07/11/89
```

AIM OF THE PROJECT :
The purpose of the project is to build and test an integrated pilot processing plant for the liquefaction of natural gas, which has been the subject of a design and feasibility study within the framework of contract nr. TH 12.09/85. This new process for natural-gas processing is a new processing technique requiring a single stage for performing a series of natural-gas processing operations upon production - dehydration, adjustment of the hydrocarbon dew point with the possible production of NGL and sweetening.
The study has shown that, compared to conventional processes of dehydration, NGL extraction and sweetening the proposed integrated process leads to a reduction in investments and to greater compactness (weight and base-level surface area). The current development phase involves the testing of the industrial qualification of the process required for its subsequent commercialization.

PROJECT DESCRIPTION :
The principle of the integrated process is based on the implementation of a physical solvent at low temperature. The lowering of the temperature causes the absorption of water (inhibiting the formation of ice and hydrates) and acid gases (CO_2, H_2S) in the solvent together with the partial condensation of the gas, which results in two liquid phases that are separated at low temperature. the process makes use of new concepts that have given rise to patent applications:
- **Processing**: because of the way the solvent is implemented, only a minimal amount is needed. Likewise, a new solvent-regeneration technique has been developed so as not to have recourse to distillation as is usually the case. This new technique is entirely integrated into the process.
- **Refrigeration**: the techniques developed to provide the refrigeration required for processing are of the static type (absorption, thermocompression) and make direct use of the heat that may be supplied by thermal discharges or by combustion. These techniques do not requrie any large rotating machinery (compressors, turbines), while increasing reliability, resulting in a reduction of investments and providing greater possibilities of integration with the processing part of the process.
The project involves four phases:
Phase 1: Designing the pilot plant. Definition of the different modules, working out the construction blueprints, defining the metrology and controls, and defining and choosing the different pieces of equipment.
Phase 2: Building the pilot plant. Ordering the principal equipment, preparing the manufacturing briefs for the different modules, equipping the installation site, and assembling the modules.
Phase 3: Testing the pilot plant. Start-up and tuning, testing over a wide range of operating conditions, performance testing, and data acquisition.
Phase 4: review and conclusions. Test interpretations and conclusions, creating computing models, studies of real applied cases.

STATE OF ADVANCEMENT :
The pilot plant and the peripheral equipment have been designed and built (Phase I and II). The project is currently in the testing phase (Phase III), the interpretation of the first results has begun and some real applied cases are being studied (Phase IV).

RESULTS :
During the phase I, the definitive diagram of the pilot plant and the principal equipment have been defined; a general control system has been studied, compoting a PC unit, in view to order the metrology and the regulation; then, a set of books have been drawn up : process book, instrumentation book, bundle of hardware equipment.
The phase II consisted to realize making drafts and to arrange the control system; the equipment have been ordered and the different skids assembled, adjusted and checked before the starting up of the plant. This phase has been completed by the installation of the analysis system and some adjustments and

settings. After preliminary tests, experimentation of the pilot plant (Phase III) has started with tests on IFPEX-1 part of the process (dehydration + NGL extraction). Extended runs have been performed with steady-state conditions, allowing data acquisition.
Interpretation of results (Phase IV) has been initiated. Comparative evaluations are currently studied on real applied cases.

REFERENCES :
****
J. LARUE, "REFRIGERATED PROCESSING FOR OFFSHORE NATURAL GAS", OCEAN INDUSTRY, p. 96, APRIL 1987.

```
*****************************************************************************
* TITLE : DESIGN, BUILDING AND EXPERIMENTAL        *      PROJECT NO        *
*         QUALIFICATION OF AN AIR-WATER MODEL FOR   *                        *
*         SIMULATING THE DYNAMIC SEPARATOR.         *   TH./12015/88/FR/..   *
*                                                   *                        *
*****************************************************************************
* CONTRACTOR :                                      *   TELEPHONE NO         *
*    BERTIN & CIE                                   *                        *
*                                                   *   1 34 81 85 00        *
*    B.P. 3                                         *                        *
*    F-78373 PLAISIR CEDEX                          *                        *
*                                                   *   TELEX NO             *
*                                                   *                        *
* TECHNICAL DIRECTOR :                              *   696231               *
*    RENE BOURASSIN                                 *                        *
*                                                   *                        *
*****************************************************************************
                                                       VERSION : 01/10/89
```

AIM OF THE PROJECT :
****
The purpose of this project is to develop a system able to ring an innovative
technological solution to the problem of two-phase mixture compression on sea-
production chains. This problem is solved presently by static separation which
weight and volume generate important costs.
Taking into account this techno-economical aspect, it seems that the two-phase
centrifugal compressor including a two-phase dynamic separator should replace
progressively the present static separators, all the more so as the development
of screw compressor, the concurrent technology, has not yet been completed.
 Consequently, the development of the two-phase centrifugal compressor which of
course strongly depends on the development of its static separator, is a very
profitable investment.

PROJECT DESCRIPTION :
****
The present project deals with the hydrodynamical and mechanical design and the
qualification of the two-phase dynamic separator on an experimental air-water
apparatus. This separator will have to treat very high massic liquid rates (up
to 50 %), under high efficiency separation objectives (down to 7 micro m).
The liquid phase is supposed to be in a mist flow configuration, assuming that
liquid slugs or plugs can be erased before the separator by specific system.
Moreover, the size of solid particles is assumed to be lower that 5 micro m for
a maximum concentration of $10(-2g)/m3$.
The dynamic separator is a conical-shaped cylinder where the liquid phase is
centrifugated and transformed into an annular liquid film which is extracted by
a bladed collecting disc at the inlet.
The development programme is divided into three part :
Part 1 : Detailed design of the two-phase air-water separator model
- Aerodynamical design
- Hydrodynamical design
-Separation study
-Mechanical study
Part 2 : Fabrication of the air-water model
Part 3 : Qualification of the air-water model by laboratory testing
The purpose of this stage will be to check :
- the separator characteristics
- the liquid annular film stability
- the liquid extraction and collecting system (pressure level).

STATE OF ADVANCEMENT :
****
Ongoing
Advanced studies were performed on these two subjects :
- aerodynamical design of the swirler system to induce rotation on the moisted
air,
- hydrodynamical design of the collecting disc to extract the liquid phase.

RESULTS :
****
The rotating flow will be obtained by a straight bladed stator (.05 meter cord)
and a twisted bladed rotor (.05 meter cord). The study of the collecting disc
has shown the possibility to obtain 40 bar pressure recovery on the water with
such a system.

ENERGY SOURCES

```
***********************************************************************
* TITLE : SURFACE-INDEPENDENT UNDERWATER ENERGY        *   PROJECT NO       *
*         SUPPLY SYSTEM-DIESEL ENGINE WITH CLOSED      *                     *
*         GAS CYCLE                                    *   TH./13006/85/DE/.. *
*                                                      *                     *
***********************************************************************
* CONTRACTOR :                                         *   TELEPHONE NO      *
*    M.A.N. TECHNOLOGIE GMBH                           *                     *
*                                                      *   089 1480 3459     *
*    DACHAUER STRASSE 667                              *                     *
*    POSTFACH 50 04 26                                 *                     *
*    DE - 8000 MUECHEN 50                              *   TELEX NO          *
*                                                      *                     *
* TECHNICAL DIRECTOR :                                 *   523 211           *
*    DR. H. GEHRINGER                                  *                     *
*                                                      *                     *
***********************************************************************
                                          VERSION : 16/03/88
```

AIM OF THE PROJECT :
****
The closed-circuit diesel engine using argon as cycle medium was developed under
the completed project TH./15056/84 and successfully tested in a pilot plant with
a 32 kW MAN diesel engine.
The aim of this present development project is the construction of a prototype
surface-independent underwater energy supply system with an argon cycle diesel
engine (100kW), and its testing in a work submergible under practical operating
conditions. In this way the operational suitability of this autonomous energy
supply system as generating plant in submarines, underwater work stations, diver
habitats, etc. in the offshore technology sector is to be demonstrated.

PROJECT DESCRIPTION :
****
The energy supply unit, consisting of the diesel engine, the closed argon gas
circuit and the exhaust gas scrubber, together with the various auxiliarly
systems, was designed and manufactured as a complete system for installation in
the test machine room section of a submachine.This test section was developed by
Merss. Bruker (TH/13007/85) and its design data were oriented on the
requirements of a small submarine for commercial applications.
The diesel engine will serve as energy source for the craft's hydraulic and
propulsion system and for its generator. Envisaged is a combined operational
mode - a closed argon cycle for submerged operation and an open air cycle for
surface operation.
Design criteria and boundary conditions for the energy supply unit were strongly
influenced by the submarine's assumed operational profile. Consequently
intensive coordination with Messr. Bruker was necessary in the design phase, in
order to achieve optimal adaptation.

STATE OF ADVANCEMENT :
****
Completed

RESULTS :
****
MOTARK (MOTOR IN ARGON-KREISLAUF / engine in argon cycle) is an alternative
drive and power-supply system integrated in the offshore-working submarine
"Seahorse 2", which belongs to Messr. Bruker Meerestechnik.
The heart of the plant is a naturally aspirated diesel engine, MAN model D 2566
ME (100 kW, 1500 rpm), which can operate in a closed argon cycle independent of
the outside air while the submarine is under water, and in the conventional
manner after the vessel has surfaced.
Subsequent to cooling down carbon dioxide CO_2, which is the combustion product
of fuel and oxygen, was removed from the process gas in a chemical process with
potassium hydroxide in a dual-stage rotary disintegrator. After another cooling
cycle and a cyclonic condensate cleaning process oxygene was supplied to the
argon carrier gas in measured quantities.
Governing of the MOTARK system and acquisition of the test data are performed by
a custom-developed micro-processor unit.
The functional tests in the submarine as well as the subsequent underwater tests
at shallow sea gave convincing evidence for the fact that this prototype unit is
now ready for regular operation.

```
*******************************************************************************
* TITLE : AN AIR INDEPENDANT POWER SOURCE OF HIGH   *      PROJECT NO         *
*         ENERGY STORAGE DENSITY                    *                         *
*                                                   *   TH./13007/85/DE/..    *
*                                                   *                         *
*******************************************************************************
* CONTRACTOR :                                      *      TELEPHONE NO       *
*    BRUKER MEERESTECHNIK GMBH                       *                         *
*                                                   *   07 21/59 67-1 80      *
*    P.O. BOX 21 02 32                              *                         *
*    DE - 7500 KARLSRUHE 21                         *                         *
*                                                   *   TELEX NO              *
*                                                   *                         *
* TECHNICAL DIRECTOR :                              *   78 25 656             *
*    MR. J. HAAS                                    *                         *
*                                                   *                         *
*******************************************************************************
                                                    VERSION : 07/02/90
```

AIM OF THE PROJECT :
****
Battery powered submarines are suffering from moderate energy densitiies of
conventional battery systems and thus from limited range and endurance.
The aim of this project was to considerably increase the energy storage capacity
of power sources for autonomous work and research submarines and other subsea
installations requiring electrical, mechanical and/or thermal energy to sustain
operation.
A complementary aim of the project was to study smaller, low cost closed cycle
diesel systems to be used, for example, for auxiliary power plants of larrger
submarines or as stand alone energy sources for smaller vehicles operating in
moderate water depths.

PROJECT DESCRIPTION :
****
The aim was to be achieved by integration of a closed cycle argon diesel engine,
developed by MAN Technologie GmbH, Munich, into an engine room section ad in the
second stage into a complete operational 50 to Autonomous
Inspection/Experimental Submarine developed by Bruker Meerestechnik GmbH. The
function and reliability of the engine plant and its contrlos was to be proven
in a series of relastic dry and wet tests including sea trials with the complete
experimental submarine.
The construction of the engine room section of the submarine mentioned included
the Argon Diesel engine with electric motor/generator and hydraulic pumping
aggregate, rotary scrubber, heat exchangers, auxiliary equipment, chemical
storage installations and liquified oxygen plant. Furthermore, the section was
to be fitted out with the fuel system, parts of the ballast, trim and drainpump
systems and the related electrical installations and switchboard, all within the
responsability of Bruker Meerestechnik GmbH.
The subsystems supplied by MAN included the diesel engine, tupe D2566, rating
100 KW at 1500 rpm, the scrubber, the heat exchangers, part of the gas circuit
pipe work, auxiliary pumps, sensors and controls and the microcomputer to survey
the diesel engine plant.
Since the bench tests with the CC-Argon Diesel plant carried out by MAN at
Munich proved to be promising, it was decided to extend the project and to built
a complete, fully operational Experimental Submarine around the engine
compartment in order to prove the functioning under at sea conditions. This
extension was not part of the EEC-project.
For safety reasons, the submarine was fitted out with a lead acid battery
installed in the forward section, consisting of newly developed maintenance free
gel type cells, providing a nominal energy of 200 KWh. When the CCD-plant has
been cut off, the battery delivers sufficient energy for 5 hrs submerged
operation.
Energy transmission to the propulsion aggregates (main thruster, side and
vertical thruster) and other equipment is performed hydrostatically.
Compared to other CCD alternatives the Argon diesel uses a mixture of Argon and
Oxygne as process gas to improve the engine's efficiency.
In addition to the Experimental Submarine with Argon Diesel, Bruker
Meerestechnik GmbH developed and built a 20 KW CO_2-Diesel plant, improved it
stepwise and tested it in a test bench arrangement.
Oxygen was supplied from an H.P. gas bank. The exhaust gas was cooled down by
water injection and furthermore in a heat exchanger to condensate the water. The
gas was heated again and oxygen was added. Excess exhaust gas was pumped into a
pressure vessel simulating various ambient pressures.

STATE OF ADVANCEMENT :
****
Both parts of the project (Engine Compartment/Experimental Submarine with closed
cycle Argon Diesel and 20 kW CO_2-diesel plant) have been successfully completed,
the first by proving the systems' capabilities during the sea trials, the latter
by performing longterm test runs under full load.

RESULTS :
****
The experimental submarine incorporating the engine compartment with closed

cycle Argon Diesel was thoroughly tested in three stages : workshop tests, shallow water trials and sea trialsm all supervised by the Germanischer Lloyd. All systems proved to be functional and reliable. The Argon Diesel plant has an exceptional high safety standard.
Fuelling the sub with liquified oxygen and chemicals was carried out repeatedly without any problems. The envisaged improvement figures in energy storage density compared to conventional batteries were met. Further improvements with regard to more economical chemical consumption are nevertheless desirable.
The Argon CCD-plant with the ambient during operation. Therefore, there is an equilibrum achieved with regard to the weight balance; only the trim of the sub has to be taken care of. The power requirement for auxiliary eauipment, pumps etc.. is depth independent as well. Transition from open to closed circuit operation and vice versa is easily possible after a short stop of the engine. Starting of the engine when the submarine is submerged can be achieved as well. The LOX tank installed has a very low self-evaporation rate. Pressure built-up from 1 bar until opening of the relief valves takes several weeks.
The experimental submarine in total proved to be outstandingly handly and manoevrable in all modes of operation, Due to the degree of automatization and the remote operation of all technical installations a crew of four proved to be sufficient to operate the submarine in two shifts.
The experimental submarine SEASHORE-KD is the ever first one operating on an absolutely closed cycle diesel engine principle.
The 20 kW closed cycle CO2-Diesel developed by Bruker worked very satisfactorily as well. As expected, the closed cycle efficiency was inferior to the operation in open circuit mode or the Argon Diesel.
Consequently, the oxygen consumption per kWh is higher than with the latter. Furthermore, a small part of the oxygen is lost by dumping some of the exhaust gas over board.
A part of the engine's power output is needed to drive the exhaust gas compressor. The power required increases with the operating depth of the submarine or power supply unit.
At a closer look, the CO2-plant proved to be astonishingly favourable: The compressor input rises underproportionql with the depth. The Argon (and all other CCD-variations) also require some power to drive auxiliary equipment.
The most important factor in favour öf the Bruker CO2-pldiesel is the almost complete absence of scrubbing installations, related machinery and agent storage facilities. With regard to the net energy storage density this largely overcompensates the increased fuel and oxygen consumption. In fact, the enrgy storage density in kWh/kg is about twice compared to the Argon Diesel at the present state. Furthermore, the operating costs of the CO2-Diesel per kWh are considerably inferior to the other alternatives.

REFERENCES :
★★★★★★★★★★★★
J. HAAS : "THE BRUKER MAN ARGON-DIESEL AND THE BRUKER CO2-DIESEL" PAPER PRESENTED ON A WORKSHOP AT GKSS/GEESTHACHT, SEPTEMBER 1987.
J. HAAS : "AIR INDEPENDENT POWER SOURCES OF HIGH ENERGY STORAGE DENSITY. THE BRUKER-MAN ARGON-DIESEL AND THE BRUKER CO2-DIESEL" PAPER PRESENTED AT ECC-SYMPOSIUM AT LUXEMBOURG IN MARCH 1988.
J. HAAS : "EXPERIMENTAL SUBMARINE WITH CLOSED CYCLE DIESEL ENGINE SEASHORE-KD" TECHNICAL DESCRIPTION; UNPUBLISHED; FEBRUARY 1988.
BRUKER/J.HAAS : "BRUKER SUBMARINE WITH CLOSED CYCLE ARGON DIESEL NEARS COMPLETION" PRESS RELEASE AUGUST 1988.
J. HAAS : "MEERESTECHNIK AUS DEM BINNENLAND" PAPER INCLUDING THE EXPERIMENTAL SUBMARINE WITH ARGON CCD-PLANT,UNIV. OF BREMEN,JULY 1989.

```
****************************************************************************
* TITLE : POWER PACKAGE FOR REMOTE OPERATED          *      PROJECT NO       *
*          SUBMARINE VEHICLE WITH HIGH SPECIFIC      *                       *
*          ENERGY POWER R.5 TO 10 KW                 *   TH./13009/87/FR/..  *
*                                                    *                       *
****************************************************************************
* CONTRACTOR :                                       *    TELEPHONE NO       *
*    BERTIN & CIE                                    *                       *
*                                                    *   (1) 34.81.85.00     *
*    BP 3                                             *                       *
*    FR - 78373 PLAISIR CEDEX                         *                       *
*                                                    *    TELEX NO           *
*                                                    *                       *
* TECHNICAL DIRECTOR :                               *    696 231            *
*    MR VERNEAU                                       *                       *
*                                                    *                       *
****************************************************************************
                                              VERSION : 26/09/89
```

AIM OF THE PROJECT :
****
The major problem in the use of deep sea (down to 1.000 m, for example) remote
operated submarine vehicles of a few kilowatts (for pipeline surveying,
cartography, etc.) is the short duration of possible missions (typically no more
than 12 hours and 150 km) due to the low specific energy of existing energy
sources. Increasing significantly the cruising radius, for example up to 300-450
km, needs to design a prime over system with a high specific energy
(approximately 200 wh/kg) and a very high reliability as well. The objective of
the project is to develop a high specific energy system including a combustion
chamber burning methyl alcohol and oxygen with condensation of the exhaust gas
and an Organic Rankine Cycle.

PROJECT DESCRIPTION :
****
The proposed system is an external combustion engine made up of a combustion
chamber with condensation of the exhaust gas and a Rankine cycle with expansion
of an organic vapour in a turbine. The power is in the range of 5 to 10 kw.
The combustion chamber burns methyl alcohol and oxygen. The design pressure of
60 bars allows to condensate the exhaust gas and store it in a small tank. This
avoids a gas compressor, heavy and powerful for outside rejection in deep sea,
and the total mass of the system remains a constant.
The down scaling (40 to 80 kW instead of 400 kW which are presently designed for
submarine application TH 15.55.84) of the combustion chamber will take into
account the results obtained on the test bench currently on construction.
The hot gas exhausting from the combustion chamber passes through the
evaporation of the Rankine Cycle.
The organic vapour so produced expands in a turbine driving a generator.
In order to meet the objectives of high reliability, high specific energy (200
Wh/kg), no material rejection outside the vehicle, adaptation to various depths,
the main technical options of the system include :
- an intermediary loop between hot gas and working fluid circuit to avoid hot
spots and prevent thermal decomposition of the working fluid.
- a sealed turboalternator ensuring no leakage of the circuit and nearly no
maintenance.
The present contract comprises two parts :
FIRST PART : a project of a 5-10 kW engine with the definition of the hot-gas
system (combustion chamber-heat exchanger-condenser) and the study of the
Rankine Cycle (selection of the fluid and thermodynamic cycle. Definition of
main components : turboalternator, feed pump, heat exchangers).
SECOND PART : design and experiment of a sealed turboalternator of 5-10 kW. For
cost reasons the combustion chamber will be replaced, at this stage, by a
standard gas burner. This does not entail any modification of the technology of
the turboalternator.

STATE OF ADVANCEMENT :
****
Ongoing. The design of the feed pump is achieved. The design of the sealed T.A
will be achieved in September. We keep or the study of the high efficiency
Rankine engine (H.E.R.E.).

RESULTS :
****
The engines for remote operated submarine vehicules have a power range of 2 to
60 kW and the most frequent power is 10 kW (electrical power). The main
components of the test bench of the Rankine cycle are designed :
- the sealed turbo-alternator (22 000 rpm, permanent magnet)
- the feed pump (3 000 rpm, pitot type)
- the condenser
- the evaporator
- the temperature of H.E.R.E. would be about 400 deg.

REFERENCES :
****
N.61.88.88.

```
********************************************************************************
* TITLE : AUTONOMOUS UNDERWATER ENERGY MODULE :        *      PROJECT NO      *
*          HEAT GENERATION LOOP                        *                      *
*                                                      *   TH./13012/89/FR/..  *
*                                                      *                      *
********************************************************************************
* CONTRACTOR :                                         *    TELEPHONE NO      *
*    BERTIN & CIE                                      *                      *
*                                                      *    34.81.85.00       *
*    B.P.3                                             *                      *
*    FR - 18373 PLAISIR CEDEX                          *                      *
*                                                      *    TELEX NO          *
*                                                      *                      *
* TECHNICAL DIRECTOR :                                 *                      *
*    MR. D. GROUSET                                    *                      *
*                                                      *                      *
********************************************************************************
                                                    VERSION : 31/03/90
```

AIM OF THE PROJECT :
****
The main objective of the project is the tuning up of a heat generation loop
with a thermal power of 400 kW, cooled by pressurized water. The loop is
composed of :
- a 60 bar aeronautical combustion chamber, working with methanol and oxygen,
and cooled by injection of recirculating combustion gases at around 200 deg
- a pressurized water boiler to cool the combustion gases down to 200 deg
- a water cooled exchanger to condense the combustion products (CO_2, H_2O)
- a pump for recirculating the gases to the combustion chamber.
In the future, this heat generation loop is to be the heat generation device of
a Rankine cycle power module using an organic fluid.
Thanks to the condensation of the combustion products, no exhaust system is
required. So, the mass of the module is constant and there is no limit of the
diving depth.

PROJECT DESCRIPTION :
****
The project began in October 1989 and has 3 years duration. It is divided into 4
phases :
Phase 1 : Design, sizing of the heat generation loop, specifications,
construction drawings of each component of the test bench.
Phase 2 : Procurement and fabrication, erection of test facilities : modified
combustion chamber, fluid circuits, electrical circuits for control, regulation
and safety. Implementation of a data acquisition programme for measurement and
control.
Phase 3 : Tests of the heat generation loop, including ignition, thermal
behaviour, thermal balance and efficiency under various operating conditions.
Phase 4 : Synthesis and integration.

STATE OF ADVANCEMENT :
****
Ongoing. Started in October 1989. Phase 1 in progress.

RESULTS :
****
- the combustion chamber has been adapted to the new configuration
- the flow-sheet of the test bench is realized
- the regulation principles are defined.

STORAGE

```
****************************************************************************
* TITLE : STORAGE OF LIQUID AND LIQUEFIED        *    PROJECT NO        *
*         HYDROCARBONS IN LINED HARDROCK CAVERNS  *                      *
*                                                 *    TH./14020/85/DE/..  *
*                                                 *                      *
****************************************************************************
* CONTRACTOR :                                    *    TELEPHONE NO      *
*    SALZGITTER AG                                *                      *
*                                                 *    030 88 42 97 13   *
*    ABTEILUNG FORSCHUNG UND ENTWICKLUNG          *                      *
*    POSTFACH 15 06 27                            *                      *
*    D - 1000 BERLIN 15                           *    TELEX NO          *
*                                                 *                      *
* TECHNICAL DIRECTOR :                            *    185 655           *
*    DR G. KLAUS                                  *                      *
*                                                 *                      *
****************************************************************************
```

VERSION : 01/07/89

AIM OF THE PROJECT :
****
Rock caverns have until now been constructed and utilized according to a storage
concept developed in Scandinavia for the storage of crude oil, mineral oil
products such as heating oil, liquid gas (LPG). The application of this storage
concept with a hydrodynamic water seal is severely restricted in the European
inland, as well as in other countries for reasons of environmental and
groundwater protection. In addition contaminations of the storage product, which
fundamentally impair the product stability occur with this storage principle due
to the direct contact between groundwater, rock and storage product.
This problem could be avoided if the storage were to occur in lined caverns. The
aim of this project is to develop storage concepts, which stand out above the
present-day technologies of their greater geotechnical safety potential, greater
geographical utilization possibilities and more diverse employment possibilities
with respect to different products stored.

PROJECT DESCRIPTION :
****
The aim of this project is the preparation of a largescale workable construction
concept for lined rock caverns for storage of a variety of hydrocarbon products.
An evaluation of the stability and tightness of cavern storage facilities (with
internal lining) is interconnected with questions of geology, rock mechanics,
geohydraulics, geothermics, construction technology, construction material
technology, belowground construction and operation technology. The stability,
tightness and economic viability of this storage type are predominantly
dependent on the geological underground conditions, the construction process and
the product-specific stress characteristics.
In the project's initial phase the bases are prepared in the individual subject
areas of underground construction. Using rock mechanical and numerical
calculations the rock caverns' stress/ strain conditions are examined and
criteria for design and dimensioning prepared. Laboratory tests on material laws
and material characteristics of the construction, sealing and heat insulation
materials are carried out and incorporated in the theoretically calculated
simulation.
The preparation of the construction concepts is carried out giving due
consideration to concrete location possibilities but the concepts are designed
in such a way that an adaptation to various different geological conditions is
possible. A definite pilot project, with respect to location selection is only
possible with regard to the construction concept and the cost efficiency
analysis and after detailed preparation of project proposals with the future
operators of the project. Construction concepts developed for lined cavern
storage facilities are the subject of cost efficiency analysis in comparison to
existing alternative storage concepts. Any concepts which do not appear feasible,
 wether for economical and/or technical reasons are dropped.
The last stage of this project is the planning of a pilot plant. After previous
preliminary talks with possible operators the following projects are carried out
to set up pilot plants.
Storage Type 1: liquid hydrocarbons
Storage Type 2: compressed liquefied hydrocarbons
                location: areas of consumption, large towns.
In the preliminary phase of the planning itemization of the proposed project
will be carried out with the operators with respect to storage product, cavity
volumes, locations, etc.

STATE OF ADVANCEMENT :
****
The project is completed

RESULTS :
****
The liquid and liquefied hydrocarbons intented for storage have been described
and specificated. The compilation of relevant safety data packages and the
determination and description of the process and of significant process elements
have been completed. The planning principles for the specified products were
also recorded diagrammatically and laid down in basic process flow charts.

A general system of classification for rock is presented and discussed by means of examples. Subsequently, six types of rock were selected as being representative for the geological formations under consideration in Central Europe. The types of rock were described in detail.
The types of rock are very different in their natural state and in their mechanical behaviour. Models in terms of rock mechanics were worked out, which contain the characteristics properties of each rock in a clear form. They are basis for the numerical calculation of the stress and deformation behaviour.
For the economic dimensioning of measures for sealing, drainage, influence of the goundwater on the structure and the environment, the geohydraulic parameters are determined: permeability, groundwater relationship, chemistry. The method of testing in order to determine the individual parameters in the rock plays an important and essential role.
General statements are made with reference to the thermal characteristics of each type of rock. Values for the six types of rock found in relevant literature were listed in tabular form.
In the section " thermomechanics" a calculation model and calculation method to determine the stress deformation behaviour as a result of thermal conduction were described.
As a result of the calculations one receives temperature distribution, thermal balances and static loads for prestated time periods, such as construction stages and differing operational conditions. The forces of the individual calculations can be used as load cases in the static design calculation.
The internal sealing systems are checked.

REFERENCES :
****
H.J. SCHNEIDER, S. SEMPRICH
REQUIREMENTS MADE OF LINED ROCK CAVERNS FOR STORAGE OF FUELS AND LIQUEFIED GASES;
INTERNATIONAL SYMPOSIUM "LARGE ROCK CAVERNS", HELSINKI, AUGUST 1986

398

```
*******************************************************************************
* TITLE : INDUSTRIAL PILOT PROJECT FOR          *      PROJECT NO        *
*         UNDERGROUND CRYOGENIC STORAGE OF       *                        *
*         LIQUEFIED GASES                        *   TH./14021/86/BE/..   *
*                                                *                        *
*******************************************************************************
* CONTRACTOR :                                   *   TELEPHONE NO         *
*    DISTRIGAZ SA                                *                        *
*                                                *   237.72.11            *
*    AVENUE DES ARTS 31                          *                        *
*    BE - 1040 BRUXELLES                         *                        *
*                                                *   TELEX NO             *
*                                                *                        *
* TECHNICAL DIRECTOR :                           *   63 738               *
*    MR. J. DERVILLE                             *                        *
*                                                *                        *
*******************************************************************************
```
VERSION : 06/09/89

AIM OF THE PROJECT :
****
The objective of the proposed project is to build a pilot facility and operate
it under industrial working conditions and over a sufficiently long length of
time in order to prove in practice the reliability of all the innovative
facilities ie:
- Design of the storage cavity and its equipment to resist the constraints due
to cryogenic conditions and depth (specially in the case of clay)
- Operating equipment - and its maintenance - working under such conditions
- Operating process. The pilot should demonstrate the technical feasability and
the advantages of the technology in terms especially of safety and lower capital
andoperating costs.

PROJECT DESCRIPTION :
****
This industrial pilot facility will have a capacity of the order of 6000 m3 and
exhibit all the specific features of a full-scale industrial unit and the
equipment needed for its operation.
The Distrigz LNG Terminal at Zeebrugge has been chosen as the project site. It
provides the LNG needed for the initial cooling down phase and all the
facilities for the subsequent testing and operating phase.
Moreover there exists in the terminal underground a clay layer which seems
adequate for the projected gallery.
The programme will break down into three phases :
1. Feasibility study :
   Its content will of course depend on information available but must
   necessarily include :
      1.1. Drilling to achieve a full geological survey of the clay layer
           and take fresh samples;
      2.2. Laboratory testing - to determine the clay properties at ambient
           and cryogenic temperatures;
      2.3. The feasibility study in itself;
      2.4. General design, including for example tunneling methods and the
           main items of operating equipment, as well as capital cost
           estimates.
2. Construction :
Will involve driving a single tunnel of 8 m diameter with a length of
approximately 150 m at a depth of roughly 150 m below ground level in a clay
formation. The access shaft will have a diameter of +/- 3,20 m and will be
supported with concrete segments for mechanical integrity during the excavation
work.
This equipment will include :
- LNG pump at ground level;
- Spraying line along the tunnel roof;
- Submersible pump for lifting LNG to surface;
- Boil-off treatment unit;
- Operating instrumentation and boil-off return lines;
- Monitoring and control equipment and safety systems.
3. Cooling-down and operation :
The timing of the cooling down phase will last 5 - 6 months at least. Once
cooled down, the unit will be operated for at least one year to provide an
opportunity for simulating various operating phases, observing cavity response,
checking the performance of equipment and control systems and optimising
operating peformances.

STATE OF ADVANCEMENT :
****
A first drilling has been made down to - 200 meters in 1986.
Two new exploration drillings have been achieved in 1989. A full range of
samples have been taken and are currently been tested in several laboratories in
France and Belgium. The operating process of the storage has been defined,
taking into account the existing facilities offered by the Distrigaz terminal,
in order to minimise the investment and operating costs.

RESULTS :

****
Laboratory tests completed first assessment of the excavating equipment, done. The laboratory tests at ambient and cryogenic temperatures on the samples taken during the two drillings achieved in 1989, have been completed. The results of the laboratory tests one can broadly characterize the Yper clay as follow :
. at ambient temperature, increase of the geotechnical characteristics with depth, more particularly, values of shear strength and shear modulus are acceptable for excavating from 170 m depth;
. at cryogenic temperature, anisotropy of values if the samples are chosen perpendicular or parallel to stratification. Furthermore, values of conductivity are consistent with the previous experiments.
The excavating conditions of the gallery have been analysed with the assistance of a similation model, starting from the parameters defining the properties of the clay at ambient temperature determined by the laboratory tests.
A first assessment of the excavating equipment required and of the size and type of the lining has been done.
The results of the above are now being fed in the simulation model representing the situation under cryogenic condition, after cooling, in order to check the feasibility under those conditions.
This step will likely lead to necessary adaptations and a recycling of the stimulation model under ambient temperature conditions before being able to draw a final conclusions on the technical feasibility.
The results of the first runs of the simulation model under ambient temperature have led to the conclusion that we can envisage the excavation in three steps of a 8 m diameter gallery, lined with a first shotcrete and then with a final lining of concrete at the depth of +/- 180 m below ground level. One has to wait for the results of the simulation under cryogenic conditions to check whether the added contraints are acceptable.

REFERENCES :
****
"CAVITE PILOTE DE STOCKAGE CRYOGENIQUE DE SCHELLE" A. BOULANGER, P.V. DE LAGUERIE & W. LUYTEN
(EUROP COMM. SYMPOSIUM, LUXEMBOURG, 5-7/12/84)
"LPG AND LNG TERMINAL ASSOCIATED WITH UNDERGROUND STORAGE" J.P. LAGRON, A. BOULANGER & W. LUYTEN
(GASTECH, NICE, 12-15/11/85)

MISCELLANEOUS

```
*********************************************************************************
* TITLE : DIMENSIONAL VERIFICATION TO BE CARRIED    *    PROJECT NO        *
*         OUT AT OPEN SEA ON OFFSHORE STRUCTURES     *                      *
*         BY MEANS OF PHOTO GRAMMETRY.               *    TH./15044/83/IT/.. *
*                                                    *                      *
*********************************************************************************
* CONTRACTOR :                                       *    TELEPHONE NO       *
*    AGIP SPA                                         *                      *
*                                                    *                      *
*    C.P. 12069                                       *    62 520 27805      *
*    IT - 20120 MILANO                                *                      *
*                                                    *                      *
*                                                    *    TELEX NO           *
*                                                    *                      *
* TECHNICAL DIRECTOR :                                *    310246 ENI        *
*    MR. BOZZOLATO                                    *                      *
*                                                    *                      *
*********************************************************************************
```

VERSION : 15/09/89

AIM OF THE PROJECT :
****
Phase 1 : Prefeasibility study and design of equipment, special instruments and
softwares perfecting. Construction of a special metal support beam (nicknamed S.
E.R.) to be installed on a medium-large size helicopter. Manufacture of small
ancillary instruments.
Phase 2 : Experimental tests of photographing from helicopter and subsequent
tests of digital stereoplotting and mathematical processing.

PROJECT DESCRIPTION :
****
Phase 1 : Prefeasibility study and design of equipment, special instruments and
softwares perfecting. Construction of a special metal support truss (nicknamed
"SER") to be installed on a medium-large size helicopter. Manufacture of small
ancillary instruments. Experimental tests of photographing from moving supports
(helicopter, floating craft) and subsequent tests of digital stereoplotting and
mathematical processing.
Phase 2 : In case of successful results in the 1st phase, proceed with the
perfecting of means, equipments, instruments and methods in order to make the
procedure fully operational and economical.

STATE OF ADVANCEMENT :
****
The Heliborn Pivoting Support (S.E.R. = Supporto Eliportato Ruotabile) is now
operational on the AB 412 AGUSTA helicopter. Such a helicopter equipped with S.E.
R. received its domestic Certificate of Airworthiness from the Italian Air
Registration Board on June 27, 1988. This official document allows the use of
the S.E.R. system in the most diverse operative situations.
Several survey tests were carried out in the second half of 1988, both on-shore
and off-shore, with successfull results.

RESULTS :
****
Practical results in very large scales up to 1:25 and in digital measurements
proved to be reliable and accurate in dimensional controls ans as built survey
field. Possible use range from civil/industrial engineering to geology and
shallow water hydrography even when the subject to be surveyed is moving.
The S.E.R. consists of a balanced beam which just out of the sides of an AB412
Agusta helicopter. At each end of the beam, contained in a specially designed
pod, is a UMK 10/1318 film metric camera. The two cameras are synchronized and
controlled by an impulse emitter placed on the control-console inside the craft
itself. There are also two TV training cameras which are integral to the
photographic cameras, and two monitors to check the accuracy of the training
operations.
The beam has a nominal base of 6500 mm and may rotale freely around its
longitudinal axis thus allowing downward vertical, forward horizontal and tilted
stereo-photogrammetric takes, through any angle and at a distance between 30 and
100 metres.
The installation of an S.E.R. on board the AB412 takes about 2 hours and a half.
on average. Removing the system takes about 1 and a half hours.
At the sea the S.E.R. installed on the helicopter while hovering allows the
survey of complex industrial installation, offshore petroleum platforms and
structures, metal structures already loaded on barges, the same while they are
launched at the sea, overall takes of ships, or details of the same, even during
navigation.
On land instead possible performances could be in civil protection and civil
engineering. For instances inventory of damages caused by eartquakes or other
natural events, nearly immediate survey (max. 4 hours, weather and light
condition permitting) of disasters occuring in any part of territory, for use in
judiciary inquests, survey of natural phenomena in rapid evolution, dimensional
checks and surveys on complex and inaccessible constructions, checks on
monumental buildings towering over densely built city centres, surveys on steep
and rocky coastal areas where there are or will be construction works.

REFERENCES :

****
THE SYSTEM WAS OFFICIALLY PRESENTED DURING TECHNICAL MEETING IN VENICE (XXXII CONGRESSO NAZIONALE SIFET, 1987), LUXEMBOURG (III E.C.SYMPOSIUM, 1988), HOUSTON (OFFSHORE TECHNOLOGY CONFERENCE, 1988), KYOTO (16TH INTERNATIONAL CONGRESS I.S.P. R.S.), MILAN (14TH EUROPEAN ROTORCRAFT FORUM) AND VIAREGGIO (XXXII CONVEGNO NAZIONALE, 1988).
ARTICLES AND QUOTES APPEARED IN "OCEAN INDUSTRY" (USA) AND "THE PHOTOGRAMMETRIC RECORD" (UK).

```
*******************************************************************************
* TITLE : FREE SWIMMING RISER PIPE INSPECTION          *    PROJECT NO       *
*         TOOLS.                                        *                     *
*                                                       *    TH./15052/84/NL/.. *
*                                                       *                     *
*******************************************************************************
* CONTRACTOR :                                          *    TELEPHONE NO     *
*    R.T.D.  B.V.                                       *                     *
*                                                       *    010 - 4150200    *
*    DELFTWEG 144 POSTBUS 10065                         *                     *
*    NL - 3004 AB - ROTTERDAM                           *                     *
*                                                       *    TELEX NO         *
*                                                       *                     *
* TECHNICAL DIRECTOR :                                  *    23366            *
*    J.TH. EERING, J.A. DE RAAD                         *                     *
*                                                       *                     *
*******************************************************************************
                                                     VERSION : 11/09/89
```

AIM OF THE PROJECT :
****
To develop ultrasonic inspection tools which can be applied under on stream
conditions of oil and gas risers. Major design criteria for oil risers which
will be taken in consideration for the design of the ultrasonic inspection tools
are :
*pipe sizes.............: 16", 20" and 24"
*overall length.......: 2,3 m for 16" tool
*pressure..............: 150 bar
*speed.................: up to 4 m/sec
*measuring distance.....: 300 metres
*travelling distance..: 50 Km
*wall thickness range..: up to 40 mm
*accuracy..............: +/- 1 mm
*bi- directional, and capable of passing 3 DF - 90 bends, 1 D-T joints, valves,
15% dents and 10 % diameter reductions.
*simultaneously internal pipe profile and wall thickness measuring in bends as
well as in straight pipe sections.
The innovative character of the project is illustrated by the fact that when the
project started only one comparable type of tools for pipe inspection existed.

PROJECT DESCRIPTION :
****
Ultrasonic riser pipe inspection tools had to be developed for several pipe
diameters, according to the above specifications. The 16" version is mainly
intended for extensive testing in a test loop of this diameter.
The 20" and 24" versions will be the tools for most practical applications.
Design is as much as possible of modular character. Experiments and design work
in the fields of ultrasonics, electronics and mechanics were needed. Frequent
interactions between the three field were essential, as well as regular
discussions with pipeline operators from an oil company and incorporation of
testing actions at many stages of the development.
The present project is concerned with design, manufacturing and testing of
inspection tools for oil risers. In addition, for gas risers non contact
ultrasonic transducers had to be developed which are considered suitable for
implementation in a gas riser pipe inspection tool. The implementation itself is
beyond the scope of this project.

STATE OF ADVANCEMENT :
****
Tools for 16" and 20" pipelines have been designed and built. The 16" version
has been extensively tested in q test loop and in a pressure tank. It passed all
tests successfully. The 20" version was built and applied in one field trial in
the North Sea in 1987. Two problems arose, one of which has been solved and
solutions for the other have been designed and partly tested. Since at present
no second field trial is planned the work has come to a break.

RESULTS :
****
With the inspection tool for oil risers internal pipe profile and wall thickness
can be mpeasured simultaneously. For the 16" tool 36 probes are in use. They
measure at intervals of 2.5 mm or a multiple of this interval. For a 20" tool 48
probes are used. All data are stored in a 6 Mbyte solid state memory, sufficient
to store all values over a pipelength of 300 metres. Data reduction facilities
enable the measurement of longer lengths of pipeline as well.
During the development solutions were engineered with respect to construction of
the inspection tool in order to cope with the rather severe mechanical
requirements. To pass 3 D bends the tool consists of 3 articulated high pressure
resistant containers which are for that purpose connected with eachother by
universal joints. In these containers all electronics and high energy batteries
are housed.
In particular a good solution was found for the hinging of the probes which
should be in a concentric and perpendicular position with respect to the
pipewall.
Spacecraft technology was introduced to design and construct very compact and

energy saving electronics. C-Mos electronics are used to save energy.
Electronics are compactly packed partly as hydbrids in order to fit in the
scarce space of the containers.
Special ultrasonic probes were designed and built in the frame of the project.
These high pressure resistant ultrasonic sensor modules contain the ultrasonic
probe, its transmitter and preamplifier electronics in one unit. The 6 Mbyte
solid state memory consisting of 32 Kbyte RAMs including a powerful set
processors was constructed to store all values at full speed of the inspection
tool.
After the data retrieved from the memory these are transferred to a powerful HP
9836 C desk top computer. With this device supported by appropriate software
several presentation modes of results can be generated. Colours are used to
enhance these results. Many function tests of parts and assembled part were
carried out to prove proper functioning of the integrated inspection tool.
Loop tests with the 16" version were completed successfully. A field test with
the 20" version resulted in two problems. The first was malfunctioning of the
system which had to switch the tool from stand-by mode into active mode at the
required location during its travel through the pipe. The reason for this was
found and the problem was solved.
The second problem was that a too high percentage of the measurements was
invalid, due to the presence of wax precipitation in the crude oil. Since this
precipitation is a continuous process, this problem could not be solved by
cleaning the pipe prior to inspection. Several possibilities to cope with this
problem have been proposed and partly tested. There is a good chance of success,
possibly at cost of some concessions in the field of the mechanical tool
specifications.

REFERENCES :
****
DE RAAD, J.A. ETAL
HOLLAND MARITIME, VOL 12, APRIL 1986, P. 14-17
HOLLAND INDUSTRIAL, VOL 10, SEPTEMBER 1986, P 18-21
DE RAAD, J.A.
COMPARISON BETWEEN ULTRASONIC AND MAGNETIC FLUX PIGS
1ST INTERNL SUBSEA PIGGING CONF., SEPTEMBER 23-25, 1986, HAUGESUND, NORWAY. WILL
BE PUBLISHED IN PIPE & PIPELINE INTERNATIONAL, JAN 1987, VOLUME 32, NR 1
VAN DEN BERG, W.H.ETAL
DEVELOPMENT OF AN ELECTROMAGNETIC TRANSDUCER
15TH SYMPOSIUM OF ACOUSTIC IMAGING.
 HALIFAX, CANADA 14-16 JULY 1986.
DE RAAD, J.A.
DEVELOPMENT OF TOOLS FOR ON-STREAM INSPECTION OF OIL RISERS USING ULTRASONICS.
WEST-EUROPEAN CONF. ON MARINE TECHN, ADVANCES IN OFFSHORE TECHN, AMSTERDAM,
NOVEMBER 25-27, 1986.

```
****************************************************************************
* TITLE : PARAFIL ROPE DEEPWATER MOORING STUDY        *      PROJECT NO    *
*                                                     *                    *
*                                                     *  TH./15053/84/UK/.. *
*                                                     *                    *
****************************************************************************
* CONTRACTOR :                                        *    TELEPHONE NO    *
*    BRITISH UNDERWATER ENGINEERING LTD               *                    *
*                                                     *                    *
*    PREMIER HOUSE                                    *                    *
*    10 GREYCOAT PLACE                                *                    *
*    UK LONDON SW1 1SB.                               *    TELEX NO         *
*                                                     *                    *
* TECHNICAL DIRECTOR :                                *                    *
*                                                     *                    *
*                                                     *                    *
****************************************************************************
                                              VERSION : 12/12/89
```

AIM OF THE PROJECT :
****
The aim of the project is to prove the feasibility of 'Parafil' ropes for
deepwater moorings both for tension leg free floating structures; guyed
structures; and semi-submersible facilities.

PROJECT DESCRIPTION :
****
"Parafil" is a parallel laid rope with high axial stiffness and lightweight. The
project will identify suitable structures and analyse the behaviour of mooring
systems for them made of "Parafil". Methods of installing, maintaining and
monitoring the moorings will be investigated properties of the material, methods
of on-site manufacture and transportation will be defined.

STATE OF ADVANCEMENT :
****
'Parafil' testing has required a longer duration than originally planned as the
material is generally outliving preliminary expectations. An approved extension
for the project period has been granted to accommodate these factors.

RESULTS :
****
"Parafil" type A, F and G are suitable for mooring oil field structures. The two
first are most suited to catenary systems and the last is better for vertically
tethered systems. For catenary systems, Parafil is best used for long term
installations than for exploration vessels due to cost and complexity of
handling operations. As displacement increases, the mooring element costs for
vertical Parafil G system becomes progressively more expensive than for
conventional steel tubular solution. Parafil should be design on a stiffness
basis rather than a strength basis. A termination geometry was generated for 60
tonnes testing. A termination suitable for 3000 T is to be further tested. From
the tests, it appears that safety factors might be lower than recommended by DNV.
 Extensive tests have been carried out which showed the following results :
- creep failure is not the dominant failure made
- tension bending showed that significant failure occured at the point when the
deflection was applied but not at the termination
- shave bending provide that handling parameters can be reduced considerably.

REFERENCES :
****
A 'PRELIMINARY DESIGN PREMISE'DOCUMENT COVERING DESIGN PARAMETERS INVOLVED IN
THE APPLICATION OF 'PARAFIL'MOORINGS TO THE MARINE STRUCTURES DESCRIBED IN
SECTION 4 ABOVE.
TO DATE THIS DOCUMENT HAS ONLY BEEN PRESENTED TO THE PARTICIPANTS IN THE PROJECT.

```
*********************************************************************
* TITLE : RANKINE CYCLE IMMERSED ENERGY SOURCE      *    PROJECT NO     *
*          WITH HIGH-PRESSURE COMBUSTION CHAMBER     *                   *
*          AND CONSTANT MASS OPERATION               *   TH./15055/84/FR/.. *
*                                                    *                   *
*********************************************************************
* CONTRACTOR :                                       *   TELEPHONE NO    *
*    BERTIN & CIE                                    *                   *
*                                                    *   34.81.85.00     *
*    B.P. 3  78373 PLAISIR CEDEX                     *                   *
*    FR - 40220 TARNOS                               *                   *
*                                                    *   TELEX NO        *
*                                                    *                   *
* TECHNICAL DIRECTOR :                               *   570 026         *
*    MR. D. GROUSET                                  *                   *
*                                                    *                   *
*********************************************************************
                                              VERSION : 29/03/89
```

AIM OF THE PROJECT :
****
The main objective of the current project is the tuning-up of a 60 bar
aeronautical combustion chambre with a thermal power of 400 kW, working with
methanol and oxygen and cooled by injection of combustion products at 25 deg C.
In the future, this new combustion chamber is to be the heat generation device
of a Rankine cycle power module using an organic fluid. By working at high
pressure (60 bar), the combustion products can be condensed in a sea water
cooled heat-exchanger and stored in a tank so that no exhaust is required and
the module mass is constant. So, this means there is no diving depth limit for
such an energy source.

PROJECT DESCRIPTION :
****
The project began in July 1985 and had a 3 year duration. It was divided into 4
phases :
PHASE 1: Design, sizing of the combustion chamber. construction drawings,
procurement and fabrication of each component of the test bench.
PHASE 2: Erection of test facilities: combustion chamber, fluids circuits,
electrical circuits for control, regulation and safety. Implementation of a data
acquisition programs for measurement and control.
PHASE 3: Tests of the combustion chamber including injection, ignition, thermal
behaviour, flame stability, combustion efficiency and control under various
operating conditions.
PHASE 4: Synthesis and integration
The final report gives the progress of the work and a synthesis of the results.
the integration of the chamber into the global system is examined for various
ranges of power corresponding to future industrial needs.
The main characteristics of the combustion chamber are:
- nominal thermal power =:
- nominal thermal power   = 400 kW
- mass flowrates   = 20 g/s methanol
                                       = 30 g/s oxygen
                                       = 335 g/s carbon dioxyd
tube flame   = 0.132 m length
                                       = 0.062 m diameter

  - operating pressure = 6MPa
- specific power   = 20 MW/m3/atm
- - range of working power   = 20 to 120 %
- injection by a pneumatic nozzle
- flame stability induced by a swirl at the oxygen inlet
- cooling of the tube flame by external flow and film cooling of CO2
- dilution of the combustion gases by injection of cold CO2
- ignition by a plasma spark plug and flame control by a photoelectric cell
- internal pressure maintained by a sonic valve during the open loop combustion
tests.

STATE OF ADVANCEMENT :
****
Completed. At this time, plan 1 is finished, and plan 2 has been initiated
according to schedule.
The test facilities are being set up, the data acquisition programs are being
written. The specific test bench for atomization measurements under high
pressure is designed. The first trials are planned in June or July 87. Finally,
contracts are taken with industry for potential applications.

RESULTS :
****
Calculation and design
a) design
The design of the chamber was carried out using iterative calculations with
several independent codes: drop size distribution, pressure drops, adiabatic gas
temperature, combustion kinetics, and wall temperatures were calculated. These
first iterative calculations resulted in the sizing of the chamber. A 3D

reactive flow code, FLUENT, was used to validate the sizing and provide more information about internal flows: recirculating zones, cooling films, dilution jets, droplet trajectories, and local wall and gas temperatures were studied.

b)atomization tests:

Atomization tests under high pressure (60 bar) were carried out on a specific test bench. Comparison between experimental and calculation results (droplet diameter and trajectory, spray geometry) permitted the design and validation of nozzle geometry.

c)

ignition trials

A number of ignition trials were performed at various mass flowrates and pressures (1/4 to 1/2 of full power, 5 to 20 bar). A reduced dilution rate is needed for successfull ignition and the CO_2 flowrate has to be increased rapidly just after ignition to maintain a wall temperature of less than 900 deg.C.

d) Combustion trials

Combustion trials at various thermal powers (80 to 400 kW, 60 bar) were successfully carried out, giving the following results :

- combustion is quite stable and silent
- wall cooling is more efficient at high power than at reduced power (maximum wall temperature is about 900 deg.C at 100 kW and about 700 deg.C at 400 kW)
- combustion efficiency, for which the calculations-are based on gas analysis measurement data, is better than 0.997 as soon as the oxygen excess is more than 7 % at 100 kW and 9 % at 400 kW.

In conclusion, the technical programme was carried out progressively and successfully. A new project consisting in the constructing of a sealed Rankine has been underway at Societe BERTIN since mid 1988. The next step in the development of the global engine will be to fit a high pressure heat generation loop and to couple it with the Rankine cycle.

```
***********************************************************************
* TITLE : DIESEL ENGINE WITH ARGON CYCLE.        *      PROJECT NO     *
*                                                 *                     *
*                                                 *   TH./15056/84/DE/.. *
*                                                 *                     *
***********************************************************************
* CONTRACTOR :                                    *    TELEPHONE NO     *
*     M.A.N. TECHNOLOGIE GMBH                      *                     *
*                                                 *    089 14803459      *
*     POSTFACH 500620                             *                     *
*     DE - 8000 MUENCHEN 50                        *                     *
*                                                 *    TELEX NO          *
*                                                 *                     *
* TECHNICAL DIRECTOR :                            *    52321             *
*     MR. H. GEHRINGER                             *                     *
*                                                 *                     *
***********************************************************************
                                             VERSION : 17/11/86
```

AIM OF THE PROJECT :
****
In the field of offshore technology, there is a great need for autonomous energy systems, particularly for use in deeper waters. Efficient diesel engines in closed cycle operation have considerable advantages against batteries for that problem. Aim of the project is therefore the development of a closed diesel engine cycle with argon as cycle medium and its test on a pilot plant.

PROJECT DESCRIPTION :
****
If argon is used as the cycle medium, the efficiency of the diesel engine can be increased by approx. 25%. In closed cycle operation the recycle gas argon has to be scrubbed free of the combustion products H_2O and CO_2 before enriched with pure oxygen leading back to the diesel engine.
One point main effort therefore is the design, test and optimization of the CO_2 scrubber system under operating conditions. Further points were the design of the overall system including all of the auxilliary components, and the erection of the closed argon cycle pilot plant with a 32 kW diesel engine. During the tests of the pilot plant the function of the overall system was to be demonstrated.

STATE OF ADVANCEMENT :
****
The project was completed on the 30th of April 1986.

RESULTS :
****
In the preliminary phase of the project parallel tests were carried out with an M.A.N. diesel engine operating in open argon/oxygen mode, during which 42% efficiency was achieved and, with a CO_2 scrubber using a caustic potash solution, a CO_2 reduction from 7.5% to 1% obtained.
The closed argon cycle pilot plant with a 32 kW M.A.N. diesel engine was tested in numerous trials, the sampled results and experiences led to an improved CO_2 scrubber system by using a self-developed 2-stage rotational scrubber. The advantages of this 2-stage M.A.N. rotational scrubber are its compact design and its very high lye absorption rate.
The test carried out with the pilot plant confirmed the function of the closed argon diesel engine cycle with all of its auxiliary components, and the increase of the engine efficiency.
The next step of the project is to produce an underwater energy supply section for submarines with a 100 kW M.A.N. diesel engine by using the sampled results and experiences and to test it under operation conditions.

```
***********************************************************************
*  TITLE : DEVELOPMENT OF NEW TECHNIQUES OF        *    PROJECT NO      *
*          OIL/WATER SEPARATION.                   *                    *
*                                                  *    TH./15057/84/NL/..  *
*                                                  *                    *
***********************************************************************
*  CONTRACTOR :                                    *    TELEPHONE NO     *
*     TECHNISCHE HOGESCHOOL DELFT, AFD MIJNBOUWKUNDE *                   *
*                                                  *    015 781617       *
*     POSTBUS 5028                                 *                     *
*     NL - 2600 GA DELFT                           *                     *
*                                                  *    TELEX NO         *
*                                                  *                     *
*  TECHNICAL DIRECTOR :                            *    38151            *
*     IR. W.M.G.T. VAN DEN BROEK                   *                     *
*                                                  *                     *
***********************************************************************
                                                    VERSION : 31/12/88
```

AIM OF THE PROJECT :
****
The aim of the project is the development of a small, light and efficient oil-
water separator suitable for a large range of oil-water mixtures.

PROJECT DESCRIPTION :
****
The following project phases can be distinguished :
-1. theoretical study of existing oil-water separation techniques;
2. design and construction of laboratory test equipment followed by experiments
on different separation methods;
3. testing the effects of compositional changes of the oil-water mixture;4.
testing combinations of separation methods;
5. selection of the best separation technique;
6. design and construction of a prototype separator based on the selected
technique;
7. prototype testing and evaluation.

STATE OF ADVANCEMENT :
****
The phases (1.) up to and including (5.) have been completed. Phase (6.) has
been partly completed.

RESULTS :
****
Laboratory experiments have been carried out on plate separation, membrane
filtration, adsorption, flotation and coalescence. In combination with plate
separation a model of a stream divider has been developed and tested.
Furthermore a transparent flow box with a contents of 0.6 m3 was constructed for
plate separation experiments on a larger scale. Separation using centrifugal
effects (centrifuge, hydrocyclone) was not incorporated in the study.
From th investigated methods plate separation and coalescence gave the most
promising results. These techniques attack different types of oil droplets.
Plate separation is very effective, but only for oil droplets with sizes larger
than about 30 micro.Coalescence can be used for smaller oil droplets. The flow
regime in the plate separator was studied theoretically (with the numerical
simulation program Hydro-Sepran) as well as experimentally (with laser-doppler
anemometry). The efficiency of both plate separation and coalescence was tested
with, among other equipment, a particle size measurement apparatus.
The first phases of the project took much more time than was foreseen at the
start of the project. Therefore it was decided not to complete the project, but
to abandon it in the course of phase (6.). The project was concluded with the
design of a prototype separator, consisting of a stream divider/plate separator
combination followed by a coalescer. The construction of the prototype and the
planned tests with the prototype on oil-water mixtures from North Sea Oil fields
(second part of phase (6.) and phase (7.), respectively) were not carried out.

REFERENCES :
****
- W.M.G.T. VAN DEN BROECK : FINAL TECHNICAL REPORT OF EEC-PROJECT NR TH/15057/84
"DEVELOPMENT OF NEW TECHNIQUES OF OIL-WATER SEPARATION", REPORT DELFT UNIVERSITY
OF TECHNOLOGY, FACULTY OF MINING AND PETROLEUM ENGINEERING, SEPTEMBER 1988.
- W.M.G.T. VAN DEN BROEK : SOME THEORETICAL ASPECTS OF DE-OILING OF WATER BY
PLATE SEPARATION, DELFT PROGRESS REPORT, 13 (1988) 87.

```
*****************************************************************************
* TITLE : DEVELOPMENT OF A SUBSEA CONNECTOR.      *      PROJECT NO        *
*                                                 *                        *
*                                                 *   TH./15058/84/IT/..   *
*                                                 *                        *
*****************************************************************************
* CONTRACTOR :                                    *      TELEPHONE NO       *
*     M.I.B. ITALIA SPA                           *                        *
*                                                 *    049 643099          *
*     C.P. 5                                      *                        *
*     IT - 35020 CASALSERUGO (PADOVA)             *                        *
*                                                 *    TELEX NO            *
*                                                 *                        *
* TECHNICAL DIRECTOR :                            *    430214             *
*     MR. G. BORMIOLI/MR. R. MASON                *                        *
*                                                 *                        *
*****************************************************************************
                                                  VERSION : 31/12/86
```

AIM OF THE PROJECT :
****
To develop a subsea emergency disconnector system designed for the special
problems associated with the use of flexible risers and flexible flowlines. The
device to be capable of connection/reconnection at the ocean floor.

PROJECT DESCRIPTION :
****
This development is in two phases : the development of mechanical
disconnection/connector itself and the investigation/development of the
associated handling system if necessary. This involves discussion with oil
companies/end users to establish the operational/design criteria and the
development of a design.
One prototype mechanical disconnection/connector has been manufactured and shop
tested following which, arrangements will be made to undertake a subsea test
together with the development of a handling system. The unit developed is
pressure compensated and will automatically release when subjected to a
preditermined external load. In addition it can ben hydraulically activated so
as to allow hydraulically activated so as to allow connection/reconnection or
disconnection.

STATE OF ADVANCEMENT :
****
Phase executed:
A design has been developed, drawings completed and one 3" prototype unit built.
 Discussions with possible users have allowed to agree on technical solutions,
test procedure and result analysis. Preliminary and detailed shop test have been
carried out, and a report completed.

RESULTS :
****
The test confirmed that all the basic principles were correct, and the unit had
good resistance to applied loads. It was estimated that the "weak bolt" feature
did not offer any particular advantage and could be eliminated.
FOLLOWING TESTS HAVE BEEN SUCCESSFULLY PERFORMED:
- Hydrostatic pressure test and cyclic pressure tests.
- Load test with and without "weak shear bolts."
- Load test with torsion and bending moments.
- Manual disconnection test.
- Cyclic axial load.
- Cyclic bending moment load.
- Disconnection with bending moment.
SUMMARY OF RESULTS
- No unexpected leakage occurred during the tests.
- The pressure compensating principle was correct.
- No damage to the unit occurred during the tests.
- No advantage was gained by the weak bolts.
- High pressure metallic seals showed good performance.
 A test report was issued detailing the test procedure, the results obtained,
the problems encountered and possible improvements or solutions.

```
*************************************************************************
* TITLE : UNDERWATER DRILL FOR LARGE CAPACITY        *   PROJECT NO     *
*          PILES.                                    *                  *
*                                                    *   TH./15059/84/BE/..  *
*                                                    *                  *
*************************************************************************
* CONTRACTOR :                                       *   TELEPHONE NO   *
*     BELGIAN OFFSHORE SERVICES NV                   *                  *
*                                                    *   03 231 87 70   *
*     SCHERMERSTRAAT 46                              *                  *
*     BE - 2000 ANTWERPEN                            *                  *
*                                                    *   TELEX NO       *
*                                                    *                  *
* TECHNICAL DIRECTOR :                               *   34129          *
*     MR. S. DECKERS                                 *                  *
*                                                    *                  *
*************************************************************************
```

VERSION : 01/10/89

AIM OF THE PROJECT :
****
This project intends to study and build a subsea drilling machine for the
excavation of very large capacity anchoring piles in medium to hard soil. This
drill will be operated from a standard drilling ship in several hundred metres
water depth (3 000 feet maximum). The main advantage of this down hole tool is
that no torque is transmitted from surface, limiting string failures as
encountered in big hole rotary drilling.

PROJECT DESCRIPTION :
****
This equipment is based on the HYDROFRAISE a hydraulically driven drilling tool
developed by SOLMARINE's mother Company SOLETANCHE to excavate trenches and
reinforced piles for Civil Works. The subsea drill consists of :
- a 150 to 200 kN guiding frame held by a string set to the drilling ship.
- 2 parallel cutter wheels, rotating on horizontal axes in opposite directions,
set at bottom of frame.
- a pump set above the wheels to remove cuttings.
- a jack to monitor weight or speed during drilling.
- a casing equipped with reentry funnel is pulled with the drilling tool to
cover overburden layers.
- a 400 HP power pack supplied hydraulic power from the surface through
hydraulic lines.
- deflection is monitored by an inclinometer and maintained below 0.3%.
This machine will be able to drill rectangular shafts (2.40 m x 1.00 m section)
down to 100 metres in soil with a simple compressive strength up to 100 MPa.

STATE OF ADVANCEMENT :
****
A full and detailed set of drawings has been made following a technical study of
every component and the whole drilling tool.

RESULTS :
****
Description of the Rectangular Marine Hydrofraise.
For these studies, we have based the concept on the following fundamentals :
- implementation of simple solutions.
- provisioning of usual components aboard the drill ship.
These fundamentals have led us to an easy and simple conception based on the use
of seawater as power fluid.
The primary energy is converted with a turbine driving an underwater hydraulic
power pack, the whole being mounted inside the A frame.
Description of the Circular Marine Hydrofraise
The Circular Marine Hydrofraise has to answer two operational criteria compared
to rectangular marine Hydrofraise :
- soil stability meanwhile drilling,
- compatibility with the present concept for offshore foundation.
These two constraints have led us to design the Circular Marine Hydrofraise on
the following principles :
- Pile drilling protected with a casing on the whole height of the shaft. This
casing is the definitive casing which is installed into the borehole in one
operation prior to the cementation. This technique is based on the pile driving
principle.
- Undereaming to help casing lowering. The motors and the mills are mounted on
an opening system to allow few centimers over drilling around the casing.
The technical file has been completed. However every contact made with main
contractors and oil companies to analyse their interest in developing the tool
with us has been negative. The main reasons are :
- the time which would be required to build and test the equipment up to an
operational level, which may be estimated as 3 years.
- the uncertainty about the market, which does not allow to analyse the
potential recovery on such an investment over the next 10 years.
- the reluctance of clients to use new equipments. An example is the tender sent
by an oil company recently which prohibits the use of non standard drilling
tools and impose conventional rotary rig even though fist there is none

available to-day for the diameter required, and second the work is to take place
in more than three years.
Considering the reductance of clients of users to share the investment with us,
and our impossiblity to invest alone, we have decided not to proceed with the
construction of the portotype even though the construction file is finalised.

REFERENCES :
****
"UN NOUVEL OUTIL FOND DE TROU NOMME HYDROFRAISE MARINE" PRESENTED IN OCTOBER
1985 TO THE CONFERENCE "ACTUALITE ET AVENIR DE L'HYDRAULIQUE MARITIME", FRANCE.
PAPER WILL BE PUBLISHED IN THE NUMBER 4/5 OF LA HOUILLE BLANCHE".
"UNDERWATER DRILL FOR LARGE CAPACITY PILES" PRESENTED IN MARCH 1988 TO THE
SYMPOSIUM ON "NEW TECHNOLOGIES FOR THE EXPLORATION AND EXPLOITATION OF OIL AND
GAS RESOURCES", LUXEMBOURG.

```
*********************************************************************************
* TITLE : AN UNDERWATER MOTION COMPENSATED SPM        *      PROJECT NO        *
* *                                                   *                        *
* *                                                   *   TH./15062/84/UK/..   *
* *                                                   *                        *
*********************************************************************************
* CONTRACTOR :                                        *   TELEPHONE NO         *
*   FLOATECH                                          *                        *
* *                                                   *   01 5752341           *
*   GREENFORD HOUSE                                   *                        *
*   309 RUISLIP ROAD EAST, GREENFORD                  *                        *
*   UK - UB6 9BQ MIDDLESEX                            *   TELEX NO             *
* *                                                   *                        *
* TECHNICAL DIRECTOR :                                *   23417               *
*   MR. M. CONWAY                                     *                        *
* *                                                   *                        *
*********************************************************************************
```

VERSION : 07/12/89

AIM OF THE PROJECT :
****
The objective and scope of this development programme is to complete the pre-
engineering design of an underwater motion compensated Single Point Mooring (SPM)
.

PROJECT DESCRIPTION :
****
The Submerged Piston Mooring System has a novel feature of a motion compensator
within a spring buoy. The motion compensator is fitted within the buoy in the
form or a large diameter axial piston arrangement and utilises the ambient
external pressure as the motive hydraulic force. The tanker mooring is attached
directly to the top of the piston thereby transmitting all mooring loads from
the tanker to the buoy via the motion compensator piston. This allows the
dynamics of the mooring buoy to be altered by selecting the piston diameter,
weight and buoy operating depth below the surface. The response of the overall
system to dynamic motions of the moored ship are such that the dynamic peak
loads are less than those encountered in a conventional mooring.
With the currently proposed system the buoy is placed well below the surface and
hence away from the effects of wave action. This has the additional advantages
for Arctic regions of keeping it below pack-ice and reduces the dynamic mooring
forces between the buoy and the tanker.
Project consists of two phases.
Phase 1 : Preliminary design and detailed assessment of forces in all parts of
the system and model testing to establish system response.
Phase 2 : Creation of a complete design package and specification of the entire
system, including all components and the approval of a certifying authority.

STATE OF ADVANCEMENT :
****
Abandoned after phase 1.

RESULTS :
****
Two variants of the mooring system have been developed :
i) Seabed mounted buoy version (for shallow water applications)
ii) Mid-water buoy version (for deep water applications).
Hydrodynamic analysis were done to analyse the quasistatic, slow frequency and
wave frequency behaviour of the system with a tanker moored.
During the model tests it has been observed that the limiting seastates for
operation of the system is higher than originally envisaged.
In view of the increase in the design limiting sea state further model testing
is to be undertaken to investigate in more detail the performance near to the
operating limit of the system.
Results obtained from the computer analysis and the model test programme confirm
that the submerged piston mooring is a viable option and has potential to exceed
the operating sea state threshold for other proven mooring concepts.
Project has been abandoned for organisational reasons.

```
*****************************************************************************
* TITLE : EXPENDABLE VEHICLE FOR MEASURMENT IN        *     PROJECT NO      *
*          HIGHLY DEVIATED WELLS.                     *                     *
*                                                     *  TH./15063/84/FR/.. *
*                                                     *                     *
*****************************************************************************
* CONTRACTOR :                                        *    TELEPHONE NO     *
*     SYMINEX                                          *                     *
*                                                     *    91 73 90 03      *
*     BOULEVARD DE L'OCEAN 2                           *                     *
*     FR - 13275 MARSEILLE CEDEX 9                     *                     *
*                                                     *    TELEX NO         *
*                                                     *                     *
* TECHNICAL DIRECTOR :                                *    400563           *
*     MR. C. CRIADO                                   *                     *
*                                                     *                     *
*****************************************************************************
                                                      VERSION : 07/11/89
```

AIM OF THE PROJECT :
****
The vehicle is a measurement probe with an internal spool of insulated copper
wires. Data acquisition, transmission of temperature, pressure, casing collar
location, etc... is performed by a low power electronic module. The aim of the
project is to assess the limitation of the probe owing to the deviation of the
well and to study a self propelled vehicle able to perform measurements in the
deviated parts of the well.

PROJECT DESCRIPTION :
****
Phase 1 : Feasibility study of a propelling system, laboratory tests.
Phase 2 : Prototype of measuring probe equipped with the propulsion system
developed in phase 1, laboratory and field tests.
Phase 3 : Extension of the principle to more sophisticated measurements,
evaluation of the needs.

STATE OF ADVANCEMENT :
****
The prototype has been completed : propellers/electronics : and body mechanics
have been successfully tested.

RESULTS :
****
The prototype tests give good results. Speed of the probe is about 1 m/s with 48
VDC/10A. The electronic motor drive is now ready for use in the operational
system. Difficulties are : to find highly deviated well in France.
Speed about 1 m/s with 48 VDC/10A are confirmed for the probe. Some difficulties
have appeared concerning the XBT internal wire spool for big length of
transmission insulated copper wires.
This technical critical issue has been satisfactorilly solved for wells of up to
2000 meters deep.
The final prototype offers to the geologists and drillers an efficient tool, at
low cost for drilling and top ciment localisation measurements.
The measure send by a differential RS 422 type transmission enable the
superposition of several signals : temperature with a 10 mv/deg.C sensitivity or
any other sensor plus an inductive magnetic weld detector which enables an easy
localisation.

```
***********************************************************************
* TITLE : DEVELOPMENT OF A MOBILE LASER SYSTEM        *    PROJECT NO      *
*         FOR THE DETERMINATION AT A DISTANCE OF      *                    *
*         AVERAGE CONCENTRATIONS OF METHANE AND       *   TH./15071/85/IT/.. *
*         ETHANE                                      *                    *
***********************************************************************
* CONTRACTOR :                                        *    TELEPHONE NO     *
*    AZIENDA ENERGETICA MUNICIPALE                    *                     *
*                                                     *   02 7720/3459      *
*    CORSO DI PORTA VITTORIA, 4                       *                     *
*    IT - 20122 MILANO                                *                     *
*                                                     *    TELEX NO         *
*                                                     *                     *
* TECHNICAL DIRECTOR :                                *   334170            *
*    ING. BONFIGLI                                    *                     *
*                                                     *                     *
***********************************************************************
                                           VERSION : 29/03/89
```

AIM OF THE PROJECT :
****
The aim of the project is the development of an IR LIDAR system able to detect
small methane concentrations. The system is based on a correlation technique and
utilizes as laser source an optical parametric oscillator pumped by a NV-YAG
laser. The system is designed to be a small, compact and reliable device
suitable to perform measurement in an urban environment from movable platform,
such as small lorries, using topographic targets as retroreflectors. Sensitivity
goal is a minimum detectable concentration of about 50 ppm*m, which corresponds
to the natural background concentration integrated over a double pass path of
about 30 metres, i.e. at working distance from topographic target of 15 metres.
The system has been designed to operate on ranges from 5 to 20 metres.
In case of successful results, it will be sought a cooperation with an
electronic firm, which already operates in the laser field.

PROJECT DESCRIPTION :
****
The detection method is based on the measurement of the atmospheric absorption
of a laser beam by the gas to be detected; this technique is called Differential
Absorption LIDAR (DIAL) and is used because it allows to cancel systematic
errors due to target reflectivity and atmospheric scattering.
Conventional DIAL techniques may lead to important errors when used to make
measurements on moving platforms, since two laser shots are required; infact,
time varying parameters, such as topographical target reflectance, may
noticeably change during the time elapsed between the two shots, since the pulse
rate is typically not grater than 10 Hz.
The gas correlation method makes it possible to self-normalize the return signal
obtained from a single laser pulse, thus leaving any time dependent
environmental effect. Main disadvantages are a lower sensitivity and a greater
effect of the measurement of interfering molecules.

STATE OF ADVANCEMENT :
****
System prototype has been constructed and tested.

RESULTS :
****
Field tests have been carried out during routine leak detection activity in the
city of Milan and also on simulated leaks for comparison with conventional
instrumentation.
Sensitivity has been measured of the order of 150 ppm* m, corresponding to a
uniform concentration of 5 ppm distributed on a double pass path of 30 metre (i.
e. at a distance from topographic target of 15 metre).
To improve this performance up to the needed one it will be necessary a review
of the detector preamplifier design in order to reduce noise and a further
optimization of the receiving optics.
Field experience has shouwn the importance of a careful design of the moveable
laboratory on the truck to make easier the use of the system in the urban
environment, with problems arising from car traffic and parking.
Further development is necessary to obtain a fully operative system.

REFERENCES :
****
H.EDNER, S.SVANBERG, L. UNEUS, W. WENDT:
GAS CORRELATION LIDAR, OPT. LETT., VOL.9, 493-495, (1984).
S.DRAGHI, E.GALLETTI, R. PETRONI, M. GARBI, E. ZANZOTTERA: METHANE GAS DETECTIN
WITH INFRARED LIDAR SYSTEM, PROC. SPIE, VOL. 701, 230-3 (1987).
3) E. GALLETTI: DETECTION OF METHANE LEAKS WITH A CORRELATION LIDAR, TOPICAL
MEETING ON LASER AND OPTICAL REMOTE SENSING, CAPE COD (MA), 1987.

```
********************************************************************
*  TITLE : IMPROVED METHOD OF MEASUREMENT AND          *    PROJECT NO     *
*          CHARACTERIZATION OF THE POROUS              *                   *
*          MICROSTRUCTURE OF OIL RESERVOIR ROCKS       *  TH./15073/85/HE/.. *
*                                                      *                   *
********************************************************************
*  CONTRACTOR :                                        *   TELEPHONE NO    *
*      INSTITUTE OF CHEMICAL ENGINEERING AND HIGH TEMP.*                   *
*                                                      *   30 61 992396    *
*                                                      *                   *
*      PO BOX 1239                                     *                   *
*      GR - 261 10 PATRAS                              *                   *
*                                                      *   TELEX NO        *
*                                                      *                   *
*  TECHNICAL DIRECTOR :                                *   312447          *
*      PROF. AC PAYATAKES                              *                   *
*                                                      *                   *
********************************************************************
                                                       VERSION : 08/11/89
```

AIM OF THE PROJECT :
****
The main objectives of the project are the following :
a) Development of a laboratory apparatus and technique for the serial
microtoming of double pore casts of reservoir rocks and data processing with
image analysis.
b) Development of a computer-aided simulator of mercury intrusion and retraction
in a network of chambers and throats.
c) Development of a code for microcomputers, which will give the characteristics
oof the pore microstructure of reservoir rocks based on data from serial
microtoming and mercury intrusion-retraction.
d) Demonstration of the system through application to core samples from Greek
oil reservoirs.

PROJECT DESCRIPTION :
****
The project is divided in four stages :
Stage 1
Task : Development of the apparatus and technique described in objective A.
Method : The equipment consists of : (i) a specially designed apparatus for the
preparation of double pore casts, using epoxy resins with appropriate properties;
 (ii) a microtome; (iii) microscope with high resolution TV camera; (iv) image
analyzer. Double prorecasts are prepared, using two different epoxy resin with
regular of fluorescent dyes, and then are subjected to social microtoming at
approximately 10 m intervals. The image analyzer is used to digitize and store
the images of the section faces. A computer code, especially developed for the
purpose, is used to analyze these data in order to obtain information about the
pore (chamber) size distribution, the skeleton of the pore network, the genus
per unit volume, and the coordination number distribution.
Stage 2
Task : Development of the simulator of mercury porosimetry described in
objective B.
Method : Information about the throat size distribution can by obtained by
deconvolving the intrusion-retraction curves. To this end, a reliable mercury
intrusion-retraction simulator for three dimensional chamber and throat networks
is developed.
Stage 3
Task : Development of a computer code for the determination of the
characteristics of the pore microstructure of reservoir rocks, as described in
objective C.
Method : A computer code (compatible with the IBM PC-AT, or similar
microcomputer) will be developed, which will be able to determine the throat
size distribution that giives the best fit to the mercury intrusion-retraction
data. To this end the code will accept as input the following; (i) chamber size
distribution; (ii) porosity (from maximum mercury intrusion); (iii) the range of
the throat-to-chamber coordination number and genus per unit volume (from stage
1); (iv) digitized data from mercury porosimetry curves (the initial drainage
and imbibition curves). A suitable parametrized throat size distribution will be
assumed, and the optimal values of the parameters will be determine using a fast
non-linear least squares method.
Stage 4
Task : Realization of objective D.
Method : The practicability of the method consisting of the results of stages 1.
2 and 3 will be demonstrated by applying the technique to core samples from
Greek reservoirs and drilling sites. Once the method has been validated and
streamlined, a systematic analysis of samples from Greek reservoirs will be
performed, in close cooperation with the Public Petroleum Corp. of Greece, and a
library of results will be established.

STATE OF ADVANCEMENT :
****
Stage 1: Experimental part almost fully developed; full development anticipated
within six months. Computer code almost fully developed.
Stage 2: First generation code for the simulation of mercury intrusion-
retraction is completed. Second generation code is under development.

Stage 3: The computer code is under development. First version expected within six months.
Stage 4: In progress.

RESULTS :
****
Stage 1: The experimental technique is developed to the point that serial microtoming of simple pore casts is done on a routine basis. Double pore casts are also produced, but we are still trying to identify dyes that give satisfactory contrast.
We have developed a computer code that converts data from serial microtoming and image analysis into information concerning the chamber size distribution, the specific genus (genus per unit volume), and the coordination number.
Stage 2: We developed an experimental apparatus for the visual and quantitative study of mercury intrusion-retraction in planar pore networks. We also developed a simulator for this process, see ref (1). The objective of this work is to study mercury porosimetry under well defined conditions.
We developed a simulator of mercury intrusion-retraction in a three dimensional chamber-and-throat network, see ref. (3) and (5).
Currently, we are using the simulator to make a systematic study of the effects of various geometric, topological and statistical parameters on the mercury porosimetry curves.
We completed a parametric experimental study of immiscible displacement in model porous media, see ref. (4). This work reveals the mechanisms of immiscible displacement, and thus it acts as a guide for the development of theoretical simulators. We also developed a network-type computer simulator of immiscible imbibition which takes into account the role of the precursor wetting frim in the disconnexion of the non-wetting fluid, see ref. (2).
Stage 3: We have the first version of a computer code that predicts mercury intrusion and retraction curves, if the pore structure is known. We are currently developing a computer code that will be used to solve the reverse problem.
Stage 4: Over a hundred core samples from Greek drilling sites have been analyzed to the extent permissible by the current state of advancement of out technique.

REFERENCES :
****
(1) TSAKIROGLOU C.D. AND A.C. PAYATAKES, "AN EXPERIMENTAL AND THEORETICAL STUDY OF MERCURY POROSIMETRY IN A PORE NETWORK MODEL", AICHE 1988 ANNUAL MEETING, PAPER NR 102 L, WASHINGTON DC, NOV. 27 - DEC. 2, 1988.
(2) VIZIKA O. AND A.C. PAYATAKES, "THEORETICAL MODELING OF THE ROLE OF THE ADVANCING WETTING FRIM IN THE DISCONNECTION OF THE NON-WETTING FLUID DURING IMMISCIBLE DISPLACEMENT," AICHE 1988 ANNUAL MEETING, PAPER NR 95 C, WASHINGTON DC, NOV. 27 - DEC. 2, 1988.
(3) CONSTANTINIDES G.N. AND A.C. PAYATAKES, "A THREE-DIMENSIONAL NETWORK MODEL FOR CONSOLIDATED POROUS MEDIA. BASIC STUDIES". CHEM. ENG. COMM. 60, IN PRESS, 1989.

```
*****************************************************************************
* TITLE : APPLICATION OF NON-DESTRUCTIVE        *      PROJECT NO          *
*         MONITORING METHODS ON COMPOSITE TUBULAR *                        *
*         EQUIPMENT                              *    TH./15077/86/FR/..    *
*                                                *                          *
*****************************************************************************
* CONTRACTOR :                                   *     TELEPHONE NO         *
*    GERTH                                       *                          *
*                                                *     1 47.52.61.39        *
*    4, AVENUE DE BOIS PREAU                     *                          *
*    FR - 92502 RUEIL-MALMAISON                  *                          *
*                                                *     TELEX NO             *
*                                                *                          *
* TECHNICAL DIRECTOR :                           *     203 050              *
*    MR. J.J. MASSOT                             *                          *
*                                                *                          *
*****************************************************************************
                                                   VERSION : 29/03/89
```

AIM OF THE PROJECT :
****
Development of the utilisation of tubular equipment made of composite materials
is presently faced with the lack of quality guaranteed by the equipment that is
implemented. This is of importance in the case of vertical tubings, which are at
the same time very heavy mechanically and for which intervention cost is the
highest.
Therefore, the object of this programme is to create, adapt or finalize non-
destructive control methods based on different physical phenomena (acoustic,
thermal, electrical,etc) so to ensure the reproducibility of the produced
equipment.
However, the non-homogeneous nature of these materials, implies in practice,
that the adaption of operational methods be done in direct with the usage value
of the equipment, in this case the loss of tightness of the pipes and of the
ancillary equipment. Hence, this programme is followed by a study on the
incidence of faults on the utilisation properties of the tubings.

PROJECT DESCRIPTION :
****
Non-destructive control methods for composite materials have been developed
particularly for the aeronautical domain, and concern top quality materials,
most often based on carbon fibers. In the case of more industrial epoxy-glass
fiber tubings, the detection of faults and the estimate of their critical point
must be evaluated simultaneously.
Furthermore, the diversity of existing equipment and methods imply a number of
choices possible for each type of development, as well as the performance
evaluation of reference equipment. The programme has been divided in two main
phases:
a) Feasibility
In this preliminary phase, two different studies will be carried out.
- evaluation of real performances in time of tubings selected as test elements.
This evaluation will be carried out under internal pressure, with and without
axial traction, in the presence of water and under temperature. This evaluation
will also include an analysis of their creeping properties.
- the inventory and evaluation of the performances of possible methods of non-
destructive control, in relation with faults currently observed and with their
possibility of being used at the factory or on worksites. This task can be
carried out without problem in laboratory test tubes.
This phase will end with the precise definition of applicable methods and their
utilisation range. Obviously, only a small number of methods will be chosen for
the following phase. Also most of the faults encountered will depend on them.
Hence, this phase is an ending point for which a reorientation of the programme
could be envisaged.
b) Development of specific methods
This phase will be entirely carried out on tubular elements, as similar as
possible with industrial tubings. It will include the following tasks:
- Creation of tubings with calibrated artificial faults similar to those
encountered previously
- Evaluation of the limit performances of chosen control methods (size, number,
etc)
- Evaluation of the incidence of these faults on the lifespan of tubings. This
incidence will be measured by long duration tests, under static and eventually
under fatigue conditions. Considering the special feature of the mechanical
operating mode of the tubings, this evaluation will not only be carried out
under internal pressure, but also under traction, and anyhow under temperature.
Attempt will also be made, when possible, to follow the progress of faults
observed initially by the chosen control methods.
Thus, it will be possible, at the end of this phase, to define a coherent
control scheme for the tubular elements made of composite materials, which will
include the maximum size of the admissible fault and the minimum performances of
the control methods to be implemented for an appropriate detection.

STATE OF ADVANCEMENT :
****
Works related to the Feasibility Phase were completed on 30 June 1988. Main

results are described here below.

RESULTS :
****

Reproducibility of faults related to the manufacturing process of composite material pipes (for instance porosity) was not possible. Thus, it is not envisaged to pursue works along this orientation as it is not possible to evaluate the incidence of these faults on the behaviour of pipes during service. On the other hand, reproducibility and characterization of defects related to shocks was possible thanks to the use of appropriate machines using the weight dropping principle. Furthermore, a simple non-destructif control method based on the principle of luminous transmission has proven a good aptitude to detect these faults. Furthermore, tests benches were manufactured to determine in a realistic manner the ageing of composite material tubings. A number of commercial supplies was tested at 90 C at internal pressure with or without superposed axial traction, allowing to define the limit solicitations to be respected so to obtain satisfactory behaviour of these products during service. However, this experimental method presents the disadvantage of a very long and costly implementation. Also, the dispersion of experimental results, phenomenon inherent to the method and the material, would not facilitate in the future, a comparative study of the behaviour of good pipes and faulty pipes so evidence the incidence of faults on the behaviour of these pipes during service.These results do not impell to pursue works beyond the feasibility study phase. Nonetheless, it is possible to envisage favourably the development of several punctual actions which could be performed independently to this project.

REFERENCES :
****

NO REFERENCES ARE YET AVAILABLE

```
****************************************************************************
* TITLE : METAL-ELASTOMER PAIRING OF MOINEAU      *       PROJECT NO       *
*         DRIVES FOR HIGH TEMPERATURES IN DEEP    *                        *
*         BOREHOLES                               *    TH./15078/86/DE/..   *
*                                                 *                        *
****************************************************************************
* CONTRACTOR :                                    *      TELEPHONE NO       *
*    EASTMAN CHRISTENSEN GMBH                      *                        *
*                                                 *      05141 203-1        *
*    POSTFACH 309                                 *                        *
*    CHRISTENSSTRASSE 1                            *                        *
*    DE - 3100 CELLE                              *      TELEX NO           *
*                                                 *                        *
* TECHNICAL DIRECTOR :                            *      925149            *
*    DR. V. KRUEGER                               *                        *
*                                                 *                        *
****************************************************************************
                                                      VERSION : 10/10/89
```

AIM OF THE PROJECT :
****
Due to the decreasing availability of fossil energy especially in Central Europe,
more and more wells are drilled down to greater depths.
Thus the demand for temperature resistant downhole direct bit drives is growing.
The more critical requirements cannot be met by the tools which are currently
available.
It is the aim of the proposed research project to develop positive displacement
motors following the Moineau principle for high temperatures. Presently these
motors allow permanent usage up to temperatures of about 140 deg.C. For very
deep wells, or wells exhibiting temperature anomalies, maximum values of about
250 deg.C have to be considerd. The critical parts of the Moineau drive are the
flexible lining of the stator tube and the metal surface of the rotor. Both
materials are subject to mechanical, chemical and temperature stresses.
Developing the new tools major emphasis has to be placed on considering a
sufficient reliability for long operating times.

PROJECT DESCRIPTION :
****
By means of LITERATURE AND PATENT RESEARCHES the state of the art regarding the
usability of positive displacement motors incorporating elastomer parts for high
temperature use will be evaluated. Starting from the collected literature
information, any commercial know-how as well as manufacturing technology will be
classified and investigated. The results of this MARKET STUDY will give
indications as to which elastomer and thermoplastics materials will be included
in the subsequent experimental evaluation. Possible materials will be
investigated regarding their temperature resistance, abrasion resistance,
tensile strength, stability in drilling mud, binding behaviour, workability and
manufacturing procedures.
THE DEVELOPMENT OF MATERIALS AND MANUFACTURING TECHNOLOGIES shall be performed
in close cooperation with research institutes and experienced manufacturers.
Regarding the reliable functions of the stator tube under high temperatures two
basic types of materials shall be investigated, elastomers for temperatures from
150 deg.C to 180 deg.C and thermoplastics for temperatures of 180 deg.C to 250
deg.C. The kind of metal coating to the rotor has to be adapted equally to the
borehole conditions and to the lining of the stator. Basically the pairing of
metal rotor and non-metal stator inner part forms a critical component of the
high temperature positive displacement motor. Also, additional components of the
motors have to be designed regarding the high temperature application. This is
especially true if rubber-elastic parts are involved such as the by pass valves
and transmission parts.
In general, experimental testing will be performed in the laboratory as well as
in the oilfield. Regarding LABORATORY TESTING tests on smaller samples, model
motors, and pilot motors have to be distinguished.
Testing on small-size samples will allow to evaluate materials roughly.
Later short segments of motors with already pre-selected materials will be
laboratory tested. Finally, complete full size downhole motors of a progressive
development stage will be thoroughly tested on the downhole motor test stand.
Modifications of the existing laboratory test equipment have to be carried out
in order to allow the application of extended temperatures. With the prototype
motors FIELD TESTING will be performed at last.

STATE OF ADVANCEMENT :
****
Major activities within the reporting period have been in the area of further
selection of high temperature elastomers, and both laboratory testing of these
materials and field testing of motors with selected rubbers.

RESULTS :
****
The investigation of literature has been continued. Major efforts have been
undertaken on the selection of high temperature elastomers for application to
downhole motors. A variety of physical parameters for five different classes of
elastomers has been determined by extensive laboratory testing. Based on the
results of these laboratory tests and on molding tests, the modification of the

elastomers and a subsequent new development of compounds is ongoing. As a result of this investigation full size motors have been manufactured and have also been tested in the field without success. Further compounds will at first be tested in model motors in the high temperature test stand which will be completed very soon.

Besides the elastomer development the principle motor design has been optimized in order to reduce the friction and internal losses of these motors and for keeping the dissipated heat as small as possible. Areas for modification and alternation of other components have been identified. Design and prototype of these components has been started.

```
************************************************************************
* TITLE : THE DEVELOPMENT OF INDUSTRIAL DIVING      *    PROJECT NO     *
*          WITH HYDROGENATED BREATHING MIXTURES     *                   *
*                                                   *  TH./15085/86/FR/..*
*                                                   *                   *
************************************************************************
* CONTRACTOR :                                      *   TELEPHONE NO     *
*    COMEX SA                                        *                   *
*                                                   *   91.23.50.00      *
*    BLD DES OCEANS 36                               *                   *
*    FR - 13275 MARSEILLE CEDEX 9                    *                   *
*                                                   *   TELEX NO         *
*                                                   *                   *
* TECHNICAL DIRECTOR :                              *   410985           *
*    MR. C. GORTAN                                  *                   *
*                                                   *                   *
************************************************************************
                                            VERSION : 09/02/90
```

AIM OF THE PROJECT :
****
The aim of the project is to perfect the techniques necessary for the
development of industrial diving with hydrogenated breathing mixtures :
- Research carried out on laboratory animals for definition of the maximum
acceptable hydrogen concentrations to avoid narcotic effects.
- Design and construction of a gas deshydrogenation system.
- Study of compression and decompression tables for hydrogenated gas mixtures.
- Accomplishment of an experimental dive for selectiion and training of divers
in view of a future actual dive at sea.
- Chemical decompression of the hyperbaric chambers by deshydrogenation of the
diving breathing mixtures.

PROJECT DESCRIPTION :
****
Traditional helium diving is limited by two factors :
- The High Pressure Nervous Syndrome (H.P.N.S.) resulting from hydrostatic
pressure on the central nervous system, which provokes motorial disorders.
- The gas mix density which increases in proportion to depth and makes breathing
increasingly difficult for the diver.
The cumulated effects of these two factors diminish thus considerably the
diver's work capacity.
The narcotic power of hydrogen tends to strongly counteract the development of H.
P.N.S.
As concerns its density, the lowest amongst gases, it allows an important
reduction of the voluminal mass of gas mixtures under pressure which induces
much easier breathing.
The programme will take the folowing four phase :
PHASE 1 : TOXICOLOGICAL RESEARCH
Its purpose is to experimentally determine the maximum acceptable hydrogen
content in respect of narcosis. The research will be conducted on small
laboratory animals under extreme pressure conditions, equivalent to 1200 to 2000
metres. A part of the animals will be sacrified for histological study.
PHASE 2 : DESIGN AND CONSTRUCTION OF GAS DESHYDROGENATION SYSTEM
The system will be designed after a bibliographic study of present state of the
art in this subject. The actual operating system will be determined on the basis
of laboratory tests results, before a full scale prototype is completed in order
to decompress the chambers during the manned simulation dive to 520 metres.
PHASE 3 : STUDY OF COMPRESSION AND DECOMPRESSION TABLES FOR HYDROGENATED GAS
MIXTURES
PHASE 4 : SELECTION AND TRAINING OF DIVERS AND REFINING OF DIVE PROCEDURES
Ten divers (two of whom serving as substitutes) will be chosen amongst very
experienced professional divers.Eight of them will perform the experimental dive
at 520 metres depth.
During this dive, they will be subjected to a complete physiological examination,
under normal conditions as well as during their activity in water.

STATE OF ADVANCEMENT :
****
Completed

RESULTS :
****
PHASE 1 :
Twelve dives were carried out at depths of from 1300 to 2000 m with 110 mice
divided up into batches of 5 or 10.
   - 5 helium dives (He-O2) to 10, 1500 and 2000 m.
   - 7 hydrogen dives (H2-He-O2) to 1300, 1400, 1500, 1800 and 2000 m.
The scientific study was chiefly concerned with :
   - The animals' behaviour
   - ponderal analysis
   - oxygen consumption
   - tissue histology
Using a slow compression speed (around 50 m/hour), a mixture with a good dose of
hydrogen (40 to 50 %) and a slow decompression speed (16m/hour), the mice can be

compressed and decompressed without serious disorders or accidents down to depths of 1400 or 1500 metres.
Hydrogen does have an anti-HPNS effect, making it possible for living creatures to withstand very high pressure, but the dosage must be exact.
The histological investigation revealed :
a) Renal and cardiac parenchyma :
   No observable change.
b) Hepatic parenchyma :
   Hepatocytic alterations of the "clarification" type in one batch of mice and of the "ballooning" type in another batch.
   As the parenchyma of a third batch was perfectly normal, these alterations can not be related to the experimental conditions.
c) Pulmonary parenchyma :
   Similar modifications :
   - congestion of the alveolar parietes varied according to the animals
   - hypertrophy and hyperplasia of type II pneumocytes.
   These modifications did not always affect the entire pulmonary parenchyma.
PHASE 2 :
The tests in laboratory of dehydrogenation by heterogen catalysis has shown the superiority of catalysis by Palladium in comparison of this by Platine.
An industrial system of gas dehydrogenation by catalytic oxydation has been developped for selective elimination of hydrogen during the decompression phases following saturated diving. This installation has been used during the decompression of HYDRA VI and VII dives to insure the elimination of hydrogen at the speed of divers' physiological decompression without any limit found.
PHASE 3 :
After completing the onshore diver selection and training dive to 520 metres, we ran a hydrogen-sensitivity test on four other divers at the beginning of 1987.
This test consisted on an onshore saturation dive to 260 metres, using a hydrox (H2-O2) mixture, with a partial pressure hydrogen of 25 bar.
The results of this test are in the process of being analysed and will be published in a separate report in the near future.

```
*********************************************************************************
* TITLE : ENVIRONMENTAL MONITORING SYSTEM WITH          *      PROJECT NO      *
*         INTERACTIVE CONTROL                           *                      *
*                                                       *   TH./15086/86/UK/.. *
*                                                       *                      *
*********************************************************************************
* CONTRACTOR :                                          *    TELEPHONE NO      *
*    SIEGER LTD                                         *                      *
*                                                       *    676161            *
*    NUFFIELD ESTATE 31                                 *                      *
*    UK - POOLE DORSET BH17 7RZ                         *                      *
*                                                       *    TELEX NO          *
*                                                       *                      *
* TECHNICAL DIRECTOR :                                  *    41138             *
*    DR. D. BALFOUR                                     *                      *
*                                                       *                      *
*********************************************************************************
                                                        VERSION : 20/09/89
```

AIM OF THE PROJECT :
****
Development of a distributed gas monitoring system in which individual
instruments and sensors can be interogated to determine their status, adjusted
and calibrated by means of coded infra red radiation generated by a hand held
compact transmitter and receiver. This development will enable a single engineer
to isolate, adjust and calibrate individual instruments and sensors on routine
basis with enhanced operational efficiency, improved reliability and reduced
cost of operation.

PROJECT DESCRIPTION :
****
The proposed programme concerns in the first instance the development of a
handheld treansmitter/receiver incorporating a keyboard and LCD display coupled
to a microprocessor which will interact via an encoder/decoder unit with an
infra red transmitter so that interrogating and correcting information can be
transmitted to the address of specific enclosures to effect, for example,
adjustment of such parameters as zero setting, span, linearisation data, range
and alarm settings and checking of diagnostic details.
Parallel development will concern the design of flameproof enclosures which in
addition to incorporating sensors or control instrumentation will incorporate a
receiving window, similar infra red transmitting and receiving circuitry to that
utilised within the handheld transmitter/receiver.
The necessary encoder/decoder circuits will also be developed to provide
effective interaction with the control processor unit within the flameproof
housing.
Meanwhile software will be developed to enable specific instruments or sensors
to be addressed to determine their status and to provide means of diagnosing
defective operation and of effecting adjustments including calibration.
Following completion of a number of development units an extensive programme of
laboratory testing will be necessary to ensure that all problems, and software
problems in particular, are resolved. Tests will cover the range of
environmental conditions anticipated in field use and the equipment will be
subjected to radiated interference, vibration and shock tests.
When designs have been finalised a pre-production batch of approximately ten
units of each new product will be manufactured. These units will be configured
into various forms of monitoring systems and subjected to extend testing both in
the laboratory and in selected locations in the field to gain initial
operational experience so that any problems arising can be corrected before
finalising the manufacturing drawings.

STATE OF ADVANCEMENT :
****
The hand-held transmitter/receiver is now ready for finalisation of the
production design. A prototype incorporating a key pad and LCD display has
demonstrated reliability of communication and is undergoing, with specially
developed software, tests of calibration and alarm setting routines initiated in
a remote station by a transmitted infra red signal.

RESULTS :
****
Still in development stage.

```
**********************************************************************************
* TITLE : DEVELOPMENT OF MECHANISED ULTRASONIC        *      PROJECT NO        *
*         FLAW DETECTION TECHNOLOGY FOR                *                        *
*         INSPECTION OF COMPLEX GEOMETRIES IN OFF-     *   TH./15089/86/DK/..   *
*         SHORE STRUCTURES                             *                        *
**********************************************************************************
* CONTRACTOR :                                         *    TELEPHONE NO        *
*    THE DANISH WELDING INSTITUTE                      *                        *
*                                                      *    45 42 96 88 00      *
*    PARK ALLE 345                                     *                        *
*    DK 2605 BROENDBY                                  *                        *
*                                                      *    TELEX NO            *
*                                                      *                        *
* TECHNICAL DIRECTOR :                                 *    33388              *
*    JENS ROEDSTED CHRISTENSEN                         *                        *
**********************************************************************************
```

VERSION : 11/09/89

AIM OF THE PROJECT :
****
This research project seeks to demonstrate that mechanised ultrasonic testing
can be successfully applied to the testing of welds in joints of complex
geometry : such configurations may be found in a wide variety of welded
constructions, but the most extreme examples are to be found at fabricated nodes
in steel tubular lattice structures of the types used for off-shore
installations, where a number of tubular members of different sizes intersect at
a variety of angles.
The main objectives of the project are :
- Development of modelling routines to predict the scanning pattern to maintain
full coverage of any given geometry;
- Development of signal processing and display software;
- Determination of optimum ultrasonic test parameters and demonstration of test
system on laboratory scale samples;
- Design and production of a prototype scanner for large scale samples;
- System trials in air and in water on large scale samples.

PROJECT DESCRIPTION :
****
Application of ultrasonic testing is essential to ensure freedom from
significant defects of welds in high quality fabrications.
A severe limitation of ultrasonic testing is the difficulty which the technician
experiences in manipulating the ultrasonic probe to ensure adequate coverage of
the welded joints, in combination with difficulties of understanding the
interaction of the ultrasonic beam with any defect present. This problem is
exacerbated by complex or varying weld geometries, which are common in large
tubular off-shore structures.
Thus despite the fact that generally suitable computer controllable and
mechanised ultrasonic testing equipment has already been developed, neither the
ultrasonic techniques nor the equipment is applicable to the off-shore sector.
The develpments required to achieve the objectives of this project will be based
on a typical off-shore node and comprises computer simulation and modelling of
geometry variations and the ultrasonic pulse-echo technique, laboratory
investigations, field tests and outlining requirements and specifications for
technique and equipment for underwater ultrasonic inspection of such complex
structures.
The project is divided into two main parts : Part 1 consists of laboratory
studies culminating in a demonstration that the approach is valid using
laboratory scale samples of a variety of complex geometries.
Part 2 will consists of procurement of number of large scale samples, production
of scanner device to enable them to be tested and the carrying out of full
inspection on the samples obtained to determine the performance of the complete
system.
The various activities in each part is described below :
Part 1 :
   Task 1 : Industrial studies and projectplanning
   Task 2 : Procedure development and manufacture of test samples
   Task 3 : Modelling and simulation of geometry changes and ultrasonic
inspection
   Task 4 : Development of testing software
   Task 5 : Examination of test samples
   Task 6 : Full scale node samples
   Task 7a: Automatic scanner concept.
Part 2 :
   Task 7b: Automatic scanner design and manufacture
   Task 8 : Interfacing of scanner device and ultrasonic equipment
   Task 9 : Examination of full scaloe test node welds
   Task 10: Underwater testing
   Task 11: Recommandation for an underwater scanner.

STATE OF ADVANCEMENT :
****
The project is in its 12th month, and still in the design phase.
During the first year a detailed project planning have taken place and

laboratory experiments have been commenced to determine sensitivity to weld
defects with the aim of optimising test procedures.
Modelling of ultrasonic wave propagation is commencing in conjunction with the
procedure development in order to predict scanning patterns on surface of
tubular joints to ensure adquate coverage.

RESULTS :
****

During the first year of research work has been concentrated on ultrasonic
procedure development and simulation and modelling of ultrasonic inspection and
geometry changes.
Within procedure developments initial work has been carried out using a 900 T-
node specimen containing extensive fatigue crack and the P-scan system
(Projection Image scanning, a computerised ultrasonic system for weld inspection,
 developed and marketed by The Danish Welding Institute) in its A-scan
Collection mode.
Using manual scanners with different angle probes valuable information for
determining a mechanised scan pattern, was collected with regard to minimising
the number of probe positions necessary to obtain full coverage and reduce scan
time.
In combination with the above, work was carried out to determine the number of
axis of freedom for a scanning mechanism which is able to cope with all probe
positions and orientations.
Development of a software for modelling and simulation of ultrasonic wave
propagation has begun. During the first year of research, work has concentrated
on producing an overall description of the system, stipulating requirements to
hardware/software and logical modelling of the system.
At present a first prototype coping with simple geometries has been completed
with the purpose of verifying the concept of the system.
The final version will be capable of depicting images of the centre beam as well
as the dispersal of the beam in a cross section of a complex structures based on
the echo pulse technique. The system will be able to handle typical standard
constructional elements representative for the off-shore industry, where tubular
elements are the main constructional part.
The system will if possible be a real time system, which means that the images
of the cross section and the A-scans will be presented in parallel while moving
the probe on the structural element.
The system will include two various way of use, one is to prepare NDT-procedures
before inspection in the fields to ensure reliable and reproducible examinations.
The second is to evaluate results gathered from examinations in the fields.

Q

```
***********************************************************************
* TITLE : PREDICTABILITY OF GRANULOMETRY :       *    PROJECT NO      *
*         COALESCENCE                            *                    *
*                                                *                    *
*                                                *    TH./15092/87/FR/.. *
*                                                *                    *
***********************************************************************
* CONTRACTOR :                                   *    TELEPHONE NO     *
*    INSTITUT NATIONAL POLYTECHNIQUE DE GRENOBLE  *                    *
*                                                *    76.82.50.00      *
*    E.N.S.H.M.-G.                               *                    *
*    46 AVENUE FELIX VIALLET                     *                    *
*    FR-38031 GRENOBLE CEDEX                     *    TELEX NO         *
*                                                *                    *
* TECHNICAL DIRECTOR :                           *    980 668          *
*    MME M. PIAU                                 *                    *
*                                                *                    *
***********************************************************************
                                             VERSION : 15/06/89
```

AIM OF THE PROJECT :
****
The main objective of the project is to prepare a computer code (CROCODIL)
capable of predicting changes in the granulometry of fluid/fluid dispersion.
This computer code requires knowledge of collision efficiencies for various
pairs of fluids. Thus the second objective of the project is to devise and to
use a specific set-up (Couette coalescer equipped with an attenuation probe)
allowing an experimental identification of these efficiencies. The current state
of knowledge in the research field is characterized by a lack of formulae
suitable for industrial use and very little experimental data.

PROJECT DESCRIPTION :
****
The first phase has two steps : I/1 Assessment of the attenuation probe : II/1
Creation of a Couette coalescer. The second phase has three steps : I/2
Adaptation of the attenuation probe (assessed in II/1) to the Couette coalescer :
 II/2 Identification code for the efficiency of coalescence: III/2 Improvment of
the computer code (CROCODIL). The third phase has two steps : II/3 Effective
identification of efficiency of coalescence for specific pairs of fluids : III/3
Runs of CROCODIL for the latter pairs.

STATE OF ADVANCEMENT :
****
Ongoing. The first phase (I/1 & II/1) have been completed. Before addressing
step I/2 a new attenuation probe has been devised.

RESULTS :
****
Two main results have been obtained :
- A variable axis cylindrical Couette coalescor has been designed. It has a 40
mm diameter rotating inner cylinder with a 1 mm annular gap to a rotating outer
cylinder 200 mm high. The electrical motor equipped with a variator can provide
9 to 1500 rev min-1 (precision loss than 1 %) to the inner or the outer
cylinders. A great care has been paid to the optical and geometrical properties
of both cylinders : the inner one is reflecting and stainless steel made and the
outer one is transparent and glass made.
- An optical probe meauring interfacial area ( ) by light attenuation has been
designed with a special emphasis en flows with sub-millimetric particles. It
permits measurements in liquid-liquid or gas-liquid dispersions without need of
introducing empirical correcting factors for the standard exponential decay law
of light intensity while keeping an extended application range. This probe was
successfully tested with an air-glass particle flow, the parameters of which
were carefully determined basically by hold-up methods. The volume fraction of
the dispersed phase was varied between 0.05 % and 5 % and the particle size
between 10   m and 300   m.

```
***********************************************************************
* TITLE : DEEPSEA THRUSTER                         *      PROJECT NO    *
*                                                  *                    *
*                                                  *  TH./15093/87/DE/.. *
*                                                  *                    *
***********************************************************************
* CONTRACTOR :                                     *    TELEPHONE NO    *
*    BOENKE GMBH                                   *                    *
*                                                  *    4102-32808      *
*    HAMBURGERSTRASSE 250                           *                    *
*    D - 2070 AHRENSBURG                           *                    *
*                                                  *    TELEX NO        *
*                                                  *                    *
* TECHNICAL DIRECTOR :                             *    2189880         *
*    HELMUT BOENKE                                 *                    *
*                                                  *                    *
***********************************************************************
                                                    VERSION : 26/10/89
```

AIM OF THE PROJECT :
****
Deepsea Thrusters can be used in the Deep Sea Drilling to position the Blow Out
Preventer on the seafloor and also allow easier handling of the Re-Entry of the
Riser System. This concerns waterdepths beginning below 1000 m.
A further aspect ae deep diving vessels to avoid the unsafety of hydraulic
aggregates to be used as well as the so fair not tested rubber sealings of
turning parts which behave critical under high pressure conditions according to
the material's floating point.
The innovation is seen in the substitution of complete hydraulic aggregates and
drives or electrical components, that have to be pressure compensated and
withstand only certain depths.
A new concept was seen in preparing the thrster to work as a down-hole pump.

PROJECT DESCRIPTION :
****
The aim of the project was the construction and prototype manufactoring of an
electric deepsea thruster without rotating sealings.
The project was separated into three main parts :
a) Mathematical calculation and Mechanical Construction of the prototype
The motor will be a completely new developement, that requires complete
mathematical and electrical calculation of stator and rotor, as well as a new
mechanical layout concerning water tightness of the stator windings, an optimal
slot between rotor and stator and a nex squirrel cage with integrated propellers.
b) Manufactoring of the prototype
c) Dry-, wet- and pressure tests of the manufactored propeller system
The dry tests shall give the mathematical basis to recalculate the electrical
values, that can be reached in sea water.
Wet tests under normal conditions in a basin gave first force and water flow
results.
High pressure tests in a pressure chamber are obtained to prove the
watertightness and resistance of the materials used.

STATE OF ADVANCEMENT :
****
Completion of Assembling and long-time testing

RESULTS :
****
An electric motor was calculated, constructed and built, that shall be able to
work completely open in seawater and under maximum seawater pressure, which is
concerned to be up to 1000 bar.
The implemented rotor has an cylindric opening and holds on it's axis directly
the propeller blades.
Problems that araised while testing the motor where the power losses according
to the magnetic saturation in the rotor.

```
*********************************************************************************
* TITLE : PROTOTYPE FOR A LASER WELDING SYSTEM TO      *      PROJECT NO       *
*         FACILITATE THE TRANSPORT IN THE              *                       *
*         HYDROCARBONS SECTOR BOTH ON AND              *   TH./15095/87/UK/..  *
*         OFFSHORE.                                    *                       *
*********************************************************************************
* CONTRACTOR :                                         *    TELEPHONE NO       *
*     N.I.S. LIMITED, LASER DIVISION                   *                       *
*                                                      *    02572-65656        *
*     ACKHURST ROAD                                    *                       *
*     CHORLEY                                           *                       *
*     PR7 1NH                                           *    TELEX NO           *
*                                                      *                       *
* TECHNICAL DIRECTOR :                                 *    677302             *
*     M. M L FERGUSSON#GROUP MANAGING DIRECTOR         *                       *
*                                                      *                       *
*********************************************************************************
```

VERSION : 21/12/88

AIM OF THE PROJECT :
****
The objectives of the project is as follows :
a) The project is to exploit a new application of laser welding techniques for
which the research stage has been only partially investigated.
b) The project offers the prospect of industrial economic, commercial and
ecological viability.

PROJECT DESCRIPTION :
****
The prototype laser welding facility is designed to reduce costs, improve safety
and reduce the risk of ecological pollution. The facility will considerably
increase the efficiency of hydrocarbon tranportation both on land and off-shore
at considerable depth and will utilise a semi-automatic pipe laying system
The project is :
the investigation of laser parameter development and base line welds (laser
power, speed, focusing and shielding provisions) and the comparison between
single and dual pass welding procedures post and pre-heatment and the
demonstration of repair procedures, to ensure the continuing feasability of the
prototype laser welding facility.
The new process offers the potential to carry out significant improvements in
pipe welding. The existing processes use multi-work station welding in the "S"
mode only,involving considerable numbers of highly paid specialists. The
technical feasability of the process has been demonstrated in a number of
alternative components and applications but there is as yet little experience of
producing welds in pipes of the dimensions utilised in on and off-shore welding.

STATE OF ADVANCEMENT :
****
Ongoing. The project has commemced with the investigation of laser weld trials
to determine the selection of equipment such as laser power, necessary optics,
whether post or pre-heat be necessary and the study of gravity effect on weld
flow.

RESULTS :
****
The initial results cast serious doubt on the economic , commercial and
ecological viability of the project. The new technical director of the project
has suspended the laser weld trials pending further evaluation of the initial
results.
A decision on the continuation or abandonment of the project will be made in the
near future.

```
****************************************************************************
* TITLE : DEVELOPMENT OF NEW TYPES OF GAS AND        *     PROJECT NO      *
*         FIRE DETECTORS FOR UNMANNED OFFSHORE       *                     *
*         PLATFORMS                                  *  TH./15099/87/UK/..  *
*                                                    *                     *
****************************************************************************
* CONTRACTOR :                                       *    TELEPHONE NO      *
*    SNAMPROGETTI LTD                                *                     *
*                                                    *    0256 461211       *
*    SNAMPROGETTI HOUSE                              *                     *
*    BASING VIEW                                     *                     *
*    BASINGSTOKE                                     *    TELEX NO          *
*                                                    *                     *
* TECHNICAL DIRECTOR :                               *    858816            *
*    DR. ING. P. GRILLO                              *                     *
*                                                    *                     *
****************************************************************************
                                             VERSION : 06/02/90
```

AIM OF THE PROJECT :
****
The aim of the Project is to develop new types of gas and fire detectors to be
used on unmanned offshore oil and gas platforms, which ensure safe, reliable
operation with the minimum maintenance requirements. In order to set the
performance targets for the new detectors with regard to
reliability/availability, it is believed necessary to perform the risk analysis
of an optimised platform design incorporating all the latest features used in
equipment and component design confining flammable and toxic gases. The
development Project is based on the Infra-Red Spectrophotometric techniques,
which are highly innovative in this field and promise to solve the major
detection problems for such an environment being characterised by windy open
areas with fog, mist, dust deposition and adverse climatic conditions.

PROJECT DESCRIPTION :
****
The project will consists of the following main phases :
2.1 Safety Analysis of a typical offshore platform designed for unmanned
operation in order to :
- characterise the potential sources of release of flammable gases
- characterise the availability requirements of gas/fire detection system
- assess the risk levels expected to be associated with the platform operation
2.2 Technological development of the fire/gas detection system comprising the
following main phases :
2.2.1. Collection and Assessment of the existing information about the
performance of fire and gas detectors for offshore plants, used up to date
2.2.2. Development of the project, by the definition of the experimental
parameters representing the offshore platform situation and the realisation of
the innovative types of detectors
2.2.3. Project realisation, consisting of :
* proposals and design of innovative detection system;
* construction of the new detectors prototypes;
* experimental gas dispersion studies in a wind tunnel to define gas dispersion
parameters;
* set up of a testing facility (environmental simulator);
* testing of the new detectors in the test facility and offshore;
2.3. Feedback of the results obtained from the technological activities into the
Safety Analysis, in order to verify the risk levels achievable.
The risk analysis developed as indicated in Para 2.1. will be reviewed in order
to implement the results obtained. The suitability of the risk levels will be
analysed and discussed in comparison with those allowed and expected for the
manned platforms.

STATE OF ADVANCEMENT :
****
Ongoing - The activities in sections 2.1 and 2.2 have been partially performed.
The reference situation has been characterised, in terms of platform optimised
design. The scenario of expected sources of gas release have been defined. The
gas transfer function, with relation to the reliability of the detectors has
been analysed. Failure data of the detectors used up to date has been collected
and analysed. The fire and gas detector design specification has been
established.

RESULTS :
****
The main result of the Project will be the development of prototypes of fire and
gas detectors based on the Infra-Red Spectrophotometric techniques. With
comparison to detectors employed to date, the new ones are expected to have
peculiar important features consisting in Fast Response, Broad Range of
Sensitivity, High Accuracy and Reliability, Low Maintenance and Testing
Requirements and Immunity from spurious signals. These features, quite relevant
for the performance in Offshore conditions, are lacking in the conventional
types.
The technology developed will have feedback from full scale tests on wind tunnel
and on offshore platforms. Possible maloperations will be recorded and analysed

in view of obtaining improvements.

Furthermore the optimised design of topside components linked to the risk analysis results will define requirements for control and management systems of the platforms. The risk analysis results will also be considered in terms of feedback into the design specifications of mechanical components and equipment confining hazardous fluids and the policy of their maintenance/inspection. The success of this development project will improve the reliability/availability of the fire/gas detection leading to improved operational and safety levels for automatic production systems. The project is believed to give improvements also in the general aspects related to gas/fire detection technology, expecially in relation to the performance requirements associated with the detectors sensitivity and fast response. There is, however, a quite relevant level of risk associated with the content of the results that will be obtained. In particular, the new safety levels associated with the unmanned operation, may not meet the acceptance criteria of all of the control/certifying authorities.

```
*************************************************************************
* TITLE : MULTIPHASE METERING EQUIPMENT       *       PROJECT NO        *
*         DEVELOPMENT                          *                        *
*                                              *   TH./15101/87/UK/..    *
*                                              *                        *
*************************************************************************
* CONTRACTOR :                                 *      TELEPHONE NO       *
*    TEXACO LIMITED                            *                        *
*                                              *   01-584-5000           *
*    1 KNIGHTSBRIDGE GREEN                     *                        *
*    UK-LONDON SW1X 7QJ                        *                        *
*                                              *      TELEX NO           *
*                                              *                        *
* TECHNICAL DIRECTOR :                         *      8956681            *
*    F B WALTER                                *                        *
*                                              *                        *
*************************************************************************
                                              VERSION : 26/02/90
```

AIM OF THE PROJECT :
****
To develop and field prove two multiphase flow meters which are expected to
enhance the economics of marginal field developments. The first meter is a two
phase liquids fiscal meter designed to continuously measure a primarily
oil/water flow from a subsea or riser separation system upstream of production
separators to nominally +/- 1 %. This meter will permit the comingling of the
product flow with other production in platform separation trains. The second
meter is a subsea three phase meter designed to meter individual well flows
subsea remote from a host facility to nominally +/- 5 % for production
allocation and reservoir management. The meter will perform the function of a
subsea test separator and allow all production to be combined into a single bulk
production pipeline. Once proven, operating companies will be able to
confidently employ the meters in new subsea field developments to significantly
reduce capital costs.

PROJECT DESCRIPTION :
****
The project involves combining basic research in multiphase metering conducted
by Texaco Inc. Exploration and Production Technology Division with offshore and
subsea expertise from Texaco Limited. Texaco Inc. originated the metering
concepts. Texaco Limited and Jiskoot Autocontrol Limited are now undertaking the
development of these concepts into prototype metering systems and will prove
them in service on hydrocarbons.
In stage I, prototypes of both the two phase liquids fiscal meter and the subsea
three phase meter will be built and tested topsides on the Tartan "A" Platform.
They will be fitted to lines carrying well fluids in series with existing
platform test separators. The performance of the meters will be evaluated using
the test separators for reference measurements.
Texaco Inc. has continued to perform related research into multiphase metering.
Texaco Limited and Jiskoot Autocontrol Limited have been informed of their work.
It has been possible to use some of their results to optimise the Stage I
prototype designs.
At the conclusion of Stage I, the tests on the two phase liquids fiscal meter
will be complete and the subsea three phase meter will have been thoroughly
tested on hydrocarbons. Reports will be prepared analysing the measurement
accuracy of each meter.
Stage II of the project will be undertaken if the subsea three phase meter
performs acceptably on Tartan. In Stage II, the subsea three phase meter will
be fitted to the Highlander Subsea Template where individual well flows can be
directed through it. A dedicated test production system exists at Highlander
which will take fluids from the meter to a dedicated test separator on the
Tartan "A" Platform.
The performance of the meter will be evaluated against this test separator and a
report on the accuracy of measurement will be prepared.
It is intended that both meters will be left installed in their final test
locations until an alternative use of curcumstance justifies the expense of
decommissioning and removal. This significantly reduces project costs,
particularly for the subsea meter, and allows the collection of data on long
term reliability. The meters will also be available for inspection by potential
clients to assist marketing efforts.

STATE OF ADVANCEMENT :
****
Theubsea three phase meter has been fabricated and tested for ten weeks on
Texaco's tartan platform.
Additional onshore testing and design work is in progress to define possible
system modifications to be made before completion of platform tests.
The two phase liquids fiscal meter has been held pending resolution of a process
flow problem which preculeds the meter's installation in the designated test
location.

RESULTS :
****
Platform tests of the subsea three phase meter have highlighted a number of
minor modifications required to the system and a basic liquid/gas separation
problem. Ongoing work is focused on these issues.

```
***************************************************************************
* TITLE : S.I.C.H.A.#(SUBSEA INTELLIGENT COMPACT    *    PROJECT NO      *
*         HYDRAULIC ACTUATORS)                      *                    *
*                                                   *  TH./15103/87/IT/.. *
*                                                   *                    *
***************************************************************************
* CONTRACTOR :                                      *   TELEPHONE NO      *
*    DRASS S.P.A.                                   *                    *
*                                                   *   35 882104         *
*    VIA VENEZIA 9                                  *                    *
*    IT-24040 ZINGONIA (BG)                         *                    *
*    ITALY                                          *   TELEX NO          *
*                                                   *                    *
* TECHNICAL DIRECTOR :                              *   300696            *
*    G CORTINOVIS                                   *                    *
*                                                   *                    *
***************************************************************************
```

VERSION : 31/12/89

AIM OF THE PROJECT :
****
DRASS S.P.A., with the collaboration of European hydraulic and robotics
specialist wants to develop and design a range of "SUBSEA INTELLIGENT COMPACT
HYDRAULIC ACTUATORS".
If successfull it will bring a major contribution to the development of second
generation underwater robotics (more autonomy decision, more high level
communication with operator, more versatily) needed to install, maintain, next
to come subsea automatic production systems.
Difficulties are important due to the technical and economic constraints (must
be compact, resistant, rreliable, modular, intelligent and evolutive, price
competitive ...).
The state of the technology is very encouraging : nothing similar is really
exisiting for subsea applications. Industrial ground produts are beginning to
appear but the level of intelligence is still very low and robotics experts have
not yet been involved. Very interesting research workds are being made on this
subject in European Universities.

PROJECT DESCRIPTION :
****
These Subsea Intelligent Compact Hydraulic Actuators will be composed of 4
different parts integrated in one package :
- the actuator itself
- the amplifier (valve)
- the sensors
- the control and communication box
This project will be developped in 4 phases themselves divided in activities,
over a period of 3 years.
First for preliminary investigation and specification of 2 demonstrative
actuators
PHASE I : preliminary studies
Then, to solve the basic R & D problems, 2 phases in parallel :
- PHASE II : development of the hydraulic components
- PHASE III : development of the electronics and associated software
Ultimately, to prove the validity of the concept and initiate the marketing
actions :
PHASE IV : integration and tests.

STATE OF ADVANCEMENT :
****
At 15 october 1989 when the project was abandoned phase I had been fully and
phase II and III partially completed.

RESULTS :
****
Technical and economical evaluation has been finalised. Underwater use to
maximum 1000 m. Maximum hydraulic pressure 350 bar, flow according to request
for use with simpler power pack.
Electric/electronic power 24 VDC, current limited, possibility of connection to
external computer, standard signal transmission.
It was decided to select only one demonstrative application to start with and to
concentrate on concretely solving the practical problems. The linear intelligent
actuationor was chosen for the particular application of making an intelligent
manipulator to be used especially on board ROV4s.
Different position sensors have been tested together with a complete linear
actuator. When the project was stopped a first laboratory model was available
and operational. The problems remaining to be solved were the final integration
and suitable packaging for underwater environments, together with the
finalisation of the software. Technically this project was very interesting and
also very promising. The project was stopped for organisational reasons.

```
***********************************************************************
* TITLE : IMPROVEMENT AND TESTING ON A GAS WELL    *   PROJECT NO    *
*         OF A FIRE FLAMING DEVICE : BIL FIRE      *                 *
*         PREVENTER                                *  TH./15105/87/FR/..  *
*                                                  *                 *
***********************************************************************
* CONTRACTOR :                                     *  TELEPHONE NO   *
*    BERTIN & CIE                                  *                 *
*                                                  *  59.64.86.48.   *
*    CENTRE DE BAYONNE                             *                 *
*    ZONE INDUSTRIELLE                             *                 *
*    F-40220 TARNOS                                *  TELEX NO       *
*                                                  *                 *
* TECHNICAL DIRECTOR :                             *  570 026        *
*    J.C. MULET                                    *                 *
*                                                  *                 *
***********************************************************************
                                                   VERSION : 15/07/89
```

AIM OF THE PROJECT :
The objective of this project is to improve a water injection system (SPOOL) fitted on a gas well to extinguish a fire after an accidental eruption on a drilling platform. This process has been developed by BERTIN and is fully successful for fire without structure above the gas well.
But the presence of a platform induces impingement and separation of the gas water mixture. Under these conditions it is not possible to extinguish fire with the current configuration of the injection system, although the temperature of the structure and the radiation of the flame are considerably reduced.

PROJECT DESCRIPTION :
The project, with a total duration of two years, is divided into 4 main phases :
PHASE 1 : SIZING AND LABORATORY TESTS
The first step of this work program consists of the analysis of the behaviour of the droplets in a gas flow with impingement on a plate and deviation by various obstacles : trajectory and vaporization.
This study must lead to the determination of the water mass flowrate and the drop size distribution required to obtain a sufficient water vapour concentration anywhere under the platform in order to reach the lower limit of inflammability of the gas.
The adequate atomizers will be tested on a specific test bench. Then, preliminary tests with a cold gas will be carried out to qualify the water and gas flows.
Finally, some extinguished tests will validate the results of the analysis.
PHASE 2 : DESIGN AND FABRICATION OF AN EXPERIMENTAL SPOOL
Design,sizing, construction drawings, procurement and fabrication of the new water injection SPOOL equiped with the new atomizers defined during the Phase 1.
PHASE 3 : TESTS ON SITE
Tests in real configurations will be carried out on a GAZ DE FRANCE facility in CHEMERY, near BLOIS in France, with natural gas extracted from underground storages.
The operating conditions will be :
- using a INTRAFOR drilling platform
- gas flowrate   : between 15.000 and 50.000 m3/h
- water flowrate : up to 150 to 200 m3/h
The tests will involve :
- injection and ignition of the gas
- stabilization of the flame and gas flowrate at a given value
- increase of the water flowrate up to the extinguishment of the flame
- tests of re-ignition and stop
The test will be analyzed with the measurement of :
- gas and water flowrates
- radiation fluxes around the platform
- temperature of the structure
PHASE 4 : SYNTHESIS AND DETERMINATION OF THE INDUSTRIAL SPOOL - ANALYSIS OF THE EXPERIMENTAL RESULTS
Inquiries to the potential users in the gas and petroleum industries to determine their needs, and so, a range of industrial spools.
Evaluation of the market.
Final report.

STATE OF ADVANCEMENT :
Phase 1 (laboratory tests) is completed. Report is in progress. Phase 2 is delayed by 12 months to perform marketing studies and gather complementary fundings.

RESULTS :
Analysis of the gas flow and water droplets trajectory and vaporisation has shown that a complementary water injection system is required to improve the effect of steam dilution in the flames attached to the structures. Tests on a 400 mm platform with 20 g/s propane flame jet have been carried out.

Extinguishment is achieved with a water to gas mass flow rate ratio of about 8, provided spray is injected into the gas flow and mean droplet diameter is smaller than 100 m. The next Blowout Spool System will incorporate an external circuit with sprayers for the platform beams and rig floor.
Contact with Oil Companies has shown interest for the system for application like :
- work-over of gas wells
- shallow gas hazards.

```
*************************************************************************
* TITLE : MODULAR TRANSFER AND SERVICE SYSTEM FOR    *    PROJECT NO    *
*         APPLICATION IN OFFSHORE FIELD              *                  *
*         DEVELOPMENT AND PRODUCTION                 * TH./15106/87/DE/..*
*                                                    *                  *
*************************************************************************
*                                                    *   TELEPHONE NO   *
* CONTRACTOR :                                       *                  *
*    TELEFUNKEN SYSTEM TECHNIK                        *                  *
*                                                    *    040/8825-0    *
*    BEHRINGSTR. 120                                 *                  *
*    D-2000 HH 50                                    *                  *
*                                                    *    TELEX NO      *
*                                                    *                  *
* TECHNICAL DIRECTOR :                               *    211925        *
*    DR ING. K. WILKE                                *                  *
*                                                    *                  *
*************************************************************************
                                            VERSION : 07/06/89
```

AIM OF THE PROJECT :
****
The oil industry presently bases its intervention scenarios on modified and
adapted maintenance methods developed for onshore or calmwater offshore
intervention which seldom show optimum and economic solutions.
In order to increase efficiency and economic for deepwater intervention, new
methods have to be developed.

PROJECT DESCRIPTION :
****
The objective of this project is to develope a Modular Transfer and Service
System on the basis of the know-how gained from the SOLS development project
(Subsea Oil Loading System).
The system to be developed is to serve not only for loading, but also for
unloading liquids or similar substances from tankers or other vessels. For
unloading applications, the following will be considered during the Concept
Study/MTS I:
- Production auxiliary materials to be pumped into injection wells.
- Pressurized water to be provided for cleaning the subsea facilities.
- Ballast materials to be transported and pumped into foundations or covering of
pipelines or other subsea facilities.
In addition, the concept of the SOLS system will be modified and fitted with
adapters for further transportation, installation, inspection and maintenance
work like through flowline servicing and pig launching.
The feasibility of the SOLS principle for the applications suggested here will
therefore be examined in the framework of the Concept Study. From this a modular
system should be developed, which could easily be adapted for the different
transfer and service tasks by modification of the individual components.

STATE OF ADVANCEMENT :
****
Completed the Concept Study/MTS I.
Ongoing with Sesign Study + Test/MTS II.

RESULTS :
****
By evaluating the possible market potential and the technical features of the
SOLS, the following groups of systems - MTS component families - have been
selected which seem relatively logical to be performed by a system based on the
SOLS concept.
- Workover/Production/Transportation
- Downhole operations/Maintenance
- Subsea intervention on tree and template
- Pipeline related tasks.
The valuation of the features of the SOLS compared with the a.m. groups showed
that methods of subsea operation, totally different from the SOLS principle,
require so many modifications that a totally new development would be necessary.
This was not the intention of this study.
The conclusion of the beforesaid is that the MTS consortium concentrated their
development work on those MTS concepts justified by criterias as :
- SOLS features adaptable to proposed MTS,
  multipurpose functional capacity of one system,
- market potential and
- economic operation of the system.
The loading of live crude, the injection, the pipeline related tasks and an
installation and retrieval system have been identified to be of most interest.
The following ranking shows the priority given to all before mentioned concepts
during the following examination :
1. Pipeline Intervention System : PINS
2. Module Installation and Retrieval System : MIRS
3. Live Crude Loading System : LCLS
   (subsea storage concept)
4. Injection System : INS
   (combined live crude loading).
In order to concentrate on the most promising concepts in the future subsea

market the MTS consortium decided to focus their ongoing development works on the concepts 1 to 3.
These system including their subsystems were conceptually defined and presented. It is envisaged to focus ongoing development works on these three systems within a second R&D phase (MTS Ph.II) which includes the design and testphase.

REFERENCES :
****
MARKET STUDY
OFFSHORE ENGINEERING RESOURCES A/S, APRIL/MAY 1988.
STUDY OF POTENTIAL APPLICATION OF THE SOLS-CONCEPT
NORWEGIAN UNDERWATER TECHNOLOGY CENTER A/S, MAI/JUNI 1988.

```
***********************************************************************
* TITLE : DEVELOPMENT OF AN AUTOMATED DEVICE FOR    *    PROJECT NO      *
*         THE IN-SITU MEASUREMENT OF METHANE IN     *                    *
*         SEAWATER                                  *  TH./15107/87/DE/.. *
*                                                   *                    *
***********************************************************************
* CONTRACTOR :                                      *    TELEPHONE NO     *
*     GKSS                                          *                    *
*                                                   *                    *
*     MAX-PLANCK-STRASSE                            *                    *
*     DE 2054 GEESTHACHT                            *                    *
*                                                   *    TELEX NO         *
*                                                   *                    *
* TECHNICAL DIRECTOR :                              *                    *
*                                                   *                    *
*                                                   *                    *
***********************************************************************
                                                     VERSION : 01/06/89
```

AIM OF THE PROJECT :
****
The aim of the project is the development, dry test, simulated ocean test, and
field test of an automated device for the in-situ detection and measurement of
methane in seawater. Methane in water around pipelines is indicative of
mechanical damage which can lead to leaks and failure of the pipeline. The
device is likely to offer operational and cost advantages over existing schemes
of inspection, both in-line (moles) and external (sonic and TV examination), and
provides qualitatively new information on pipeline integrity.

PROJECT DESCRIPTION :
****
A project is proposed that consists in the development of an automated system
for the measurement of methane in seawater. Operated in the immediate vicinity
of underwater pipelines, the system will allow the detection of cracks and other
leaks in oil or gas carrying pipelines at an early stage.
The idea underlying the technique relies on the use of a natural tracer, methane,
 present in mineral oil and gas and highly mobile because of its small molecular
weight and relative inertness. Brought into contact with water at elevated
pressure as is found at the bottom of the sea, methane forms bubbles to a much
lesser extent than at atmospheric pressure. Rather, a large fraction of the gas
dissolves in the water. An increased methane content of the water surrounding a
pipeline can thus be indicative of a defect through which the gas leaked into
the sea.
The physical and physico-chemical problems in connection with the project have
already been investigated. The results show that no unsurmountable obstacles
will probably be encountered in the technical implementation.
After the initialization and specification definition phase, optical, mechanical,
 and electronic components will be procured or developed, and the overall system
will be designed and built. A remotely operated vehicle has to be chosen.
Components and system will subsequently be tested under dry, wet, and simulated
pressure conditions, followed by field tests under realistic conditions in the
North sea. The final phase will consist in the documentation and assessment of
the viability of the system for routine underwater pipeline inspection.
Total cost of the project is estimated at DM 5.08 million, to be distributed
over four years. If funded, the project will be carried out by GKSS in
cooperation with French partners.
The successful completion of the project will provide an early-detection system
of damage to underwater oil and gas pipelines. It will thus help to reduce
operational losses, contribute to the safety of the supply of hydrocarbons to
the European Community, and ease the protection of marine ecosystems in
endangered bodies of water such as the North Sea.

STATE OF ADVANCEMENT :
****
Ongoing. The scientific basis has been worked out, results are published in the
literature cited. Present activities concentrqte on the elaboration of contracts
with prospective partners and preparation of the text to be negociated with the
EEC.

RESULTS :
****
The project is presently in the definition phase. No results of technical
activities can as yet be reported.

REFERENCES :
****
W. MICHAELIS, C. WEITKAMP : GERMAN PATENT P 33 02 656.4, FILED 27 JANUARY 1983.
G. BALSEN : REPORT GKSS 86/E/37 (1986), 103 PP.
G. BALSEN, W. MICHAELIS, C. WEITKAMP : GKSS JAHRESBERICHT 1985, PP. 18-31 (1986).

```
**************************************************************************
* TITLE : DEVELOPMENT OF AN IMPROVED FLAMMABLE      *    PROJECT NO      *
*          GAS DETECTION SYSTEM TO INCREASE         *                    *
*          OFFSHORE SAFETY AND TO REDUCE THE RISK   *    TH./15108/87/UK/.. *
*          FROM FIRE/EXPL                           *                    *
**************************************************************************
* CONTRACTOR :                                      *    TELEPHONE NO     *
*     EDINBURGH SENSORS LIMITED                     *                    *
*                                                   *    031-449-5844     *
*     RICCARTON                                     *                    *
*     CURRIE                                        *                    *
*     EDINBURGH EH14 4AP                            *    TELEX NO         *
*                                                   *                    *
* TECHNICAL DIRECTOR :                              *    72553           *
*     MR. P. BRAMLEY                                *                    *
*                                                   *                    *
**************************************************************************
```

VERSION : 17/10/89

AIM OF THE PROJECT :
****
To develop a flammable gas sensor for use in the hostile offshore environment,
based on infra-red technology; a product capable of operation on a remote basis
and of obtained the necessary safety certification.

PROJECT DESCRIPTION :
****
Based on infra-red interference technology, some progress has already been made
(before this application) in developing cheap devices for gas sensing purposes
to replace the existing cumbersome and expensive instruments.
In particular, innovative steps have been taken in the areas of the source,
thermostating of the filter assemblies, the guiding of light by a simple tube
mechanism and the introduction of a microprocessor. But, considerable further
development is required to make this prototype product suitable for the offshore
environment. Much effort will be applied to -
1/ design of tube length
2/ gas sampling methods
3/ weather protection
4/ design of emitter housing
5/ design of detector housing
6/ anti-condensation
7/ support housing
8/ certification (BASEFA/NEMKO)
9/ performance certification
10/ batch evaluation.

STATE OF ADVANCEMENT :
****
The conventional head feasibility study is largely study is largely complete and
two instruments based on our Guardian product range have now been manufactured
(one for 0 -10 % and one for 0 - 100 % methane). Performance evaluation is in
progress, including evaluation by potential users.
Work on the detailed design of a conventional IR head capable of full state
sensitivity of between 10 % and 100 % methane has now commenced.

RESULTS :
****
The current conventional IR technology known to Edinburgh Sensors is capable of
full scale sensitivity down to 10 % methane (200 % L.E.L.) with a
resolution/accuracy of 0.1 % methane (2 % L.E.L.). Zero drift after 6 months
continuous operation was found to be 0.1 % methane (2 % L.E.L.).

```
********************************************************************
* TITLE : A SYSTEMATIC STUDY OF ENVIRONMENTAL    *   PROJECT NO    *
*         EFFECTS ON GAMMA-RAY DENSITY SONDES    *                 *
*         INCLUDING ROUGH WALLED BOREHOLES.      *  TH./15109/87/UK/.. *
*                                                *                 *
********************************************************************
* CONTRACTOR :                                   *  TELEPHONE NO   *
*    WINFRITH PETROLEUM TECHNOLOGY               *                 *
*                                                *  0305 251888 EXT.3396*
*    DORCHESTER                                  *                 *
*    UK - DORSET DT2 8DH                         *                 *
*                                                *  TELEX NO       *
*                                                *                 *
* TECHNICAL DIRECTOR :                           *  41231          *
*    DR J LOCKE                                  *                 *
*                                                *                 *
********************************************************************
                                        VERSION : 23/10/89
```

AIM OF THE PROJECT :
****
The main objective of this project is to establish and quantify the causes of
apparently anomalous logs from gamma density tools. To do this requires the
development and validation of a special version of the Mc BEND code, called
McDUFF, which will afford the efficient prediction of individual detector count
rates to a very high accuracy. Experimental benchmark data will be provided
ranging from the determination of basic detector response functions in
simplified geometries to experimental logging tool measurements in full scale
testblocks. The latter will also be used to simulate variations in the borehole
environment paying particular attention to rugosity of different amplitudes and
frequency.

PROJECT DESCRIPTION :
****
The project has mbeen planned in phases in order to achieve the following main
objectives :
1) Computer Code Development
A critical review of the accuracy and efficiency of Winfrith Radiation Shielding
Codes will be carried out which will identify those aspects which can be
enhanced. This will not involve any major change in the codes but will
concentrate on the validation of nuclear data sets and, in particular, on
detector response functions.
2) Code Validation
Initial validation will be accomplished using measurements made with an
effectively infinite plane slab of aluminium in an existing shield test facility
at Winfrith. A series of measurements will then be made in a number of standard
density test blocks, varying the instrumental parameters in the Laboratory sonde
over the working range.
3) Study of Environmental Effects
The validated computer model will be used to predict the effects of variations
in experimental boreholes including mudcake thickness, borehole size and shape
etc. Rugosity and caving will be simulated in the test blocks by means of liners
with sinuosidal or other variations along the length of the borehole. These
measurements will involve dynamic tests in which the sonde pads will be moved up
through the test region at typical logging speeds.
4) Demonstration Using Operational Sondes
The experimental facilities and software will be available to participating
companies under preferential terms for a period of one year follo<wing
completion of the project. These will allow logging companies to calibrate
sondes and investigate their performance in a range of primary standards with
simulated mud-cake and rough-walled borehole liners. In addition, "back-to-back"
comparisons of different service company sondes could be made which would allow
major oil companies to acquire greater confidence in measurements being made on
their behalf.
5) Management and Timescale
The work will be undertaken within the Radiation Physics & Shielding Group at
Winfrith, over a period of two years from 1 February 1989.

STATE OF ADVANCEMENT :
****
The computer code development and validation are now well advanced. Further
benchmark tests are underway to test various aspects of the codes performance.
Design work for the experimental logging tool and test blocks is nearing
completion and manufacture expected to begin shortly.

RESULTS :
****
In order to achieve the objectives of this project it is necessary for the code
to predict individual detector count rates to a high accuracy. A major obstacle
to achieving this was that modelling of the scintillation detector response had
yet to be adequately demonstrated. Two approaches have been developed for
calculating the pulse height distribution (PHD) for the detector.
a) The gamma flux spectrum calculated within the vicinity of the detector is
multiplied by an experimentally determined response function to obtain the PHD.

442

b) A special algorithm developed at Winfrith is used to score directly the photon energy deposition within the crystal.
Both methods have been compared with measurements for simple benchmark conditions and found to be successful. The second approach has been adopted for use within this project as it involves less data manipulation.
A specially designed experimental rig which simulates the essential physics of a gamma-density tool has been fully commissioned. Test work carried out with this rig has enabled various aspects of the sensitivity of tool response to source/detector geometry and the nature of the formation to be assessed. Apart from providing benchmark data for validating the computer code, these results have also assisted in the design of future experimental work involving an experimental tool and full scale test blocks.

```
*****************************************************************************
* TITLE : DEVELOPMENT OF INDUSTRIAL EQUIPMENT FOR    *    PROJECT NO       *
*          HYDROGEN DEEP DIVING                      *                     *
*                                                    *    TH./15110/87/FR/..*
*                                                    *                     *
*****************************************************************************
* CONTRACTOR :                                       *    TELEPHONE NO      *
*     COMEX S.A.                                     *                     *
*                                                    *    91.23.50.00      *
*     BD DES OCEANS 36                               *                     *
*     F-13275 MARSEILLE CEDEX 9                      *                     *
*                                                    *    TELEX NO         *
*                                                    *                     *
* TECHNICAL DIRECTOR :                               *    41410.985        *
*     MR CLAUDE GORTAN                               *                     *
*                                                    *                     *
*****************************************************************************
                                          VERSION : 14/03/90
```

AIM OF THE PROJECT :
****
The purpose of the project is to develop the specific industrial equipments
designed for hydrogen diving.
This equipment will be evaluated in actual diving conditions during the sea
trials at the depth of 520 meters.

PROJECT DESCRIPTION :
****
Research carried out since 1983 has demonstrated that divers breathing
hydrogenated gas mixtures are less affected by pressure :
- No clinical signs of the High Pressure Nervous Syndrome appears.
- The respiratory resistances are considerably reduced.
To allow industrial application of this new diving technology the specific
diving equipment should be developed.
The programme will be carried out in three phases :
PHASE 1 :
Engineering study and adaptation to hydrogen of individual diving equipments.
- Open-circuit breathing apparatus.
- Semi-closed-circuit breathing apparatus.
- Closed-circuit breathing apparatus.
- Hyperbaric tests of selected equipments on ventilatory simulator.
- Evaluation of emergency breathing system prototype and tests in hyperbaric
  conditions.
- Divers thermic protection. Optimization of diving clothes with reduced flow
  of hot water, and cold water tests in atmospheric and hyperbaric pools.
PHASE 2 :
Automatization of gas dehydrogenator :
- System definition.
- System components research and selection :
     * pressure sensors
     * electronic regulators
     * microprocessors
     * regulation valves
- Design and construction of the system prototype based on differential
  pressure control and monitoring.
- Design and construction of the system prototype based on hydrostatic or
  absolute pressure control and monitoring.
PHASE 3 : test dives in the sea to 520 meters'depth to evaluate the equipments
and technics in actual conditions gas mixtures :
- Divers selection and training
- Diving site selection
- Work platform construction
- Mobilisation on board of the DSV ORELIA of specific materials and
  equipments and gases
- HYDRA VIII dives at 520 meters depth.
  The purpose of this operation is to accomplish at least 60 hours cumulated
  with two teams of 3 divers at a rate of one dive per day for each team.
The total duration of HYDRA VIII is approximatively 30 days including of 6 to 9
days spent at the maximum depth.

STATE OF ADVANCEMENT :
****
Ongoing.

RESULTS :
****
PHASE 1 : The breathing adaptation to hydrogen of individual equipments
The breathing apparatus has been adapted without changing the basic equipment
structure. Modifications mainly concern gas distribution circuits, and the
second regulation chamber which delivers the breating mixture inside the diving
helmet through the breathing mask. Injection valves optimisation allowed to
bring breathing resistance very low rates. The helmet is then very comfortable.
The emergency breathing system, the "ball out system" type BOS II of COMEX PRO
has been adapted to hydrogen, enabling a 5 to 10 minutes autonomy at a 520

meters depth, (function of the breathing frequency).

For divers'thermic protection, a diving suit with reduced hot water flow has been developed from a neoprene-silicon composite material. Tests in hyperbaric chamber revealed the importance of the diving-suit design. This one has to be adjusted to divers mensurations to ensure an efficient thermic protection with a hot water flow of 5/6 liters per minute. In case of water supply failure, inertiae is important and ensures a larger protection to the emergency breathing system autonomy.

PHASE 2 : automatisation of gas dehydrogenator.

The selective hydrogen elimination is based on a dehydrogenator especially developed for decompression chamber in hydrogen saturation. A dehydrogenator monitoring system has been studied to increase safety and ease the monitoring process.

Two regulation systems are tested. The first one was based on the differential pressure measurement between a standard capacity and the saturation chambers. This method of measurement ensures a high accuracy of decompression. In normal function the variation between theoretical pressure and absolute pressure is lower than +/- 0,025 kPa. The system accuracy however causes problem because of the high regulaiton loop sensitivity to environment temperature variations. The second one was based on the system of direct measurement of chambers'pressure. It was significantly less sensitive to the temperature variations. The regulation does not reach the level of accuracy of the first system, but stays within the physiological constraints imposed.

PHASE 3 : offshore diving tests at 520 and 534 meters depth.

The offshore validation operation took place between the 22th of February and the 22th of March 1988, offshore Marseille.

After hydrogen adaptation of the D.P.DSV "ORELIA", with mobilisation on board of the required equipment :

Six divers (4 belonging to Comex and 2 to the French Navy) have been selected for this HYDRA VII ofshore operation.

OPERATION SCHEDULE :

Works performed (26 hours underwater in total) :

a) Connections of two 12" pipe-joints with a flanged spool piece.

b) Connection of a flexible "COFLEXIP 6" with a GRAYLOCK type flange.

Conclusion were:

- Divers easy evolution on the workplatform underline proves the previous onshore physiological comfort observed.

```
*******************************************************************************
* TITLE : EXPERT MONITORING SYSTEM FOR DEEP        *      PROJECT NO          *
*         DIVING OPERATIONS                        *                          *
*                                                  *    TH./15111/87/FR/..    *
*                                                  *                          *
*******************************************************************************
* CONTRACTOR :                                     *      TELEPHONE NO        *
*    COMEX S.A.                                     *                          *
*                                                  *      91.23.50.48         *
*    36 BLD DES OCEANS                              *                          *
*    FR-13275 MARSEILLE CEDEX 9                     *                          *
*                                                  *      TELEX NO            *
*                                                  *                          *
* TECHNICAL DIRECTOR :                             *      410985              *
*    MR J.P. IMBERT                                 *                          *
*                                                  *                          *
*******************************************************************************
                                                       VERSION : 30/06/89
```

AIM OF THE PROJECT :
****
Deep diving operations are carried out on a routine basis to 300m in Brazil,
Gulf of Mexico and Norway. This activity will extend to 450m in the years to
come.
The aim of the project is to add a new dimension to diving safety and efficiency
by introducing an Expert System (ES) to assist the diving supervisor during deep
diving operations. The system will be based on existing monitoring system and
will provide condition monitoring, data consistency, trend assessment and
diagnosis of abnormal situations for the working diver.

PROJECT DESCRIPTION :
****
The Expert System will focus on the problem of the working divers. It will
interface the monitoring system - that collects the information relevant to the
dives - to the Diving Supervisor who is in charge of the operations. The ES will
assist the diving Supervisor but will not control the operations : the
Supervisor will always be able to ride over its recommendations.
Tasks to be performed by the ES will include :
- Quality control of the incoming data
- Condition monitoring of the objects according to the various phases of dives
- Alarm generation in case of abnormal states, including early detection of
trends that might lead to such states
- Diagnosis of the causes for abnormal states
- Assistance in selecting appropriated procedures or setting the equipment to
correct the situation
- Educational programs for personnel training
Information from the knowledge base will be provided by COMEX operational and
diving personnel : Safety Officers, Methods Engineers and Medical Doctors,
Equipment Engineers and Technicians, Barge Superintendents and Diving
Supervisors, Life Support Technicians and Divers.
Operating and emergency procedures will be derived from the Company procedures
and manuals. The ES software will be selected according to the following
criteria :
- Capacity to cope with fast evoluting signals and to differentiate levels of
priorities according to the emergencies
- Ease of developing/adapting/modifying the knowledge base with experience
- Possibility of implementing the ES on a small computer, compatible with
offshore conditions of operation.
The project will be carried out in three phases :
a) FEASIBILITY STUDY AND SPECIFICATIONS : Review of existing monitoring systems,
establishment of a simple demonstration system and functional specifications of
ES
b) DEVELOPMENT : Interface with monitoring system, interface with operators,
prototype expert and system test of prototype expert system in onshore
facilities
c) EVALUATION : Adaptation of the expert system to the actual case of a deep
diving vessel, offshore testing of the complete system during one diving season,
evaluation of the performances of the expert system, modifications if required
and documentation.

STATE OF ADVANCEMENT :
****
The work package nr 1 of the project is completed. This included a feasibility
study and specification of the system, a classic approach to expert system
development.

RESULTS :
****
The feasibility of the project has been demonstrated on the simple example of
the thermal balance of the diver and specifications written on this experience.
A subcontractor, ITMI, has been selected that can provide an expert system
software with a time response compatible with the safety of the diver. In
addition, the system can be installed on an IBM type micro-computer suitable for
the offshore environment.

The Cybernetic monitoring was selected that complies with our diver's monitoring specifications. The different modules of the expert system have been identified and the thermal model of the diver finally designed and successfull implemented on the micro computer. This model is the key to the deductions of the expert system. Its first validation tests with information available from the literature are very promising.

```
*********************************************************************
* TITLE : PROCESS OF DIGITAL AND AUTOMATIC      *      PROJECT NO      *
*          UNDERWATER OPTICAL METROLOGY         *                      *
*                                               *   TH./15112/87/FR/.. *
*                                               *                      *
*********************************************************************
* CONTRACTOR :                                  *   TELEPHONE NO       *
*    CYBERNETTY                                 *                      *
*                                               *   91252500           *
*    36, BOULEVARD DE L'OCEAN                   *                      *
*    FR - 13009 MARSEILLE                       *                      *
*                                               *   TELEX NO           *
*                                               *                      *
* TECHNICAL DIRECTOR :                          *   401339             *
*    MR. BARAONA                                *                      *
*                                               *                      *
*********************************************************************
                                                  VERSION : 14/09/89
```

AIM OF THE PROJECT :
****
The underwater metrology system studied by CYBERNETIX will be able to carry out
automatically accurate distance measurements to the depth of 300 metres.
The major innovative aspect of the system consists in using an optical device to
take measurements.
The main applications of this development will be :
- pipeline/riser connection.
- subsea inspection and intervention on platform structures.

PROJECT DESCRIPTION :
****
The metrology system is based on the principle of optical triangulation.
It includes :
* A pinpoint light placed on the point to be measured.
* A two metre wide base divided into two identical parts.
Each part is equiped with a CCD camera and a miror.
* One of the miror is fix and gives the direction of sight. The other one is
mobile to enable the two beams to cross at the considered point.
* The base itself is mobile on its vertical axis to enable pan and tilt of the
system.
The pinpoint light is shown on a CCD camera and then analyzed with an image
processing card, to calculate the pinpoint data.
The system includes an automatic tracking of the pinpoint light in order to
automatize the measures as much a possible.

STATE OF ADVANCEMENT :
****
The principle was validated on a prototype during sea trials down to 30 meters
at IFREMER.
The prototype of the system is in the process of achievement and will be
avaiable by beginning of 1990.

RESULTS :
****
Expected performances
MEASUREMENT DISTANCE : 6 to 30 meters (more if turbidity permits)
MEASUREMENT DEPTH : 300 metres maximum
FIELD OF CAMERA : automatic positioning
- automatic option,
. at 30 m : 630 x 473 mm
. at 5 m : 135 x 100 mm
. semi-automatic option,
. at 30 m : 125 x 109 mm
. at 5 m : 35 x 25 mm
MEASUREMENT PRECISION
A change in water rate (n=1,336 +/- 0,005) can effect the measurement precision
from 0 to 25 mm.
This precision depends on geometrical configuration of points to be measured.
DIMENSIONS
Length x width x height
2100 mm x 400 mm x 1000 mm
Weight ~ 500 kg

```
******************************************************************************
* TITLE : SPOOLPIECE MEASUREMENT SYSTEM                    *      PROJECT NO     *
* *                                                        *                     *
* *                                                        *   TH./15115/87/UK/.. *
* *                                                        *                     *
******************************************************************************
* CONTRACTOR :                                             *    TELEPHONE NO     *
*   BROWN & ROOT CONSTRUCTION(UK)LTD-SURVEY DIVISION       *                     *
* *                                                        *    0224-724855      *
*   WELLHEADS PLACE                                        *                     *
*   WELLHEADS INDUSTRIAL ESTATE                            *                     *
*   UK-DYCE ABERDEEN AB2 0GG                               *    TELEX NO         *
* *                                                        *                     *
* TECHNICAL DIRECTOR :                                     *    73260            *
*   MARK VORENKAMP                                         *                     *
* *                                                        *                     *
******************************************************************************
                                                          VERSION : 05/10/89
```

AIM OF THE PROJECT :
****
To produce an Acoustic Spoolpiece Measurement System which can be deployed at
the flange and riser portions by means of a Remotely Operated Vehicle. The
crucial part of the system is the design and manufacture of the magnetic clamp
attachment.

PROJECT DESCRIPTION :
****
PHASE 1 - SYSTEM OUTLINE DESIGN
PHASE 2 and 3 - EQUIPMENT PROCUREMENT
PHASE 4 - SYSTEM DETAIL DESIGN
PHASE 5 - ELECTRONIC AND SOFTWARE DESIGN AND PROTOTYPE
PHASE 6 - SYSTEM ASSEMBLY
PHASE 7 - COMMISSIONING

STATE OF ADVANCEMENT :
****
Ongoing. Basic system has been used by deploying with divers. However, no work
has been carried out under the specification and scope of work of this project.

RESULTS :
****
No results available to date.

```
**************************************************************************
* TITLE : SECOND GENERATION DOPPLER SONAR          *    PROJECT NO       *
*                                                  *                     *
*                                                  *  TH./15118/87/FR/..  *
*                                                  *                     *
**************************************************************************
* CONTRACTOR :                                     *  TELEPHONE NO       *
*    REMTECH S.A.                                  *                     *
*                                                  *  33 1 39 46 59 58   *
*    AVENUE DE L'EUROPE 2-4                         *                     *
*    BP 159                                         *                     *
*    F-78143 VELIZY                                 *  TELEX NO           *
*                                                  *                     *
* TECHNICAL DIRECTOR :                             *  698 227            *
*    MR Jean-Michel FAGE                           *                     *
*                                                  *                     *
**************************************************************************
                                                    VERSION : 20/09/89
```

AIM OF THE PROJECT :
****
The aim of this project is to develop a commercial Doppler Sonar for measuring current speed and direction within the ocean to a depth of one thousand meters. Signal processing techniques and acoustic research already developed for atmospheric measurements with the same kind of instrument will be used to produce a so-called "Second generation Doppler Sonar". The innovations are an electronically steerable transducer and very sophisticated signal processing. This will improve measurement reliability and accuracy in "hostile" environments and the sensitivity to increase the range of measurement.

PROJECT DESCRIPTION :
****
The work will include the following phases :
1. PROTOTYPE DEVELOPMENT :
The prototype of the Doppler Sonar used in the first study (September 86) will be modified to be more efficient and flexible. The future test program will utilize different frequencies.
2. TRANSDUCER OPTIMIZATION :
Different transducers and beamforming techniques will be tested and modified to find, for each frequency, the optimum one according to phased array antenna emitted power and directivity. These tests will be performed in a shallow basin.
3. FREQUENCY OPTIMIZATION :
An optimization of the frequency for each suitable range will be examined from both theory and existing measurements. These results will be validated in sea experiment where different frequencies will be tested according to range.
4. INTERCOMPARISON DOPPLER SONAR - CURRENTMETER :
An intercomparison will be done between a line of currentmeters and the Doppler Sonar prototype in a specific configuration (transducer mounted on sea floor measuring upward).
5. SOFTWARE OPTIMIZATION :
During and after the intercomparison, the software will be optimized in relation to the results.
6. DOPPLER SONAR ON BOARD A SHIP :
This phase will examine all problems related to the use of Doppler Sonar and more specifically measurement corrections as a function of ship movements.
7. ADAPTATION OF THE SYSTEM TO SEA OPERATION :
The two versions of the Doppler Sonar (ship mounted and self contained on sea floor) have to be modified to work properly in a sea environment. For the self contained version the major effort will be done to achieve low power consumption.

STATE OF ADVANCEMENT :
****
Phase 2.1 and the theoretical part of phase 2.2 are already done. Some difficulties in the design of the phased array transducer delayed instrument construction until the beginning of 89.
The transducer will be ready for the first tests in September 1989.

RESULTS :
****
Intermediate results of this project will be available at the end of 1989.

REFERENCES :
****
- PINKEL, R., "ON THE USE OF DOPPLER SONAR FOR INTERNAL WAVE MEASUREMENTS", DEEP SEA RESEARCH, V. 28a NO 3, 269-289,1981.
- RD INSTRUMENTS : ACOUSTIC DOPPLER CURRENT PROFILERS
   RD-DR SERIES.
   RD INSTRUMENTS 10035 CARROLL CANYON RD, SUITE G. SAN DIEGO, CA 92131
- REMTECH SA : JEAN-FRANCOIS SOULIER, RAPPORT SONAR DT 87/055, OCTOBRE 1987.

```
********************************************************************************
* TITLE : DEVELOPMENT OF HYDROGEN DIVING IN LONG      *      PROJECT NO        *
*         PERIOD SATURATION                           *                        *
*                                                     *   TH./15120/88/FR/..   *
*                                                     *                        *
********************************************************************************
* CONTRACTOR :                                        *   TELEPHONE NO         *
*      COMEX S.A.                                     *                        *
*                                                     *   33.91.23.50.00       *
*    36, BD DES OCEANS                                *                        *
*    FR-13275 MARSEILLE                               *                        *
*                                                     *   TELEX NO             *
*                                                     *                        *
* TECHNICAL DIRECTOR :                                *   410985               *
*      MR C. GORTAN                                   *                        *
*                                                     *                        *
********************************************************************************
```

VERSION : 31/01/90

AIM OF THE PROJECT :
****
\The aim of the project is to conduct a long saturation dive in order to study
physiological, biologicaql, and psycho-sociological effects on divers uring a
long stay in a hyperbaric hydrogen atmosphere.

PROJECT DESCRIPTION :
****
The research on hydrogen diving carried out by COMEX from 1983 to 1988 has
revealed that hydrogen gas mixes provide a great breathing comfort in hyperbaric
saturation diving conditions :
* important attenuation of the High Pressure Nervous Syndrome effect
* important improvement in breathing ease
* stress reduction caused by the confined divers environment.
Hydrogen diving seems to be the only way to overpass the helium diving depth
limitations but it can also be used within actual diving depth where it should
play a major role for improving divers'productivity.
Different development axis are investigated :
* study on hydrox utilization depth range
* study on human behavior (psycho-sociological factors due to long stay in
confined environment)
* study of long stay under hydrogen hyperbaric atmosphere biological effects.
This project is divided into six phases :
Phase 1 : human factors necessary tests definition
Phase 2 : physiological, biological and medical follow-up tests definition
Phase 3 : long saturation procedures definition
Phase 4 : saturation diving equipment mobilization
Phase 5 : long saturation dive
Phase 6 : data processing and analysis divers' medical follow-up.

STATE OF ADVANCEMENT :
****
Ongoing-Phases 1 to 5 completed
Phase 6 in progress.

RESULTS :
****
a) Hydrox (H2O2) gas mix utilisation depth :
* minimum depth : around 70 meters using a Hydrox gas mix with 2,5 % oxygen
representing 200 hPa of partial pressure. Divers can be submitted to workload
and the low O2 partial pressure does not seem to have major effects.
* Maximum depth : from 240 meters and down, the divers show first signs of
narcosis with worsten at 270 meters and become major at 300 meters.
All datas are not processed yet but we can already define the maximum HYDROX
utilisation depth to be around 240 meters for short saturation dives (a few days
and around 180 to 200 meters for longer stays.
A low Helium percentage added in the HYDROX mix seems to reduce the neurological
effects of hydrogen.
b) Long term expose biological effects :
The first medical results following the dive do not show any toxic effects due
to HYDROX.
c) Human factors related to long term confinement :
No result for the moment; the datas are being processed.